Plant-Environment Interactions

Third Edition

BOOKS IN SOILS, PLANTS, AND THE ENVIRONMENT

Editorial Board

Agricultural Engineering	Robert M. Peart, University of Florida, Gainesville
Crops	Mohammad Pessarakli, University of Arizona, Tucson
Environment	Kenneth G. Cassman, University of Nebraska, Lincoln
Irrigation and Hydrology	Donald R. Nielsen, University of California, Davis
Microbiology	Jan Dirk van Elsas, Research Institute for Plant Protection, Wageningen, The Netherlands
Plants	L. David Kuykendall, U.S. Department of Agriculture, Beltsville, Maryland Kenneth B. Marcum, Arizona State University, Tempe
Soils	Jean-Marc Bollag, Pennsylvania State University, University Park Tsuyoshi Miyazaki, University of Tokyo, Japan

Soil Biochemistry, Volume 1, edited by A. D. McLaren and G. H. Peterson

Soil Biochemistry, Volume 2, edited by A. D. McLaren and J. Skujins

Soil Biochemistry, Volume 3, edited by E. A. Paul and A. D. McLaren

Soil Biochemistry, Volume 4, edited by E. A. Paul and A. D. McLaren

Soil Biochemistry, Volume 5, edited by E. A. Paul and J. N. Ladd

Soil Biochemistry, Volume 6, edited by Jean-Marc Bollag and G. Stotzky

Soil Biochemistry, Volume 7, edited by G. Stotzky and Jean-Marc Bollag

Soil Biochemistry, Volume 8, edited by Jean-Marc Bollag and G. Stotzky

Soil Biochemistry, Volume 9, edited by G. Stotzky and Jean-Marc Bollag

Organic Chemicals in the Soil Environment, Volumes 1 and 2,
    edited by C. A. I. Goring and J. W. Hamaker
Humic Substances in the Environment, M. Schnitzer and S. U. Khan
Microbial Life in the Soil: An Introduction, T. Hattori
Principles of Soil Chemistry, Kim H. Tan
Soil Analysis: Instrumental Techniques and Related Procedures,
    edited by Keith A. Smith
*Soil Reclamation Processes: Microbiological Analyses and
    Applications,* edited by Robert L. Tate III and Donald A. Klein
Symbiotic Nitrogen Fixation Technology, edited by Gerald H. Elkan
Soil–Water Interactions: Mechanisms and Applications, Shingo Iwata
    and Toshio Tabuchi with Benno P. Warkentin
Soil Analysis: Modern Instrumental Techniques, Second Edition,
    edited by Keith A. Smith
Soil Analysis: Physical Methods, edited by Keith A. Smith
    and Chris E. Mullins
Growth and Mineral Nutrition of Field Crops, N. K. Fageria,
    V. C. Baligar, and Charles Allan Jones
Semiarid Lands and Deserts: Soil Resource and Reclamation,
    edited by J. Skujins
Plant Roots: The Hidden Half, edited by Yoav Waisel, Amram Eshel,
    and Uzi Kafkafi
Plant Biochemical Regulators, edited by Harold W. Gausman
Maximizing Crop Yields, N. K. Fageria
Transgenic Plants: Fundamentals and Applications, edited by
    Andrew Hiatt
*Soil Microbial Ecology: Applications in Agricultural and Environmental
    Management,* edited by F. Blaine Metting, Jr.
Principles of Soil Chemistry: Second Edition, Kim H. Tan
Water Flow in Soils, edited by Tsuyoshi Miyazaki
Handbook of Plant and Crop Stress, edited by Mohammad Pessarakli
Genetic Improvement of Field Crops, edited by Gustavo A. Slafer
Agricultural Field Experiments: Design and Analysis,
    Roger G. Petersen
Environmental Soil Science, Kim H. Tan
*Mechanisms of Plant Growth and Improved Productivity: Modern
    Approaches,* edited by Amarjit S. Basra
Selenium in the Environment, edited by W. T. Frankenberger, Jr.
    and Sally Benson
Plant–Environment Interactions, edited by Robert E. Wilkinson
Handbook of Plant and Crop Physiology, edited by
    Mohammad Pessarakli

Handbook of Phytoalexin Metabolism and Action, edited by M. Daniel and R. P. Purkayastha

Soil–Water Interactions: Mechanisms and Applications, Second Edition, Revised and Expanded, Shingo Iwata, Toshio Tabuchi, and Benno P. Warkentin

Stored-Grain Ecosystems, edited by Digvir S. Jayas, Noel D. G. White, and William E. Muir

Agrochemicals from Natural Products, edited by C. R. A. Godfrey

Seed Development and Germination, edited by Jaime Kigel and Gad Galili

Nitrogen Fertilization in the Environment, edited by Peter Edward Bacon

Phytohormones in Soils: Microbial Production and Function, William T. Frankenberger, Jr., and Muhammad Arshad

Handbook of Weed Management Systems, edited by Albert E. Smith

Soil Sampling, Preparation, and Analysis, Kim H. Tan

Soil Erosion, Conservation, and Rehabilitation, edited by Menachem Agassi

Plant Roots: The Hidden Half, Second Edition, Revised and Expanded, edited by Yoav Waisel, Amram Eshel, and Uzi Kafkafi

Photoassimilate Distribution in Plants and Crops: Source–Sink Relationships, edited by Eli Zamski and Arthur A. Schaffer

Mass Spectrometry of Soils, edited by Thomas W. Boutton and Shinichi Yamasaki

Handbook of Photosynthesis, edited by Mohammad Pessarakli

Chemical and Isotopic Groundwater Hydrology: The Applied Approach, Second Edition, Revised and Expanded, Emanuel Mazor

Fauna in Soil Ecosystems: Recycling Processes, Nutrient Fluxes, and Agricultural Production, edited by Gero Benckiser

Soil and Plant Analysis in Sustainable Agriculture and Environment, edited by Teresa Hood and J. Benton Jones, Jr.

Seeds Handbook: Biology, Production, Processing, and Storage, B. B. Desai, P. M. Kotecha, and D. K. Salunkhe

Modern Soil Microbiology, edited by J. D. van Elsas, J. T. Trevors, and E. M. H. Wellington

Growth and Mineral Nutrition of Field Crops: Second Edition, N. K. Fageria, V. C. Baligar, and Charles Allan Jones

Fungal Pathogenesis in Plants and Crops: Molecular Biology and Host Defense Mechanisms, P. Vidhyasekaran

Plant Pathogen Detection and Disease Diagnosis, P. Narayanasamy

Agricultural Systems Modeling and Simulation, edited by Robert M. Peart and R. Bruce Curry

Agricultural Biotechnology, edited by Arie Altman

Plant–Microbe Interactions and Biological Control, edited by Greg J. Boland and L. David Kuykendall

Handbook of Soil Conditioners: Substances That Enhance the Physical Properties of Soil, edited by Arthur Wallace and Richard E. Terry

Environmental Chemistry of Selenium, edited by William T. Frankenberger, Jr., and Richard A. Engberg

Principles of Soil Chemistry: Third Edition, Revised and Expanded, Kim H. Tan

Sulfur in the Environment, edited by Douglas G. Maynard

Soil–Machine Interactions: A Finite Element Perspective, edited by Jie Shen and Radhey Lal Kushwaha

Mycotoxins in Agriculture and Food Safety, edited by Kaushal K. Sinha and Deepak Bhatnagar

Plant Amino Acids: Biochemistry and Biotechnology, edited by Bijay K. Singh

Handbook of Functional Plant Ecology, edited by Francisco I. Pugnaire and Fernando Valladares

Handbook of Plant and Crop Stress: Second Edition, Revised and Expanded, edited by Mohammad Pessarakli

Plant Responses to Environmental Stresses: From Phytohormones to Genome Reorganization, edited by H. R. Lerner

Handbook of Pest Management, edited by John R. Ruberson

Environmental Soil Science: Second Edition, Revised and Expanded, Kim H. Tan

Microbial Endophytes, edited by Charles W. Bacon and James F. White, Jr.

Plant–Environment Interactions: Second Edition, edited by Robert E. Wilkinson

Microbial Pest Control, Sushil K. Khetan

Soil and Environmental Analysis: Physical Methods, Second Edition, Revised and Expanded, edited by Keith A. Smith and Chris E. Mullins

The Rhizosphere: Biochemistry and Organic Substances at the Soil–Plant Interface, Roberto Pinton, Zeno Varanini, and Paolo Nannipieri

Woody Plants and Woody Plant Management: Ecology, Safety, and Environmental Impact, Rodney W. Bovey

Metals in the Environment, M. N. V. Prasad

Plant Pathogen Detection and Disease Diagnosis: Second Edition, Revised and Expanded, P. Narayanasamy

Handbook of Plant and Crop Physiology: Second Edition, Revised and Expanded, edited by Mohammad Pessarakli

Environmental Chemistry of Arsenic, edited by William T. Frankenberger, Jr.

Enzymes in the Environment: Activity, Ecology, and Applications, edited by Richard G. Burns and Richard P. Dick

Plant Roots: The Hidden Half, Third Edition, Revised and Expanded, edited by Yoav Waisel, Amram Eshel, and Uzi Kafkafi

Handbook of Plant Growth: pH as the Master Variable, edited by Zdenko Rengel

Biological Control of Major Crop Plant Diseases edited by Samuel S. Gnanamanickam

Pesticides in Agriculture and the Environment, edited by Willis B. Wheeler

Mathematical Models of Crop Growth and Yield, , Allen R. Overman and Richard Scholtz

Plant Biotechnology and Transgenic Plants, edited by Kirsi-Marja Oksman Caldentey and Wolfgang Barz

Handbook of Postharvest Technology: Cereals, Fruits, Vegetables, Tea, and Spices, edited by Amalendu Chakraverty, Arun S. Mujumdar, G. S. Vijaya Raghavan, and Hosahalli S. Ramaswamy

Handbook of Soil Acidity, edited by Zdenko Rengel

Humic Matter in Soil and the Environment: Principles and Controversies, edited by Kim H. Tan

Molecular Host Plant Resistance to Pests, edited by S. Sadasivam and B. Thayumanayan

Soil and Environmental Analysis: Modern Instrumental Techniques, Third Edition, edited by Keith A. Smith and Malcolm S. Cresser

Chemical and Isotopic Groundwater Hydrology, Third Edition, edited by Emanuel Mazor

Agricultural Systems Management: Optimizing Efficiency and Performance, edited by Robert M. Peart and W. David Shoup

Physiology and Biotechnology Integration for Plant Breeding, edited by Henry T. Nguyen and Abraham Blum

Global Water Dynamics: Shallow and Deep Groundwater: Petroleum Hydrology: Hydrothermal Fluids, and Landscaping, , edited by Emanuel Mazor

Principles of Soil Physics, edited by Rattan Lal

Seeds Handbook: Biology, Production, Processing, and Storage, Second Edition, Babasaheb B. Desai

Field Sampling: Principles and Practices in Environmental Analysis,
  edited by Alfred R. Conklin

Sustainable Agriculture and the International Rice-Wheat System,
  edited by Rattan Lal, Peter R. Hobbs, Norman Uphoff,
  and David O. Hansen

Plant Toxicology, Fourth Edition, edited by Bertold Hock
  and Erich F. Elstner

*Drought and Water Crises: Science, Technology, and Management
  Issues*, edited by Donald A. Wilhite

Soil Sampling, Preparation, and Analysis, Second Edition, Kim H. Tan

Climate Change and Global Food Security, edited by Rattan Lal,
  Norman Uphoff, B. A. Stewart, and David O. Hansen

Handbook of Photosynthesis, Second Edition, edited by
  Mohammad Pessarakli

*Environmental Soil-Landscape Modeling: Geographic Information
  Technologies and Pedometrics*, edited by Sabine Grunwald

Water Flow In Soils, Second Edition, Tsuyoshi Miyazaki

Biological Approaches to Sustainable Soil Systems, edited by
  Norman Uphoff, Andrew S. Ball, Erick Fernandes, Hans Herren,
  Olivier Husson, Mark Laing, Cheryl Palm, Jules Pretty, Pedro
  Sanchez, Nteranya Sanginga, and Janice Thies

Plant–Environment Interactions, Third Edition, edited by Bingru Huang

Plant-Environment Interactions

Third Edition

edited by
Bingru Huang

Taylor & Francis Group
Boca Raton   London   New York

CRC is an imprint of the Taylor & Francis Group,
an informa business

Published in 2006 by
CRC Press
Taylor & Francis Group
6000 Broken Sound Parkway NW, Suite 300
Boca Raton, FL 33487-2742

© 2006 by Taylor & Francis Group, LLC
CRC Press is an imprint of Taylor & Francis Group

No claim to original U.S. Government works
Printed in the United States of America on acid-free paper
10 9 8 7 6 5 4 3 2 1

International Standard Book Number-10: 0-8493-3727-5 (Hardcover)
International Standard Book Number-13: 978-0-8493-3727-7 (Hardcover)
Library of Congress Card Number 2005044887

This book contains information obtained from authentic and highly regarded sources. Reprinted material is quoted with permission, and sources are indicated. A wide variety of references are listed. Reasonable efforts have been made to publish reliable data and information, but the author and the publisher cannot assume responsibility for the validity of all materials or for the consequences of their use.

No part of this book may be reprinted, reproduced, transmitted, or utilized in any form by any electronic, mechanical, or other means, now known or hereafter invented, including photocopying, microfilming, and recording, or in any information storage or retrieval system, without written permission from the publishers.

For permission to photocopy or use material electronically from this work, please access www.copyright.com (http://www.copyright.com/) or contact the Copyright Clearance Center, Inc. (CCC) 222 Rosewood Drive, Danvers, MA 01923, 978-750-8400. CCC is a not-for-profit organization that provides licenses and registration for a variety of users. For organizations that have been granted a photocopy license by the CCC, a separate system of payment has been arranged.

Trademark Notice: Product or corporate names may be trademarks or registered trademarks, and are used only for identification and explanation without intent to infringe.

Library of Congress Cataloging-in-Publication Data

Plant-environment interactions / edited by Bingru Huang.--3rd ed.
    p. cm. -- (Books in soils, plants, and the environment)
Includes bibliographical references and index.
ISBN 0-8493-3727-5 (alk. paper)
1. Plant ecophysiology. I. Huang, Bingru. II. Series.

QK717.P53 2006
581.7--dc22   2005044887

Taylor & Francis Group
is the Academic Division of Informa plc.

Visit the Taylor & Francis Web site at
http://www.taylorandfrancis.com

and the CRC Press Web site at
http://www.crcpress.com

Preface to the Third Edition

Environmental stresses have been recognized as the most detrimental factors leading to the decline in plant productivity around the world. Understanding plant–environment interactions is becoming increasingly important in agriculture, horticulture, and natural ecosystems, as climates continue to change and natural resources are more limited due to increasing demand. Abiotic stress is the leading cause for the failure of crops to reach their potential yield in agriculture and horticulture. This book examines some of the most important and common abiotic stresses limiting plant growth and productivity in their natural and managed environments, including high temperature, low temperature, drought, waterlogging, salinity, light, and soil physical and chemical (nutrient deficiency and heavy metals) constraints.

The previous editions of *Plant–Environment Interactions* and most other books on the topic of plant and environment interactions focused on plant responses to various environmental stresses or mechanisms of stress tolerance on the whole-plant level. The fundamentals of basic plant-stress physiology continue to receive increased attention. In recent years, much progress has also been made in molecular biology and biotechnology toward understanding plant responses to environmental stress. We are now in an era of rapid discovery, when tremendous amounts of information have been generated, providing insights into the mechanisms of plant and environmental interactions at cellular and molecular levels. It is almost impossible to include all research developments in these broad subjects. Therefore, this new edition of *Plant–Environment Interactions* is an attempt to discuss most, if not all, recent advances in cellular and molecular mechanisms of plant tolerance to abiotic stresses, with consideration of whole-plant stress physiology (Chapters 1 to 10).

Recent literature on the genome (gene expression), proteome (changes in protein metabolism), and biotechnology (genetic manipulation) in relation to plant and environmental interactions are examined. New methods and techniques in stress physiology and molecular biology have been developed in recent years. This book has been expanded from the first two editions to discuss physiological and biochemical methodologies commonly used for the evaluation of plant-stress tolerance (Chapter 11). It also addresses the application of genomic and molecular approaches for the improvement of stress tolerance of plants utilizing mechanisms identified through the analysis of plant–environment interactions at whole-plant, cellular, and molecular levels (Chapter 12). Lastly, the book, through the re-examination of old literature and review of recent advances, raises many unanswered, challenging questions in plant responses to its environment for further exploration in future research.

This book is written with an aim to help readers develop a better understanding of the interaction and integration of whole-plant, cellular, and molecular mechanisms of abiotic stress tolerance of plants, as well as the application of information from cellular and molecular responses to improve whole-plant growth and productivity in suboptimal environments. This book should be a valuable resource for plant physiologists, breeders, molecular biologists, agronomists, horticulturists, and crop scientists at the graduate and research level in both the academic and industrial sectors.

Bingru Huang

Preface to the Second Edition

Every environmental factor influences plant growth and development. Sometimes we recognize a stress factor and can commence experimentation to evaluate responses. These preliminary experiments are done with plants exposed to a single stress. Then after much data gathering and with much trepidation, researchers expose plants to two (or more) stress factors. The ensuing melange of plant responses often induces greater stress in the researcher than in the plant because the results can be additive, synergistic, or antagonistic.

The previous edition focused on plant responses to individual stresses. In this book we attempt to correlate and understand responses to multiple factors. How do water deficit, wind, and salinity affect plant growth when all three are present at one time? Even if it is relatively simple to solve this problem the added man-made stress of polluted water may cause major problems. What does the capacity of roots to change growth patterns in response to any or all of these stress factors such as soil physical and/or chemical constraints, water supply (quantity and depth), salinity, acidity, and alkalinity do to our understanding of the plant responses? This book is an attempt to discuss the frequently complex response of plants to multiple stress factors.

The chapters are fascinating, and they demonstrate how much we have yet to learn. I hope the reader enjoys this book as much as I have enjoyed editing it.

I would like to thank my son, Randall Wilkinson, for his help in preparing the Index.

Robert E. Wilkinson

Preface to the First Edition

Plant response to environment has been a paradigm for millennia. But the total ecosystem influencing plant growth and development has multiparameters. Science has attempted to isolate individual environmental factors so that the plant response to a single stimulus can be quantitated. The success of this attack on apparently insoluble problems can best be evaluated by the increased understanding of plant growth and development that has accumulated in the last century.

But, increasingly, evidence has shown that plant response to a single stimulus is not uniform during the life of the plant, and that a plant is an integrated whole biological entity whereby a change at one level can have a profound influence at a second tissue, organ, or process separated in time or location by some distance from the original stimulus. Thus, this text is an attempt to correlate some of these variables. And, because so many of the environmental parameters produce concomitant responses, those environmental influences produce interactions in the plant.

Basically, a large percentage of the interactions that have been reported have been studied in agricultural systems. Since agriculture is only applied ecology, the relationships between mineral nutrition, plant growth and development, plant–water relations, photoperiod, light intensity, temperature, pesticides, and plant biochemistry are closely interwoven.

Also, there is a natural progression of study and understanding that proceeds from (a) description of biological responses to (b) biochemical and biophysical mechanisms that produce the responses, to (c) genetic manipulation of DNA to understand and create new biological responses. Each portion must correlate with the other types of study.

Concepts of natural food production have evolved to an absence of pesticides that are only synthetic plant growth regulators (PGRs). True, the pesticides may not necessarily be hormone-type PGRs. But, as the various genetic mutants have shown, loss of the ability to produce a requisite component (i.e., chlorophyll) places a severe restraint on continued plant growth and development. This includes eukaryotic and prokaryotic plants. Thus, utilization of one pesticide (i.e., alar) may produce excellent apples that have a degradation product that may possibly induce cancer in x numbers of humans over a two-decade span. At the same time, that particular PGR inhibits the growth and development of "natural" pathogens whose "natural" products induce lethal responses in $200x$ humans over the same period. These applied ecology problems are the province of a vast array of biologists, chemists, biochemists, etc., and, factually, a large proportion of the biochemical and biological knowledge that has accreted in the last few decades about plant growth and development has been a direct result of the development of herbicides, fungicides, and PGRs for use in agriculture.

Definition of the mode of action of these various chemicals has permitted scientists to further isolate specific processes in plants that control plant growth and development. Examples of this progression are seen in the study of diuron as a herbicide that inhibits photosystem II (PSII). And the ability to selectively inhibit specific portions of the entire photosynthetic process by certain diuron concentrations has led to major advances in the study of photosynthesis as a biochemical process. Currently, this study is focusing on the amino acid constituents of DNA involved in the production of specific proteins utilized in PSII. Similar progressions in the development of plant biochemistry, etc., are occurring in many other areas. However, there is always the consideration that conditions, concentrations, and so forth must be carefully monitored. For example, although one diuron concentration has been utilized extensively to study PSII, greater diuron concentrations influence several other biochemical processes. Very rarely does an exogenous compound produce only one reaction regardless of the concentration. Examples of metabolic control by the "second messenger" Ca^{2+} have shown that cytosol Ca^{2+} concentration is very tightly regulated. Variation from the optimal Ca^{2+} concentration results in massively modified cellular metabolism.

Root absorption of mineral nutrients and the growth of plants in relation to the concentrations and ratios of various ions have been studied by agriculturally oriented plant scientists for decades. These studies have benefited mankind tremendously in the production of food and fiber for a constantly increasing world population. But understanding how plant nutrients are absorbed into the root has been a puzzle. Recently, studies of human heart arrhythmia have led to the development of chemicals that control the transfer of Ca^{2+} through heart cell plasma membranes. An entire scientific discipline has developed that is concerned with the biochemistry and biophysics of the ion pumps and voltage-gated ion pores that control Ca^{2+} transport through the plant plasma membrane. These studies have been extended to plant roots, and an explanation of how ions are transported through root plasma membranes may soon be more completely established. Additionally, these processes have been shown to vary between roots of cultivars of a single species, between organelles within a cell, and between (or at the interface of) specific tissues (i.e., phloem sieve tube elements or xylem elements). When confirmed and extended to different species, genomes, and so forth, these findings may help explain many correlations of plant growth and development that currently are totally inexplicable. These enzymes that control transport through membranes are also found in plant pathogens. And alteration of plant epicuticular chemistry has been shown to have a profound influence on the growth and infestation of some plant pathogens.

Natural PGRs (i.e., hormones) are present at different concentrations during the development of the plant. Factors determining concentration-response have been shown to include (a) species, (b) tissue, (c) age, (d) relative concentration of other PGRs, and (e) other *stresses* that develop in the tissue/organ/organism.

Thus, interactions and correlations of plant growth extend through a complete ecological array. One environmental parameter produces one primary response at specific growth stages, etc. However, side reactions also occur. And occasionally those side reactions have striking results. Because these various environmental stresses alter plant growth to differing degrees depending on time of stress and the particular plant response being measured, computer modeling of these factors offers hope of developing an integrated understanding of the entire process. But first the influence of individual stresses and other factors must be ascertained. This text is a compilation of a few of the correlations and interactions that are currently known. We make no claim for discussing all the known interactions, and more correlations will be discovered with additional research. Thus, we present some data for the perusal of students of plant growth and development. Extension of these concepts lies in the province of individual researchers. And, since 90–98% of the researchers who have ever worked throughout recorded history are alive and working today, we feel confident that much more will be learned about these interactions and correlations in the future.

This book constitutes a preliminary introduction to a possible study of a large and difficult subject. I hope that readers learn as much as I have learned while editing these chapters.

Robert E. Wilkinson

Acknowledgments

The chapters in this book were contributed by prominent scientists around the world in the areas of plant-stress physiology, molecular biology, genetics, and breeding, representing international and multidisciplinary perspectives. I wish to express sincere thanks to each of the contributors for their generous scholarly contributions and their enthusiastic cooperation during the process of the book development. I would like to thank all the reviewers who generously devoted their time to provide invaluable comments on the chapters: Dr. Rajeev Arora, Dr. Chee-Kok Chin, Ms. Michelle DaCosta, Dr. Thomas Gianfagna, Dr. Peter Goldsbrough, Dr. Anthony Hall, Dr. Mary Beth Kirkham, Dr. Hans Lambers, Dr. George Liang, Dr. Bobbie McMichael, Dr. Brett Robinson, Mr. Steve McCann, Dr. Shimon Rachmilevitch, Dr. Tara VanToai, and Dr. Jianhua Zhang. Special thanks go to Ms. Emily Merewitz, senior research technician in my lab, who served as a great assistant in compiling this book. Last, but not the least, I want to thank my husband, Ping, and my two lovely children, Eddie and Lisa, for their support and inspiration.

Editor

Bingru Huang is currently a professor at the Department of Plant Biology and Pathology, Rutgers University. She was an assistant professor at Kansas State University prior to joining Rutgers University (1996–2000). She received her B.S. degree (1984) from Hebei Agricultural University and M.S. degree (1987) from Shandong Agricultural University in China. Her Ph.D. degree is from Texas Tech University (1991). She has authored or co-authored over 100 refereed journal articles and 17 book chapters, and co-authored or edited two books in the areas of plant stress physiology. She is the recipient of the Young Crop Scientist Award (1997) from the Crop Science Society of America. She was elected as a Fellow of the American Society of Agronomy in 2003 and Fellow of the Crop Science Society of America in 2004.

Contributors

Jenny Ballif
Department of Plants, Soils
& Biometeorology
Utah State University
Logan, UT

Stacy Bonos
Department of Plant Biology
and Pathology
Rutgers University
New Brunswick, NJ

Kathleen Brown
Department of Horticulture
Pennsylvania State University
University Park, PA

R. Brouquisse
UMR-PCV, DRDC
CEA-Grenoble
Grenoble Cedex, France

John J. Burke
USDA Plant Stress and Germplasm
Development Unit
Lubbock, TX

Junping Chen
USDA Plant Stress and Germplasm
Development Unit
Lubbock, TX

Vishwanathan Chinnusamy
Water Technology Centre
Indian Agricultural Research
Institute
New Delhi, India

Michelle DaCosta
Department of Plant Biology and
Pathology
Rutgers University
New Brunswick, NJ

Dion G. Durnford
Department of Biology
University of New Brunswick
Fredericton, NB, Canada

I. Gharbi
Laboratoire de Physiologie Végétale
Département de Biologie
Faculté des Sciences
Tunis, Tunisia

Bingru Huang
Department of Plant Biology and
Pathology
Rutgers University
New Brunswick, NJ

Penny L. Humby
Department of Biology
University of New Brunswick
Fredericton, NB, Canada

Mary B. Kirkham
Department of Agronomy
Kansas State University
Manhattan, KS

M.S. Liphadzi
Water Research Commission
Gezina, Pretoria, South Africa

Jonathan P. Lynch
Department of Horticulture
Pennsylvania State University
University Park, PA

Shimon Rachmilevitch
Department of Plant Biology
  and Pathology
Rutgers University
New Brunswick, NJ

C.B. Rajashekar
Department of Horticulture,
  Forestry and Recreation
  Resources
Kansas Sate University
Manhattan, KS

Bérénice Ricard
Institut de Biologie Végétale
  Moléculaire
Villenave d'Ornon, France

R.K. Sairam
Division of Plant Physiology
Indian Agricultural Research Institute
New Delhi, India

S. Aschi-Smiti
Laboratoire de Physiologie Végétale
Département de Biologie
Faculté des Sciences
Tunis, Tunisia

Aruna Tyagi
Division of Biochemistry
Indian Agricultural Research Institute
New Delhi, India

Hong Wang
Graduate School of Bioagricultural
  Sciences
Nagoya University
Chikusa, Nagoya, Japan

Yajun Wu
Department of Plants, Soils &
  Biometeorology
Utah State University
Logan, UT

Akira Yamauchi
Graduate School of Bioagricultural
  Sciences
Nagoya University
Chikusa, Nagoya, Japan

Contents

Chapter 1 Cellular Membranes in Stress Sensing and Regulation of Plant Adaptation to Abiotic Stresses 1

Bingru Huang

Chapter 2 Changes in Cellular and Molecular Processes in Plant Adaptation to Heat Stress 27

John J. Burke and Junping Chen

Chapter 3 Molecular Responses and Mechanisms of Plant Adaptation to Cold and Freezing Stress 47

C.B. Rajashekar

Chapter 4 Photoacclimation: Physiological and Molecular Responses to Changes in Light Environments 69

Penny L. Humby and Dion G. Durnford

Chapter 5 Molecular Mechanisms in Hormonal Regulation of Drought Tolerance..........101

Yajun Wu, Jenny Ballif, and Bingru Huang

Chapter 6 Salinity Tolerance: Cellular Mechanisms and Gene Regulation 121

R.K. Sairam, Aruna Tyagi, and Vishwanathan Chinnusamy

Chapter 7 Cellular and Molecular Mechanisms of Plant Tolerance to Waterlogging 177

B. Ricard, S. Aschi-Smiti, I. Gharbi, and R. Brouquisse

Chapter 8 Whole-Plant Adaptations to Low Phosphorus Availability 209

Jonathan P. Lynch and Kathleen Brown

Chapter 9 Physiological Effects of Heavy Metals on Plant
Growth and Function 243

M.S. Liphadzi and M.B. Kirkham

Chapter 10 Growth and Function of Roots under Abiotic
Stress in Soils 271

Hong Wang and Akira Yamauchi

Chapter 11 Physiological and Biochemical Indicators
for Stress Tolerance 321

Shimon Rachmilevitch, Michelle DaCosta, and Bingru Huang

Chapter 12 Breeding and Genomic Approaches to Improving
Abiotic Stress Tolerance in Plants 357

Stacy A. Bonos and Bingru Huang

Index 377

1 Cellular Membranes in Stress Sensing and Regulation of Plant Adaptation to Abiotic Stresses

Bingru Huang

CONTENTS

1.1 Introduction 2
1.2 Membrane Structure and Composition 3
1.3 Membrane Lipids and Plant Adaptation to High
    and Low Temperatures 3
    1.3.1 Changes in Membrane Structure and Lipid Properties
          in Response to Temperature Changes 4
          1.3.1.1 Heat Stress 4
          1.3.1.2 Low Temperature 7
    1.3.2 Lipid Metabolism and Temperature Stress Tolerance 8
          1.3.2.1 Regulation of Lipid Saturation or Unsaturation
                  in Temperature Stress Adaptation 8
          1.3.2.2 Membrane Sensing and Signaling
                  of Temperature Stress 10
1.4 Membrane Lipids and Plant Adaptation to Drought Stress 11
    1.4.1 Composition and Saturation of Membrane Lipids
          in Relation to Drought Tolerance 11
    1.4.2 Lipid Metabolism and Drought Stress Signaling 12
1.5 Membrane Lipids and Plant Adaptation to Salinity Stress 13
    1.5.1 Change in Membrane Lipids during Salinity Stress 13
    1.5.2 Lipid Metabolism in Response to Salinity Stress 15
1.6 Major Cellular Processes Associated with Membrane
    Damages during Abiotic Stresses 16
1.7 Future Research Perspectives 17
References 18

1.1 INTRODUCTION

Plants growing in natural environments are constantly subjected to various environmental stresses. Environmental stresses can be from the soil, including soil physical constraints and chemical stress such as nutrient deficiency and toxicity, and can also be from the atmosphere, such as temperature, moisture, air pollutants, and light. High temperature is a primary stress limiting the growth of cool-season plant species during the summer. The problem of high-temperature stress is expected to accentuate in the future as global temperature is projected to rise 1 to 4.5°C in the next 50 years [1]. It seems likely that throughout most of the temperate zone there will be longer warm periods and hotter extremes, which could intensify heat-stress injury in many cool-season plant species. Low temperatures typically occur during winter, late fall, and early spring in cold climatic regions. Chilling or freezing is a major stress threatening the growth of warm-season plants. Drought stress has a global impact, which limits the growth of both cool-season and warm-season plant species. Water deficit due to lack of water supply or drought stress is often associated with salinity, winter desiccation, and heat stress. Water stress can become more significant as water becomes increasingly limited for irrigation.

Unlike animals, plants are by and large incapable of escape from stresses that affect their activities. However, plants adapted to their growing environments have developed some resistance mechanisms in order to survive in those conditions. Although numerous processes change during plant adaptation to environmental stresses, cellular membranes are found to be the critical sites of stress injury, including chilling, freezing, heat, drought, and salinity stress. Membranes define the outer boundary of plant cells and the structure of internal organelles. Membranes regulate the flow of materials between cells and their environment, as well as their internal compartments and provide key metabolic interfaces between the plant and its environment. As the selective barrier between living cells and their environments, plasma membranes may be the first structures involved in the perception and transmission of external stress signals, and protecting a cell from the stress, and thus changes in membrane structure and property may constitute the initial responses of a plant cell to environmental stresses among other processes [2]. A variety of reports suggests that specific membrane lipids may be involved in the early stress signal perception and transduction [3].

Maintenance of membrane integrity in anhydrobiotic organisms has been found to play a critical role in desiccation tolerance [4]. The role of membranes in survival of bacteria at extreme temperatures, salinity, and water deficit has been reported [5,6]. Larcher [2] attributed high temperature resistance in plants to highly resistant cell membranes, among other things. Membrane lipids are important for maintaining photosynthetic activity and photosynthetic integrity of plants exposed to high temperature stress, contributing to improved heat tolerance [7]. There is ample evidence for the role of membrane lipids in cold acclimation and freezing tolerance [8]. Temperature, light quality and intensity, degree of hydration, ionic interactions, pH, and phytohormones affect membrane fluidity, and thus cell functions [9]. It is clear from the literature that modifications in membrane structure and composition play a major role in plant adaptation to diverse environmental stresses, including extreme temperatures, drought, and salinity.

This chapter reviews some recent advances in the involvement of cellular membrane structure and composition changes in plant adaptation and tolerance to heat, cold, drought, and salinity stress. Membrane structure and composition are introduced briefly here, inasmuch as more basic information on membrane structure and function can be found in many textbooks. The alteration in membrane lipid composition, saturation, content, and metabolism during plant adaptation to extreme temperatures, drought, and salinity are discussed. Other physiological and molecular mechanisms of plant tolerance to these major environmental stresses are discussed in Chapters 2 through 5.

1.2 MEMBRANE STRUCTURE AND COMPOSITION

Plasma membranes are a vital parts of a cell, which separate the cell from the external environment. Most of the functional units or organelles within a cell, such as chloroplasts, mitochondria, nuclei, vacuoles, and endoplasmic reticula, have one or more membrane layers. Cellular membranes are composed of lipids, peripheral and integral proteins, sterols, and carbohydrates. The major components of membranes are lipids and proteins. The proportion of each component varies with the type of membrane. The bilayer lipid arrangement defines the plane of the membrane with which proteins are associated. The lipid portion of the membrane represents a substantial barrier to material flow, but the proteins associated with the membrane impart the means for selective transport and accumulation of solutes.

One of the principal types of lipids in cellular membranes is the phospholipid. Phospholipids consist of a polar head group (hydrophilic) and two hydroxyl groups esterified to long-chain fatty acids (hydrophobic) The fatty acids in membrane lipids may either be saturated (no double bond between carbon atoms) or unsaturated (one or more double bonds in the *cis* or *trans* configuration). The composition and saturation level of fatty acids in phospholipids gives the membrane its fluid characteristics, regulating cell membrane fluidity. Proper cellular metabolism and function requires a certain physical state or fluidity of membranes. The presence of common *cis* double bonds helps maintain membrane fluidity by introducing bends or kinks in the fatty acyl chains, which inhibits tight packing of adjacent lipid molecules [10]. Environmental stresses may modify membrane structure and composition, and thus cellular functions.

1.3 MEMBRANE LIPIDS AND PLANT ADAPTATION TO HIGH AND LOW TEMPERATURES

Membrane lipids are highly susceptible to changes in temperatures, which can consequently change membrane fluidity, permeability, and cellular metabolic functions [11]. High temperature fluidizes membranes by "melting" the lipid bilayer, which allows lipids and other membrane components to move around, rotate, and exchange places. In contrast, low temperature rigidifies membranes in which phospholipid bilayers are present in a gel-like, tightly packed state. Changes in membrane physical state or fluidity are largely dictated by the composition of lipid molecule species and the degree of saturation of fatty acids in membrane phospholipids.

A large body of literature reports membrane lipid changes during cold acclimation and the involvement of membrane lipid metabolism in cold stress signal transduction and gene expression, which is discussed in detail in Chapter 3 in this book and in other recent reviews [12]. Relatively less information is available on how changes in membrane physical structures and chemical properties regulate plant tolerance to heat stress. The following section reviews some of the recent literature on alteration of membrane lipid properties and metabolism in response to temperature changes, with emphasis on heat stress and a brief discussion on cold stress.

1.3.1 Changes in Membrane Structure and Lipid Properties in Response to Temperature Changes

1.3.1.1 Heat Stress

Cell integrity depends on the maintenance of the structural and functional integrity of cellular membranes. Heat stress alters the structural and functional integrity of cell membranes, as manifested by changes in enzyme dysfunction, membrane leakage, or other cellular damages such as enzyme inefficiency and impairment of cell metabolic activity [2]. Cell membranes become more fluid as temperature rises, increasing membrane permeability and the chance for leakage of ions and other cellular compounds from the cell.

Alteration of membrane fluidity has been found to affect plant tolerance to heat stress. The importance of maintaining proper membrane fluidity in temperature stress tolerance has been illustrated in studies using chemical modification, mutation analysis, and gene transformation by modifying membrane physical state, lipid composition, or degree of lipid saturation. For example, mutants of soybean (*Glycine max* (L.) Merr.) [13] and *Arabidopsis* [14] that are deficient in fatty acid unsaturation maintained stable membrane fluidity and showed improved tolerance to high temperature. Zaharieva et al. [15] examined the effect of short-term (5 min) heat stress (20–55°C) on energy distribution and photochemical activity of photosystem I in isolated pea (*Pisum sativum L.*) chloroplasts that were previously treated with chemicals to alter thylakoid membrane fluidity. They found that incorporation of cholesterol into thylakoid membranes reduced membrane fluidity by 50% compared with controls, which resulted in a 50% increase in photosystem I photochemical activity. However, a 20% increase in thylakoid membrane fluidity due to benzyl alcohol treatment inhibited electron transport by 10%. This study illustrated the importance of membrane fluidity in regulating thermotolerance of the photosynthesis system.

Changes in membrane fluidity can be the result of single or combined effects of the degree of saturation and composition of lipids in the membrane [11]. The degree of saturation and composition of lipids are sensitive to changes in environmental conditions, including high temperature. Lipid saturation level typically increases whereas unsaturated lipids decrease with increasing temperatures [16–20]. In *Arabidopsis* exposed to high temperatures, total lipid content decreased and the ratio of unsaturated to saturated fatty acids decreased [21]. An increase in the temperature caused a reduction in α-linolenic acid (18:3) and hexadecatrienoic acid (16:3) in some desert and evergreen plants [22,23]. The increase in lipid saturation in *Agrostis stolonifera* exposed to high temperature was primarily due to a decrease in linolenic acid (18:3) and an increase in linoleic and palmitic acids (18:2, 16:0) [20].

The increases in the levels of saturated fatty acids have been found in different cellular membranes including thylakoid membranes [24] and plasma membranes [25]. Many of the measured changes in membrane lipids during heat acclimation and stress have been done in the thylakoid membranes [19,26]. Some studies have shown that lipid changes in thylakoid membranes are more sensitive to heat stress than changes in the plasma membrane, and that the changes in lipid saturation are predominantly occurring in the thylakoid membranes [17]. However, the relative importance of lipid changes in different membranes in heat tolerance is not clear.

An increase in saturation of membrane lipids increases their melting temperature, which might retard the increase in membrane fluidity at high temperatures. Therefore, an increase in fatty acid saturation may increase the thermostability of biological membranes. However, although there are certain amounts of data supporting a direct link of lipid saturation and heat tolerance, there are also some data showing otherwise. Some studies indicate an association between increased lipid saturation and heat stress tolerance. For example, transgenic tobacco (*Nicotiana tobaccum* L.) plants with a low level of unsaturated fatty acids survived longer at high temperatures [19]. Heat-resistant wheat (*Triticum aestivum* L.) cultivars had significantly more saturated lipid membranes than sensitive cultivars under heat stress [27]. A heat-sensitive cultivar of *A. stolonifera* that had acquired thermotolerance through heat acclimation had increased concentration of saturated fatty acids [20]. Alfonso et al. [13] reported that a new herbicide-resistant D1 mutant from soybean (STR7) had much higher tolerance to high temperature than wild-type plants, which was associated with an unusually high accumulation of saturated C16:0 and reduced levels of C16:1 and C18:3 unsaturated fatty acids in the chloroplast membranes of the heat-resistant mutant compared with the wild-type plants.

In contrast, some studies have found that increasing lipid saturation is not related to improved heat tolerance [28–32]. Increasing lipid saturation did not protect photosynthesis from heat injury [29,31]. Heat-tolerant cultivars of potato (*Solanum tuberosum* L.) had slightly higher linoleic and linolenic contents than control plants (i.e., more unsaturated lipids), whereas the levels of oleic acid were lower in tolerant cultivars [30]. In the same study, the heat-tolerant cultivar had increased palmitic and decreased oleic and linolenic acids, and sensitive cultivars had decreased oleic and linolenic acids when both cultivars were exposed to heat stress [30]. As palmitic is a saturated fatty acid, the data appear to support the contention that increase in saturation enhances heat tolerance. Studies that found no correlation between heat tolerance and lipid saturation suggest that factors other than membrane stability may be limiting growth at high temperature. However, at present, it is obvious that the relationship between membrane lipid saturation and heat tolerance is not yet clear and requires further investigation.

The physical properties of a biological membrane are also affected by lipid composition, which has been found to be involved in temperature acclimation of plants [7]. In a comparison of the lipids in the chloroplasts and mitochondria between a heat-tolerant wheat (*Triticum aestivum*) cultivar and a heat-sensitive cultivar exposed to 45°C, the content of neutral lipids increased, especially in chloroplasts of the tolerant cultivar whereas content of phospholipids increased only in the heat-sensitive cultivar [33]. In wheat leaves of both a heat-tolerant mutant and its heat-sensitive parent, polar lipids constituted 80–90% of the total lipids; galactolipids, phospholipids, and neutral lipid contents were higher in leaves of heat-stressed plants of both genotypes than controls; and the mutant showed a higher content of galactolipid-bound linolenic acid and especially phospholipids-bound trans-Δ-3-hexadecanoic acid than the heat-sensitive cultivar [34]. In a heat-tolerant mutant of soybean (STR7), chloroplastic lipids, including mono-galactosyl-diacyl-glycerol (MGDG), phosphatidylglycerol (PG), and di-galactosyl-diacyl-glycerol (DGDG), synthesized via the prokaryotic pathway represented more than 75% of the total lipid classes among all the lipid classes [13]. It seems that high proportion and stability of membrane polar lipids is important for plant adaptation to heat stress.

As mentioned previously, most of the studies on changes in lipid properties during heat stress have been done primarily with chloroplast thylakoid membranes [17,19,26]. Roots, which consist of mainly endoplasmic reticulum, plasma, and mitochondrial membranes, are more sensitive to high temperatures than leaves [35–38]. This difference may be related to the difference in the lipid properties of chloroplast and root membranes. Few studies have been conducted on root lipid changes in response to increasing temperatures. Diepenbrock et al. [30] reported that palmitic acid level did not change during high-temperature treatment (30°C) in roots of a heat-sensitive cultivar of potato, but increased in roots of a heat-tolerant cultivar compared to roots at 20°C; oleic acid decreased in both genotypes at 30°C; and linolenic acid concentration decreased with time of treatment, but cultivar differences were not significant. Steyer et al. [39] reported a significant increase in the proportion of oleate (18:1) at the expenses of linoleate (18:2) for carrot (*Daucus carota L.*) cells in response to increasing temperatures. The differential change in lipid composition at high temperatures was attributed to the inhibition or loss of desaturase responsible for inserting the double bond between carbons 12 and 13 of oleate esterified to the glycerol moiety of phosphatidylcholine (PC) and phosphatidylethanolamine (PE) [39].

Liu and Huang [40] reported that the level of saturation of fatty acids in *A. stolonifera* roots increased with increases in soil temperature. Larkindale and Huang [20] also found that cultivars of *A. stolonifera* with contrasting heat tolerance differed in root lipid saturation, whereby the double bond index for lipids from roots of the heat-sensitive cultivar was significantly higher than that for the heat-tolerant cultivar. This difference was predominantly associated with a higher level of the unsaturated fatty acid linolenic acid (18:3). Results from this study suggested that lipid saturation levels of roots could be an important factor controlling root tolerance to high soil temperatures. However, how changes in lipid composition and fluidity may be associated with thermotolerance of roots is not clear, and which membrane lipid property or lipid/fatty acid species is more important in regulating root thermotolerance is unknown.

1.3.1.2 Low Temperature

In contrast to responses to heat, membranes respond to chilling or freezing stress by undergoing a phase transition from a highly fluid, liquid crystalline phase to a more rigid gel phase, in which lipids are closely packed and more highly ordered, and thereby can interfere with normal physiological functions of the membranes. Unsaturated lipids may lower the transition temperature by introducing bends in the otherwise linear fatty acyl chains, and therefore may have a pivotal role in maintaining membrane fluidity for proper functioning under low-temperature stress. Numerous studies have documented that membrane lipids become more unsaturated during cold acclimation or low-temperature stresses including chilling and freezing [41–46].

Accumulating evidence has shown that higher levels of unsaturation of membrane lipids are positively correlated to plant tolerance to low-temperature stresses [44,45,47–50]. Bertin et al. [50] examined the relationship between chilling tolerance of six rice (*Oryza sativa L.*) cultivars differing in chilling tolerance and the fatty acid composition in total lipids, phospholipids, galactolipids, and neutral lipids from leaves. They reported that higher double bond index and proportions of linolenic acid in the phospholipid and galactolipid classes accounted for the variation in chilling tolerance among cultivars, but chilling tolerance was not related to changes in total lipids and the neutral lipid class.

Cyril et al. [45] found no changes in the two saturated fatty acids, palmitic acid and stearic acid, in three genotypes of seashore paspalum (*Paspalum vaginatum* Sw.), but a significant increase in the triunsaturated linolenic acid during low-temperature exposure. The increase in unsaturated fatty acids was more pronounced in the more cold-tolerant cultivar "SeaIsle1" than in the intermediately cold-tolerant "Adalayd" or in the cold-susceptible PI 299042. Cold-tolerant bermudagrass (*Cynodon dactylon* (L.) *Pers.*) genotypes also showed a greater magnitude of increase in linolenic acid than did the relatively cold-susceptible genotypes [43,45].

Extensive unsaturation of phosphatidylglycerol, an abundant phospholipid in thylakoid membranes, correlated with improved chilling tolerance of tobacco [51].

Similarly, chilling-resistant cucumbers (*Cucumis sativus* L.) and sweet peppers (*Capsicum annuum* L.) have a greater amount of unsaturated fatty acids than saturated fatty acids [52]. Collectively, by comparing species or cultivars differing in cold tolerance, various studies present strong evidence that the increase in unsaturated fatty acids or lipids in membranes is associated with improved tolerance to low-temperature stress.

The role of lipid unsaturation in cold tolerance has been confirmed using transgenic plants. Murata et al. [47] transformed tobacco using acyl-ACP:glycerol-3-phosphate acyltransferase (GPAT) from two plant species with contrasting chilling tolerance, chilling-sensitive squash (*Cucurbita maxima Duchesne*), and chilling-tolerant *Arabidopsis*. Transgenic tobacco carrying GPAT from squash had increased levels of saturated phosphatidylglycerol, whereas those transgenic tobacco plants with the *Arabidopsis* enzyme showed decreased levels of saturated lipids. The increase in saturation level of lipids in the transgenic tobacco was correlated to lower chilling tolerance. Conversely, the decrease in lipid saturation was correlated with more chilling tolerance of the transgenic tobacco. Tobacco transformed with GPAT from *E. coli* that had more saturated phosphatidyl choline also was more cold sensitive [48]. An increase in the unsaturation level of fatty acids in thylakoid membrane lipids in transgenic *Arabidopsis* has been associated with the enhanced recovery of the photosystem II protein complex from photoinhibitory damage at low temperature, suggesting that the degree of unsaturation of fatty acids in membrane lipids was directly proportional to the chilling tolerance of photosynthesis [53–55].

1.3.2 LIPID METABOLISM AND TEMPERATURE STRESS TOLERANCE

1.3.2.1 Regulation of Lipid Saturation or Unsaturation in Temperature Stress Adaptation

The biosynthetic pathway leading to the synthesis of saturated or unsaturated fatty acids or lipids may be regulated in plant adaptation to temperature stress. Fatty acid desaturases are the enzymes that introduce the double bonds into the hydrocarbon chains of fatty acids, and thus these enzymes play an important role during the process of temperature acclimation of higher plants and other organisms [56,57]. Linolenic acid is synthesized from stearic acid (C18:0) by a series of desaturase enzymes. Stearic acid is converted to oleic acid (C18:1) by stearoyl-ACP desaturase. Formation of the second double bond is catalyzed by oleoyl-PC desaturase to form linoleic acid (C18:2). The third double bond is synthesized by linoleoyl-PC desaturase to form linolenic acid (C18:3). It is possible that either the activity of the desaturase(s) or the expression of the gene(s) responsible for the synthesis of one or more of the desaturase enzymes may be temperature regulated.

An *Arabidopsis* mutant that was deficient in the activity of chloroplast fatty acid T9 desaturase had increased saturation of chloroplast lipids, which was associated with the increase in the optimum growth temperature [29,58]. Recently, Murakami et al. [7] silenced the gene encoding the chloroplast version of T3-fatty acid desaturase enzyme (Fad7 synthesizes lipids containing three double bonds) in tobacco, which resulted in a decrease in unsaturated lipids with three double bonds and a corresponding increase in lipids with two double bonds in thylakoid membranes.

Transgenic tobacco plants contained a lower level of trienoic fatty acid than wild-type plants, and grew much better than controls at higher temperatures. Differences in growth rate were noted at 36°C, and transgenic plants survived for 2 h at 47°C, which killed their wild-type counterparts. This work illustrated that thermotolerance is a function of the lipid profile of photosynthetic membranes. Murakami et al. [7] provided strong evidence that the lower number of unsaturated lipids in the thylakoid membranes of chloroplasts was associated with increased photosynthetic capacity and the plant's ability to survive at temperatures of 35°C or higher. They pointed out that the T3-fatty acid desaturase enzyme is conservatively expressed in nearly all plant species, which may be involved in high-temperature tolerance in all plants. This research demonstrated that thermotolerance was related to membrane properties, and that the growth and survival of plants were affected by the thermostability of membranes.

Various mechanisms have been suggested for the regulation of the unsaturation of the fatty acid of membrane lipids in plant adaptation to cold temperature. Several studies indicate that genes of a number of desaturases can be activated by low temperatures (see review [59]). Linoleoyl (ω-3) desaturase regulates crucial steps in the unsaturation of membrane lipids, and has been found to play a critical role in improving cold tolerance [60]. Kodama et al. [49] reported that overexpressing a plastid localized ω-3 desaturase (*fad7*) gene from *Arabidopsis* had increased levels of two triunsaturated fatty acids in tobacco plants, and those plants had improved chilling tolerance.

The *fad8* ω-3 desaturase gene is also upregulated in response to low temperature in *Arabidopsis* [61]. In contrast, *Arabidopsis* mutants that were deficient in ω-3 desaturase activity demonstrated high chilling sensitivity [62,63]. The temperature-dependent desaturase activity could be controlled at the transcriptional, translational, or post-translational level [64]. In this aspect, isolation of the genes encoding plastid ω-3 desaturase from *Arabidopsis* (*FAD8* [61]) and *Zea mays* (*ZmFAD8* [65]), whose transcript levels increased at low temperature, supports the concept that the level of desaturase activity is regulated at the mRNA level.

Phospholipid hydrolysis or degradation occurs in response to various environmental stresses [46,66,67]. Phospholipases are major lipid-degrading enzymes in plants. They hydrolyze phospholipids to produce a free fatty acid and a lysophospholipid. As such, their activities may have a significant impact on the structure and stability of cellular membranes. These enzymes are also involved in a broad range of other functions in cellular regulation. One of the phospholipases, phospholipase D (PLD), has been identified as an important enzyme regulating a wide range of cellular processes in plants [68]. This enzyme has been found to be involved in abscisic acid- and ethylene-promoted senescence [69], wound-induced accumulation of unsaturated fatty acids [70,71], and reactive oxygen generation [72]. It has also been found to affect plant tolerance to various stresses.

Wang and Wang [73] reported that PLD was more abundant in roots, flowers, and stems than in leaves, and the amount of PLD was greater in senescent than young leaves in *Arabidopsis*.

Maarouf et al. [74] reported that gene expression of PLD in cowpea (*Vigna unguiculata L.W.Ip.*) plants was strongly upregulated by water stress in the drought-sensitive plants while remaining remarkably stable in drought-tolerant plants during stress. Li et al. [75] suggested that freezing tolerance of *Arabidopsis* could be manipulated by depletion or overexpression of PLD, suggesting the involvement of these enzymes in plant tolerance to freezing. Phospholipase D activity increased markedly during Ca^{2+} starvation for roots of cucumber, suggesting that the increase in PLD activity and the reduction in membrane-bound Ca^{2+} might be responsible for membrane structure disorganization and the changes in enzyme activities [76].

In addition to PLD, other lipases have also been found to be involved in stress responses. He et al. [77] reported that phospholipase C (PLC) activity increased in plasma membranes of maize root tips exposed to hypoxia for three days and exogenous ethylene also increased PLC activity in the apical 10 mm root zone, suggesting that increased PLC activity in the plasma membrane is an early marker for cell death in the root cortex during hypoxia. It is suggested that early changes in lipid composition of wheat roots associated with mechanical stress may also be caused by the action of lipases, in particular phospholipase A2 [78].

1.3.2.2 Membrane Sensing and Signaling of Temperature Stress

Presumably, changes in membrane fluidity begin with detection of temperature change by a sensor or sensors. There are likely multiple primary sensors that perceive the initial stress signal [79]. Sangwan et al. [80] indicated that cold and heat were sensed by structural changes in the plasma membrane that translated the signal via cytoskeleton, Ca^{2+} fluxes, and CDPKs (calcium-dependent protein kinases) into the activation of distinct MAPK cascades. Mitogen-activated protein kinases (MAPKs) appear to be ubiquitously involved in signal transduction in responses to external stimuli. Cold-induced rigidification and heat-induced fluidization of membranes may trigger the activation of stress-activated MAPK and heat-shock activated MAPK, respectively. Some reported that membrane proteins, including histidine kinases, are possible sensors for low-temperature perception [81]. There is evidence suggesting that histidine kinase Hik33 in cyanobacterium [81] and histidine kinase DesK in *Bacillus subtilis* [82] are thermosensors, which regulate desaturase gene expression in response to lowering temperatures.

Phospholipases have also been found to play an important role in regulating many critical cellular functions by generating second messengers [68]. Environmental stresses may activate various lipases releasing diacyglycerol, lipid head groups, free fatty acids, and other moieties that may be directly or indirectly involved in stress-signaling processes [68]. PLD hydrolyzes phospholipids to generate phosphatidic acid (PA), a second messenger in animal cells involved in the activation of PI-PLC and protein kinase C [83]. Wang [68] reported that PA may also serve as a messenger in plant stress responses. Some recent studies demonstrated the importance of PLD in regulating plant tolerance to freezing by producing signaling molecules [46,75]. The involvement of PLD in signaling and regulation of cold-stress responses is reviewed in detail by Rajashekar [8]. The role of PLD in heat-stress signaling and regulation is not well documented.

1.4 MEMBRANE LIPIDS AND PLANT ADAPTATION TO DROUGHT STRESS

The ability of plant survival during drought stress, or often referred to as drought tolerance, involves structural and physiological adaptations, which allow plants to survive extended periods of limited water availability. Drought injury is characterized by cell dehydration, which alters cell membranes. Drought tolerance could be partially due to changes in membrane composition and structure, which optimize membrane fluidity [84]. Membrane alteration induced by dehydration can be associated with changes in lipids and proteins. Phase property and composition of lipids are sensitive to dehydration, and changes in lipids represent early responses of the membrane to the stress. Alteration in lipid composition and saturation level may help to maintain membrane integrity and preserve cell compartmentation during cell dehydration. Changes in membrane lipid composition, such as caused by dehydration, may induce changes in the structure and function of intrinsic membrane protein complexes, and thus interrupt cell functions [85,86].

1.4.1 Composition and Saturation of Membrane Lipids in Relation to Drought Tolerance

Changes in the content, composition, and saturation of phospholipids are detected in plants and other organisms exposed to water stress [87]. Membrane lipid content generally decreases during drought stress [88,89], which has been shown to be associated with an inhibition of lipid biosynthesis and a stimulation of lipolytic and peroxidative processes [90–92]. In a recent study, *Arabidopsis* leaf membranes appeared to be remarkably resistant to water deficit, which was attributed to the capacity to maintain their polar lipid contents and the stability of their lipid composition under severe stress conditions [93]. When leaf relative water content declined to 48% during drying, total leaf lipid contents decreased, but were still maintained at 60% of the control level, and lipid class distribution remained relatively stable.

Among the polar lipids, monogalactosyldiacylglycerols (MGDG) seems to be most sensitive to dehydration, as found in *Arabidopsis* [93], the resurrection plants *Ramonda serbica* and *R. nathaliae* [94], rape (*Brassica napus* L.) [95], and cowpea [96]. It is known that MGDG in a pure lipid–water mixture forms a cylindrical inverted hexagonal structure instead of forming the bilayer configuration as for other lipids such as digalactosyldiacylglycerols (DGDG). Therefore, the proportion between MGDG and DGDG is very important for the stability of membranes.

Under stressful conditions, an increase in the ratio of DGDG:MGDG would help maintain membrane stability and fluidity necessary for biological functions [97]. An increase in the DGDG:MGDG ratio in response to drought has been reported in *Arabidopsis* [93].

Quartacci et al. [98] reported that water deficit caused significant changes in the composition and saturation level of lipids in thylakoid membranes in two wheat (*Triticum durum* Desf.) cultivars differing in drought tolerance. The level of DGDG was unaffected by water deficit in both cultivars whereas MGDG content decreased in the tolerant cultivars, but increased in the sensitive cultivars under stress. The decrease in MGDG content was associated with upregulation of a patatinlike and lipoxygenase gene controlling MGDG degradation and the downregulation of expression of AtMGD1, the gene responsible for 70% of the bulk MGDG synthesis [93,99].

Changes in fatty acid saturation level allow cells to maintain an appropriate balance of bilayer- and nonbilayer-forming lipids in the membrane, which may play a role in plant adaptation to drought [100]. Gounaris et al. [100] suggested that increasing the saturation level reduces the tendency of MGDG to form a nonbilayer structure, thus increasing the stability of membranes. Following water deficit, the level of unsaturation for MGDG increased in the drought-sensitive wheat cultivar, but decreased in the tolerant cultivar [98]. The lower unsaturation level or higher saturation level of MGDG may help maintain membrane stability during drought and protect cells from dehydration. However, other comparative studies between cultivars differing in drought tolerance showed that the unsaturation level of polar lipids decreased in sensitive cultivars, but was unchanged or increased in tolerant cultivars exposed to drought stress in cowpea [89] and coconut trees (*Cocos nucifera* L.) [101]. Apparently, a clear relationship between the unsaturation level of membrane lipids and drought tolerance is yet to be established.

Drought adaptation is not only affected by changes in lipid composition and saturation levels, but also may be related to changes in the proportion of lipids to proteins in membranes. Quartacci et al. [98] reported increased lipid-to-protein ratios in thylakoid membranes of drought-tolerant wheat cultivar, which may also be important for drought adaptation, as the lipid-to-protein ratio could contribute to the regulation of membrane fluidity. The increase in this ratio has been shown to lead to a more fluid membrane [102].

1.4.2 LIPID METABOLISM AND DROUGHT STRESS SIGNALING

It has recently been shown that molecules produced through phospholipid degradation under environmental stresses may serve as precursors for the generation of secondary messenger molecules in stress-signal transduction pathways [87]. Drought upregulates the mRNA levels for certain PI-PLC (phosphoinositide-specific phospholipase C) isoforms [103,104], contributing to increased cleavage of phosphotidylinositol 4,5-bisphosphate (PIP_2) to produce two important molecules, diacylglycerol and inositol 1,4,5-trisphosphate (IP_3). Diacylglycerol and IP_3 are second messengers that can activate protein kinase C and trigger Ca^{2+} release, respectively, initiating a stress-signal cascade. In addition to serving as precursors to IP_3 and diacylglycerol, phosphoinositides have been shown recently to actively participate in other cellular functions during plant adaptation to drought. They may regulate the dynamics of the actin cytoskeleton through the interaction with actin-binding proteins and to potentiate the activation of protein kinase C and PI-PLC [105].

Accumulating evidence suggests that PLD in plants is also involved in the transduction of dehydration stress signals. Drought stimulates activities of PLD, leading to transient increases in PA levels in plants [106–108]. PLD appears to be activated by water stress through a G-protein [106]. Several studies reported that PLD may also be involved in drought adaptation by mediating ABA-signaling for stomatal closure, which reduces water loss during water stress [72,109].

Some studies suggest that excess PLD activity may have a negative impact on plant stress tolerance. High concentration of PA, a nonbilayer lipid favoring hexagonal phase formation, may destabilize membranes [68]. Responses of PLD to water stress were studied in two cultivars of cowpea plants differing in their tolerance to drought [74]. The activities of PLD increased when plants were exposed to water stress, with the increase being more pronounced in the drought-sensitive cultivar. This study suggested that PLD in cowpea plants was regulated at the transcriptional level, and that the expression of *VuPLD1* gene in the drought-sensitive cultivar was sensitive to water stress; in contrast, the drought-tolerant plants maintained remarkable stability of PLD gene expression, suggesting that a high PLD activity may jeopardize membrane integrity [74].

1.5 MEMBRANE LIPIDS AND PLANT ADAPTATION TO SALINITY STRESS

Salinity injury in plants includes both ionic (chemical) and osmotic (physical) components. One of the important mechanisms of salt tolerance of higher plants is ion compartmentation, and compartmentation of salts depends on the permeability of cellular membranes [110]. Therefore, the regulation of ion movement across the tonoplast and plasma membranes is a key cellular factor in salinity tolerance in order to maintain a low Na^+ concentration in the cytoplasm and the accumulation of salts in vacuoles. Numerous studies demonstrated the importance of plasma membranes in plant tolerance to salinity in both glycophytes and halophytes [111–113]. This section discusses changes in membrane lipid composition and saturation in relation to plant adaptation to salinity. Changes in membrane lipid metabolism associated with salinity stress signaling are discussed briefly here. Some related literature has been reviewed in the above sections, because osmotic stress associated with salinity may share the same signaling pathways as dehydration due to drought stress or heat stress.

1.5.1 CHANGE IN MEMBRANE LIPIDS DURING SALINITY STRESS

Plants survive in saline soils by excluding salts or tolerating salt accumulated within the cells. High salt resistance of plants depends on the preservation of membrane integrity of cellular membranes. Some variations in plasma membrane sensitivity to salt in salt-sensitive and tolerant plants may be explained by differences in composition or structure of the plasma membrane [114]. High salt resistance of barley (*Hordeum vulgare L.*) was attributed to decreases in the content of galactolipids in chloroplast membranes that facilitate Cl⁻ transport across membranes. Salinity effects on membrane lipids are largely due to ionic effects rather than osmotic stress [115]. Salt accumulation in cells affects lipid-synthesizing enzymes such as galactosyl transferase, and acylase in chloroplasts.

The responses of different lipid species to salinity may vary in different plant species. Parti et al. [116] examined the effects of salinity on lipid content in Indian mustard (*Brassica juncea L. Czern.*) and found that the content of total and nonpolar lipids decreased with increasing salinity whereas polar lipid content increased. Zenoff et al. [117] found no changes in phospholipid content during salt stress, but an increase in free sterols and triglycerides in roots of two soybean cultivars differing in salt tolerance. Glycolipid content increased in root membranes of the salinity-sensitive cultivars, but did not change in the tolerant cultivar in response to salinity. A salt-tolerant cultivar of tomato (*Lycopersicon esculentum* Mill.) accumulated Na⁺ in roots without considerable modifications in lipid metabolism, whereas phosphatidylinositol 4,5-bisphosphate increased and phosphorylation of phosphatidic acid and phosphatidylcholine decreased in the sensitive cultivar [118].

In another study with *Plantago* species differing in salinity tolerance, it was found that the level of phospho-, galacto-, and sulpholipids in roots of salt-sensitive *P. media L.* decreased considerably with increasing NaCl concentration, whereas in the roots of two salt-tolerant species (*P. maritima L., P. coronopus L.*), the level of most lipid classes was unchanged or even increased at 75 mM NaCl [119]. Major membrane lipid classes remained stable in calli of a salt march grass, *Spartina patens*, cultured at different salinity levels (0, 170, 340 mmol/L), which was associated with stable membrane fluidity [120]. In general, structural and signal lipids seem to be more stable in salt-tolerant plants, which may contribute to their higher cell membrane stability under salinity stress.

The ratio of different lipid species and the saturation level of lipids also change in plant adaptation to salinity, which affect the membrane structural configuration and stability. Under salinity stress, the phospholipid constituent of the polar lipids was relatively higher than the glycolipid component in Indian mustard [116]. Zenoff et al. [117] reported that the ratio of DGDG:MGDG decreased in the salt-tolerant cultivar of soybean. Halophyte *S. patens* exhibited a decline in the membrane protein to lipid ratio in the calli exposed to elevated NaCl [120].

Kerkeb et al. [121] found that plasma membrane vesicles isolated from tomato calli tolerant to 100 mM NaCl exhibited a higher phospholipid and sterol content, as well as a higher free sterol-to-phospholipid ratio compared to NaCl-sensitive calli. Mansour et al. [122] also reported a higher free sterol-to-phospholipid ratio and a higher total sterol-to-phospholipid ratio in plasma membranes of wheat roots exposed to salt stress. Sterols play a role as stabilizers of glycerolipids, which help maintain membrane fluidity at the appropriate physiological level (not too fluid or not fluid enough). Salt-tolerant tomato calli had a lower double bond index or higher saturation of phospholipid fatty acids [121].

The sterol and saturated fatty acids packing in the membrane bilayer would reduce membrane permeability, because they are the major component contributing to membrane rigidity [123,124]. It is suggested that "more planar" sterols enhance ion exclusion, and thus enhance plant survival of high salts [125]. It can be concluded that the increases in membrane lipid saturation and sterol levels helps stabilize cellular membranes and protect cells from salinity injury.

Inorganic ions such as Ca^{2+} and Mg^{2+} in high saline soils may also directly affect membrane stability or permeability by interacting with phospholipids without changing lipid composition. Ca^{2+} reacts primarily with negatively charged or acidic phospholipid heads and causes a condensing effect on phospholipids in the membrane bilayer, resulting in rigid or gel-like membranes [9]. The water hydration associated with phospholipid heads can be displaced in high Ca^{2+}-salinity conditions, resulting in the alteration of membrane fluidity.

1.5.2 Lipid Metabolism in Response to Salinity Stress

The alteration of membrane lipid composition is related to changes in enzyme activities that control the synthesis, degradation, or hydrolysis of different phospholipid species. Most of the current literature concerns lipid hydrolysis induced by salinity stress from the stress-signaling point of view. Studies suggest that an increase in the lipoxygenase activity in salt-stressed plasma membranes may contribute to the decline in the content of some phospholipids. As discussed above, phospholipid hydrolysis provides secondary messengers, which play roles in many cellular responses to environmental stresses.

Salt stress was recently shown to induce an increase in $PtdIns(3,5)P_2$ synthesis in *Chlamydomonas moewusii* and in some higher plant cells [105]. In *Arabidopsis*, the expression of one PtdInsP kinase gene and one PI-PLC gene is increased by salt stress at 150–200 mM NaCl. Pical et al. [105] found increased synthesis of $PtdIns(4,5)P_2$ and diacylglycerol pyrophosphate (DGPP), and increased turnover of phosphatidylcholine (PtdCho) in response to salt stress. These studies suggest a possible involvement of the classical $(4,5)P_2$-inositide pathway in plant response to salinity stress. Some of $PtdIns(4,5)P_2$ may be hydrolyzed to $Ins(1,4,5)P_3$ and diacylglycerol during the early phase of stress, which may participate in salinity signaling. A gene encoding phosphatidylinositol-specific phospholipase C (AtPLC1) is induced to a significant extent in *Arabidopsis* by various environmental stresses, including salinity, which might be involved in the signal-transduction pathways of salinity stress [103].

Wu et al. [120] reported that NaCl stress and osmotic stress induced by mannitol enhanced the expression of all three different forms of PLDs (PLDα, PLDβ, PLDγ) in *Arabidopsis*. It is suggested that activation of PLD induced by salinity may be involved in the regulation of cellular stress responses by producing secondary messengers and through altering physical properties of membranes.

The increase in PLD activity may also promote membrane degradation in stress injury and senescence [126]. The role of lipid hydrolysis in plant adaptation to salinity stress is not completely understood, and deserves further investigation.

1.6 MAJOR CELLULAR PROCESSES ASSOCIATED WITH MEMBRANE DAMAGES DURING ABIOTIC STRESSES

As discussed above, damage to membranes has been documented in a number of plant species under various environmental stresses including heat, cold, drought, and salinity. Such damage may be associated with changes in membrane physical structure or chemical properties. The double bonds of the polyunsaturated fatty acids in membrane lipids not only affect membrane fluidity, but are also the target of oxidation. Cellular injuries induced by environmental stresses have been linked to increase in membrane lipid peroxidation [127]. Further information on lipid peroxidation can be found in Chapter 11. Loss of membrane integrity caused by excessive fluidity or lipid peroxidation may result in increases in membrane permeability and cell death during stress. Numerous cellular changes take place in the process of stress-induced cell damage. Release of cellular compounds such as proteins, ions, and other metabolites from the cell is a hallmark of cellular membrane damage and cell death [128].

In addition to lipids, proteins are important components of the membranes, as they play critical functions in signal transduction, ion transport, or energy generation. Abiotic stresses may directly affect the stability and functions of these proteins or indirectly exert effects on the proteins by altering the membrane fluidity. Obviously, loss of activities of proteins with critical functions, such as ion transport and energy generation, can have deleterious effects. Indeed, many stress factors have been found to involve the release of membrane-bound proteins leading to disruption of cellular activities. For example, release of cytochrome C from mitochondria into the cytosol is regarded as a hallmark of cell death. Pasqualini et al. [129] reported that ozone-induced oxidative stress triggered a cell death program in tobacco, which involved the early release of cytochrome C from mitochondria. Virolainen et al. [130] reported that Ca^{2+}-induced high amplitude swelling and cytochrome C release from wheat root mitochondria under anoxic stress. Peters and Chin [131,132] observed that an inducer of cell death, palmitoleic acid, induces cell death in part by causing the release of a number of chloroplast membrane proteins, including cytochrome f, and the release of these proteins contributes to the demise of the cells. It is probable that the stability of membrane proteins under environmental stresses is affected by the levels of membrane lipid saturation. Heat tolerance of leaves depends on the thermal stability of the photochemical reactions in the thylakoid membranes of chloroplasts. The primary site of thermal damage to the photosynthetic function is believed to be associated with photosystem II [133]. Heat stress induces the separation of PSII from light harvesting protein complex in thylakoid membranes [134,135] by reducing the stability of thylakoid membranes [133,136,137].

A change in membrane permeability can also result in ion leakage, leading to cell death. Membrane leakage of solutes including ions, soluble sugars, and free amino acids is positively correlated to the severity of stress injury. In fact, membrane leakage has become a common method for evaluating cell membrane stability and measuring the degree of injury or tolerance of plants to various environmental stresses [138–140] (also see Chapter 11). Stress-sensitive plants generally have more leaky cell membranes under stress. For example, a *Arabidopsis* mutant (*hot2-1*), deficient in acquired thermotolerance, showed higher levels of electrolyte leakage compared to wild-type plants under heat stress [141]. Conversely, PLD-deficient *Arabidopsis* plants that showed improved freezing tolerance had decreased membrane leakage [46]. In terms of temperature response, maximum leakage of membranes is seen at the phase transition temperature due to packing defects in the phase boundaries between the liquid and gel portions of the membrane.

One of the major cations that leak out of a damaged cell is K^+. The loss of K^+ induced by chilling stress can be seven- to tenfold greater than that of Ca^{++} or Mg^{++}. The efflux of ions such as K^+ could occur either by diffusion across the lipid bilayer or through channels in membranes. High temperature may cause membrane leakage by changing membrane fluidity or the activity of membrane channels and transporters [142]. K^+ channels are sensitive to changes in the lipid environment. Free fatty acids such as arachidonic acid have been shown to activate K^+ channels resulting in K^+ efflux from cells [143]. K^+ efflux has been associated with apoptosis [144], and prevention of K^+ efflux with K^+ channel inhibitors has been associated with the prevention of cell death [144]. It is evident that cell membrane stability or integrity is important for maintaining functional cells during plant adaptation to environmental stresses.

1.7 FUTURE RESEARCH PERSPECTIVES

As discussed above, a large body of literature elucidates the importance of cellular membranes in regulating stress tolerance in higher plants. However, many questions still remain as to how cellular membranes are involved in plant adaptation to environmental stresses and how lipids are involved in fine-tuning membrane stability and function during stress adaptation. For example, what are the critical thresholds of different environmental factors such as temperature and water stress that trigger changes in membrane composition and fluidity prior to inducing other physiological changes (environmental threshold)? To what extent are changes in membrane fluidity, lipid composition, or lipid saturations necessary in order to cause functional changes of membranes in stress adaptation (intrinsic threshold)? How are changes in membrane composition and fluidity involved in stress-signal perception and signal transduction? Are common or different signaling molecules sensed and transduced into an adaptational response by the alteration of membrane properties under different environmental stresses? How do different membranes (i.e., thylakoid, plasma, and mitochondrial membranes) vary in their sensitivity to environmental stresses, and to what extent are these different membranes involved in stress adaptation? Finally, what are the specific lipid species in different membranes linked to stress tolerance? Addressing these questions in future research will provide insight into the roles of cellular membranes in plant adaptation to various environmental stresses. Such information could be utilized in biotechnology and breeding techniques to improve plant tolerance to environmental stresses.

REFERENCES

1. Bengtsson, L., Climate change. Climate of the 21th century. *Agric. For. Meteor.*, 72:3, 1994.
2. Larcher, W., *Physiological Plant Ecology. Ecophysiology and Stress Physiology of Functional Groups.* Springer-Verlag, Berlin, 1995, p. 345.
3. Munnik, T. and Meijer, H.J., Osmotic stress activates distinct lipid and MAPK signaling pathways in plants. *FEBS Lett.*, 498:172, 2001.
4. Crowe, J.H. and Crowe, L.M., Membrane integrity in anhydrobiotic organisms: Toward a mechanism for stabilizing dry cells. In *Water in Life: Comparative Analysis of Water Relationships at the Organismic, Cellular, and Molecular Level.* Somero, G.N., Osmond, C.B., and Bolis, C.L., Eds., Springer-Verlag, Berlin, 1992, pp. 87–103.
5. Russell, N.J. and Fukunga, N., A comparison of thermal adaptation of membrane lipids in psychrophilic and thermophilic bacteria. *FEMS Microbiol.* 75:171, 1990.
6. Oliver, A.E., Crowe, L.M., and Crowe, J.H., Methods for dehydration-tolerance: Depression of the phase transition temperature in dry membranes and carbohydrate vitrification. *Seed Sci. Res.*, 8:211, 1998.
7. Murakami, Y.M. et al., Trienoic fatty acids and plant tolerance of high temperature. *Science,* 287:476, 2000.
8. Rajashekar, C.B., Molecular responses and mechanisms of plant adaptation to cold freezing stress. In *Plant–Environment Interactions*, Huang, B., Ed., CRC, Boca Raton, FL, 2006.
9. Nilsen E.T. and Orcutt D.M., *Physiology of Plants under Stress. Abiotic Factors*, Wiley, New York, 1996, pp. 62–78.
10. Lehninger, A.L., *Biochemistry,* 2d ed. Worth, New York, 1977.
11. Singh, S.C., Sinha, R.P., and Hader, D.P., Role of lipids and fatty acids in stress tolerance in cyanobacteria. *Acta Protozool.*, 41:297, 2002.
12. Sung, D. et al., Acquired tolerance to temperature extremes. *Trends Plant Sci.*, 8: 179, 2003.
13. Afonso, M.I. et al., Unusual tolerance to high temperatures in a new herbicide-resistant D1 mutant from *Glycine max* (L.) Merr. cell culture deficient in fatty acid desaturation. *Planta,* 212:573, 2001.
14. Hugly, S. et al., Thermal tolerance of photosynthesis and altered chloroplast ultrastructure in a mutant of *Arabidopsis* deficient in lipid desaturation. *Plant Physiol.*, 90:1134, 1989.
15. Zaharieva, I., Velitchkova, M., and Goltsev, V., Effect of cholesterol and benzyl alcohol on prompt and delayed chlorophyll fluorescence in thylakoid membranes. In *Photosynthesis: Mechanisms and Effects*, Garab, G., Ed., Kluwer Academic, Dordrecht, The Netherlands, 1998, pp. 1827–1830.

16. Whitaker, B.D. et al., Influence of pre-storage heat and calcium pre-treatments on lipid metabolism in Golden Delicious apples, *Phytochem.*, 45:465, 1997.
17. Horvath, I. et al., Membrane physical state controls the signaling mechanism of the heat shock response in *Synechocystis* PCC 6803: Identification of hsp17 as a "fluidity gene." *Proc. Natl. Acad. Sci.,* 95:3513, 1998.
18. Nishiyama, Y., Los, D., and Murata, N., PsbU, a protein associated with PSII is required for the acquisition of cellular thermotolerance in *Synechococcus* species PCC7002. *Plant Physiol.*, 120:301, 1999.
19. Grover, A. et al., Production of high temperature tolerant transgenic plants through manipulation of membrane lipids. *Current Sci.,* 79:557, 2000.
20. Larkindale, J. and Huang, B., Changes of lipid composition and saturation level in leaves and roots for heat-stressed and heat-acclimated creeping bentgrass. *Exp. Environ. Bot.* 51:57, 2004.
21. Somerville, C. and Browse, J., Plant lipids, metabolism and membranes. *Science*, 252:80, 1991.
22. Pearcy, R., Effects of growth temperature on the fatty acid composition of the leaf lipids in *Atriplex lentiformis* (Torr.) Wats. *Plant Physiol.*, 61:484, 1978.
23. Raison, J.K., Pike, C.S., and Berry, J.A., Growth temperature-induced alterations in the thermotropic properties of *Nerium oleander* membrane lipids. *Plant Physiol.*, 70:215, 1982.
24. Vigh, L. et al., Saturation of membrane lipids by hydrogenation induces thermal stability in chloroplast inhibiting the heat dependent stimulation of Photosystem I mediated electron transport. *Biochem. Biophysic. Acta*, 979:361, 1989.
25. Vigh, L. et al., The primary signal in the biological perception of temperature: Pd-catalyzed hydrogenation of membrane lipids stimulated the expression of the desA gene in *Synechocystis* PCCC6803. *Proc. Natl. Acad. Sci.* 90:9090, 1993.
26. Sharkey, T. Some like it hot. *Science*, 287:435, 2000.
27. Yang, J.F., Cheng, B.S., and Wang, H.C., Influence of high temperature and low humidity on the fatty acid composition of membrane lipids in wheat. *Acta Bot. Sinica*, 26:386, 2001.
28. McCourt, P. et al., The effects of reduced amounts of lipid unsaturation on chloroplast ultrastructure and photosynthesis in a mutant of *Arabidopsis*. *Plant Physiol.*, 34:353, 1987.
29. Kunst, L., Browse, J., and Somerville, C., Enhanced thermal tolerance in a mutant of Arabidopsis deficient in palmitic acid unsaturation. *Plant Physiol.*, 91:401, 1989.
30. Diepenbrock, W., Muller-Rehbehn, A., and Sattelmacher, B., Fatty acid composition of root membrane lipids from two potato (*Solanum tuberosum* L.) genotypes differing in heat tolerance as affected by superaoptimal root-zone temperature. *Agrochimica*, 33:478, 1989.
31. Gombos, Z., Wada, H., and Murata, N., Direct evaluation of effects of fatty acid unsaturation on the thermal properties of photosynthetic activities, as studied by mutation and transformation of *Synechocystis* PC6803. *Plant Cell Physiol.*, 32:205, 1991.
32. Rikin, A., Dillwith, J.W., and Bergman, D.K., Correlation between the circadian rhythm of resistance to extreme temperatures and changes in fatty acid composition in cotton seedlings. *Plant Physiol.* 101:31, 1993.
33. Musienko, N.N. et al., Lipid composition of the subcellular factions in leaves of wheat varieties differing in heat resistance. *Fiz. Bio. Kul. Ras.* 1:36, 1984.

34. Behl, R.K., Heise, K.P., and Moawad, A.M., High temperature tolerance in relation to changes in lipids in mutant wheat. *Tropenlandwirt* 97:131, 1996.
35. Nielsen, K.F., Roots and root temperatures. In *The Plant Root and Its Environment*. Carson, E.W., Ed., Univ. Press Virginia, Charlottesville, 1974, p. 293.
36. McMichael, B.L., Soil temperature and root growth. *HortScience*, 33:947, 1998.
37. Xu, Q. and Huang, B., Growth and physiological responses of creeping bentgrass to differential shoot and root temperatures. *Crop Sci.,* 40:1363, 2000.
38. Xu, Q., and Huang, B., Effects of air and soil temperature on carbohydrate metabolism in creeping bentgrass. *Crop Sci.,* 40:1368, 2000.
39. Styer, E.H. et al., Lipid composition of microsomes from heat-stressed cell suspension culture. *Phytochemistry* 41:187, 1996.
40. Liu, X. and Huang, B., Changes in fatty acid composition and saturation level for creeping bentgrass exposed to high soil temperature. *J. Am. Soc. Hort.*, 129: 795–801, 2004.
41. Latsague, M. et al., Frost resistance and lipid composition of cold-hardened needles of Chilean conifers. *Phyochemistry*, 31:3419, 1992.
42. Palta, J.W., Whitaker, B.D., and Weiss, L.S., Plasma membrane lipids associated with genetic variability in freezing tolerance and cold acclimation of *Solanum* species. *Plant Physiol.*, 103:793, 1993.
43. Samala, S., Yan, J.Y., and Baird, W.V., Changes in polar lipid fatty acid composition during cold acclimation in "Midiron" and "U3" bermudagrass. *Crop Sci.,* 38:188, 1998.
44. Zhang, J. et al., Genetic engineering for abiotic stress resistance in crop plants. *In Vitro Cell. Dev. Biol.*, 36:108, 2000.
45. Cyril, J. et al., Changes in membrane polar lipid fatty acids of seashore paspalum in response to low temperature exposure. *Crop Sci.*, 42:2031, 2002.
46. Welti, R. et al., Profiling membrane lipids in plant stress response: Role of phospholipase Dδ in freezing-induced lipid change in *Arabidopsis*. *J. Biol. Chem.*, 277:31994, 2002.
47. Murata, N. et al., Genetically engineered alteration in the chilling sensitivity of plants. *Nature*, 356:710, 1992.
48. Wolter et al., Chilling sensitivity of *Arabidopsis thaliana* with genetically engineered membrane lipids. *EMBO J.*, 11:4685, 1992.
49. Kodama, H. et al., Genetic enhancement of cold tolerance by expression of a gene for chloroplast [omega]-3 fatty acid desaturase in transgenic tobacco. *Plant Physiol.*, 105:601, 1994.
50. Bertin, P. et al., Somaclonal variation and chilling tolerance improvement in rice: Changes in fatty acid composition. *Plant Growth Reg.*, 24:31, 1998.
51. Murata, N. et al., Composition and positional distribution of fatty acids in phospholipids from leaves of chilling-sensitive and chilling-resistant plants. *Plant Cell Physiol.* 23:1071, 1982.
52. Wang, C.Y. and Baker, J.E., Effects of two freeze radical scavengers and intermittent warming on chilling injury and polar lipid composition of cucumbers and sweet pepper fruits. *Plant Cell Physiol.*, 20:243, 1979.
53. Gombos, Z., Wada, H., and Murata, N., Unsaturation of fatty acids in membrane lipids enhances tolerance of the cyanobacterium *Synechocystis* PCC6803 to low-temperature photoinhibition. *Proc. Natl. Acad. Sci. USA*, 89:9959, 1992.

54. Moon, B.Y. et al. Unsaturation of the membrane lipids of chloroplasts stabilizes the photosynthetic machinery against low-temperature photoinhibition in transgenic tobacco plants. *Proc. Natl. Acad. Sci. USA*, 92:6219, 1995.
55. Tasaka, Y. et al., Targeted mutagenesis of acyl-lipid desaturases in *Synechocystis*: evidence for the important roles of polyunsaturated membrane lipids in growth, respiration, and photosynthesis. *EMBO J.*15:6416, 1996.
56. Sato, N. and Murata, N., Temperature shift-induced responses in lipids in the blue-green alga, Anabaena variabilis: The central role of diacylmonogalactosylglycerol in thermo-adaptation. *Biochem. Biophys. Acta.* 619;353, 1980.
57. Wada, H. and Murata, N., Temperature-induced changes in the fatty acid composition of the cyanobacterium, *Synechocystis* PCC6803. *Plant Physiol.*, 92:1062, 1990.
58. Raison, J.K., Alterations in the physical properties and thermal response of membrane lipids: correlations with acclimation to chilling and high temperature. In *Frontiers of Membrane Research in Agriculture*, St John, J.B. and Jackson, P.C., Eds., Rowman and Allanheld, Totowa, NJ, 1986, p. 383.
59. Buchanan, B.B., *Biochemistry and Molecular Biology of Plants*. Am. Soc. Plant Physiol., Maryland, 2000.
60. Murata, N. and Wada, H., Acyl-lipid desaturases and their importance in the tolerance and acclimatization to cold of cyanobacteria. *Biochem. J.*, 308:1, 1995.
61. Gibson, S. et al., Cloning of a temperature-regulated gene encoding a chloroplast w-3 desaturase from *Arabidopsis thaliana*, *Plant Physiol.*, 106:1615, 1994.
62. Browse, J. et al., A mutant of *Arabidopsis* deficient in the chloroplast, 16:1/18:1 desaturase. *Plant Physiol.*, 90:522, 1989.
63. Miquel, M.D. et al., *Arabidopsis* requires polyunsaturated lipids for low temperature survival. *Proc. Natl. Acad. Sci. USA*, 90:6208, 1993.
64. Williams, J.P. et al., The effect of temperature on the level and biosynthesis of unsaturated fatty acids in diacylglycerols of *Brassica napus* leaves. *Plant Physiol.* 87:904, 1988.
65. Berberich, T. et al., Two maize genes encoding -3 fatty acid desaturase and their differential expression to temperature. *Plant Mol. Biol.* 36:297, 1998.
66. Chapman, K.D., Phospholipase activity during plant growth and development and in response to environmental stress. *Trends Plant Sci.*, 11:419, 1998.
67. Zhu, J.K., Salt and drought stress signal transduction in plants. *Ann. Rev. Plant Physiol.*, 53:247, 2002.
68. Wang, X., The role of phospholipase D in signaling cascades. *Plant Physiol.* 120;645, 1999.
69. Fan, L., Zheng, S., and Wang, X., Antisense suppression of phopspholipase D retards abscisic acid- and ethylene-promoted senescence of post harvest *Arabidopsis* leaves. *Plant Cell*, 9:2916, 1997.
70. Wang, C. et al., Involvement of phospholipase D in wound-induced accumulation of jasmonic acid in *Arabidopsis. Plant Cell*, 12:2237, 2000.
71. Zien, C.A. et al., In-vivo substrates and the contribution of the common phospholipase D, PLD, to wound-induced metabolism of lipids in *Arabidopsis. Biochem. Biophys. Acta.*, 1530:236, 2001.
72. Sang, Y. et al., Regulation of plant water loss by manipulating the expression of phospholipase D. *Plant J.*, 28:135, 2001.
73. Wang, C. and Wang, X., A novel phospholipase D of *Arabidopsis* that is activated by oleic acid and associated with the plasma membrane, *Plant Physiol.*, 127:1102, 2001.

74. Maarouf, H.E. et al., Enzymatic activity and gene expression under water stress of phospholipase D in two cultivars of *Vigna unguiculata* L. Walp. differing in drought tolerance. *Plant Mol. Biol.,* 39:1257, 1999.
75. Li, W. et al., The plasma membrane-bound phospholipase D enhances freezing tolerance in *Arabidopsis thaliana*. *Nature Biotech.*, 22:427, 2004.
76. Yapa, P.A.J., Kawasaki, T., and Matsumoto, H., Changes of some membrane-associated enzyme activities and degradation of membrane phospholipases in cucumber roots due to Ca^{2+} starvation. *Plant Cell Physiol.,* 27:223, 1986.
77. He, C.J. et al., Ethylene signal transduction pathway in cell death during aerenchyma formation in maize root cells: Role of phospholipases. In *Biology and Biotechnology of the Plant Hormone Ethylene*, Charng, Y.Y. et al., Eds., Kluwer Academic, Dordrecht, The Netherlands, 1998, pp. 103–104.
78. Lygin, A.V. and Gordon, L.K., Change in lipid composition of wheat roots on early phase of mechanical stress. In *Root Demographics and Their Efficiencies in Sustainable Agriculture, Grasslands and Forest Ecosystems*, Kluwer Academic, Dordrecht, The Netherlands, 1998, p. 565.
79. Xiong, L., Schumaker, K., and Zhu, J., Cell signaling during cold, drought, and salt stress. *Plant Cell*, supplement, 165, 2002.
80. Sangwan, V. et al., Opposite changes in membrane fluidity mimic cold and heat stress activation of distinct plant MAP kinase pathways. *Plant J.*, 31:629, 2002.
81. Suzuki, I. et al., The pathway for perception and transcription of low-temperature signals in *Synechocystis*. *EMBO J.*, 19:1327, 2000.
82. Aguilar, P.S. et al., Molecular basis of thermosensing: A two-component signal transduction thermometer in *Bacillus subtilis*. *EMBO J.*, 20:1681, 2001.
83. English, D., Phosphatidic acid: A lipid messenger involved in intracellular and extracellular signaling, *Cell. Signal*, 8:341, 1996.
84. Navari-Izzo F., et al., Lipid composition of plasma membranes isolated from sunflower seedlings grown under water-stress. *Physiol. Plant*, 87:508, 1993.
85. Caldwell, C.R. and Whitman, C.E., Temperature-induced protein conformational changes in barley root plasma membrane-ennriched microsomes. I. Effect of temperature on membrane protein and lipid mobility. *Plant Physiol.,* 84:918, 1987.
86. Horvath, I. et al., The role of phospholipids in regulating photosynthetic electron transport activities: treatment of chloroplasts with phospholipase A2. *J. Photochem. Photobiol B Biol,* 3:515, 1989.
87. Munnik, T., Irvine, R.F., and Musgrave, A., Phospholipid signaling in plants. *Biochim. Biophy. Acta*, 1389:222, 1998.
88. Pham Thi, A.T. et al., Effects of water stress on lipid metabolism in cotton leaves. *Phytochemistry*, 24:723, 1985.
89. Monteiro de Paula, F. et al., Effects of water stress on the molecular species composition of polar lipids from *Vigna unguiculata* L. leaves. *Plant Sci.*, 66:188, 1990.
90. Sahsah, Y. et al., Enzymatic degradation of polar lipids in *Vigna unguiculata* leaves and influence of drought stress. *Physiol. Plant.*, 104:577, 1998.
91. El Maarouf, H. et al., Enzymatic activity and gene expression under water stress of phospholipadse D in two cultivars of *Vigna unguiculata* L. Walp. Differing in drought tolerance. *Plant Mol. Biol.*, 39:1257, 1999.
92. Matos, A.R. et al., A novel patatin-like gene stimulated by drought stress encodes a galactolipid acyl hydrolase. *FEBS Lett.,* 491:188, 2001.

93. Gogon, A. et al., Effect of drought stress on lipid metabolism in the leaves of *Arabidopsis thaliana* (Ecotype Columbia). *Ann. Bot.*, 94:345, 2004.
94. Stevanovic, B., et al., Effects of dehydration and rehydration on the polar lipid and fatty acid composition of *Ramonda* species. *Can. J. Bot.*, 70:107, 1992.
95. Benhassaine-Kesri, G. et al., Drought stress affects chloroplast lipid metabolism in rape (*Brassica napus*) leaves. *Physiol. Plant.*, 115:221, 2002.
96. Monteiro de Paula, F. et al., Effect of water stress on the biosynthesis and degradation of polyunsaturated lipid molecular species in leaves of *Vigna unguiculata*. *Plant Physiol. Biochem.*, 31:707, 1993.
97. Dorman, P. and Benning, C., Galactolipids rule in seed plants. *Trends Plant Sci.*, 7:112, 2002.
98. Quartacci, M. et al., Lipid composition and protein dynamic in thylakoids of two wheat cultivars differently sensitive to drought. *Plant Physiol.*, 108:191, 1995.
99. Jarvis, P. et al., Galactolipid deficiency and abnormal chloroplast development in the Arabidopsis MGD synthase 1 mutant. *Proc. Natl. Acad. Sci. USA*, 97:8175, 2000.
100. Gounaris, K. et al., Polyunsaturated fatty acyl residues of galactolipids are involved in the control of bilayer/non-bilayer lipid transition in higher plant chloroplasts. *Biochem Biophys Acta.*, 732:229, 1983.
101. Repellin, A. et al., Leaf membrane lipids and drought tolerance in young coconut palms (*Cocos nucifera* L.). *Eur. J. Agron.*, 6:25, 1997.
102. Chapman, D.J., De-Felice, J., and Barber, J., Growth temperature effects on thylakoid membrane lipid and protein content of pea chloroplast. *Plant Physiol.*, 72:225, 1983.
103. Hirayama, T. et al., A gene encoding a phosphatidylinositol-specific phospholipase C is induced by dehydration and salt stress in Arabidopsis thaliana. *Proc. Natl. Acad. Sci. USA,* 92:3903, 1995.
104. Kopka, J. et al., Molecular and enzymatic characterization of three phosphoinositide-specific phospholipase C isoforms from potato. *Plant Physiol.*, 116:239,1998.
105. Pical, C. et al., Salinity and hyperosmotic stress induce rapid increases in phosphatidylinositol 4,5,-bisphosphate, diacylglycerol pyrophosphate, and phosphatidylcholine in *Arabidopsis thaliana* cells. *J. Biol. Chem.*, 274:38232, 1999.
106. Frank, W. et al., Water deficit triggers phospholipase D activity in the resurrection plant *Craterostigma plantagineum*. *Plant Cell,* 12:111, 2000.
107. Munnik, T. et al., Hyperosmotic stress stimulates phopholipase D activity and elevates the levels of phophatidic acid and diacylglycerol pyrophosphate. *Plant J.,* 22:147, 2000.
108. Katagiri, T., Takahashi, S., and Shinozaki, K., Involvement of a novel *Arabidopsis* phopholipase D, AtPLDδ, in dehydration-inducible accumulation of phosphatidic acid in stress signaling. *Plant J.,* 26:595, 2001.
109. Jacob, T. et al., Abscisic acid signal transduction in guard cells is mediated by phopholipase D activity. *Proc. Natl. Acad. Sci. USA*, 9:12192, 1999.
110. Yeo, A., Molecular biology of salt tolerance in the context of whole-plant physiology. *J. Exp. Bot.,* 49:915, 1998.
111. Douglas, T.J. and Walker, R.R., Phospholipids free sterols and adenosine triphophatase of plasma membrane enriched preparation from roots of citrus genotypes differing in chloride exclusion capacity. *Physiol. Plant.,* 62:51, 1984.
112. Ben-Hayyim, G. and Ran, U., Salt-induced cooperativity in ATPase activity of plasma membrane-enriched fractions from cultured citrus cells: Kinetic evidence. *Physiol. Plant,* 80:210, 1990.

113. Blits, K.C. and Gallagher, J.L., Effect of NaCl on lipid content of plasma membranes of the dicot halophyte *Kosteletzkya virginica* (L.) Presl. *Plant Cell Environ.,* 7:601, 1990.
114. Mansour, M.M.F. and Salama, K.H.A., Cellular basis of salinity tolerance in plants. *Environ. Exp. Bot.,* 52:113, 2004.
115. Müller, M. and Santarius, K.A., Changes in chloroplast membrane lipids during adaptation of barley to extreme salinity. *Plant Physiol.,* 62:326, 1978.
116. Parti, R.P.S., Gupta, S.K., and Chandra, N., Effect of salinity on plant growth and lipid components of Indian mustard (*Brassica juncea* L.). *Cruciferae Newslett.,* 24:45, 2002.
117. Zenoff, A.M. et al., Changes in root lipid composition and inhibition of the extrusion of protons during salt stress in two genotypes of soybean resistant or susceptible to stress. *Plant Cell Physiol.* 35:729, 1994.
118. Racagni, G. et al., Effect of short-term salinity on lipid metabolism and ion accumulation in tomato roots. *Biologia Plantarum,* 47:373, 2003/2004.
119. Erdei, L., Stuiver, C.E., and Kuiper, J.C., The effect of salinity on lipid composition and on activity of Ca2+ - and Mg2+ stimulated ATPases in salt-sensitive and salt-tolerant *Plantago* species. *Physiol. Plant.,* 49:315, 1980.
120. Wu, J., Seliskar, D., and Gallagher, J., The response of plasma membrane lipid composition in callus of the halophyte *Spartina patens* (*Poaceae*) to salinity stress. *Am. J. Bot.,* 92:852, 2005.
121. Kerkeb, L. et al., Tolerance to NaCl induces changes in plasma membrane lipid composition, fluidity and H+-ATPase activity of tomato calli. *Physiol. Plant.,* 113:217, 2001.
122. Mansour, M.M.F. et al., Effect of NaCl and polyamines on plasma membrane lipids of wheat roots. *Biol. Plant.,* 45:235, 2002.
123. Chapman, D., Recent studies of lipids, lipid cholesterol and membrane system. In *Biological Membranes,* Chapman, D. and Wallach, D.F., Eds., Academic, New York, 1973, p. 91.
124. Van Blitterswijk, W.J., Van Hoeven, R.P., and Van Der Meer, B.W., Lipid structural order parameters (reciprocal of fluidity) in bio-membranes derived from steady-state fluorescence in polarization measurements. *Biochem. Biophys. Acta,* 644:323, 1981.
125. Douglas, T.J. and Walker, R.R., 4-Desmethylsterol composition of citrus rootstocks of different salt exclusion capacity. *Physiol. Plant.,* 58:69, 1983.
126. Wang, X., Multiple forms of phospholipase D in plants: The gene family, catalytic and regulatory properties, and cellular functions. *Prog. Lipid. Res.,* 39:109, 2000.
127. Dhindsa, R.S., Plumb-Dhindsa, P., and Thorpe, T.A., Leaf senescence: Correlated with increased levels of membrane permeability and lipid peroxidation, and decreased levels of superoxide dismutase and catalase. *J Exp Bot.,* 32:93, 1981.
128. Maeno, E. et al., Normotonic cell shrinkage because of disordered volume regulation is an early prerequisite to apoptosis. *Proc. Natl. Acad. Sci. USA,* 97:9487, 2000.
129. Pasqualini, S. et al., Ozone-induced cell death in tobacco cultivar Bel W3 plants. The role of programmed cell death in lesion formation. *Plant Physiol.,* 133:1122, 2003.
130. Virolainen, E., Blokhina, O., and Fagerstedt, K., Ca^{2+}-induced high amplitude swelling and cytochrome c release from wheat (*Triticum aestivum* L.) mirochondria under anoxia stress. *Ann. Bot.* 90:509, 2002.

131. Peters, J.S. and Chin, C.K., Inhibition of photosynthetic electron transport by palmitoleic acid is partially correlated to loss of thylakoid membrane proteins. *Plant Physiol. Biochem.*, 41:117, 2002.
132. Peters, J.S. and Chin, C.K., Evidence for cytochrome f involvement in eggplant cell death induced by palmitoleic acid. *Cell Death Diff.*, 12:405, 2005.
133. Havaux, M. and Tardy, F., Temperature-dependent adjustment of the thermal stability of photosystem II in vivo: Possible involvement of xanthophylls-cycle pigments. *Planta*, 198:324, 1996.
134. Sundby, C. et al., Temperature dependent changes in the antenna size of photosystem II. Reversible conversion of Photosystem II" to Photosystem $. *Biochem. Biophys. Acta*, 851:475, 1986.
135. Pastenes, C. and Horton, P., Effect of high temperature on photosynthesis in beans. 1. Oxygen evolution and chlorophyll fluorescence. *Plant Physiol.* 112:1245, 1996.
136. Havaux, M. et al., Thylakoid membrane stability to heat stress studied by flash spectroscopic measurements of the electrochromic shift in intact potato leaves: influence of xanthophylls content. *Plant Cell Environ.* 19:1359, 1996.
137. Tardy, F. and Havaux, M., Thylakoid membrane fluidity and thermostability during the operation of the xanthophylls cycle in higher-plant chloroplasts. *Biochem. Bio. Acta*, 1330:179, 1997.
138. Blum, A. and Ebercon, A., Cell membrane stability as a measure of drought and heat tolerance in wheat. *Crop Sci.*, 21:43, 1981.
139. McKay, H., Electrolyte leakage from fine roots of conifer seedlings: A rapid index of plant vitality following cold storage. *Can. J. For Res.*, 22:1317, 1992.
140. Marcum, K., Cell membrane thermostability and whole-plant heat tolerance of Kentucky bluegrass. *Crop Sci.* 38:1214, 1998.
141. Hong, S., Ung, L., and Vierling, E., Arabidopsis *hot* mutants define multiple functions required for acclimation to high temperature. *Plant Physiol.*, 132:757, 2003.
142. Levitt, J., *Responses of Plants to Environmental Stresses I. Chilling, Freezing, and High Temperature Stresses*. Academic, New York, 1980, p. 407.
143. Fink, M. et al., A neuronal two P domain K+ channel stimulated by arachidonic acid and polyunsaturated fatty acids. *EMBO J.*, 17:3297, 1998.
144. Bortner, C.D., Hughes, F.M.J, and Cidlowski, J.A., A primary role for K$^+$ and Na$^+$ efflux in the activation of apoptosis. *J. Biol. Chem.*, 272:32436, 1997.

2 Changes in Cellular and Molecular Processes in Plant Adaptation to Heat Stress

John J. Burke and Junping Chen

CONTENTS

2.1 Introduction 27
2.2 Heat Tolerance during Seed Germination and Stand Establishment 29
    2.2.1 Developmentally Regulated Protein Changes 29
    2.2.2 Metabolite Adjustment 31
2.3 Inducible Processes Associated with Heat Tolerance 32
2.4 Heat Tolerance during Reproductive and Vegetative Stages 35
    2.4.1 Reproductive Heat Responses 35
    2.4.2 Vegetative Heat Responses 38
2.5 Concluding Remarks 40
References 41

2.1 INTRODUCTION

Plant adaptation to temperature occurs constantly as temperature patterns modulate diurnally and seasonally. These adaptations entail qualitative or quantitative metabolic and physiological changes that often provide a competitive advantage, affect adjustment to new environments, and affect the survival of the species. Changes in isozymes or allozymes, changes in enzyme concentration, modification by substrate and effectors, and metabolic regulation of enzyme function without changing enzyme composition are all possible strategies for adaptation to changes in temperature [1].

To better understand plant responses to elevated temperatures it is essential that the optimal temperature for the plant first be identified. The concept of the thermal kinetic window (TKW) arose from a desire to investigate temperature stresses in plants and the realization of a lack of knowledge about how to identify the optimal temperatures for metabolism. The thermal kinetic window of optimal enzyme function was defined as the temperature range over which the value of the apparent Km was within 200% of the minimum apparent Km value observed for the enzyme being assayed [2]. The 200% cutoff value was used because previous studies [3,4] had reported that enzymes could function optimally with Km values below 200% of the minimum Km value. The purpose of the thermal kinetic window was to provide a general indicator of the range of temperatures in which the optimal temperature for metabolism was located.

The temperature ranges comprising the thermal kinetic windows for wheat (*Triticum aestivum*) and cotton (*Gossypium hirsutum*) were 17.5 to 23°C and 23.5 to 32°C, respectively. Although the TKWs were 5 to 8°C in breadth, it was shown that these plants were only within the optimal temperature range of their TKWs for a fraction of the growing season [2]. These initial observations called for a re-evaluation of our understanding of the temperature stresses experienced by plants in the field. To date, thermal kinetic windows have been reported for numerous species [2,5–10].

The best correlative evidence of the validity of the TKWs comes from the determination of the temperature-dependence of the reappearance of photosystem II variable fluorescence following illumination [5,6]. Chlorophyll fluorescence functions as a natural indicator of the in vivo temperature characteristics of the plant. An early example supporting the validity of the TKW concept was an analysis of the minimum apparent Km for NADH for hydroxypyruvate reductase from "Norgold M" potatoes (*Solanum tuberosum*) [6]. The study demonstrated that the minimum apparent Km occurred at 20°C, with the thermal kinetic window falling between 15 and 25°C. The optimum temperature for variable fluorescence (Fv) reappearance (expressed as the ratio Fv/Fo where Fo is the initial fluorescence) was defined as the temperature providing the maximum Fv/Fo ratio and the minimum time in darkness required to reach this ratio. The optimal temperature identified from the fluorescence reappearance was also 20°C for the "Norgold M" potato. Similar correlations between the temperatures of the TKW and the temperatures at which maximum fluorescence reappeared have been reported for cucumber (*Cucumis sativus*), wheat, cotton, soybean (*Glycine max*), tomato (*Lycopersicon esculentum* Mill.), and bell pepper (*Capsicum annuum*) [3–5].

Any discussion of heat tolerance should therefore take into account the optimal temperature for the plant species being studied. Some or all of the negative impacts of damaging high temperatures can be ameliorated by prior exposure to elevated but nonlethal temperatures, a process called acquired thermotolerance. Like most organisms, plants respond to an elevation in temperature above their optimum temperature by the synthesis of heat shock proteins (HSPs). Temperatures required for the induction of a heat shock response and thermotolerance vary with plant species and are often associated with environmental adaptation [11–16].

The pioneering work of Key et al. [17] in 1981 first introduced the plant research community to heat shock proteins and the related stress adaptation process. In a review of heat shock proteins, molecular chaperones, and the stress response, Feder and Hofmann [18] stated, "The relevant literature on HSPs and molecular chaperones is huge, now comprising more than 12,000 references.

Even a review of the relevant reviews is difficult." For this reason, we direct the reader to some of the excellent reviews on the topic (see [18–26]). In this chapter, we describe some cellular and molecular aspects of heat stress tolerance and susceptibility at different stages of plant development. Specifically, we discuss (1) the contribution of metabolites and developmentally regulated heat shock proteins to heat tolerance during seed germination and seedling establishment; (2) inducible processes that enhance heat tolerance; and (3) the natural sensitivity of reproductive and vegetative tissues to injury from high temperature.

2.2 HEAT TOLERANCE DURING SEED GERMINATION AND STAND ESTABLISHMENT

Plants are most tightly linked to their surrounding thermal environments from planting to the establishment of fully functional photosynthetic and water transport systems. This period in a plant's life ultimately determines survival and competitiveness across diverse environments. Germinating seeds may experience a 25°C fluctuation in temperature throughout the course of a day under dry land conditions. Figure 2.1 shows changes in soil temperature measured at 5 cm depth over a two-day period in the month of July. Nonirrigated soil temperatures (solid circles) may range from a minimum of 22°C to a maximum of 47°C over a 12 h period, with the cycle repeating the second day. Irrigated soils (open circles) that had not received water for two weeks showed temperatures similar to those of the nonirrigated soils. Soil temperatures of the irrigated plots following a two-week drying responded immediately to an irrigation event (arrow) with a 14°C drop in the maximum temperature. Clearly, germinating seeds may experience high-temperature stress before emergence. It is interesting to speculate that there would be an evolutionary advantage for seeds to develop mechanisms that metabolically regulate seed responses to the large fluctuation in soil and air temperatures, and thereby optimize emergence and stand establishment.

2.2.1 DEVELOPMENTALLY REGULATED PROTEIN CHANGES

How do germinating seeds protect themselves from abiotic stresses? Seeds have evolved certain mechanisms to deal with the potential threat of environmental stress, including the induction of certain proteins such as late embryogenesis abundant (LEA) proteins and heat shock proteins (HSPs) [27–40]. LEA proteins accumulate to high levels in desiccation-tolerant tissues, especially seeds. They are overwhelmingly hydrophilic and have been shown to protect two mitochondrial matrix enzymes, fumarase and rhodanase, during drying in an in vitro assay [28]. Dehydrins are also proteins with high water solubility that accumulate during late seed development, which are considered a subset of the LEA proteins. Although the specific function of the dehydrins remains unclear, a positive correlation exists between dehydrins and adaptation to cellular dehydration [29].

Many HSPs are developmentally controlled and expressed in the absence of heat stress in maturing seeds [32–40]. Although some of these developmentally

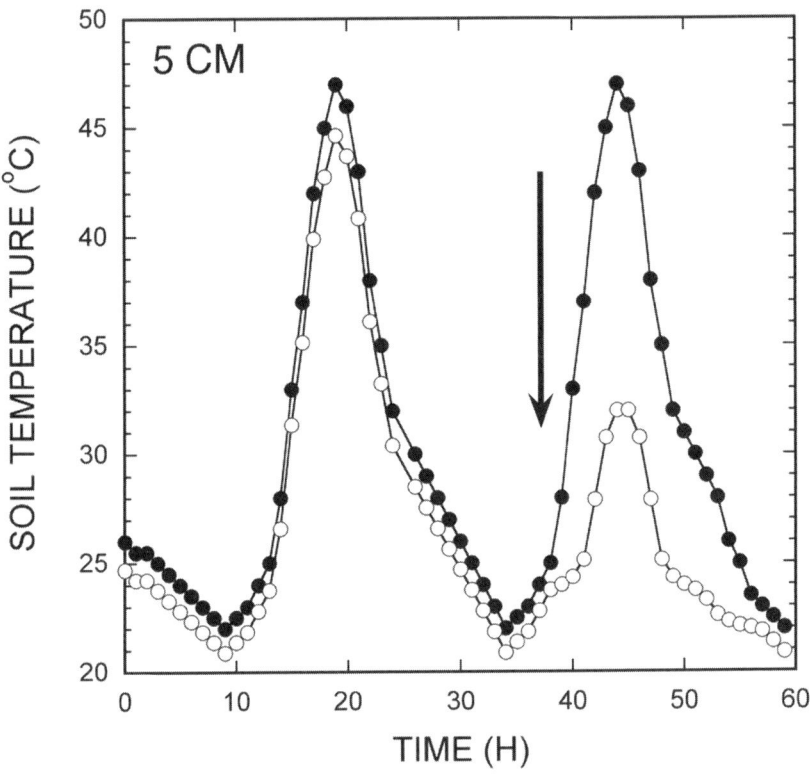

FIGURE 2.1 Diurnal changes in soil temperatures at 5 cm depth of nonirrigated (solid circles) and irrigated (open circles) soils beneath a cultivated cotton crop in the month of July in 1992 in Lubbock, TX. The arrow indicates the occurrence of an irrigation event.

dependent heat shock proteins are also responsive to elevated temperatures, many appear in the seed during the onset of desiccation and are still present several days after seedling emergence [39]. This suggests that HSPs possibly have roles in desiccation tolerance or dormancy during seed development and perhaps during germination. Clearly, seeds have evolved desiccation protection mechanisms composed of numerous hydrophilic proteins and heat shock proteins.

It is interesting to speculate that the developmentally induced HSPs might also contribute to seedling heat tolerance during germination. A report of an investigation into the impact of water stress on the heat protection system of cotton seedlings [41] showed that though the disappearance of a high molecular weight and a low molecular weight heat shock protein was delayed by up to three days in water-stressed seedlings, inherent and acquired thermotolerance in the stressed plants was not enhanced when compared to well-watered seedlings. The results of this study suggest that the developmentally induced HSPs do not contribute to seedling heat tolerance. If these native heat shock proteins in seeds

Changes in Cellular and Molecular Processes

do not enhance early seedling heat tolerance, then what mechanism(s) might be utilized in young seedlings to enhance tissue heat tolerance?

2.2.2 METABOLITE ADJUSTMENT

The presence of elevated sugar levels and other metabolites during seed-reserve mobilization may play a role for temperature tolerance in young plants. Cucumber seedlings exposed to low temperatures reduced seedling growth and resulted in the accumulation of sugars in the cotyledonary tissues [42]. The accumulation of sugars broadened the temperature range at which chlorophyll accumulation occurred [43]. Similar responses are observed in cotton (Figure 2.2) with chlorophyll accumulating over a broad range of temperatures (15°C to 40°C) during the period of rapid mobilization of stored reserves (three days after planting). The temperature range of chlorophyll accumulation (25°C to 35°C) narrowed with the depletion of the stored reserves at seven and nine days after planting. These findings suggest that the mobilized sugars contribute either directly or indirectly to the heat tolerance of the germinating seedling.

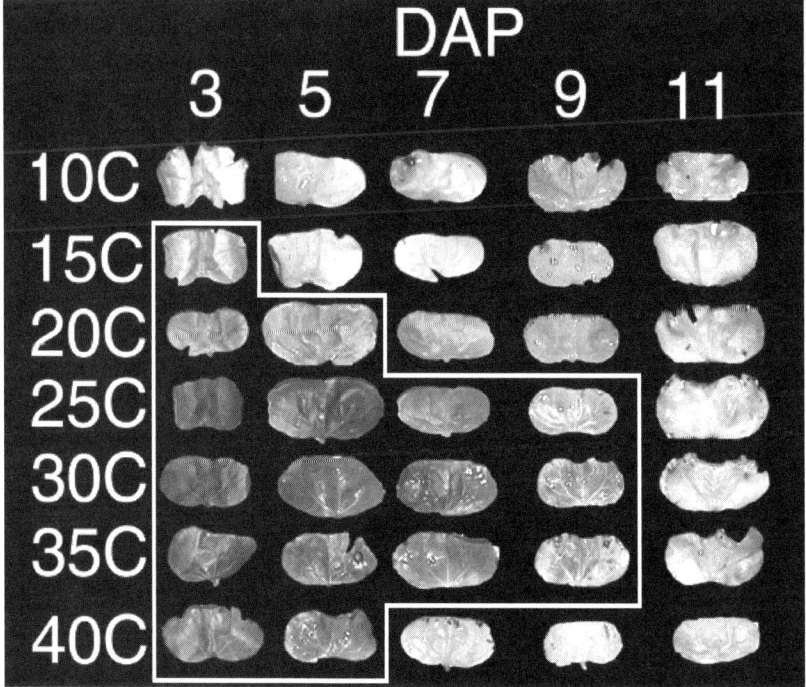

FIGURE 2.2 Cotton cotyledons from dark-grown seedlings harvested on 3, 5, 7, 9, and 11 days after planting (DAP) and exposed to 24 h of continuous light at 10, 15, 20, 25, 30, 35, and 40°C. The cotyledons within the white outline exhibited chlorophyll accumulation and the cotyledons outside the outline failed to accumulate chlorophyll.

Plants often respond to abiotic stresses by accumulating organic solutes such as sugars, polyols, and amino acids [44–53]. These osmolytes have been reported to aid in the stabilization of membranes and minimize protein denaturation under stress conditions [46–55]. It is therefore possible that the very nature of the mobilization of seed reserves during germination may serve to protect the seedling from thermal injury.

2.3 INDUCIBLE PROCESSES ASSOCIATED WITH HEAT TOLERANCE

Following seedling emergence, various metabolic changes occur under heat stress conditions. Metabolic adjustments play important roles in protecting plants from heat injury. Changes in metabolism include alteration of the lipid composition of the endoplasmic reticulum [56,57], inhibition of the release of signal recognition particles from the endoplasmic reticulum [58], the disruption of cap and poly (A) tail function during translation [59], the coordinated loss of translational efficiency and an increase in mRNA stability [57,59], elevation of the level of xanthophyll lutein, and other processes such as induction of circadian rhythms, and changes in morphogenesis [60].

Numerous studies have identified genetic diversity in HSP induction in response to heat shock among various organisms or in different environments [61–72]. For example, a comparison of heat shock response in two sexually compatible species of tomato, *L. esculentum* and *L. pennellii*, revealed a number of species-specific differences with regard to unique HSPs [62]. The response of the F1 generation to heat shock was not intermediate to the parental responses: the F1 generation induced only half of *L. esculentum*-specific HSPs, and all of the *L. pennellii*-specific HSPs. Differences in the induction of low molecular weight HSPs have been observed between cell lines of carrot (*Daucus carota* cv. Danvers Half-Long) in response to heat shock; those lines were derived from different seedlings within a seed stock or derived from a single parent cell line.

These observations highlight the potential for variation in such a highly conserved response and gene set as the heat shock genes [64]. Leaves of maize (*Zea mays*) inbred lines B73 and Mo17 were analyzed for intraspecific genetic variability in heat shock response [73]. Mo17 synthesized 12 unique HSPs in the 15–18 kD range, but B73 synthesized only three unique HSPs in the same range. In another study, heat shock response of wheat cultivars "Mustang" (heat-tolerant) and "Sturdy" (heat-susceptible) that differ in heat tolerance revealed that Mustang synthesized low molecular weight HSP mRNA earlier during exposure to heat shock and at a higher level than "Sturdy" [66,72]. This was especially true for the chloroplast-localized HSP.

Although mutational analysis of thermotolerance has been very effective in identifying genes contributing to heat tolerance in other organisms, in higher plants it had been limited primarily due to the limited number of HSP mutants [74–76]. The HSP104 gene of yeast has been demonstrated to play an essential role in thermotolerance by virtue of the fact that HSP104-minus yeasts are unable to acquire thermotolerance [77]. Plant homologues to yeast HSP104 from soybean (GmHsp101) [77] and *Arabidopsis* (AtHsp101) [78] are capable of complementing HSP104-minus yeast. In a similar study, bacterial thermotolerance was enhanced by synthesis of a plant 16.9 kDa heat shock protein [79].

The association between HSP expression and thermotolerance is also revealed by studies on a heat shock factor, an essential transcription factor for the expression of many HSP genes. The constitutive expression of a heat shock transcription factor HSF3 resulted in HSP synthesis at 25°C and increased the level of thermotolerance in transgenic *Arabidopsis* without prior exposure to elevated temperatures [80]. Transgenic *Arabidopsis* plants expressing less than normal amounts of HSP101, a result of either antisense inhibition or co-suppression, grew at normal rates but had a severely diminished capacity to acquire heat tolerance [81]. Conversely, plants constitutively expressing HSP101 tolerated sudden shifts to extreme temperatures better than the controls. Although a correlation between the development of acquired thermotolerance and the synthesis of HSPs has been noted, a cause-and-effect relationship between the two is only now coming forward with the analysis of knock-out and EMS mutants.

Plant adaptation to heat stress is a complex process and involves many mechanisms in addition to the induction of HSPs. Unfortunately, contributions to the development of acquired thermotolerance in plants from other genes and metabolic processes are not well understood. Using a chlorophyll accumulation assay in an attempt to identify genes and the various mechanisms contributing to thermotolerance, we have isolated a series of *Arabidopsis* EMS mutants that are defective in acquired thermotolerance. Genetic analysis of these mutants has revealed additional genes/mechanisms other than HSPs that play an important role in thermotolerance.

Mapping and cloning of one thermosensitive mutant, AtTS02, identified the mutation to digalactosyldiacylglycerol synthase 1 (DGDG synthase; also known as MGDG:DGDG galactosyltransferase) located on Chromosome 3 [82]. A second allele of DGDG1 was identified in the AtTS100 and AtTS104, two mutants in the same complementation group of AtTS02. DGD1 catalyzes the conversion of monogalactosyldiacylglycerol (MGDG) to digalactosyldiacylglycerol (DGDG) in chloroplast. The DGD1 mutation in AtTS02 resulted in a reduced level of DGDG lipid and decrease in DGDG to MGDG ratio.

The finding that the mutation in DGD1 affects acquired thermotolerance is consistent with the findings of Suss and Yordanov [83], Goncharova et al. [84], Di Baccio et al. [85], Murakami et al. [86], and Lee [87] on the importance of digalactosyldiacylglycerol in heat tolerance. Suss and Yordanov [83] reported that thermotolerance of photosynthetic light reactions in vivo was correlated with an increase in the ratio of DGDG to MGDG and an increased incorporation into thylakoid membranes of saturated digalactosyldiacylglycerol species. This and other recent studies indicate that the ratio of DGDG and MGDG and its change in response to thermoenvironments play a crucial role in maintaining the stability of the chloroplast membrane, effective electron, and ATP transport [83,88]. Lipid profiles of AtTS02 and AtTS104 mutants are consistent with these findings. The reduced thermotolerance of AtTS02 and AtTS104 mutants is correlated with decreases in DGDG lipid level and in the ratio of DGDG to MGDG under both normal and elevated temperatures.

The gene responsible for thermosensitivity of a second-group EMS mutant, AtTS244, also has been identified. A single base pair mutation in AtTS244 results in an amino acid change from glycine to aspartic acid at position 564 in the At5g53170 gene that encodes a putative FtsH protease. Genetic confirmation that the identified mutation in At5g53170 is responsible for the thermosensitive phenotype came from characterizing two independent SALK T-DNA knock-out lines of the At5g53170 gene. Both T-DNA lines display the expected thermosensitive phenotype in chlorophyll accumulation assay. Additional confirmation of the importance of At5g53170 in acquired thermal tolerance came from a previously identified mutant from Burke's lab, AtTDNA-5, originally found within the Feldmen T-DNA tagged population.

Although the T-DNA tag in AtTDNA-5 (kanamycin resistance) was found to segregate from the thermosensitive phenotype of the mutant line, subsequent crosses and linkage analysis mapped the AtTDNA-5 lesion to the same chromosomal region as the AtTS244 mutation. Crosses between AtTDNA-5 and AtTS244 indicated that both mutations belong to the same complementation group. Sequencing of the AtTDNA-5 mutant plant identified a 1670 bp deletion, covering the At5g53170 promoter region plus 530 bp of coding sequence from the first exon of the gene. These results provide strong evidence that the putative FtsH protease (At5g53170) is the gene responsible for the thermosensitivity identified in both the AtTS244 and AtTDNA-5 mutants.

FtsH protease is an ATP-dependent metalloprotease that belongs to an AAA (ATPase associated with diverse cellular activities) protease subfamily [89]. There are total of 12 FtsH protease homologues identified in the *Arabidopsis* genome [90]. Among them, nine are chloroplast targeted and three are located in mitochondria [92]. The FtsH protein identified in the thermosensitive mutants is targeted to the chloroplast and is FtsH11 according to the nomenclature [91]. Several homologues of FtsH11 with a high degree of similarity to the *Arabidopsis* gene have been identified in rice (*Oryza sativa*), tomato, and pea (*Pisum sativum*), suggesting an evolutionarily important role for FtsH11 in plants.

The involvement of FtsH in heat shock regulation was reported initially in *E. coli* [89,92]. Fischer et al. [94] found that FtsH was also involved in the control of heat shock response and in the development and cell cycle control in *Caulobacter crescentus*. Mutants deprived of the FtsH protease are sensitive to elevated temperatures or high light in plants and other organisms [93,94]. Recent studies show that FtsH protease plays an important role in the repair of photodamaged proteins in thylakoid membranes [91,94–98]. Enhanced FtsH gene expression in plant tissues subjected to heat stress or high light acclimation was also observed in gene expression studies [99,100]. Our findings that defects in the FtsH protease gene are detrimental to acquired thermotolerance in *Arabidopsis* agrees with recent reports on the involvement of FtsH protease in heat shock, and other general stress responses in *Arabidopsis*, *Synechocystis* sp PCC 6803, and *Pisum sativum* L. [91,96–98] and provide independent genetic evidence for the involvement of FtsH in high temperature tolerance in higher plants.

The identification of non-hsp mutants [DGDG synthase and FtsH protease] is impetus for the study of additional *Arabidopsis* sequence indexed T-DNA insertional mutants in genes representing a range of metabolic functions. A cursory analysis of phenotypes with T-DNA insertions in ten lipid biosynthetic proteins using the chlorophyll accumulation bioassay revealed six additional genes contributing to thermotolerance. The analysis of phenotypes of T-DNA insertions in 155 proteolytic proteins using the chlorophyll accumulation bioassay revealed 40 additional genes contributing to thermotolerance. Clearly, the availability of genetic stocks will greatly advance the process of identifying genes associated with thermotolerance.

The chlorophyll accumulation assay does not identify all of the genes contributing to heat tolerance, although it is very effective in identifying some genes associated with heat tolerance. An analysis of the HOT1 and HOT3 mutants, kindly provided by Hong and Vierling (University of Arizona), showed that the chlorophyll accumulation bioassay could detect HOT1, but not the HOT3. Similarly, the AtTS02 mutant failed to show a mutant phenotype using the hypocotyl elongation assay described by Hong and Vierling [74]. Hong et al. [101] reported that although the *HOT1* locus was shown previously to encode a major heat shock protein (Hsp), Hsp101, chromosomal map positions indicate that *HOT2, 3,* and *4* did not correspond to major Hsp or heat shock transcription factor genes. Measurement of thermotolerance at different growth stages revealed that the mutants have growth stage–specific heat sensitivity. Analysis of Hsp accumulation showed that *HOT2* and *HOT4* produce normal levels of Hsps, whereas *HOT3* shows reduced accumulation. Thermotolerance evaluated by luciferase activity and ion leakage also varied in the mutants. These data provide additional genetic evidence that distinct functions, independent of Hsp synthesis, are required for thermotolerance, including protection of membrane integrity and recovery of protein activity/synthesis.

2.4 HEAT TOLERANCE DURING REPRODUCTIVE AND VEGETATIVE STAGES

2.4.1 REPRODUCTIVE HEAT RESPONSES

The majority of literature on heat tolerance focuses on vegetative responses of plants. Less is known about heat tolerance mechanisms during reproductive stages of a plant's life. In a report by Boyer [102], a comparison of average crop yields with reported record yields showed that the major crops grown in the United States exhibit annual average yields three- to sevenfold lower than record yields because of unfavorable environments. Analysis of yields from maize, wheat, soybeans, sorghum (*Sorghum bicolor*), oats (*Avena sativa*), barley (*Hordeum vulgare*), potatoes, and sugar beets (*Beta vulgaris*) revealed that the average yield represented only 22% of the mean record yield. The greatest discrepancy between average and record yields was observed in crops with economically valuable reproductive structures. Those crops having marketable vegetative structures exhibited approximately threefold reductions in yield [102]. Environmental stresses during the reproductive stage can have a major impact on crop yield and seed quality.

Reproductive tissues are very sensitive to environmental stresses such as drought and heat. Defective maize pollen development was associated with an inefficient induction of heat shock gene expression [103]. Upon heat shock treatment, mRNAs for HSPs were induced but accumulated only to low levels in germinating pollen. Conversely, a high level of RNA for alpha-tubulin, a representative normal transcript [103], was observed in the pollen. Duck and Folk [104] reported that pollen of angiosperms lacked the ability to respond to heat stress by synthesizing HSPs. In tomato, they found heat shock protein cognate HSC70 (HSP70 expressed in the absence of heat stress) present throughout development of the microspores. HSC70 mRNA transcripts and proteins were detected in nonstressed sporogenous tissues, microspores, and in pretapetal layers during early pollen development.

The HSC70 proteins were detected in mature pollen, however, HSP70 was not induced by heat stress at the latest stages of pollen development [104]. The expression pattern of HSP70 in tomato pollen is consistent with the results of Gagliardi et al. [105]. They analyzed the expression of HSP70 and hsf genes throughout pollen development. There were at least three hsf genes expressed in maize. One hsf gene, whose expression was independent of temperature, was expressed at similar levels throughout pollen development. The two other heat-inducible hsf genes in maize vegetative tissues were not significantly increased in expression levels after heat shock at any stage of pollen development. Gagliardi et al. [105] concluded that the loss of hsp gene expression during the late stages of pollen development was not due to a modification of hsf gene expression at the mRNA level and that hsf gene expression was differentially regulated in vegetative and microgametophytic tissues. These results indicate that pollen may not exhibit the same heat responses as the vegetative tissues and HSP and HSF in pollen may have different functions from those in vegetative tissues.

Genetic diversity of heat tolerance exists in pollen. Rodriguez-Garay and Barrow [106] identified that a greater heat tolerance existed in cotton pollen of a *Gossypium barbadense* breeding line 7456 than in the *G. hirsutum* commercial cultivar Paymaster 404 following a 35°C, 15 h treatment of flowers the day before anthesis. Emasculated flowers of a tester stock, DHNE, were pollinated with the heat-treated pollen from line 7456 and Paymaster 404. The number of pollen tube penetrations in the DHNE styles was determined with Paymaster 404 having an average of 4.3 pollen tubes per style, and line 7456 averaged 59.3 pollen tubes per style.

Rodriguez-Garay and Barrow [106] analyzed a backcrossed population and concluded that a few genes were responsible for the transferred pollen heat tolerance, and heat tolerance expressed in the gametophyte could be transferred by pollen selection. In another study, the influence of selection of F1 heat-tolerant pollen on the structure of BC1 segregating populations was studied for a number of agronomically important traits in oil flax (*Linum usitatissimum*) [107]. Liakh et al. [107] found that heat treatment of a heterogeneous pollen population considerably changed the genetic composition of plant populations not only for tolerance to drought, but also for the duration of seedling emergence, flowering, and quantity and inclination angle of side shoots. They hypothesized that genes that confer these traits are at least partially linked to the genes responsible for pollen sensitivity to high temperature.

Our laboratory has evaluated genotypic diversity in heat tolerance of five cotton lines during reproduction by analyzing pollen dehiscence under elevated temperatures. For the comparison between genotype, a fertility rating was applied in the analysis: 1 = 100% anther dehiscence, 2 = 75% of the anthers dehisced, 3 = 50% of the anthers dehisced, 4 = 25% of the anthers dehisced, and 5 = 0% of the anthers dehisced (Figure 2.3). Fertility ratings varied with breeding lines: Stoneville 474 (2.84), Suregrow 248 (2.92), NM 67 (3.46), PHY 72 (3.62), and Acala Maxxa (3.77). A similar ranking for these lines was obtained in a field

FIGURE 2.3 Cotton flower anthers showing pollen dehiscence on all anthers in the photograph labeled with 100% (left) and no pollen dehiscence from the photograph labeled 0% (right). The flower showing 100% dehiscence came from a cotton plant grown in a greenhouse under a 28°C day/25°C night regime and the flower showing 0% dehiscence was grown under a 42°C day/28°C night regime.

stress study in Arizona [R. Percy, personal communication]. The metabolic mechanisms for the genetic diversity in heat tolerance of cotton during reproduction are unknown and warrant further investigation.

2.4.2 Vegetative Heat Responses

Stem or root cell growth begins slowly in the meristematic region, rapidly increases through the "zone of cell elongation," and ceases once the cell wall has become rigid. The switch from meristematic cell growth to rapidly expanding cell growth is accompanied by higher expression of aquaporins [108] and genes encoding for cell wall synthesis and wall metabolizing enzymes [109,110].

Plants grow by physically expanding their cell walls while maintaining mechanical integrity in the presence of high turgor pressure. This occurs by a coordinated balance of cell wall loosening agents (expansins, endotransglycosylases, glucanases) and wall synthesis machinery. Expansins are cell wall proteins that induce pH-dependent wall extension and stress relaxation in a characteristic and unique manner [111–114]. Expansins appear to weaken glucan–glucan binding, although the detailed mechanisms of action are not well established. McQueen-Mason and Cosgrove [115] provided evidence for expansin action in its weakening effect on pure cellulosic paper without evidence of hydrolytic activity and in the action of chaotropic agents like 2M urea, which can partially mimic and synergize the effect of expansins on cell walls. Although no enzymatic activities have been detected in expansin preparations, they act catalytically to induce cell wall expansion in vitro.

Because endogenous expansin activity is rate limiting for the enlargement of cells, Cosgrove et al. [111] concluded that they might serve as a potential control point for cell growth. Other enzymes found in cell walls include xyloglucan endotransglycosylase (XET) and endo-1, 4-β-D-glucanase (EG). XET cleaves the xyloglucan mid-chain, forming a covalently bonded xyloglucan–enzyme intermediate that can transfer xyloglucan to the nonreducing end of another xyloglucan chain [116]. EG is thought to cleave xyloglucans and may act as secondary wall loosening agent [117,118].

Clearly, cell growth is a coordinated process allowing for expansive growth while maintaining mechanical integrity of the cell. Disruption of this coordination could, in theory, have devastating effects on cell survival. High temperatures affect multiple biochemical and physiological processes related to cell expansion. A sudden increase in temperature of more than 15°C above ambient level should more than double the rates of enzymatic activity within cells and cell walls. The impact of a temperature-induced rapid increase in the activity of cell wall loosening enzymes in the zone of elongation may be severe because the cell wall is under considerable tensile stress due to the hydrostatic pressure of the cell.

We investigated the impact of a rapid increase in temperature on the survival of cells residing within the zone of cell elongation to determine if expansive cell growth predisposes elongating tissues to injury. Cotton seedlings were grown in

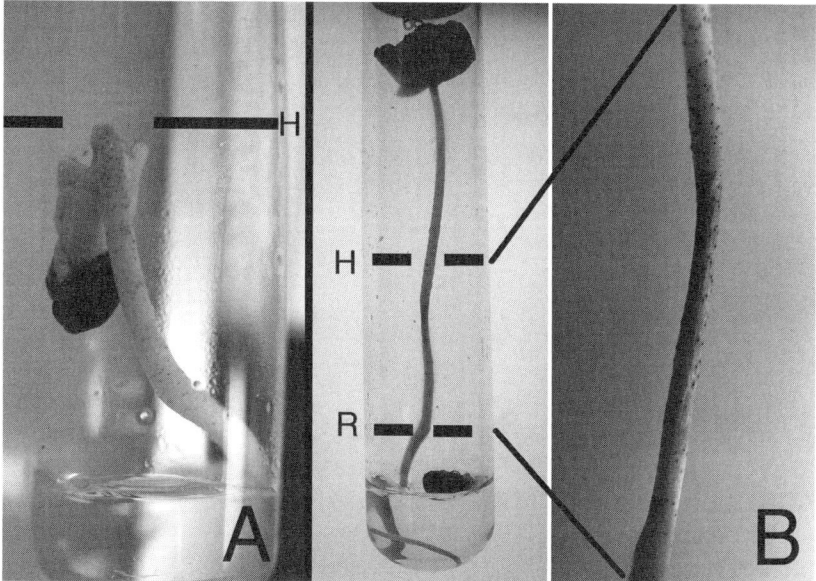

FIGURE 2.4 (A) A representative cotton seedling that was heat-treated for 90 min at 48°C. The horizontal black line labeled "H" is the location of the hypocotyl hook at the time of the heat treatment. (B) A seedling treated for 60 min at 48°C in dark and then exposed to light for 48 h. The horizontal black line labeled "H" is the location of the hypocotyl hook at the time of the heat treatment and the horizontal black line labeled "R" is the location of the hypocotyl–root junction at the time of the heat treatment.

capped culture tubes under sterile conditions in the dark prior to heat treatment. Internal temperature of the culture tube changed from ambient level (approximately 24°C, cooling within the tubes occurred in transit from a 30°C incubator to a water bath) to 48°C within 10 min in a water bath. Following 10 min of 48°C treatment the seedlings were moved to a 30°C incubator for 24 h prior to analysis of changes in hypocotyl elongation.

A photograph of a representative seedling treated at 48°C for 90 min in the dark is shown in Figure 2.4A. The hypocotyl of the seedling shown in Figure 2.4A did not elongate following the 90 min heat treatment. A brown coloration appeared just below the hook and continued down the hypocotyl for approximately 1.5 cm. There was no visible injury to the cotyledons or to the lower regions of the hypocotyl. When moved to light for 48 h the cotyledons accumulated chlorophyll, however, no additional elongation was observed.

Seedlings that were heat-treated for 60 min showed additional hypocotyl elongation that was not observed following the 90 min treatments (Figure 2.4B). The cotton seedling above the "H" line appeared normal, whereas the region of the hypocotyl between the "H" and "R" lines (formerly the zone of cell elongation) showed brown coloration and some constriction of the hypocotyl in the

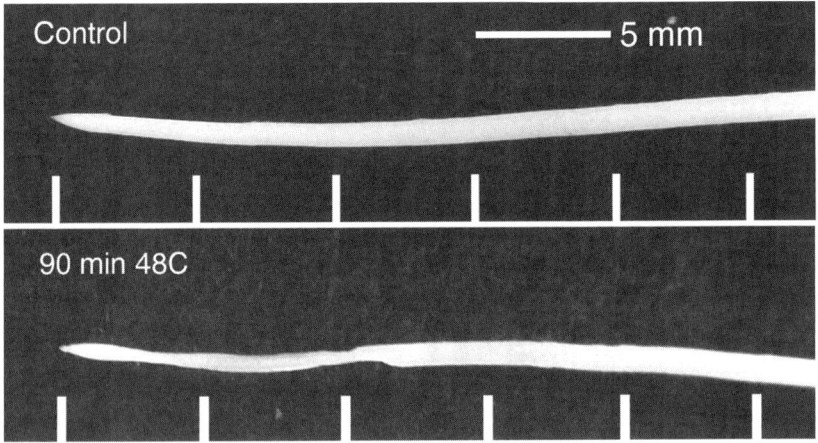

FIGURE 2.5 Roots of cotton seedlings grown at 30°C (control, top) and treated with 48°C for 90 min (bottom) in a 25 × 150 mm culture tube containing 8 ml of Stewart's medium and 0.2% phytogel.

injured region. Figure 2.5 shows photographs of cotton roots from control and heat-treated (90 min at 48°C) seedlings. Damage to the heat-treated root is apparent in the elongation zone, 3 mm to 12 mm behind the tip. These data suggest that the regions of expansive cell growth are predisposed to injury from a rapid elevation of temperature.

2.5 CONCLUDING REMARKS

The literature on plant responses to high temperatures is quite extensive and has been the subject of numerous reviews. The first 20 years of plant research into acquired thermotolerance was focused on the protein changes that occurred during the heat shock response, the sequences of the genes coding for the HSPs, regulation of the response, and identification of genetic diversity in the heat shock response. The Genomics Revolution has provided new tools to help us gain insight into the complexity of plant metabolic responses to adverse environments. Today, research is evaluating wholesale changes in mRNA expression patterns in response to heat stress. This information provides clues about the role of numerous gene products involved in heat tolerance. The availability of sequence-indexed t-DNA insertional mutants for most genes will facilitate the functional analysis of the contribution that specific genes make to overall thermotolerance. The development of plants more tolerant to adverse environments will occur through continued research on tissue-specific responses to heat and alternative heat protection mechanisms for those tissues that exhibit inherent sensitivity to heat.

REFERENCES

1. Hochachka, P.W. and Somero, G.N., *Biochemical Adaptation*. Princeton University Press, Princeton, NJ, 1984.
2. Burke, J.J., Mahan, J.R., and Hatfield, J.L., Crop-specific thermal kinetic windows in relation to wheat and cotton biomass production. *Agron. J.*, 80:553, 1988.
3. Burke, J.J., Variation among species in the temperature dependence of the reappearance of variable fluorescence following illumination. *Plant Physiol.*, 93:652, 1990.
4. Burke, J.J. and Oliver, M.J., Optimal thermal environments for plant metabolic processes (*Cucumis sativis* L.). *Plant Physiol.*, 102:295, 1993.
5. Ferguson, D.L. and Burke, J.J., Influence of water and temperature stress on the temperature dependence of the reappearance of variable fluorescence following illumination. *Plant Physiol.*, 97:188, 1991.
6. Somero, G.N. and Low, P.S., Temperature: A "shaping force" in protein evolution. *Biochem. Soc. Symp.*, 41:33, 1976.
7. Teeri, J.A., Adaptation of malate dehydrogenase to temperature variability. In *Adaptation of Plants to Water and High Temperature Stress*, Turner, N.C. and Kramer, P.J., Eds., Wiley, New York, 1978, p. 251.
8. Kidambi, S.P., Mahan, J.R., and Matches, A.G., Purification and thermal dependence of glutathione reductase from two forage legume species. *Plant Physiol.*, 92:363, 1990.
9. Mahan, J.R., Burke, J.J., and Orzech, K.A., Thermal dependence of the apparent Km of glutathione reductases from three plant species. *Plant Physiol.*, 93:822, 1990.
10. Anderson, J.V., Chevone, B.I., and Hess, J.L., Seasonal variation in the antioxidant system of eastern white pine needles. Evidence for thermal dependence. *Plant Physiol.*, 98:501, 1992.
11. Duck, N., McCormick, S., and Winter, J., Heat shock protein hsp70 cognate gene expression in vegetative and reproductive organs of *Lycopersicon esculentum*. *Proc. Natl. Acad. Sci.*, 86:3674, 1989.
12. Friedrich, K.L. et al., Interactions between small heat shock protein/substrate complexes. *J. Biol. Chem.*, 279:1080, 2004.
13. Gallie, D.R. et al., Translation initiation factors are differently regulated in cereals during development and following heat shock. *Plant J.*, 14:715, 1998.
14. Gehring, W.J. and Wehner, R., Heat shock protein synthesis and thermotolerance in Cataglyphis, an ant from the Sahara desert. *Proc. Natl. Acad. Sci.*, 92:2994, 1995.
15. Giese, K.C. and Vierling, E., Changes in oligomerization are essential for the chaperone activity of a small heat shock protein in vivo and in vitro. *J. Biol. Chem.*, 277:46310, 2002.
16. Gifford, D.J. and Taleisnik, E., Heat-shock response of *Pinus* and *Picea* seedlings. *Tree Physiol.*, 14:103, 1994.
17. Key, J.L., Lin, C.Y., and Chen, Y.M., Heat shock proteins of higher plants. *Proc. Natl. Acad. Sci.*, 78:3526, 1981.
18. Feder, M.E. and Hofmann, G.E., Heat-shock proteins, molecular chaperones, and the stress response: evolutionary and ecological physiology. *Ann. Rev. Physiol.*, 61:243, 1999.
19. Boston, R.S., Viitanen, P.V., and Vierling, E., Molecular chaperones and protein folding in plants. *Plant Mol. Biol.*, 32:191, 1996.
20. Miernyk, J.A., Protein folding in the plant cell. *Plant Physiol.*, 121:695, 1999.

21. Nover, L. and Scharf, K.-D., Heat stress proteins and transcription factors. *Cell. Mol. Life Sci.*, 53:80, 1997.
22. Schöffl, F., Prändl, R., and Reindl, A., Molecular responses to heat stress. In *Molecular Responses to Cold, Drought, Heat and Salt Stress in Higher Plants*, R.G. Landes, 1999, Austin, Texas, p. 81.
23. Vierling E., The roles of heat shock proteins in plants. *Ann. Rev. Plant Physiol. Plant Mol. Biol.*, 42:579, 1991.
24. Sorger, P.K., Heat shock factor and the heat shock response. *Cell*, 65:363, 1991.
25. Baniwal, S.K. et al., Heat stress response in plants: A complex game with chaperones and more than twenty heat stress transcription factors. *J. Biosci.*, 29:471, 2004.
26. Sung, D.Y. et al., Acquired tolerance to temperature extremes. *Trends Plant Sci.*, 8:179, 2003.
27. Finkelstein, R., Abscisic acid-insensitive mutations provide evidence for stage specific signal pathways regulating expression of an *Arabidopsis* late embryogenesis-abundant (LEA) gene. *Mol. Gen. Genet.*, 238:401, 1993.
28. Grelet, J. et al., Identification in pea seed mitochondria of a late-embryogenesis abundant protein able to protect enzymes from drying. *Plant Physiol.*, 137:157, 2005.
29. Ingram, J. and Bartels, D., The molecular basis of dehydration tolerance in plants. *Ann. Rev. Plant Physiol. Plant Mol. Biol.*, 47:377, 1996.
30. Kermode, A., Approaches to elucidate the basis of desiccation-intolerance in seeds. *Seed Sci. Res.*, 7:7595, 1997.
31. Rodriguez, E.M. et al., Barley *Dhn13* encodes a KS-type dehydrin with constitutive and stress responsive expression, *Theor. Appl. Gene.*, 110:852, 2005.
32. Helm, K.W., Petersen, N.S., and Abernethy, R.H., Heat shock response of germinating embryos of wheat. Effects of imbibition time and seed vigor. *Plant Physiol.*, 90:598, 1989.
33. Wehmeyer, N. et al., Synthesis of small heat-shock proteins is part of the developmental program of late seed maturation. *Plant Physiol.*, 112:747, 1996.
34. Almoguera, C., Prieto-Dapena, P., and Jordano, J., Dual regulation of a heat shock promoter during embroygenesis: Stage-dependent role of heat shock elements. *Plant J.*, 13:437, 1998.
35. Bettey, M. and Finch-Savage, W.E., Stress protein content of mature Brassica seeds and their germination performance. *Seed Sci. Res.*, 8:347, 1998.
36. Bettey, M., et al., Irrigation and seed quality development in rapid-cycling brassica: Accumulation of stress proteins. *Ann. Bot.*, 82:657, 1998.
37. Sung, Y., Cantliffe, D.J., and Nagata, R.T., Seed development temperature regulation of thermotolerance in lettuce. *J. Am. Soc. Hort. Sci.*, 123:700, 1998.
38. Carranco, R., Almoguera, C., and Jordano, J., An imperfect heat shock element and different upstream sequences are required for the seed-specific expression of a small heat shock protein gene. *Plant Physiol.*, 121:723, 1999.
39. Wehmeyer, N. and Vierling, E., The expression of small heat shock proteins in seeds responds to discrete developmental signals and suggests a general protective role in desiccation tolerance. *Plant Physiol.*, 122:1099, 2000.
40. DeRocher, A. and Vierling, E., Cytoplasmic HSP70 homologues of pea: differential expression in vegetative and embryonic organs. *Plant Mol. Biol.*, 27:441, 1995.
41. Burke, J.J. and O'Mahony, P., Protective role in acquired thermotolerance of developmentally regulated hest shock proteins in cotton seeds. *J. Cotton Sci.*, 5:174, 2001.

42. Trelease, R.N. et al., Microbodies (glyoxysomes and peroxisomes) in cucumber cotyledons; correlative biochemical and ultrastructural study in light- and dark-grown seedlings. *Plant Physiol.*, 48:461, 1971.
43. Burke, J.J., Integration of acquired thermotolerance within the developmental program of seed reserve mobilization. In *Biochemical and Cellular Mechanisms of Stress Tolerance in Plants*, Cherry, J.H., Ed., Springer-Verlag, Berlin, 1994.
44. Tanner, J.J., Hecht, R.M., and Krause, K.L., Determinants of enzyme thermostability in the molecular structure of *Thermus aquaticus* d-glyceraldehyde-3-phosphate dehydrogenase at 2.5 A resolution. *Biochemistry*, 35:2597, 1996.
45. Yancey, P.H. et al., Living with water stress: Evolution of osmolyte systems. *Science*, 217:1214, 1982.
46. Crowe, J.H., Crowe, L.M., and Chapman, D., Preservation of membranes in anhydrobiotic organisms: The role of trehalose. *Science*, 223:701, 1984.
47. Somero, G.N., Protons, osmolytes and fitness of internal milieu for protein function. *Am. J. Physiol.*, 251:197, 1986.
48. Sola-Penna, M. et al., Polyols that accumulate in renal tissue uncouple the plasma membrane calcium pump and counteract the inhibition by urea and guanidine hydrochloride. *Z. Naturforsch.*, 50c:114, 1995.
49. Sola-Penna, M. et al., Carbohydrate protection of enzyme structure and function against guanidinium chloride treatment depends on the nature of carbohydrate and enzyme. *Eur. J. Biochem.*, 248:24, 1997.
50. Crowe, L.M. et al., Effects of carbohydrates on membrane stability low water activities. *Biochem. Biophys. Acta*, 769:141, 1984.
51. Crowe, J.H. et al., Stabilization of dry phospholipid bilayers and proteins by sugars. *Biochem. J.*, 242:1, 1987.
52. Carpenter, J.F. and Crowe, J.H., An infrared spectroscopy study of the interactions of carbohydrates with dried proteins. *Biochemistry*, 28:3916, 1989.
53. De Virgilio, C. et al., Acquisition of thermotolerance in Saccharomyces cerevisae without heat shock protein hsp 104 and in the absence of protein synthesis. *FEBS Lett.*, 288:86, 1991.
54. De Virgilio, C. et al., The role of trehalose synthesis for the acquisition thermotolerance in yeast. I. Genetic evidence that trehalose is a thermoprotectant. *Eur. J. Biochem.*, 219:179, 1994.
55. Hottiger, T. et al., The role of trehalose synthesis for the acquisition of thermotolerance in yeast. II. Physiological concentrations of trehalose increase the thermal stability of proteins in vitro. *Eur. J. Biochem.*, 219:187, 1994.
56. Campbell, J.D., Fielding, L.A., and Brodl, M.R., Heat shock temperature acclimation of normal secretory protein synthesis in barley aleurone cells. *Plant Cell Environ.*, 20:1349, 1997.
57. Grindstaff, K.K., Fielding, L.A., and Brodl, M.R., Effect of gibberellin and heat shock on the lipid composition and endoplasmic reticulum in barley aleurone layers. *Plant Physiol.*, 110:571, 1996.
58. Chu, B., Brodl, M.R., and Belanger, F.C., Heat shock inhibits the release of the signal recognition particle from the endoplasmic reticulum in barley aleurone layers. *J. Biol. Chem.*, 272:7306, 1997.
59. Gallie, D.R., Caldwell, C., and Pitto, L., Heat shock disrupts cap and poly(A) tail function during translation and increases mRNA stability of introduced reporter mRNA. *Plant Physiol.*, 108:1703. 1995.

60. Beator, J., Potter, E., and Kloppstech, K., The effect of heat shock on morphogenesis in barley. Coordinated circadian regulation of mRNA levels for light-regulated genes and of the capacity for accumulation of chlorophyll protein complexes. *Plant Physiol.*, 100:1780, 1992.
61. DiMascio, J.A., Sweeny, P.M., and Denneberger, T.K., Analysis of heat shock response in perennial ryegrass using maize heat shock protein. *Crop Sci.*, 34:798, 1994.
62. Fender, S.E. and O'Connell, M.A., Expression of the heat shock response in a tomato interspecific hybrid is not intermediate between the two parental responses. *Plant Physiol.*, 93:1140, 1990.
63. Gifford, D.J. and Taleisnik, E., Heat-shock response of *Pinus* and *Picea* seedlings. *Tree Physiol.*, 14:103, 1994.
64. Hwang, C.H. and Zimmerman, J.L., The heat shock response of carrot. Protein variations between cultured cell lines. *Plant Physiol.*, 91:552, 1989.
65. Jinn, T.L. et al., Tissue-type-specific heat-shock response and immunolocalization of class I low-molecular-weight heat-shock proteins in soybean. *Plant Physiol.*, 114:429, 1997.
66. Krishnan, M., Nguyen, H.T., and Burke, J.J., Heat shock protein synthesis and thermal tolerance in wheat. *Plant Physiol.*, 90:140, 1989.
67. Porter, D.R., Nguyen, H.T., and Burke, J.J., Genetic control of acquired high temperature tolerance in winter wheat. *Euphytica*, 83:153, 1995.
68. Trent, J.D. et al., Acquired thermotolerance and heat shock proteins in thermophiles from the three phylogenetic domains. *J. Bacteriol.*, 176:6148, 1994.
69. Vayda, M.E. and Yuan, M.L., The heat shock protein response of an Antarctic alga is evident at 5°C. *Plant Mol. Biol.*, 24:229, 1994.
70. Vierling, E., Harris, L.M., and Chen, Q., The major low-molecular-weight heat shock protein in chloroplasts shows antigenic conservation among diverse higher plant species. *Mol. Cell. Biol.*, 9:461, 1989.
71. Vierling, R.A. and Nguyen, H.T., Heat-shock protein gene expression in diploid wheat genotypes differing in thermal tolerance. *Crop Sci.*, 32:370, 1992.
72. Weng, J. and Nguyen, H.T., Differences in the heat-shock response between thermotolerant and thermosusceptible cultivars of hexaploid wheat. *Theor. Appl. Gene.*, 84:941, 1992.
73. Jorgensen, J.A. et al., Genotype-specific heat shock proteins in two maize inbreds. *Plant Cell Rep.*, 11:576, 1992.
74. Hong, S.-W. and Vierling, E., Mutants of *Arabidopsis thaliana* defective in the acquisition of tolerance to high temperature stress. *Proc. Natl. Acad. Sci.*, 97:4392, 2000.
75. Ludwig-Müller, J., Krishna, P., and Forreiter, C., A glucosinolate mutant of Arabidopsis is thermosensitive and defective in cytosolic Hsp90 expression after heat stress. *Plant Physiol.*, 123:949, 2000.
76. Mullarkey, M. and Jones, P., Isolation and analysis of thermotolerant mutants of wheat. *J. Exp. Bot.*, 51:139, 2000.
77. Sanchez, Y. and Lindquist, S.L., HSP104 required for induced thermotolerance. *Science*, 248:1112, 1990.
78. Lee, Y.-R.J., Nagao, R.T., and Key, J.L., A soybean 101-kD heat shock protein complements a yeast HSP104 deletion mutation in acquiring thermotolerance. *Plant Cell*, 6:1889, 1994.

79. Yeh, C.H. et al., Expression of a gene encoding a 16.9-kDa heat-shock protein, Oshsp16.9, in Escherichia coli enhances thermotolerance. *Proc. Natl. Acad. Sci.*, 94:10967, 1997.
80. Prändl, R. et al., HSF3, a new heat shock factor from *Arabidopsis thaliana*, derepresses the heat shock response and confers thermotolerance when overexpressed in transgenic plants. *Mol. Gen. Genet.*, 258:269, 1998.
81. Queitsch, C. et al., Heat shock protein 101 plays a crucial role in thermotolerance in *Arabidopsis*. *Plant Cell*, 12:479, 2000.
82. Burke, J.J., O'Mahony, P.J., and Oliver, M.J., Isolation of *Arabidopsis thaliana* mutants lacking components of acquired thermotolerance. *Plant Physiol.*, 123:575, 2000.
83. Suss, K.-H. and Yordanov, I.T., Biosynthetic cause of in vivo acquired thermotolerance of photosynthetic light reactions and metabolic responses of chloroplasts to heat stress. *Plant Physiol.*, 81:192, 1986.
84. Goncharova, S.N., Sanina, N.M., and Kostetsky, E.Y., Role of lipids in molecular thermoadaptation mechanisms of seagrass *Zostera marina*. *Biochem. Soc. Trans.*, 28:887, 2000.
85. Di Baccio, D. et al., Bleaching herbicide effects on plastids of dark-grown plants: lipid composition of etioplasts in amitrole and norflurazon-treated barley leaves. *J. Exp. Bot.*, 53:1857, 2002.
86. Murakami, Y. et al., Trienoic fatty acids and plant tolerance of high temperature. *Science*, 287:476, 2000.
87. Lee, A.G., Membrane lipids: It's only a phase. *Curr. Biol.*, 10:R377, 2000.
88. Dörmann, P. and Benning, C., Galactolipids rule in seed plants. *Trends Plant Sci.*, 7:112, 2002.
89. Tomoyasu, T. et al., *Escherichia coli* FtsH is a membrane-bound, ATP-dependent protease which degrades the heat-shock transcription factor sigma 32. *EMBO J.*, 14:2551, 1995.
90. Sokolenko, A. et al., The gene complement for proteolysis in the cyanobacterium *Synechocystis* sp. PCC 6803 and *Arabidopsis thaliana* chloroplasts. *Curr. Genet.*, 41:291, 2002.
91. Sakamoto, W., Coordinated regulation and complex formation of yellow variegated 1 and yellow variegated 2, chloroplastic FtsH metalloproteases involved in the repair cycle of photosystem II in Arabidopsis thylakoid membranes. *Plant Cell*, 15:2843, 2003.
92. Tatsuta, T., Heat shock regulation in the FtsH null mutant of *Escherichia coli*: Dissection of stability and activity control mechanisms of sigma32 in vivo. *Mol. Microbiol.*, 30;583, 1998.
93. Fischer, B. et al., The FtsH protease is involved in development, stress response and heat shock control in *Caulobacter crescentus*. *Mol. Microbiol.*, 44:461, 2002.
94. Lindahl, M. et al., The thylakoid FtsH protease plays a role in the light-induced turnover of the photosystem II D1 protein. *Plant Cell*, 12:419, 2000.
95. Adam, Z. and Clarke, A.K., Cutting edge of chloroplast proteolysis. *Trends Plant Sci.*, 7:451, 2002.
96. Bailey, S. et al., A critical role for the Var2 FtsH homologue of Arabidopsis thaliana in the photosystem II repair cycle in vivo. *J. Biol. Chem.*, 277:2006, 2002.
97. Ostersetzer, O. and Adam, Z., Light-stimulated degradation of an unassembled Rieske FeS protein by a thylakoid-bound protease: The possible role of the FtsH protease. *Plant Cell*, 9:957, 1997.

98. Silva, P. et al., FtsH is involved in the early stages of repair of photosystem II in *Synechocystis* sp PCC 6803. *Plant Cell*, 15:2152, 2003.
99. Hihara, Y. et al., DNA microarray analysis of cyanobacterial gene expression during acclimation to high light. *Plant Cell*, 13:793, 2001.
100. Rizhsky, L. et al., When defense pathways collide. The response of *Arabidopsis* to a combination of drought and heat stress. *Plant Physiol.*, 134:1683, 2004.
101. Hong, S.-W., Lee, U., and Vierling, E., *Arabidopsis* hot mutants define multiple functions required for acclimation to high temperatures. *Plant Phsiol.*, 132:757, 2003.
102. Boyer, J.S., Plant productivity and environment. *Science*, 218:443, 1982.
103. Hopf, N., Plesofsky-Vig, N., and Brambl, R., The heat shock response of pollen and other tissues of maize. *Plant Mol. Biol.*, 19:623, 1992.
104. Duck, N.B. and Folk, W.R., Hsp70 heat shock protein cognate is expressed and stored in developing tomato pollen. *Plant Mol. Biol.*, 26:1031, 1994.
105. Gagliardi, D. et al., Expression of heat shock factor and heat shock protein 70 genes during maize pollen development. *Plant Mol. Biol.*, 29:841, 1995.
106. Rodriguez-Garay, B. and Barrow, J.R., Pollen selection for heat tolerance in cotton. *Crop Sci.*, 28:857, 1988
107. Liakh, V.A., Mishchenko, L.I., and Soroka, A.I., Improvement of a series of agriculturally valuable traits by pollen selection in *Linum usitatissimum* L. *Tsitol Gene.*, 34:30, 2000.
108. Weig, A., Deswarte, C., and Chrispeels, M.J., The major intrinsic protein family of *Arabidopsis* has 23 members that form three distinct groups with functional aquaporins in each group. *Plant Physiol.*, 114:1347, 1997.
109. Pear, J.R. et al., Higher plants contain homologs of the bacterial celA genes encoding the catalytic subunit of cellulose synthase. *Proc. Natl. Acad. Sci.*, 93:12637, 1996.
110. Shimizu, Y. et al., Changes in levels of mRNAs for cell wall-related enzymes in growing cotton fiber cells. *Plant Cell. Physiol.*, 38:375, 1997.
111. Cosgrove, D.J. et al., The growing world of expansins. *Plant Cell Physiol.*, 43:1436, 2002.
112. Cosgrove, D.J., Enzymes and other agents that enhance cell wall extensibility. *Ann. Rev. Plant Physiol. Plant Mol. Biol.*, 50:391, 1999.
113. Cosgrove, D.J., Expansive growth of plant cell walls. *Plant Physiol. Biochem.*, 38:109, 2000.
114. Darley, C.P., Forrester, A.M., and McQueen-Mason, S.J., The molecular basis of plant cell wall extension. *Plant Mol. Biol.*, 47:179, 2001.
115. McQueen-Mason, S. and Cosgrove, D.J., Disrupton of hydrogen bonding between wall polymers by proteins that induce wall extension. *Proc. Natl. Acad. Sci.*, 91:6574, 1994.
116. Steele, N.M. and Fry, S.C., Purification of xyloglucan endotransglycosylases (XETs): a general applicable and simple method based on reversible formation of an enzyme-substrate complex. *Biochem. J.*, 340:207, 1999.
117. Cosgrove, D.J. and Durachko, D.M. Autolysis and extension of isolated walls from growing cucumber hypocotyls. *J. Exp. Bot.*, 45:1711, 1994.
118. Sulova, Z. et al., Xyloglucan endotransglycosylase: Evidence for the existence of a relatively stable glycosyl-enzyme intermediate. *Biochem. J.*, 330:1475, 1998.

3 Molecular Responses and Mechanisms of Plant Adaptation to Cold and Freezing Stress

C.B. Rajashekar

CONTENTS

3.1 Introduction 47
3.2 Cold Signal Transduction 48
3.3 Is Water Stress an Integral Part of Cold Acclimation? 50
3.4 Cold-Responsive Genes 52
3.5 Functions of COR Proteins and Osmolytes 55
3.6 Membranes and Phospholipase D 56
3.7 Bioengineering Plants for Freezing Tolerance 59
3.8 Summary 61
References 62

3.1 INTRODUCTION

The ability of plants to tolerate low temperatures is one of the key factors in their successful growth, productivity, survival, and geographical distribution. Crop plants, especially in temperate and subtropical regions, are adversely affected by low temperatures on a routine basis. Hence, there is a need to understand the nature and mechanism of tolerance to low temperatures in plants, and to develop plants with greater tolerance. As temperate perennial plants have to survive winters, they need to tolerate not only low temperatures but also the freezing of their tissue water. Freezing of plant tissue usually involves the presence of ice in the tissue and accompanying cell dehydration. Typically, plant tissues can tolerate extreme low temperatures by freezing extracellularly, that is, by allowing ice growth to occur only in intercellular spaces [1], which entails extreme cellular dehydration, particularly in plants that survive moderate to extreme low temperatures. Although typically herbaceous plants may not be very freezing tolerant, many woody species can survive extremely low temperatures, some even below –196°C.

Plants that survive temperatures below −40°C have all or most of their freezable water removed from the cells through freeze-induced dehydration and thus, such plants need to and do tolerate near complete cell dehydration [2]. Thus, overwintering plants have to cope with at least two key stresses, namely, low temperatures and cell dehydration.

Although most temperate plants have low tolerance to freezing during their active growth in summer, they typically acquire an ability to tolerate freezing during the fall in response to environmental cues. These environmental cues can trigger a distinct set of adaptive changes leading to the induction of freezing tolerance in plants, commonly referred to as cold acclimation. The common environmental cues that have been associated with inducing freezing tolerance in plants are low temperature and short photoperiod [1]. The extent of cold acclimation varies greatly among species and even within a plant among various plant parts. Many woody species can acclimate to survive below −196°C, however, most herbaceous plants, by comparison, have limited ability to acclimate, although some can survive as low as −38°C. Thus, plants transform themselves to cope with low temperatures and freezing through cold acclimation.

Plants respond to acclimating conditions by going through many physical, biochemical, and physiological changes at molecular, cellular, and whole-plant levels. Recently, Thomashow and his group [3] profiled changes in metabolites in response to cold acclimation in *Arabidopsis*. They found extensive changes in metabolite levels: 325 metabolites accumulated in response to cold. Although we understand the role played by a few of these in relation to freezing tolerance, the relationship of most of them to freezing tolerance is not known. However, it has been long known that cold acclimation leads to many responses in plants, including morphological changes, dehydration, accumulation of sugars and other osmolytes, changes in membrane physical structure and its lipid composition, and modulation of gene expression [1,2,4–7].

Considering the complex nature of cold acclimation in plants, it is not surprising that we do not clearly understand how this transformation takes place, although in recent years we have begun to gain some insights into the cold-induced molecular responses, how they fit into the signaling pathway, and their role in freezing tolerance induction. Nevertheless, it is largely intriguing that all these changes in plants can result in freezing tolerance to such remarkable levels as those which often occur in extremely freezing-tolerant plants in response to cold acclimating conditions. In this chapter, we consider some of the recent developments in plant molecular responses to cold, putative adaptation mechanisms, and emerging potential strategies for plant improvement with regard to freezing tolerance.

3.2 COLD SIGNAL TRANSDUCTION

One of the intriguing aspects of cold response, and possibly, the induction of adaptation mechanisms in plants is how the low-temperature signal is perceived by the cells, which has to be subsequently transduced into a series of many biochemical and physiological events resulting eventually in cold acclimation.

The plant response to cold is surprisingly rapid; for example, downstream gene activation can occur within only a few minutes [8,9], although full cold acclimation can take from a few days to a few weeks, depending on the plant species [1]. Considering the various plant responses, and a large number of cold-responsive genes and their varied regulation, it is reasonable to assume that cold acclimation is complex, and that plants may indeed use more than one pathway to acquire freezing tolerance [10]. Although presently we understand some of the downstream elements of the signaling pathway induced by cold, very little is known about the mechanisms of signal perception and the immediate initial targets. However, we will consider some of the known downstream targets later in the subsequent sections.

The primary sensor of low temperature in plants is assumed to be the plasma membrane. Because temperature can directly modulate the fluidity of the lipid bilayer, the plasma membrane turns out to be an ideal candidate for sensing temperature fluctuations. Thus, it could act as a primary switch in turning on the signaling cascade. Indeed, low temperatures can induce membrane rigidity, which has been shown to activate cold-responsive genes [11–13]. The temperature-induced changes in membrane fluidity are likely to affect the cytoskeleton organization, which in turn could lead to the alteration of gene expression, including mitogen-activated protein kinases, an ubiquitous element in the signaling cascade of eukaryotes, and the downstream cold-responsive genes [12,14]. This can lead to influx of Ca^{2+} which is also considered a common mediator of stress response in a wide range of stresses including cold [15,16]. The transient increase in cytosolic Ca^{2+} has been known to mediate low-temperature response and is required for inducing freezing tolerance in plants [17].

Thus, to summarize the sequence of events in the pathway, the low-temperature signal is perceived via temperature-induced changes in the plasma membrane, which can affect the cytoskeleton organization, and subsequently result in transient cytosolic Ca^{2+} influx. The elevated Ca^{2+} can lead to activation of kinases and other cold-responsive genes. However, it is not clear how these signaling elements such as Ca^{2+}, which is a common mediator of many diverse stress responses in plants, can target specifically the cold-inducible genes and downstream events.

In addition to the direct effects of temperature on the membrane fluidity, there are other possible events that can trigger the signaling cascade in response to cold. One such event may be the loss of turgor in cells in response to cold, which has not garnered as much attention as the temperature-induced fluidity changes in the membrane. Many plants subjected to cold acclimating conditions show clear symptoms of water stress, including transient wilting. However, the loss of turgor is much more pronounced and is well documented in chilling-sensitive plants and occurs immediately after exposure to low temperatures [18,19]. Similarly, chilling- and freezing-tolerant plants may show the symptoms of wilting immediately after exposing them to low temperatures, but only transiently, as they usually tend to recover from these symptoms over the long term [20]. The loss of turgor in *Arabidopsis* occurs within minutes of exposure to low temperature.

Interestingly, although plants do recover from the wilting symptom, they experience water stress as indicated by reduced leaf water potential throughout the cold acclimation period. It is unclear how plants are able to eventually regain turgor despite the persistent low water status in leaves during cold acclimation. More importantly, it is not known whether the transient loss of turgor in plants in response to cold, which involves distinct changes not only in the plasma membrane but also in other cell structures and functions, is involved in perception of the low-temperature signal in plants. However, it is clear that plant responses to water stress and cold do have overlapping signaling pathways with the expression of some of the same target genes in response to these stresses. In the following section, we discuss the possible role of water stress during cold acclimation in inducing freezing tolerance in plants.

The other possible initial signal that could arise in response to low temperature is the oxidative burst which is known to occur in response to both biotic and abiotic stresses [21,22]. Plants exposed to low temperatures respond with a transient increase in hydrogen peroxide levels [23]. Hydrogen peroxide can trigger catalase synthesis, which helps in scavenging the accumulated hydrogen peroxide, and has been shown to induce chilling tolerance in maize (*Zea mays*) [24]. Furthermore, hydrogen peroxide is also known to induce an increase in the cytosolic Ca^{2+} levels [25], which has been associated with the activation of cold-responsive genes. Thus, from these observations, it is tempting to conclude that hydrogen peroxide may be involved in signaling cold response in plants, as it does in the case of other stresses. However, presently it is unclear as to whether oxidative burst is a part of the signaling pathway in cold response or is involved in inducing freezing tolerance in plants.

3.3 IS WATER STRESS AN INTEGRAL PART OF COLD ACCLIMATION?

Dehydration has long been known to be a part of cold acclimation and is likely to play a key role in inducing freezing tolerance in plants [1]. However, the specific role played by water stress and its role in inducing adaptive responses during cold acclimation have not been characterized in plants. Similar to chilling-sensitive plants, which readily exhibit water stress in response to low temperatures, many cold-tolerant plants including *Arabidopsis* and strawberry (*Fragaria* × *ananassa*) show water stress in response to low temperatures [20,26,27]. In these plants, there was an immediate decrease in leaf water potential after exposing them to cold [20]. The plants showed visible symptoms of loss of turgor or even wilting, but they remained turgid with longer exposure to cold.

Similar observations have been made in chilling-sensitive maize in which plants partially regained their turgor and leaf water status over a period of time in response to low temperatures [19]. Surprisingly, however, in *Arabidopsis*, the leaf water potential continued to decrease up to ten days of cold acclimation, reaching approximately −2 MPa [20]. One of the key factors that can affect plant water status is the absorption of water by roots, which can be severely depressed by the declining root hydraulic conductivity at low temperatures [18]. Thus typically, cold acclimating conditions can provide low temperatures, which in turn can induce water stress in plants.

In a number of plants, dehydration has been shown to induce freezing tolerance, and interestingly, it has been found to be as effective as cold acclimating conditions [26,28,29]. However, in some species such as barley (*Hordeum vulgare*) and oats (*Avena sativa*), although water stress could increase freezing tolerance, it was less effective than the cold acclimating conditions [28]. Similar results have been observed in *Arabidopsis* and strawberry where water stress could not completely substitute the cold acclimating conditions in inducing freezing tolerance [20,27]. Although it was nearly as effective as the cold acclimating conditions in inducing freezing tolerance, still low temperatures were needed for inducing the maximum freezing tolerance [20]. Low temperature alone could contribute to less than 20% of the increase in freezing tolerance induced during cold acclimation, and the rest was perhaps due to water stress. Thus, surprisingly, low temperatures alone, without water stress, contribute to a minor increase in freezing tolerance in *Arabidopsis* and strawberry. However, it is often assumed that cold acclimation entails only low temperatures without the consideration of the accompanying water stress and its possible effects on the induction of freezing tolerance.

The two components of cold acclimation may act via disparate pathways in inducing freezing tolerance in plants. However, there is increasing evidence to show that there is a great deal of similarity in molecular responses induced by low temperature and water stress. This is rather expected because both low temperature and water stress can trigger the induction of freezing tolerance in plants. Thus, one may expect overlapping signaling pathways with regard to these stresses. Indeed, there is overwhelming evidence to show that a number of cold-regulated (COR) genes which have conserved C-repeat (CRT)/dehydration responsive elements (DRE) are activated by both cold as well as dehydration [30–32]. Thus, both low temperatures and water stress may eventually activate the same set of target genes, although the pathways involved often are different.

For example, two transcription activators, DREB1A (also referred to as CBF3, a homologue of CBF1 and CBF2) and DREB2A, were identified in *Arabidopsis* that bind to the DRE regulatory element. The DREB1A gene is primarily inducible by low temperature whereas DREB2A is inducible by dehydration, however, both transcription factors can activate the same target genes [10,33]. Thus, *Arabidopsis* plants overexpressing cold-responsive DREB1A induced not only significant freezing tolerance but also dehydration tolerance. Heterologous expression of CBF1 in tomato plants (*Lycopersicon esculentum*) was found to increase their drought resistance [34]. Similarly, heterologous expression or expressions of CBF orthologues have been shown to increase tolerance to both freezing and water stress in an increasing number of crop plants [35–37]. Kasuga et al. [38] showed that overexpression of CBF3 of *Arabidopsis* in tobacco (*Nicotiana tabacum*) increased not only tolerance to water stress but also conferred freezing tolerance to these chilling-sensitive plants. They also found that the activated target genes of CBF3 were identical, regardless of whether the stimulus was water stress or low temperature, illustrating the similarity in the adaptation pathways for these two stresses.

Unlike the cold-responsive CBFs, CBF4, a homologue of CBFs/DREB1, was identified in *Arabidopsis* to be responsive only to water stress [39]. However, the transgenic plants overexpressing CBF4 resulted in the expression of cold- and drought-responsive genes and thus, led to an increase in both freezing tolerance and drought resistance, similar to the overexpression of other CBFs [39]. Thus, DREB2, a nonhomologue of CBFs, and CBF4, a homologue of CBFs, although both activated by water stress, apparently use separate pathways in activating the cold and water stress-responsive genes containing the C repeat/DRE element. This resulted in improved tolerance to both freezing and dehydration [33,38].

In addition to COR proteins, osmolytes like proline, sucrose, raffinose, and galactinol also accumulate in response to both cold and water stress [40,41]. In *Arabidopsis*, Taji et al. [41] found three distinct galactinol synthase genes, which are involved in raffinose biosynthesis; out of these one was responsive to cold and the others to drought. Thus, it is reasonable to assume that osmolytes are likely to accumulate in response to either water or cold stress. Therefore, as water stress can induce freezing tolerance, low temperatures can likewise induce drought resistance in plants. Indeed, it has long been known that a positive correlation exists between freezing tolerance of plants and their drought resistance [1].

In fact, recently it has been shown that *Arabidopsis* plants acquire drought resistance in response to cold acclimation treatment [20]. However, in this study plants were subjected to typical cold acclimating conditions, which were expected to include both low temperature and water stress, and thus, it was not conclusive as to whether the observed increase in drought resistance was due to low temperatures alone. Nonetheless, one can conclude that cold acclimating conditions may provide complex environmental stimuli, which may act additively through disparate but overlapping pathways to induce both freezing tolerance and drought resistance in plants. Furthermore, it is worth noting that either one of these stresses can trigger, at least to some degree, the adaptation mechanisms for both stresses.

3.4 COLD-RESPONSIVE GENES

Over the last two decades, there has been overwhelming evidence that demonstrates that plants respond to cold by activating a wide array of genes [7]. Of the diverse group of genes that are induced by cold, only a few have been studied in greater detail and are the basis of our understanding of the cold-induced signaling pathway, including gene regulation and functions of gene products [10]. Thus, as this forms only a tiny fraction of a vast array of cold-inducible genes, with further studies one would undoubtedly expect to gain better insights into the scope of cold response and adaptation mechanisms in plants.

A number of cold-responsive genes have been identified and characterized in both monocots and dicots [7,42]. However, most of our current understanding of plant molecular responses and adaptation to cold comes from the extensive work on *Arabidopsis*. When *Arabidopsis* plants are exposed to cold acclimating conditions, within minutes, they accumulate transcription activators, which activate a number of cold-responsive genes. Transcription activators, CBF1, CBF2, and CBF3 (also referred to as DREB1B, DREB1C, and DREB1A, respectively), having AP2/EREB-DNA binding motif can bind to the CRT/DRE domain of the cold-responsive genes to activate them [10,33]. The binding affinity of transcription factors may be temperature-modulated, with higher affinity at lower temperatures [43]. These proteins share considerable similarity with regard to their amino acid sequence and are coded by genes in tandem on Chromosome 4 which are also regulated by cold, but the exact mechanism by which this is accomplished is yet to be determined. However, recently a transcription factor of CBF3, ICE1, has been identified which is a MYC-like bHLH protein and can activate CBF3, and also, two cold-responsive promoter regions of CBF2 gene have been identified [44,45]. Furthermore, CBF2 has been found to exercise some control over the expression of the other cold-responsive CBF genes [46].

Cold-responsive transcription factors appear to be conserved among a wide range of plant species in both monocots and dicots. Orthologues of CBFs, which are also regulated by cold, have been identified in diverse plant species including strawberry, sour cherry (*Prunus cerasus*), canola (*Brassica juncea*), wheat (*Triticum aestivum*), barley (*Hordeum vulgare*), rye (*Secale cereale*), and even some chilling-sensitive crops [9,37,43,47,48]. In many of these species CBF orthologues function in a similar manner as in *Arabidopsis* by activating cold-responsive genes, which typically leads to stress tolerance.

Furthermore, it has been shown that CBF1, 2 and 3, which are inducible by cold, appear to have wide-ranging but similar effects on plants [49]. A number of regulatory genes in both dicots and monocots have been shown to share the AP2 DNA binding domain [50–52], suggesting that CBF transcription factors may have a broad effect on plant function beyond the stress response. Thus, under the broad umbrella of CBF regulon, the transcription activators can activate target genes, which are associated with responses to low temperature and drought, and may even affect the growth and development in plants [33,49,53].

The cold-responsive CBF transcription factors have been shown to activate a number of cold- and drought-responsive genes such as KIN1, COR6.6, COR15a, COR47, and COR78 in *Arabidopsis* [54]. These genes typically encode hydrophilic proteins, which may have protective functions against freezing and dehydration stresses in plants. The significance of COR proteins has been demonstrated in mutants of *Arabidopsis* (*sfr*). These mutants do not accumulate COR genes and are found to be deficient in cold acclimation despite their unimpaired CBF1, 2, and 3 expression [55]. In addition to these cold-regulated proteins, CBF regulon may modulate the accumulation of cryoprotective osmolytes such as sugars and proline, which are known to play a central role in inducing freezing tolerance in plants [1,56,57].

Furthermore, in recent microarray studies a number of CBF-regulated genes have been identified. Many of the CBF-targeted genes cover a broad spectrum of plant functions, including stress response and adaptation, metabolism, transport facilitation, and even synthesis of other transcription factors [53,61]. Transcriptome analysis of *Arabidopsis* in response to cold and in CBF-overexpressing plants has found that a number of transcription factors such as ZAT12 and RAV1 are cold-induced in parallel pathways along with CBFs, and some are actually the targets of CBFs [53]. Although AP2-domain proteins such as RAP2.1 and RAP2.6 may be activated by CBFs and are a part of CBF regulon, a number of CBF-regulated genes do not have a CRT/DRE domain and are perhaps induced through a secondary cascade action of these transcription factors. It is also interesting that there may be more than 60 cold-responsive genes that are independent of CBF regulon [53]. Furthermore, among the CBF-targeted gene products are a number of unidentified proteins, which raises the possibility that the CBF pathway may have a larger role than we assume at this point.

However, what is staggering is the extent of impact of the CBF pathway on the overall metabolism of plants. Over 400 metabolites were profiled in response to cold acclimation in *Arabidopsis*; 75% of the metabolites monitored increased in response to cold, suggesting that cold acclimation constitutes a major metabolic shift in plants that enables them to adapt to low temperatures [3]. Interestingly, most of the metabolic changes that occur during cold acclimation appear to occur through the CBF pathway. It is likely that the changes in the metabolome in response to cold are not only due to the accumulation of numerous CBF-targeted gene products but also perhaps to their indirect effects on the overall metabolism leading to a shift in the metabolic homeostasis. Thus, although CBF-nexus plays an important role in cold acclimation, it may have a greater impact on plant function and metabolism than we presently understand. Also, it is worth noting that cold adaptation in plants may also involve a complex array of changes, beyond the accumulation of thus-far characterized COR proteins and the cryoprotective substances.

Furthermore, in addition to CBF regulon, there may be multiple pathways by which plants acquire freezing tolerance [10]. Xin and Browse [58] identified an *Arabidopsis esk1* mutant, which is markedly freezing tolerant but did not involve the expression of COR genes. Their superior freezing tolerance was attributed to the accumulation of high levels of osmolytes, particularly proline. These findings suggest that there may be alternative pathways in acquiring freezing tolerance in plants. Similar, conclusions were reached in freezing-tolerant PLDδ-expressing *Arabidopsis* plants, where neither COR genes nor osmolytes appear to play a role in their freezing tolerance [62]. In addition, there may be pathways or genes that may have an effect on the CBF regulon, thus ultimately affecting the CBF-regulated plant responses involving stress tolerance and other characteristics in plants. Examples of genes that may regulate the CBF pathway negatively are HOS1 and HOS2 [63,64], which may affect the cold response, and ACG1, which may affect flowering in *Arabidopsis* [65]. Thus, cold response and adaptation strategies are complex, and perhaps cannot be explained by a single linear pathway, but instead, may involve separate or at least overlapping pathways in plant species capable of tolerating freezing.

3.5 FUNCTIONS OF COR PROTEINS AND OSMOLYTES

Of the COR family of proteins, COR15a has been shown to have a modest cryoprotective activity in chloroplast membranes [66]. When COR15a was overexpressed in *Arabidopsis*, the chloroplasts were more freezing tolerant by 1 to 2°C. This 15-kDa polypeptide coded by the COR15a gene is targeted to the stroma of the cholorplast and helps stabilize the membrane by reducing its propensity to form a hexagonal II phase during freezing [66,67].

However, this modest effect did not explain the typical increase in freezing tolerance one observes during cold acclimation in *Arabidopsis*. It was subsequently shown that the increase in freezing tolerance could be comparable to that resulting from cold acclimation when a number of COR genes were expressed by overexpressing CBF1 [68]. Thus, *Arabidopsis* mutant (*sfr6*), which was deficient in its ability to acclimate to cold, did not express COR genes, confirming the role of COR genes in freezing tolerance [55]. Similarly, overexpression of DREB1A was shown to induce freezing tolerance and drought resistance [33] and also these plants, in addition to expressing COR genes, showed a higher accumulation of sucrose, raffinose, glucose, fructose, and proline [69]. Among the CBF-targeted genes that are associated with freezing tolerance are those involved in synthesis of raffinose, sucrose, proline, and trehalose [3,41,53,58]. These compatible osmolytes protect membranes, organelles, and macromolecules against dehydration and freezing [70,71].

Cold acclimation in plants has long been associated with accumulation of compatible osmolytes [1,57], and exogenous application of osmolytes has been found to increase both freezing tolerance and drought resistance [72]. Similarly, increasing osmolytes through genetic manipulation has confirmed the role of these compatible solutes in improving stress tolerance in plants [73,74]. The freezing-tolerant *Arabidopsis* mutant, *esk1*, accumulates very high levels of osmolytes, with concentrations of free proline and sugars increasing by about 30-fold and 2-fold, respectively, but no involvement of COR genes was found [58]. In *Arabidopsis* mutant (*sfr4*), depressed accumulation of osmolytes has been attributed to their inability to fully cold acclimate [59]. Using the protoplasts of this mutant, Uemura et al. [60] found that this impairment can be overcome by providing sucrose exogenously, which demonstrates the importance of sugars and other osmolytes in the cold acclimation process in *Arabidopsis*. Antisense-phospholipase Dα1 plants of *Arabidopsis* were more freezing tolerant than wild-type, and accumulated higher levels of raffinose under both non- and cold-acclimating conditions [20]. Also in petunia (*Petunia hybrida*), *acuminata* Graham = *Nicotiana acuminata* var. *acuminata*), Pennycooke et al. [75] showed that antisense suppression of α-galactosidase led to a high accumulation of raffinose, which resulted in a significant increase in freezing tolerance of both nonacclimated and cold-acclimated plants.

Microarray analysis in *Arabidopsis* also shows that many genes involved in osmolyte metabolism were found to be cold responsive [53]. Genes involved in sucrose and proline biosynthesis are cold responsive, regulated by CBF and are activated during cold acclimation [69]. Galactinol synthase, one of the enzymes in the biosynthesis of raffinose, is a cold-responsive CBF-regulated gene and is known to contribute to raffinose accumulation in response to cold acclimation [41,53,61]. It is interesting that of the metabolome that was profiled by Cook et al. [3], the metabolites that increased by more than 25-fold were carbohydrates including fructose, galactinol, glucose, and raffinose, which constituted more than 50% of the changes induced by low temperature in *Arabidopsis*. Similarly in CBF overexpressing plants, which are freezing tolerant, microarray analysis shows that by far the most dominant transcripts were for the gene encoding galactinol synthase [49]. Taken together, these results clearly support the premise that accumulation of osmolytes plays a central role in inducing freezing tolerance in plants during cold acclimation.

3.6 MEMBRANES AND PHOSPHOLIPASE D

Membranes have been known to be the primary site of freezing injury, and thus their stability and competency play a key role in freezing tolerance [4,76]. Loss of bilayer stability and its competency, as typically indicated by leakage of cell solutes, can result from injury caused by low temperatures and freezing. Plants can be injured when membrane lipids undergo freeze-induced transition into the nonbilayer hexagonal II phase. Evidently, membrane lipid composition can play a central role in this membrane destabilizing transition [77]. In fact, the stability of protoplasts against freezing has been shown to be dependent on the lipid composition of plasma membrane [78]. For example, phosphatidylcholine (PC) may favor bilayer stability whereas unsaturated phosphatidylethanolamine (PE) and phosphatidic acid (PA) show a strong propensity to form hexagonal II phase, which can be damaging to the integrity of membranes during freezing.

As an adaptation mechanism, plant membranes undergo numerous changes, including changes in lipid compositions during cold acclimation. These changes are assumed to help in the stability of bilayer against freezing stress [79,80]. In addition to changes in membrane constitution, a number of cold-induced responses may contribute to membrane competency as well. For example, the cold-induced osmolytes [2] and COR15a polypeptide [67] are known to favor membrane stability and protect membranes against freezing stress. Overexpression of COR15a in *Arabidopsis* was shown to reduce the transformation of bilayers to hexagonal phase II in chloroplasts in response to freezing [67]. Sucrose, one of the osmolytes associated with cold acclimation and freezing tolerance, may actually interact with membrane-bound phospholipase D (PLD) in stabilizing membranes [81].

Membrane phospholipids are hydrolyzed by several types of phospholipases, which are often activated by both abiotic and biotic stresses [81–84,86] (also see Chapter 1). Phospholipases constitute a diverse group of lipolytic enzymes that interact and produce a network of lipid and lipid-derived messengers that are involved in signaling plant responses to various stresses [87–90]. In addition, phospholipases can have a direct effect on the membranes, as they are involved in the hydrolysis and turnover of phospholipids. Of the phospholipase family of hydrolytic enzymes, PLD has been shown to play an important role in plant response to stress including freezing, and has been characterized in relation to its genetic regulation, isoforms, and biochemical properties [86,91]. Direct involvement of PLD in degrading phospholipids and its activation properties in response to freezing were demonstrated by Yoshida [81].

Twelve genes encoding five classes of PLDs have been identified, with each having distinct properties with regard to its activation, regulation, and substrate-specificity [80,92]. Of these, PLDα is the most common form and is expressed in most tissues in plants. It needs moderately high levels of calcium for its activity and hydrolyzes phospholipids, with a preference for PC, into PA and free head groups. PLDα plays an important role in response to low temperature and is likely to be activated during freezing, suggesting an increased lipolytic activity associated with this stress [86,93,94]. When Arabidopsis plants were frozen to −8°C, there was a sharp decrease in membrane PC, PE, and phophatidylglycerol. Notably, a large increase in PA occurs in response to freezing. PA, like many other phopholipase-derived lipid messengers, is known to play a key role in signaling stress responses in plants [95,96]. In addition, PA is likely to have an adverse effect on membrane integrity as demonstrated by its destabilizing effect on bilayers [97].

Also, PA can stimulate the production of active oxygen species such as superoxide [98] which can have a damaging effect on the membranes as well. Thus, it is reasonable to assume that diminished PLDα1 activity can lead to improved freezing tolerance in plants. In fact, it has been shown that the reduced PLDα activity in *Arabidopsis* can lead to significant changes in membrane lipid composition during cold acclimation and freezing, which are likely to favor the plant's adaptation to freezing stress [86]. PLDα-deficient plants showed a much reduced level of hydrolysis of membrane PC, and relatively low PA accumulation following freezing compared to wild-type plants. Evidently, these two conditions are likely to favor the stability of membranes in PLDα-deficient plants.

Furthermore, PLDα deficient plants also generate lower amounts of oxidative species which also favors the stability of membranes [27,98]. When PLDα1-deficient plants were exposed to stress, they typically produced less oxidative species than did the wild-type (Figure 3.1). Antisense suppression of PLDα1 resulted in a significant increase in freezing tolerance of both nonacclimated and cold-acclimated *Arabidopsis* plants (Figure 3.1). The increase in freezing tolerance was 2°C and 3.5°C in nonacclimated and cold-acclimated plants, respectively [20]. Similar to antisense PLDα1-suppression, an improved freezing tolerance in plants was observed by blocking the PLD activity by using lysophospahtidylethanola-mine, a PLD-inhibitor [27]. The fact that there is an increase in freezing tolerance of cold-acclimated plants suggests that PLDα1-suppression is truly able to induce freezing tolerance but not just mimic changes that are typically associated with cold acclimation. Similar observations have been made in CBF3-overexpressing *Arabidopsis* and canola (*Brassica napus*) plants as well [47,69]. This raises the

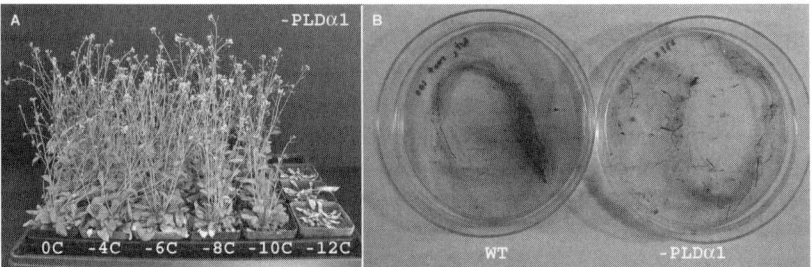

FIGURE 3.1 Freezing tolerance and production of oxidative species in antisense PLDα1 *Arabidopsis* plants. Plants were cold acclimated at 2°C with an 8 h photoperiod for up to four weeks. Cold-acclimated antisense PLDα1 plants survived −10°C (A) and wild-type plants survived −7.5°C. The PLDα1-deficient plants also produced much lower oxidative species as illustrated by root staining (B). Oxidative species in roots of wild-type and antisense PLDα1 plants were detected by using KI-starch assay [23]. Roots were incubated in 100 mm KI in 4% starch under hypoxic conditions for 12 h.

possibility that these responses may be quantitative and stronger expressions of certain genes, as in the case of CBF3, and may actually lead to better adaptation. There is also the possibility that the pathway or the mechanism of adaptation in plants that are bioengineered to overexpress/suppress a gene may respond somewhat differently from the pathway followed during natural cold acclimation.

Furthermore, in PLDα1-abrogated plants, there was a stronger expression of COR genes in response to cold acclimation and freezing. The COR gene expression was particularly enhanced in response to freezing in these plants compared to that in the cold-acclimated plants. The expression of COR genes appears to take place in partially frozen samples. This raises the possibility that the elevated COR gene expression in frozen samples is perhaps due to the mechanical perturbation that occurs during extracellular freezing. Mechanical perturbation involving cell collapse is known to be an essential feature of extracellular freezing [2]. Gilmour et al. [8] found that mechanical agitation in *Arabidopsis* could activate CBF1, CBF2, and COR15a genes transiently, and Zarka et al. [45] have identified the promoter regions of CBF2 that may be responsive to mechanical agitation. Thus, from these observations, it is tempting to propose that cell collapse during extracellular freezing may contribute to stronger expression of COR genes. Furthermore, CBF expression has also been shown to be temperature-sensitive and that the level of its expression is actually modulated by the history of plant exposure to temperature regimes. There was a progressively higher expression of CBF and COR genes with exposure to lower temperatures, even at subzero temperatures [45]. Thus, it is possible that lower temperatures may also enhance the expression of cold-responsive genes.

In addition, PLDα1-deficient plants had higher levels of osmolytes such as raffinose in a nonacclimated state compared to wild-type. Thus, PLDα*1*-suppression can lead to concerted changes in plants including membrane lipid composition, COR gene expression, and osmolyte accumulation, which may in turn result in a greater tolerance to freezing and water stress. These changes may be a result of not only the diminished catabolic PLDα1 activity, but also the signaling activity of a network of lipolytic intermediates generated by the reduced PLDα1 activity. However, it is not known whether PLDα1 and its lipolytic messengers can modulate the cold response in relation to COR genes or genes involved in osmolyte metabolism. Recently, a PLDδ knock-out mutant of *Arabidopsis* was identified to be freezing sensitive, and the overexpression of PLDδ resulted in an increase in the freezing tolerance of plants [62]. However, the genetic manipulations involving either knock-out or overexpression did not affect the expression of COR47 or COR78 or change the osmolyte accumulation. This again supports the premise that the induction of freezing tolerance in plants is complex and is brought about by multiple pathways.

As in CBF-expressing plants, PLDα1 suppression also leads to an increase in drought resistance in *Arabidopsis* [20]. Although cold acclimation induced drought resistance in both wild-type and PLDα1-suppressed plants, it was greater in PLDα1-suppressed plants. PLDα1-deficient plants had better water status during water stress, which could be attributed to some of the morphological changes induced by PLDα1 suppression. These plants have more extensive and deeper root systems, which helps them maintain a better water status. In addition, PLDα1-deficient plants showed thicker stem epidermal wax and leaf characteristics such as lower stomatal density that are likely to retard water loss and help maintain a better plant water status. Thus, PLDα1-suppression can induce wide-ranging responses in *Arabidopsis*, including activating a number of genes, osmolyte accumulation, and changes in growth and development. Clearly, some of these changes contribute to the enhancement of the stress tolerance in these plants.

3.7 BIOENGINEERING PLANTS FOR FREEZING TOLERANCE

CBF overexpression in *Arabidopsis* has provided a unique opportunity to improve freezing tolerance and drought resistance in many crops. Expression of CBF from *Arabidopsis* or its orthologues has been successfully used in a number of crop plants to increase tolerance to freezing, water stress, and salt stress [8,9,37,47]. Similar to *Arabidopsis*, overexpression of CBF in canola resulted in the activation of cold-regulated genes, which led to an increase in freezing tolerance in both nonacclimated and cold-acclimated plants. Cold-responsive orthologues of CBFs have also been identified in rice (*Oryza sativa*) and heterologous expression of these in *Arabidopsis* resulted in an increased tolerance to drought, freezing, and salt stress [35]. Recently, a homologue of CBFs was identified in chilling-sensitive maize and was found to be responsive to cold and mechanical perturbation. Its expression in *Arabidopsis* activated the cold-responsive target genes resulting in tolerance to freezing and water stress [36].

From the above observations it is clear that the function of CBFs appears to be conserved in a number of plants including monocots, dicots, and chilling-sensitive and tolerant plants. Although monocots and dicots may vary somewhat in the binding characteristics of CBF proteins to the DNA-binding domains of target genes, their overall function appears to be similar [35]. Thus, with the successful expression of CBFs heterologously, it is possible to exploit this approach readily in many plants, including those in which we do not yet clearly understand the mechanism of stress adaptation.

CBF overexpression leads to increased freezing tolerance and drought resistance in plants, and it also results in some undesirable consequences with regard to plant growth and development. The adverse effects on growth and development in *Arabidopsis* by overexpressing CBF1, 2, or 3 appear to be similar [49]. Such observations of growth retardation have also been made in plants such as tomatoes and tobacco overexpressing CBFs [34,38,99]. This is perhaps expected due to the wide-ranging and varied effects of CBFs on the metabolism and plant functions, including the activation of transcription factors involved in growth and morphogenesis [49]. The CBF-overexpressing *Arabidopsis* plants typically were smaller, with prostrate growth. The leaves were darker with higher density of chloroplast. The leaves were also thicker with two layers of palisade cells whereas the wild-type has only one palisade layer. In addition, the plants showed developmental delay at least in reproductive growth as indicated by delayed bolting [49]. The CBF-expressing plants produced less biomass and seed yield due to the production of fewer inflorescence shoots [69].

This growth retardation appeared to be associated with the level of expression of CBF3 (DREB1A) and is perhaps expected as CBF3 is one among many AP2 DNA-binding proteins, which are known to have a negative effect on growth [100]. The negative effects of CBF expression on the vegetative and reproductive growth and development are important in the overall context of the plant's performance, and the biological cost of improving freezing tolerance using such an approach has to be considered carefully, especially while attempting to extend such strategies to other plant species and crop plants. However, this problem can be mitigated by avoiding constitutive CBF expression in plants. When CBF was expressed under the control of cold-inducible promoter, the adverse effects on growth and development were minimized [101]. To demonstrate this, Kasuga et al. [38] compared the effects of expression of DREB1A in tobacco using the constitutive promoter, CaMV35S, and the cold-responsive *rd29A* promoter. The plant growth was severely retarded with the constitutive expression of DREB1A, whereas it was greatly improved when DREB1A was expressed with the *rd29a* promoter, suggesting that there is a relationship between the DREB1A transcript levels and the growth retardation. Thus, the approach of CBF-expression driven by a nonconstitutive promoter may hold a better promise for its wider application in developing stress-tolerant crops.

PLDα1 suppression, which results in increased freezing tolerance, also has a wide range of effects on growth and development of plants [20]. The transgenic plants were slightly shorter, but produced larger leaf area and a larger number of inflorescence branches than did the wild-type (Figure 3.2). Unlike CBF-overexpressing plants, the antisense PLDα1 plants bolted about six days earlier under long-day

Molecular Responses and Mechanisms of Plant Adaptation to Cold

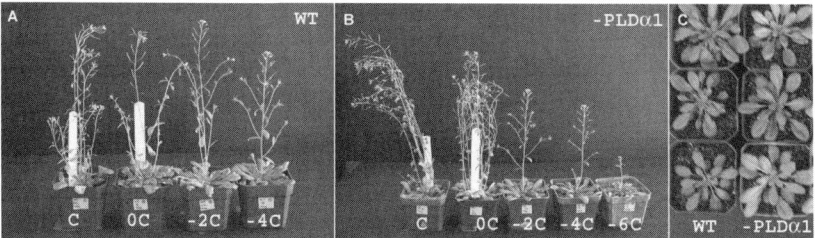

FIGURE 3.2 Growth and development of antisense PLDα1 *Arabidopsis* plants. Bolting following freezing to various temperatures in wild-type (A) and antisense PLDα1 plants (B) is presented. PLDα1-deficient plants, although slightly larger, were similar in their growth characteristics to wild-type plants (C). PLDα1-deficient plants produced more inflorescence branches (B) compared to wild-type and the growth of their inflorescence shoots was retarded by freezing.

conditions than did the wild-type plants. Interestingly, the elongation and growth of inflorescence shoots were affected by freezing in antisense PLDα1 plants. As the temperature decreased, there was progressive retardation in inflorescence growth, which was not observed in wild-type plants (Figure 3.2).

Typically, the lower the temperature, the greater was the growth retardation. Both PLDα1-deficient and wild-type plants grow at a similar rate at 23°C, however, slower growth occurs after their exposure to low temperatures in only PLDα1-deficient plants. Nonetheless, slower growth in PLDα1 plants did not appear to significantly affect their final biomass or seed yield. The PLDα1-deficient plants have a deeper and a more extensive root system and leaves with smaller stomatal apparatus, and lower stomatal density and stomatal index both on abaxial and adaxial surfaces. Despite the extensive inflorescence branching, the number of siliques produced was slightly less (95% of wild-type), with shorter siliques and fewer seeds than in the wild-type. These observations show that there were many changes in growth and development of *Arabidopsis*, induced by PLDα1 suppression. However, the negative effects of PLDα1 suppression appear to be rather minimal on the overall growth and development of plants, especially on the reproductive development and seed yield. Thus this suggests that PLDα1 suppression can be used to improve freezing tolerance without the adverse effects on the plant growth performance.

3.8 SUMMARY

In the last few years with transcriptome and metabolome analyses in response to cold and in CBF-overexpressing *Arabidopsis* plants, we have gained significant knowledge on the cold-signaling network and adaptation strategies, primarily involving CBF regulon using CBF-overexpressing *Arabidopsis* plants. Furthermore, various studies have uncovered hitherto-unknown responses of plants to low temperature in many diverse and broad areas of plant functions, showing the significant challenges ahead in understanding and relating these responses to cold adaptation.

From studies on *Arabidopsis*, it appears that there are multiple pathways by which plants acquire freezing tolerance. However, very little is known about these either in *Arabidopsis* or in more freezing-tolerant species such as woody plants. In fact, it will be certainly valuable and interesting to gain an insight into molecular responses and adaptation strategies to cold in the extremely freezing-tolerant species. Among the other areas in plant cold acclimation in which we do not yet have a clear understanding are the specific environmental cues, the mechanism of low-temperature signal perception, and the involvement of initial targets in the cold signal transduction pathway. Nevertheless, some of the recent advances over the last two decades have undoubtedly helped us gain considerable insights into cold response and adaptation mechanisms in *Arabidopsis*. In fact, presently we may have significant opportunities in applying some of the recent discoveries to develop crop plants that are tolerant to various environmental stresses.

REFERENCES

1. Levitt, J., *Responses of Plants to Environmental Stresses*, 2d ed. Academic, New York, 1980, p. 607.
2. Rajashekar, C.B., Cold response and freezing tolerance in plants. In *Plant–Environment Interactions* (II ed.), Wilkinson, R.E., Ed., Marcel Dekker, New York, 2000, pp. 321–341.
3. Cook D., Fowler, S., Fiehn, O., and Thomashow, M.F., A prominent role for the CBF cold response pathway in configuring the low-temperature metabolome of *Arabidopsis*. *Proc. Natl. Acad. Sci. USA*, 101:15243, 2004.
4. Steponkus, P.L., Role of the plasma membrane in freezing injury and cold acclimation. *Ann. Rev. Plant Physiol.*, 35:543, 1984.
5. Guy, C.L., Cold acclimation and freezing stress tolerance: role of protein metabolism. *Ann. Rev. Plant Physiol. Plant Mol. Biol.*, 41:187, 1990.
6. Ristic, Z. and Ashworth, E.N., Changes in leaf ultrastructure and carbohydrates in *Arabidopsis thaliana* (hayen) cv. Columbia during rapid cold acclimation. *Protoplasma*, 172:111, 1993.
7. Thomashow, M.F., Plant cold acclimation: Freezing tolerance genes and regulatory mechanisms. *Ann. Rev. Plant Physiol. Plant Mol. Biol.,* 50:571, 1999.
8. Gilmour, S.J. et al., Low temperature regulation of the *Arabidopsis* CBF family of AP2 transcriptional activators as an early step in cold-induced COR gene expression. *Plant J.,* 16:433, 1998.
9. Owens, C.L. et al., CBF1 orthologs in sour cherry and strawberry and the heterologous expression of CBF1 in strawberry. *Am. Soc. Hort. Sci.*, 127:489, 2002.
10. Thomashow, M.F., So what's new in the field of cold acclimation? Lots! *Plant Physiol.,* 125:89, 2001.
11. Vigh, L. et al., The primary signal in the biological perception of temperature: Pd-catalyzed hydrogenation of membrane lipids stimulated the expression of the desA gene in Synechocystis PCC6803. *Proc. Natl. Acad. Sci. USA*, 90:9090, 1993.

12. Sangwan, V. et al., Cold-activation of *Brassica napus gene* BN115 promoter is mediated by structural changes in the membrane and cytoskeleton, and requires CA^{2+} influx. *Plant J.,* 27:1, 2001.
13. Inaba, M. et al., Gene-engineered rigidification of membrane lipids enhances the cold inducibility of gene expression in Synechocystis. *J. Biol. Chem.*, 278:12191, 2003.
14. Sangwan, V. et al., Opposite changes in membrane fluidity mimic cold and heat stress activation of distinct plant MAP kinase pathways. *Plant J.*, 31:629, 2002.
15. Knight, H., Trewavas, A.J., and Knight, M.R., Cold calcium signaling in *Arabidopsis* involves two cellular pools and a change in calcium signature after acclimation. *Plant Cell,* 8:489, 1996.
16. Sanders, D., Brownlee, C., and Harper, J.F., Communicating with calcium. *Plant Cell*, 11:691, 1999.
17. Monroy, A.F., Sarahan, F., and Dhindsa, R.S., Cold-induced changes in freezing tolerance, protein phosphorylation and gene expression. Evidence for a role of calcium. *Plant Physiol.*, 102:1227, 1993.
18. Bloom, A.J., et al., Water relations under root chilling in sensitive and tolerant tomato species. *Plant Cell Environ.*, 27:971, 2004.
19. Melkonian, J., Long-Xi, Y., and Setter, T.L., Chilling responses of maize (*Zea Mays* L.) seedlings: Root hydraulic conductance, abscisic acid, and stoamatal conductance. *J. Expt. Bot.*, 55:1751, 2004.
20. Zhou, H., Studies on the effects of suppression of PLDα and PLDγ genes on freezing tolerance, morphology, and reproductive biology of *Arabidopsis thaliana.* PhD. Thesis. Kansas State University, Manhattan, KS, 2002.
21. Lamb, C. and Dixon, R.A., The oxidative burst in plant disease resistance. *Ann. Rev. Plant Physiol. Plant Mol. Biol.*, 48:251, 1997.
22. Varnova, E., Inze, D., and Van Breusegem, F., Signal transduction during oxidative stress. *J. Exp. Bot.*, 53:1227, 2002.
23. Baek, K., Role of active oxygen species in cold response in Arabidopsis and in alleviation of hypoxia in germinating bean seeds. MS Thesis, Kansas State University, Manhattan, KS, 1999.
24. Prasad, T.K., Anderson, M.D., and Stewart, C.R., Acclimation, hydrogen peroxide, and abscisic acid protect mitochondria against irreversible chilling injury in maize seedlings. *Plant Physiol.*, 105:619, 1994.
25. Rentel, M.C. and Knight, M.R., Oxidative stress-induced calcium signaling in *Arabidopsis. Plant Physiol.*, 135:1471, 2004.
26. Lang, V. et al., Alterations in water status, endogenous abscisic acid content and expression of rab18 gene during the development of freezing tolerance in *Arabidopsis thaliana. Plant Physiol.*, 104:1341, 1994.
27. Rajashekar, C.B., Unpublished results.
28. Cloutier, Y. and Andrews, C.J., Efficiency of cold hardiness induction by desiccation stress in four winter cereals. *Plant Physiol.*, 76:595, 1984.
29. Siminovitch, D. and Cloutier, Y., Twenty-four-hour induction of freezing and drought tolerance in plumules of winter rye seedlings by desiccation stress at room temperature in the dark. *Plant Physiol.*, 69:250, 1982.
30. Wang, H. et al., Promoters from *Kin1* and *cor6.6,* two homologous *Arabidopsis* thaliana genes, transcriptional regulation and gene regulation induced by low temperature, ABA, osmoticum and dehydration. *Plant Mol. Biol.,* 28:605, 1995.

31. Jiang, C., Iu, B., and Singh, J., Requirement of a CCGAC cis-acting element for cold induction of the *BN115* gene from winter *Brassica napus*. *Plant Mol. Biol.*, 30:679, 1996.
32. Stockinger, E.J., Gilmour, S.J., and Thomashow, M.F., *Arabidopsis thaliana* CBF1 encodes an AP2 domain-containing transcriptional activator that binds to the C-repeat/DRE, a *cis-acting* regulatory element that stimulates transcription in response to low temperature and water deficit. *Proc. Natl. Acad. Sci. USA*, 94:1035, 1997.
33. Liu, Q. et al., Two transcription factors, DREB1 and DREB2, with an EREBP/AP2 DNA binding domain separate two cellular signal transduction pathways in drought- and low-temperature-responsive gene expression, respectively, in *Arabidopsis*. *Plant Cell,* 10:1391, 1998.
34. Hsieh, T.H. et al., Tomato plants ectopically expressing *Arabidopsis* CBF1 show enhanced resistance to water deficit stress. *Plant Physiol.*, 130:618, 2002.
35. Dubouzet, J.G. et al., OsDERB genes in rice, *Oriza sativa* L. encode transcription activators that function in drought, high-salt, and cold-responsive expression. *Plant J.,* 33:751, 2003.
36. Qin F. et al., Cloning and functional analysis of a novel DERB1/CBF transcription factor involved in cold-responsive gene expression in *Zea mays* L. *Plant Cell Physiol.*, 45:1042, 2004.
37. Zhang, J.Z., Creelman, R.A., and Zhu, J.K., From laboratory to field. Using information to engineer salt, cold, and drought tolerance in crops. *Plant Physiol.* 135:615, 2004.
38. Kasuga, M. et al., A combination of the *Arabidopsis* DREB1A gene and stress-induced *rd29A* promoter improved drought-and low-temperature stress tolerance in tobacco by gene transfer. *Plant Cell Physiol.*, 45:346, 2004.
39. Haake, V. et al., Transcription factor CBF4 is a regulator of drought adaptation in *Arabidopsis*. *Plant Physiol.*, 130:639, 2002.
40. Yoshiba, Y. et al., Regulation of levels of proline as an osmolyte in plants under water stress. *Plant Cell Physiol.*, 38:1095, 1997.
41. Taji, T. et al., Important roles of drought-and cold-inducible genes for galactinol synthase in stress tolerance in *Arabidopsis thaliana*. *Plant J.*, 29:417, 2002.
42. Hughes, M.A. and Dunn, M.A., The molecular biology of plant acclimation to low temperature. *J. Expt. Bot.*, 47:291, 1996.
43. Xue, G.P., The DNA-binding activity of an AP2 transcriptional activator HvCBF2 involved in regulation of low-temperature responsive genes in barley is modulated by temperature. *Plant J.*, 33:375, 2003.
44. Chinnusamy, V. et al., ICE1, a regulator of cold-induced transcriptome and freezing tolerance in *Arabidopsis*. *Genes Dev.*, 17:1043, 2003.
45. Zarka, D.G. et al., Cold induction of *Arabidopsis* CBF genes involves multiple ICE (inducer of CBF expression) promoter elements and a cold-regulatory circuit that is desensitized by low temperature. *Plant Physiol.*, 133:910, 2003.
46. Novillo, F. et al., CBF2/DREB1C is a negative regulator of CBF1/DREB1B and CBF3/DREB1A expression and plays a central role in stress tolerance in *Arabidopsis*. *Proc. Natl. Acad. Sci. USA*, 101:3985, 2004.
47. Jaglo, K.R. et al., Components of the *Arabidopsis* C-Repeat/dehydration-responsive element binding factor cold-responsive pathway are conserved in *Brassica napus* and other plant species. *Plant Physiol.*, 127:910, 2001.

48. Choi, D.W., Rodriguez, E.M., and Close, T.J., Barley Cbf3 gene identification, expression pattern and map location. *Plant Physiol.*, 129:1781, 2002.
49. Gilmour, S.J., Fowler, S.G., and Thomashow, M.F., *Arabidopsis* transcriptional activators CBF1, CBF2, and CBF3 have matching functional activities. *Plant Mol. Biol.*, 54:767, 2004.
50. Elliot, R.C. et al., *AIN-TEGUMENTA*, an *APETALA2*-like gene of *Arabidopsis* with pleiotropic roles in ovule development and floral organ growth. *Plant Cell*, 8,:155, 1996.
51. Moose, S.P., and Sisco, P.H., *Glossy1*, an *APETALA2*-LIKE gene from maize that regulates leaf epidermal cell identity. *Genes Dev.*, 10:3018, 1996.
52. Buttner, M. and Singh, K.B., *Arabidopsis thaliana* ethylene-responsive element binding protein (AtEBP) and ethylene, GCC box DNA-binding protein interact with an ocs element binding protein. *Proc. Natl. Acad. Sci. USA*, 94:5961, 1997.
53. Fowler, S. and Thomashow, M.F., *Arabidopsis* transcriptome profiling indicates that multiple regulatory pathways are activated during cold acclimation in addition to the CBF cold response pathway. *Plant Cell*, 14:1675, 2002.
54. Thomashow, M.F. et al., Role of the *Arabidopsis* CBF transcription activators in cold acclimation. *Physiologia Plant.*, 112:171, 2001.
55. Knight, H. et al., The sfr6 mutation in *Arabidopsis* suppresses low-temperature induction of genes dependent on the CTR/DRE sequence motif. *Plant Cell*, 11:875, 1999.
56. Guy, C.L., Huber, J.L.A., and Huber, H.C., Sucrose phosphate synthase and sucrose accumulation at low temeperature. *Plant Physiol.*, 100:502, 1992.
57. Koster, K.K. and Lynch, D.V., Solute accumulation and compartmentation during the cold acclimation of puma rye. *Plant Physiol.*, 98:108, 1992.
58. Xin, Z. and Browse, J., *Eskimo1* mutants of *Arabidopsis* are constitutively freezing-tolerant. *Proc. Natl. Acad. Sci. USA*, 95:7799, 1998.
59. McKown, R., Kuroki, G., and Warren, G., Cold responses of *Arabidopsis* mutants impaired in freezing tolerance. *J. Exp. Bot.*, 47:1919, 1996.
60. Uemura, M., Warren, G., and Steponkus, P.L., Freezing sensitivity in the *sfr4* mutant of *Arabidopsis* is due to low sugar content and is manifested by loss of osmotic responsiveness. *Plant Physiol.*, 131:1800, 2003.
61. Maruyama, K. et al., Identification of cold-inducible downstream genes of *Arabidopsis* DREB1A/CBF3 transcriptional factor using two microarray systems. *Plant J.*, 38, 982:2004.
62. Li, W. et al., The plasma membrane-bound phospholipase Dδ enhances freezing tolerance in *Arabidopsis thaliana*. *Nat. Biotech.*, 22:427, 2004.
63. Ishitani, M. et al., HOS1, a genetic locus involved in cold-responsive gene expression in *Arabidopsis*. *Plant Cell*, 10:1151, 1998.
64. Lee, H., Cold-regulated gene expression and freezing tolerance in an *Arabidopsis thaliana* mutant. *Plant J.*, 17:301, 1999.
65. Kim, H.Y. et al., A genetic link between cold responses and flowering time through FVE in *Arabidopsis thaliana*. *Nature Gen.*, 36:167, 2004.
66. Artus, N.N. et al., Constitutive expression of the cold-regulated *Arabidopsis thaliana* COR15a gene affects both chloroplast and protoplast freezing tolerance. *Proc. Natl. Acad. Sci. USA*, 93:13404, 1996.
67. Steponkus, P.L. et al., Mode of action of the COR15a gene on the freezing tolerance of *Arabidopsis thaliana*. *Proc. Natl. Acad. Sci. USA*, 95:14570, 1998.

68. Jaglo-Ottosen, K.R., Arabidopsis CBF1 overexpression induces COR genes and enhances freezing tolerance. *Science*, 280:104, 1998.
69. Gilmour, S.J. et al., Overexpression of the Arabidopsis CBF3 transcriptional activator mimics multiple biochemical changes associated with cold acclimation. *Plant Physiol.*, 124:1854, 2000.
70. Carpenter, J.F. and Crowe, J.H., The mechanism of cryoprotection of proteins by solutes. *Cryobiology*, 25:244, 1988.
71. Santarius, K.A., Freezing of isolated thylakoid membranes in complex media. VIII. Differential cryoproytection by sucrose, proline and glycerol. *Physiologia Plant.*, 84:87, 1992.
72. Xing, W. and Rajashekar, C.B., Glycine betain involvement in freezing tolerance and water stress in *Arabidodpsis thaliana*. *Environ. Exp. Bot.*, 46:21, 2001.
73. Nanjo, T. et al., Antisense suppression of proline degradation improves tolerance to freezing and salinity in *Arabidopsis thaliana*. *FEBS Lett.*, 461:205, 1999.
74. Tarczynski, M.C., Jensen, R.G., and Bonhert, H.J., Stress protection of transgenic tobacco by production of the osmolytes mannitol. *Science*, 259:508, 1993.
75. Pennycooke, J.C., Jones, M.L., and Stushnoff, C., Down-regulating alpha-galactosidase enhances freezing tolerance in transgenic petunia. *Plant Physiol.*, 133:901, 2003.
76. Palta, J.P., Whitaker, B.D., and Weiss, L.S., Plasma membrane lipids associated with genetic variability in freezing tolerance and cold acclimation of *Solanum* species. *Plant Physiol.*, 103:793, 1993.
77. Uemura, M., Joseph, R.A., and Steponkus, P.L., Cold acclimation of *Arabidopsis thaliana*. Effect of plasma membrane lipid composition and freeze-induced lesions. *Plant Physiol.*, 109:15, 1995.
78. Steponkus, P.L. et al., Transformation of the cryobehavior of rye protoplasts by modification of the plasma membrane lipid composition. *Proc. Natl. Acad. Sci. USA*, 85:9026, 1988.
79. Uemura, M. and Steponkus, P.L., A contrast of the plasma membrane lipid composition of oat and rye leaves in relation to freezing tolerance. *Plant Physiol.*, 104:479, 1994.
80. Uemura, M. and Steponkus, P.L., Effect of cold acclimation on the lipid composition of the inner and outer membrane of the chloroplast envelope isolated from rye leaves. *Plant Physiol.*, 114:1493,1997.
81. Frank, W. et al., Water deficit triggers phospholipase D activity in the resurrection plant *Craterostigma plantagineum*. *Plant Cell*, 12:111, 2000.
82. Yoshida, S. Freezing injury and phospholipd degradation in *vivo* in woody plant cells, III. Effects of freezing on activity of membrane bound phospholipase D in microsome-enriched membranes. *Plant Physiol.*, 64:252, 1979.
83. Young, Y.A., Wang, X., and Leach, J.E., Changes in the plasma membrane distribution of rice phospholipase D during resistant interactions with *Xanthomonas oryzae* pv oryzae. *Plant Cell*, 8:1079, 1996.
84. Pinhero, R.G. et al., Modulation of phospholipase D and lipoxygenase activities during chilling. Relation to chilling tolerance of maize seedlings. *Plant Physiol. Biochem.*, 36:213, 1998.
85. Maarouf, H.E. et al., Enzymatic activity and gene expression under water stress of phospholipase D in two cultivars of *Vigna unguiculata* L. Walp. differing in drought tolerance. *Plant Mol. Biol.*, 39:1257, 1999.

86. Welti, R. et al., Profiling membrane lipids in plant stress responses—Role of phospholipase Dα in freeze-induced lipid changes in *Arabidopsis. J. Biol. Chem.*, 35:31994, 2002.
87. Munnik, T., Irvine, R.F., and Musgrave, A., Phospholipid signaling in plants. *Biochem. Biophys. Acta.*, 1389:222, 1998.
88. Wang, X., The role of phospholipase D in signaling cascade. *Plant Physiol.*, 120:645, 1999.
89. Wang, X. et al., Networking of phospholipases in plant signal transduction. *Physiol. Plant.*, 115:331, 2002.
90. Wang, X., Phospholipase D in hormonal and stress signaling. *Curr. Opinion Plant Biol.* 5:408, 2002.
91. Wang, X., Plant phospholipases. *Ann. Rev. Plant Physiol. Plant Mol. Biol.*, 52:211, 2001.
92. Wang, X., Multiple forms of phospholipase D in plants: The gene family, catalytic and regulatory properties, and cellular functions. *Prog. Lipid Res.*, 39:109, 2000.
93. Willemot, C., Rapid degradation of polar lipids in frost damaged winter wheat crown and root tissue. *Phytochemistry*, 22:861, 1983.
94. Ruelland, E. et al., Activation of phospholipases C and D in an early response to a cold exposure in *Arabidopsis* suspension cells. *Plant Physiol.*, 130:999, 2002.
95. Jacob, T. et al., Abscisic acid signal transduction in guard cells is mediated by phospholipase D activity. *Proc. Natl. Acad. Sci. USA*, 96:1292, 1999.
96. Testerink, C. et al., Isolation and identification of phosphatidic acid targets from plants. *Plant J.*, 39:527, 2004.
97. Miao, Q., Han, X., and Yang, F., Phosphatidic acid–phosphotidylethanolamine interaction and apocytochrome c translocation across model membranes. *Biochem. J.*, 354:681, 2001.
98. Sang, Y., Cui, D., and Wang, X., Phospholiplase D and phosphatidic acid-mediated generation of superoxide in *Arabidopsis. Plant Physiol.*, 126:1449, 2001.
99. Hsieh, T.H. et al., Heterology expression of the *Arabidopsis* C-repeat/dehydration responsive element factor 1 gene confers elevated tolerance to chilling and oxidative stresses in transgenic tomato. *Plant Physiol.*, 129:1086, 2002.
100. Wilson, K. et al., A dissociation insertion causes a semidormant mutation that increases expression of *TIN*, an *Arabidopsis* gene related to *APETALA2*. *Plant Cell*, 8:659, 1996.
101. Kasuga, M. et al., Improving plant drought, salt, and freezing tolerance by gene transfer of a single stress-inducible transcription factor. *Nature Biotechnol.*, 17:287, 1999.

4 Photoacclimation: Physiological and Molecular Responses to Changes in Light Environments

Penny L. Humby and Dion G. Durnford

CONTENTS

4.1 Introduction 70
4.2 Short-Term Cellular and Chloroplast Responses 72
    4.2.1 Energy Dissipation/Quenching 72
    4.2.2 Thermal Dissipation 74
    4.2.3 State Transitions 75
    4.2.4 Photoinhibition 76
4.3 Long-Term Responses 77
    4.3.1 Metabolism 77
    4.3.2 Developmental Attributes 78
    4.3.3 PSI/PSII Stoichiometry 80
    4.3.4 Light-Harvesting Antennae 81
        4.3.4.1 LHC Regulation 82
4.4 Chloroplast to Nucleus Signal Transduction 83
    4.4.1 Feedback from the Tetrapyrrole Biosynthetic Pathway 83
    4.4.2 Plastid Redox Status 84
        4.4.2.1 Plastoquinone (PQ) Pool 85
        4.4.2.2 Sugars 86
        4.4.2.3 Reactive Oxygen Species (ROS) 87
    4.4.3 Photoreceptors 87
4.5 Final Thoughts 88
References 88

4.1 INTRODUCTION

Light is a complex environmental cue that varies in intensity and quality both daily and seasonally. In addition to the predictable daily changes tracked from dawn to dusk, plants must also cope with the unpredictable changes that can occur in an instant (as seen in sun flecks, cloud cover, etc.) or that may persist for longer durations (due to events such as weather patterns, windthrow, or deforestation). Fluctuations in light intensity and quality have clear consequences on biochemical, physiological, and developmental processes in plants. If the light level is too low, there is insufficient energy to maintain growth and development. If it becomes too high, the system can overload resulting in photodamage.

Plants, must constantly monitor light conditions and initiate appropriate acclimation mechanisms to optimize growth, a process called photoacclimation [1–6]. Photoacclimatory mechanisms can be categorized as short-term responses on a time scale of seconds to minutes [2,3,6], or long-term responses that take minutes to hours to induce (Figure 4.1) [3–5]. Photoacclimation can occur at the whole-plant/leaf level and the cellular/chloroplast level [7]. This review focuses the discussion primarily on the short- and long-term cellular, chloroplast-level changes, their effects on the physiological status of the plant, and how these processes allow organisms to acclimate to changing light environments. In addition, we examine the sensing of light changes and the signaling events that initiate photoacclimation.

Photosynthesis proceeds from the capture of solar energy by light-harvesting complexes (LHCs) via bound chlorophyll and carotenoids within the thylakoid membranes of the chloroplast. This excitation energy is transferred to the reaction centers of the two photosystems (PSII and PSI) to drive an electron transport chain. Electron transport leads to the reduction of NADP+ and the

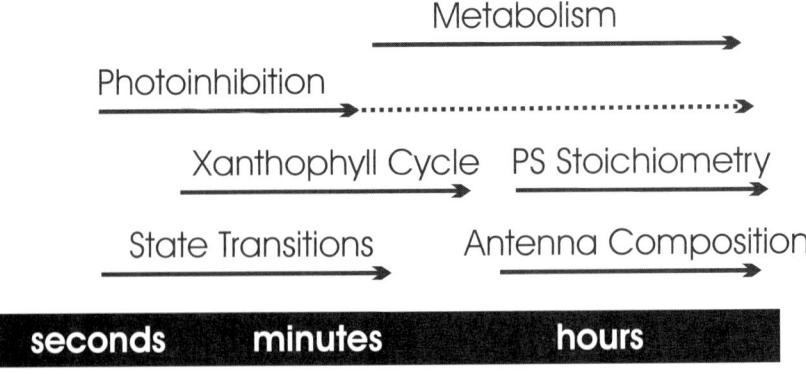

FIGURE 4.1 Effective timescale of short- and long-term photoacclimation mechanisms discussed in the text. The dashed line for photoinhibition refers to the chronic photoinhibition that can persist in high-light exposed plants.

formation of a proton gradient for the synthesis of ATP. When light harvesting outpaces a plants' ability to utilize it, the excitation energy may pass to oxygen, leading to the generation of a series of reactive oxygen species (ROS). ROS production, in the form of singlet oxygen, superoxide, and hydrogen peroxide, can lead to the oxidation of macromolecules and eventual bleaching of the chloroplast. Plants face the challenge of capturing sufficient energy to meet their metabolic needs and minimizing the absorption of excess light that would result in photodamage.

Plants and green algae appear to use similar mechanisms for coping with excess light, although the physiological importance of specific strategies may vary. Much of what we discuss in the following sections involves studies from plants and green algae owing to the considerable research on the light acclimation mechanisms in these systems. Even a casual glance at the literature will lead to the inevitable conclusion that there are many different ways of dealing with changes in light intensity and that even shared responses have different relative contributions depending on the species and the magnitude of the change in the light environment. There are studies of photoacclimation in more diverse groups, however, the mechanistic basis of the changes is less understood and we have left them out of the discussion.

One of the many possible strategies for coping with changes in light intensity is avoidance [8]. Motile aquatic protists are able to swim, and the reversal of swimming direction in response to a light shock, the so-called photophobic response, in *Chlamydomonas reinhardtii* and *Euglena gracilis* has been well described [9,10]. Plants, on the other hand, have modified this response owing to their sessile nature. Under low light, plants will reorient both their leaves and chloroplasts so that they optimize light absorption. In response to high light, however, they fold or drop their leaves to minimize exposure and reorient the chloroplasts to maximize self-shading [8,11–13]. This movement accounted for a 10% decrease in light absorption in a facultative shade plant resulting in higher tolerance to high light [14].

Arabidopsis chloroplast movement was shown to be fluence rate-dependent, whereby higher light intensities are associated with greater chloroplast movement [13]. Field studies have shown that chloroplasts move and change position in vivo and rarely remain in the same position [15]. Mutants' lacking chloroplast reorientation in response to increasing light levels exhibited greater susceptibility to photodamage [8,13]. Because these mutants showed no differences in nonphotochemical quenching (NPQ) or scavenging enzymes compared to wild-type plants, the photoinhibition observed may be the result of the mutants' inability to avoid high irradiance through chloroplast movement [8,13]. This strategy can cope with the transient changes in light level. Indeed, chloroplast phototropism appears to be quick enough to potentially minimize damages due to high light exposure, such as in the case of sun flecks [15], and it is argued that this should be considered as a significant first line of defense in the photoacclimatory process [8].

The synthesis of light-reflecting or absorbing compounds is another common avoidance response. In plants, this may include the biosynthesis of waxes [16] or anthocyanins [17]. Two maize (*Zea mays*) genotypes, HOPI (with anthocyanins) and W22 (anthocyanin-deficient), showed no differences in photochemistry, but photoinhibition occurred where the HOPI strain was more resistant to plutoinhibition than W22. This suggests that anthocyanins protect against excess light, but do not directly interfere with photosynthesis [17]. In addition to their light and UV screening abilities, anthocyanins are also hypothesized to have a role in scavenging superoxide [18]. In unicellular green algae, the biosynthesis of secondary carotenoids, such as astaxanthin, are common in some species inhabiting extreme environments and are reported to protect against photo-oxidative damage [19,20].

Often avoidance strategies are limited or insufficient to protect against photodamage, and additional responses are required to cope with light changes. A combination of avoidance and short- and long-term acclimation mechanisms, as discussed below, should allow plants to successfully photoacclimate to their environment.

4.2 SHORT-TERM CELLULAR AND CHLOROPLAST RESPONSES

Rapid photoprotective mechanisms are induced within seconds to minutes of exposure to a high-light stress [6]. These include excess energy dissipation [21], state transitions [22,23], and photoinhibition [24] (Figure 4.2). All of these strategies function to temporarily redirect excitation energy away from PSII, thus preventing the oxidation of the PSII core and reducing linear electron flow. This effectively acts to buffer photochemistry against transient increases in light or help maintain efficient light utilization until longer-term mechanisms can be initiated.

4.2.1 ENERGY DISSIPATION/QUENCHING

Under conditions of excess light, photosynthetic organisms must either utilize the light energy (photosynthesis) or dissipate the collected excitation energy in order to prevent photodamage. Photochemistry (q_P), which is the normal process of linear electron flow from water to reduced NADPH, is the preferred route for excitation energy. Even when this process is functioning at its optimum, some energy is lost during transfer between chlorophyll molecules [25–27]. Chlorophyll molecules emit the energy not used in photochemistry as fluorescence as they return to their stable, ground energy state. The fluorescence signal can be diminished (or quenched) by the induction of mechanisms whose function is to bleed off this excitation energy. The amount of fluorescence can be measured to indicate the physiological status of the organism and assess plant stress [28]. Chlorophyll fluorescence has been invaluable in assessing light responses in vivo since the amount of chlorophyll fluorescence is related to the redox state of PSII [29,30].

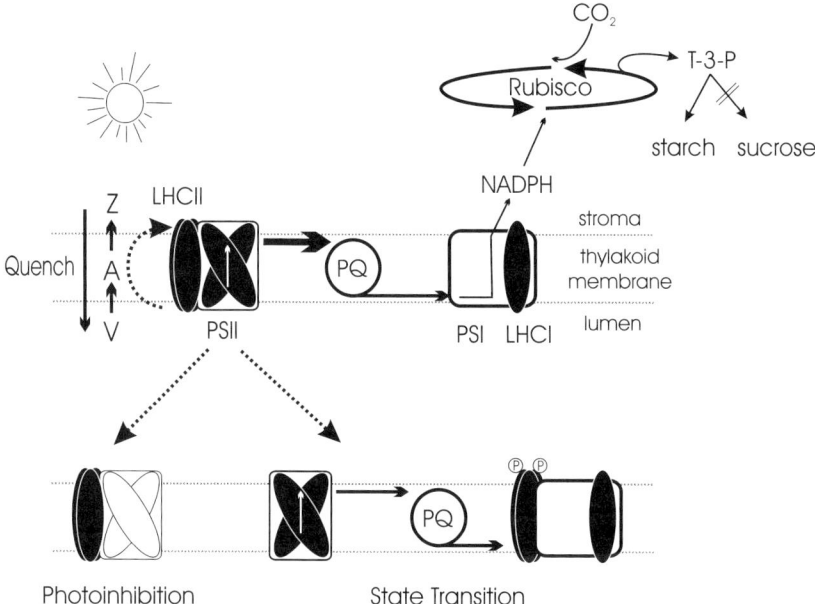

FIGURE 4.2 Short-term acclimation mechanisms existing at the thylakoid membrane. Following an increase in light intensity there is an imbalance in the processes of energy collection/generation and utilization via carbon fixation into triose-phosphates (T–3–P), starch, and sucrose, the latter of which occurs in the cytosol (top portion). Arrow thickness in and out of the plastoquinone (PQ) pool represents the relative rate of electron flow. The generation of a high *trans*-thylakoid pH under high light induces the xanthophyll cycle, which involves the enzymatic conversion of violaxanthin (V) to zeaxanthin (Z) via antheraxanthin (A). Zeaxanthin synthesis is correlated with quenching of fluorescence and represents a significant energy dissipation mechanism. The inactivation of PSII reaction centers (photoinhibition, illustrated by loss of color in PSII reaction center) is believed to participate in dissipating excitation energy as heat and reduce electron flow into the PQ pool. Finally, reduction of the PQ pool is known to activate an LHCII kinase, which phosphorylates the major LHCII polypeptides, leading to the dissociation of the complex from PSII and association with PSI, thus assisting to re-establish a balance in the electron flow between PSII and PSI (see text for details).

The degree of stress on PSII is referred to as excitation pressure [31]. When excitation pressure exceeds the plant's capacity for photochemistry (q_P), alternative forms of energy dissipation called nonphotochemical quenching (q_N) must be activated. The relationship between total absorbed irradiance (I) and fluorescence (F) is given in the following equation [32].

$$I = q_P + q_N + F$$

Nonphotochemical quenching (q_N) is composed of three distinct quenching mechanisms [26,27]: high-energy thermal dissipation (q_E) through the use of the xanthophyll cycle and other pigments [21]; state transitions (q_T) involving the phosphorylation of the light harvesting complex (LHCII) and subsequent detachment from PSII [22]; and photoinhibition (q_I) induced by damage to PSII [33]. Small amounts of q_N may also be attributed to such things as P680 functioning as a shallow trap for electrons [34] or cyclic flow around PSII [32], but these are highly speculative and require further investigation. Of the different quenching mechanisms, the xanthophyll cycle predominates; therefore, most of q_N is associated with loss of excitation energy through heat [35].

4.2.2 Thermal Dissipation

The primary component of thermal dissipation (q_E) is referred to as NPQ (non-photochemical quenching) [29,35,36]. It involves carotenoid pigments located in the light harvesting antennae in the form of β-carotenes and the xanthophyll cycle pigments [37,38]. This parameter is easy to measure in the field inasmuch as it does not require the subject to be dark-adapted unlike chlorophyll fluorescence measurements, and is often used to indicate thermal dissipation in the LHCII [35]. Most NPQ is high-energy-dependent, meaning that its induction is associated with a pH gradient across the thylakoid membrane [39].

β-carotene quenching is an extremely effective mechanism for protecting against high-light stress. It has the ability to remove excess energy and quench triplet states of chlorophyll [6] and, possibly, triplet pheophytin [40]. It may also donate electrons to oxidized P680 (reaction center of PSII) to prevent this extremely strong oxidizing agent from pulling electrons from other molecules when water is limited [41]. Increases in triplet state formation during high-light stress can be used as a mechanism for photoprotection by "bleeding off" energy to adjacent β-carotenes [42].

Although β-carotene quenches triplet state chlorophylls, the xanthophyll cycle quenches the singlet state of chlorophyll, which has a higher energy state than triplet chlorophyll [6]. The xanthophyll cycle involves the energy-dependent conversion of violaxanthin to zeaxanthin, a process that facilitates energy dissipation as heat [43]. The pool of xanthophyll pigments increases under high light and decreases in low light, reflecting its role in excess energy dissipation [44,45].

As photosynthesis continues, there is an increase in the pH gradient across the thylakoid membrane and a subsequent acidification of the lumen. This activates, via protonation, violaxanthin de-epoxidase (VDE) that de-epoxidzes violaxanthin (V) to produce antheraxanthin (A) and zeaxanthin (Z) [46]. Excited antennae chlorophylls can dissipate their energy via resonance transfer to A or Z, and this energy is then dissipated via an unknown mechanism as heat [6,21]. For NPQ to function, the xanthophylls must be in the correct location and orientation to accept the excitation energy [47]. Exposure to high light can lead to conformational changes in the antennae due to protonation of lumen-exposed portions [48] and phosphorylation through kinase activation that promote the binding of the newly synthesized zeaxanthin [39,49–50] or the reorientation of pigments so they can accept excitation energy [51].

NPQ is primarily regulated via the xanthophyll cycle although it does not account for it completely. Addition of dithiothreitol, which inhibits VDE, causes a large drop in NPQ, but then NPQ rises slowly to a steady state. This suggests that under phosphorylating conditions, the increase in NPQ is not a result of VDE but rather dissipation within the reaction centers [52].

Exactly where in the photosystem the xanthophyll cycle is functioning is under considerable debate [53] and this is a very active field of research. The minor antennae (CP29, CP26, and CP24) bind the majority of xanthophyll pigments and have protonation sites, and thus are potential sites of thermal dissipation [54]. For instance, quenching in an NPQ mutant of *Arabidopsis* appears to be associated with isoform changes in CP26 due to the binding of zeaxanthin, suggesting that this minor antenna plays a role in quenching [55]. Conversely *Arabidopsis* antisense lines for CP26 and CP29 show essentially no change in their NPQ, suggesting that the sight of thermal dissipation is not at the minor antennae [56,57].

The LHC–related PsbS protein is essential in the execution of efficient qE in higher plants [58]. *Arabidopsis npq4* and *npq4-1* mutants, which have no PsbS or a point mutation in the PsbS gene respectively, exhibited a significant reduction of NPQ in high light (80% and 50%, respectively) [58,59], however, they are still able to initiate some quenching [60]. Interestingly, although PsbS has been identified in *Chlamydomonas* [61], its role in green algae is not well understood [55]. Although several NPQ mutants have been isolated in *Chlamydomonas*, no PsbS mutant has yet been identified.

Evidence to support NPQ occurring in the peripheral antennae in green algae is seen in the qE-deficient *Chlamydomonas* mutant *npq5*. Although it shows little thermal dissipation, V de-epoxidation still occurs in high light. This mutant was found to have no *Lhcbm1* (the major trimerizing LHCII antenna protein), suggesting that reduced dissipation is due to a reduction in trimerization of the peripheral antenna, not the minor antennae [62]. A contradictory study of chl *b*-less mutants in *Chlamydomonas*, where little trimerization of the LHC occurs, showed that dissipation still occurred and therefore this study suggests that NPQ is associated not with trimerization of LHCII but likely with the minor antennae [63]. As one can see, the picture of thermal dissipation in the photosystems is unresolved and more research, especially with mutant and antisense lines in both *Arabidopsis* and *Chlamydomonas*, are needed to help elucidate the location and mechanisms of NPQ.

4.2.3 State Transitions

During periods of excess light, an imbalance in the excitation of PSII and PSI leads to photoinhibition. State transitions function to re-establish the distribution of excitation energy between the two reaction centers by regulating the antennae size in a reversible fashion [22,23,64]. Chloroplasts in State 1 have LHCII polypeptides that are primarily unphosphorylated, associated with PSII, and characterized by linear electron flow [65]. Following an imbalance between PSII and PSI, there is a transition to State 2 which is correlated with LHCII phosphorylation and the lateral migration of LHCII and cytochrome (Cyt) b_6f from appressed to nonappressed regions of the thylakoid membrane where PSI is located [66–68]. In *Chlamydomonas*, State 2 enhances cyclic flow around PSI [65], effectively reducing the number of photons PSII) absorbs whereas PSI circulates the electrons back through the plastoquinone (PQ) pool via a cyclic electron flow. This essentially stalls the system while maintaining the pH gradient and allowing for continued ATP production [65] that is potentially required for PSII repair [69].

Phosphorylation of LHC is correlated to the redox state of the PQ pool and is specifically triggered when plastoquinol binds the Q_O site of Cyt b_6f [70], although under low light, regulation may be a result of the redox state of thioredoxin [71]. Identification of the LHCII kinase involved in state transitions has been difficult, but this pathway is starting to be dissected. The thylakoid-associated kinases (TAKs) of *Arabidopsis* are responsible for phosphorylation of the LHCII and are required for state transitions [72,73]. Although TAK1 antisense plants are viable under higher light intensities, they did exhibit slowed growth and bleaching [73], indicating an important role in optimizing photosynthesis. Depege et al. [74] identified another kinase involved in state transitions by screening for *Chlamydomonas* mutants lacking the fluorescence decrease following saturating flashes, a response attributable to state transitions. These mutants are locked in state I, have reduced LHCII phosphorylation, and lack the ability to undergo state transitions. The mutation responsible for the phenotype was a defective chloroplast-localized kinase called Stt7. The exact role of Stt7 in the induction of state transitions is not known.

Prolonged or excessive light appears to prevent further phosphorylation by deactivating LHCII kinase via a ferredoxin–thioredoxin trigger [71,75,76]. Thus, the role of state-transitions may be limited to imbalances in photosystem activity under relatively low to medium light intensities [77]. Although it is an important mechanism for light acclimation, plants appear to be able to compensate for losses in the ability to undergo state transitions [78]. Nevertheless, the regulation of state transitions is complex and is in need of further investigation.

4.2.4 Photoinhibition

Photoinhibition is the result of excess light causing damage to the RC of PSII that can be described as rapidly relaxing "dynamic photoinhibition" or "chronic photoinhibition" that slowly recovers following excess light exposure [79]. Because damage to PSI is irreversible [80], photoprotective mechanisms are in place that include the quenching mechanisms described above but also the actual photoinhibition of PSII (qI) [24]. The PSII core protein D1 (where Q_a is located) is the primary site for photodamage. Under high-light stress, damaged D1 becomes targeted for destruction and D1 synthesis is upregulated within minutes as part of a complex repair mechanism [81]. The photoinhibition of PSII functions as an active photoprotection mechanism in high light; it is thought to participate in the dissipation of light energy as heat, and is not merely a result of photodamage [24,82].

4.3 LONG-TERM RESPONSES

The above-described photoacclimatory responses compensate for significant but transient changes in light intensity. In the event of light changes that are of longer duration, long-term mechanisms are induced (Figure 4.3). Of course, none of these strategies are mutually exclusive and are likely utilized concurrently in many systems.

4.3.1 Metabolism

The most obvious strategy is to adjust metabolic output in order to compensate for the change in energy input. In response to an elevation in light intensity, metabolism is stimulated to increase the sink capacity of photosynthesis [83–85], preventing an overreduced electron transport chain that can cause oxidative stress. In high light, carbon fixation and photosynthate production increase, often fueling increases in respiration and growth. The reverse typically happens in low light [86–88].

Generally, in plants, levels of Rubisco, Cyt b_6f, and ATP synthase activity increase in high light, which reflects the increases in downstream activity [86,89]. Increases in Rubisco also lead to higher P_{max} values supporting the idea that

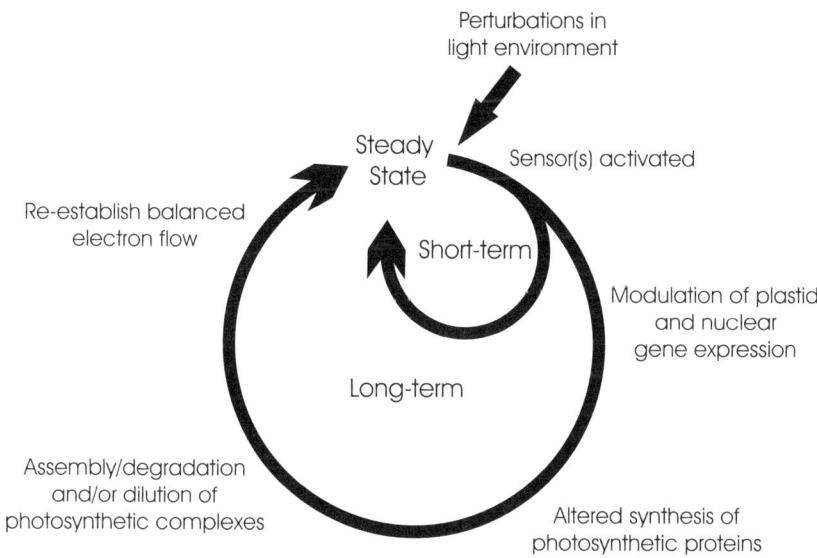

FIGURE 4.3 The integration of short- and long-term photoacclimation responses. Following a change in the light environment (short arrow), intracellular sensors recognize the fluctuation and induce both short- and long-term mechanisms to re-establish a balanced electron flow in the chloroplast. For long-term changes, this includes an alteration in chloroplast and nuclear gene expression to facilitate the long-term change in the concentration of photosynthetic complexes that is required to reach a new steady state and to reduce photo-oxidative damage to the chloroplast.

downstream carbon fixation is associated with photochemical efficiency [90]. The *Arabidopsis* mutant lacking the triose-phosphate/phosphate translocator (*ape2*), which is responsible for exporting sugar from the chloroplast, preferentially synthesizes starch in order to maintain a metabolic sink and its photosynthesis appears unaffected in low light [91]. In high light, starch synthesis can become limiting and lead to a reduction in P_{max} and quantum efficiency [91,92]. The phosphate pool also may become limiting and a subsequent reduction in growth and photosynthesis can occur [92]. When the metabolic sink is unavailable, a negative feedback occurs, affecting photosynthesis.

Krapp et al. [93] supplied an alternative sugar source to detached spinach (*Spinacia oleracea*) leaves, thus saturating the energy sink within the leaves. Although respiration rates increased in response to the increased availability of energy, downregulation both of photosynthesis rate and carbon assimilation, with a concomitant reduction in Rubisco was observed [93].

Increasing sinks for available reductants in high light is an effective way to photoacclimate; however, it may be limited by environmental or physiological constraints. Normal recovery upon exposure to high light and initiation of photoinhibition occurs over some time (hours to days) as the organism adjusts to the new light regime [86]. Nitrogen availability [94,95], temperature [96], nutrient availability [97], and water status [98] have been shown to modulate the process of photoacclimation.

The ability to photoacclimate also depends on the physiological constraints of the organism. In algae, increases in starch and sugar metabolism occur, however, mobilization and storage of photosynthates are restricted [83]. When *Chlamydomonas* is shifted into high light, photoinhibition is induced and a transient decrease in Rubisco expression is observed [99]. Microalgae rely more heavily on cell division and growth, rather than carbohydrate synthesis [83]. *Chlamydomonas*, when shifted to high light, ceases cell division for approximately six to eight hours [100,101]. During this time period cells become larger, Rubisco and Lhc expression decline, and an upregulation in D1 synthesis to repair PSII is observed. Upon the resumption of division, the Rubisco and Lhc expression return to their low-light levels and growth increases dramatically [99,101].

Although the approach to the initial light-shock is different in plants and microalgae (carbohydrate metabolism vs. cell division), the final outcome for both is the same. The increase in metabolic yield and growth balances light capture and usage. These too have their limitations, therefore, an organism must induce other long-term responses in order to maintain metabolic efficiency [102].

4.3.2 Developmental Attributes

Many plants are adapted to specific light environments and their development, photosynthetic capacity, and chloroplast structure have been fine-tuned to optimize growth under the prevailing conditions [103]. As a result, many plants and algae have limitations regarding overall tolerance to changes in light levels (sun vs. shade) and may have species-specific [104–108] or highly variable responses to altered light regimes [109–111]. There is a strong interplay between developmental attributes of a plant grown in a particular light regime and the ability to photoacclimate to changes within this context. If a plant is grown under low light, its leaves are thinner, having one palisade layer, and a reduced mesophyll, which results in more intercellular air spaces. Chloroplasts are arranged parallel to the leaf surface, large and numerous, and have higher thylakoid membrane volume and a larger number of granal stacks [109].

Murchie and Horton [7] investigated 22 species of temperate angiosperms in an attempt to quantify the continuum between sun and shade plants. They identified two strategies: changes in chlorophyll per leaf area (whole leaf alteration) or changes in chlorophyll a/b ratio and P_{max} (alterations at the chloroplast level). Obligate sun species alter their entire leaf whereas intermediate shade-tolerant plants are flexible in changing parameters at the chloroplast level. Sun plants have greater plasticity than shade plants and obligate shade plants have an overall limitation in their ability to acclimate to high light [7,107].

The absorption of light as it passes through leaves generates a light gradient, leaving chloroplasts of different acclimation states within individual leaves [112,113] adding a level of complexity. In spinach, the chlorophyll content increased from the adaxial surface into the leaf and then declined again as it reached the abaxial surface, suggesting the decreased irradiance as light passed through the leaf [114]. This is also manifested as different P_{max} capacities depending on the positions where the chloroplasts are located within a leaf [115]. Conceptually, microalgae experience a similar light gradient within a water column, the significance of which is dependent on vertical mixing and cell motility.

To cofound the issue even more plants have the ability to communicate the light conditions in one part to other areas of the plant. Karpinski et al. [109] showed that *Arabidopsis* was capable of "systemic acquired acclimation." In their study, plants were partially exposed to high light and acclimatory response of the shaded leaves was measured. They found that shaded leaves exhibited photoprotective mechanisms even though they had not directly been exposed to the light stress. Yano and Terashima [111] examined *Chenopodium album* under various conditions of partial light exposure and the morphology of new leaf development was assessed. In one treatment, the apex was shaded while the mature leaves were exposed to high light. New leaves produced by these plants morphologically resembled mature leaves acclimated to high light even though they developed in low light. However, the chloroplast morphology in these new leaves reflected the light condition at the apex (shade-acclimated). When the mature leaves were shaded while the apex was exposed to high light, the results were reversed. Consequently, leaf morphology of newly formed leaves reflected the conditions under which the mature plant was exposed, but the chloroplast acclimation state was defined by the light environment of the new leaves. This suggests two mechanisms at work: the environment of the mature leaves dictating leaf differentiation (development), and local light environment dictating chloroplast morphology (acclimation).

Many species-specific differences in plant photoacclimation have to do with leaf longevity. In general, long-lived leaves are likely to readjust photosynthesis at the chloroplast level, whereas short-lived leaves drop off and are replaced by "sun"-adapted leaves [106,109]. It can be argued that this latter case is not truly photoacclimation, as the susceptible leaf is lost, but rather a light intensity-dependent developmental response. In older leaves, even though Rubisco levels increase as predicted in high light, increases in P_{max} are limited, suggesting that age of a leaf may also play a role in its ability to photoacclimate [116].

It is important when studying photoacclimation to consider the background development within which the photoacclimatory responses are being studied, as developmental conditions have a strong influence on the plant's ability to acclimate [117]. Plants grown under a particular light environment do not necessarily respond in the same fashion as would one shifted into that light regime [118,119], nor do they have the same acclimatory responses to a shift in light as one grown under a different light level [110]. Also, physiological and morphological constraints may be placed on the organism's ability to acclimate, such as leaf development restricting chloroplast response to light changes [119]. Many studies look at long-term acclimation under steady-state light environments rather than shifting a pre-exposed organism to a different regime. Although these are valid approaches, it should be noted that these results reflect the organism's ability to grow under specific environmental conditions and do not tease out the photoacclimation mechanisms induced during the stress of a change in acclimation states.

4.3.3 PSI/PSII Stoichiometry

Within each chloroplast, photoacclimatory mechanisms are initiated to augment the changes seen in metabolism under altering light intensity. Modulation of the PSI/PSII ratio is an additional approach to dealing with alterations in light intensity [1,5,120]. This response occurs as a means to offset imbalances in light absorption between the two PSs, thus maintaining optimal electron transport and minimizing photodamage [121,122].

In low light there are fewer PSII reaction centers in relation to PSI [122,123]. This increase in the PSI/PSII ratio may support an enhanced cyclic electron flow around PSI, increasing the ATP/NADPH ratio, and providing extra ATP for housekeeping functions when the organism is close to the light compensation point following a decline in light intensity [124]. In high light the relative number of PSII reaction centers increases compared to the number of PSIs [122,123]. The decline in the PSI/PSII ratio and the concomitant photoinhibition of PSII in high light likely protects the cell against photodamage [122].

PSI/PSII stoichiometry is also altered by light quality. In nature, shady environments are enriched in red light whereas blue-light levels are reduced. PSI preferentially absorbs red light over blue light, so exposure to a shade environment results in an imbalance favoring excitation of PSI. To compensate for this imbalance an increase in the number of PSIIs is observed, effectively restoring the balance in linear electron flow between the two systems. Conversely, light that favors PSII absorption triggers an increase in PSI centers [121,125,126]. The effect of light quality clearly shows that a disruption in the balance between the photosystems results in an adjustment of the PSI/PSII stoichiometry.

4.3.4 LIGHT-HARVESTING ANTENNAE

The light-harvesting complex family (LHC) is made up of pigment–protein complexes that bind both chlorophyll and carotenoids, which often have a dual role in light harvesting and photoprotection [127]. X-ray crystallography has elucidated the structure of the LHCII in spinach [51]. The interaction of the LHCs within the photosystems is also being examined, revealing a complex architecture [47,128,129]. The major, or peripheral, LHCs of PSII are each known to bind eight chlorophyll a (chl a), six chlorophyll b (chl b), two lutein, and one neoxanthin, and a fourth site exists for xanthophyll carotenoids [51]. Energy is captured in these peripheral antennae, travels through the minor antenna (CP29, CP26, CP24) or the core antenna (CP43 and CP47), finally reaching the reaction center (D1, D2) where charge separation occurs. The core antennae are tightly associated with the reaction center and bind only chl a, however, the peripheral antennae proteins bind both chl a and chl b and are often subjected to alterations in both number and association depending on the light environment. The entire structure dimerizes in a mirror-image arrangement to make a heterodimer supercomplex. Peripheral to this, more LHC trimers are positioned creating complex, yet flexible, harvesting antennae [47,51]. The PSI complex, on the other hand, is monomeric and contains four peripheral antenna (LHCIs) that form a crescent shape around one side of the core [128].

The LHC peripheral antennae are organized as stable trimers, however, they may exist as either functional trimers or monomers [130]. In high light, LHCII trimers are unstable and monomerization is significant in plants [131]. Monomers may facilitate NPQ of singlet chl a through exposure of the carotenoid sites [51], but they can also be targeted for proteolytic degradation [132]. Monomerization may provide a means for photoprotection in high light through both quenching and reorganization of the PSII complex [131]. Although this may be the case in plants, NPQ in *Chlamydomonas* appears to be dependent on the trimerization of LHCs [62]. An intriguing observation about the importance of the trimerization of the antennae comes from *Arabidopsis* lines lacking the LHCs (*Lhcb1, 2*) involved in trimer formation. Trimerization, however, was observed owing to the upregulation of CP26 (*Lhcb5*) [133]. This indicates that trimerization is not restricted to the main LHCII proteins and that it must have an important function in PSII [131,134].

Typically, there is an inverse relationship between light and chlorophyll levels. The LHC antennae bind most of the chlorophyll in the chloroplast, changes in chlorophyll levels are symptomatic of alterations in the composition and concentration of the LHC antennae system [7,121,134]. Because the antennae bind all of the chl b, changes in the chlorophyll a/b ratio indicate alterations in the antennae system. Much of the change in total chlorophyll is due to changes in the size of the antennae per reaction center, usually being larger in low light and smaller in high light [12,124,135,136]. However, reductions in total LHC levels, and hence chlorophyll per cell, can also occur by maintaining the same LHC/PSII ratio, but reducing the density of the entire complex within the thylakoid membrane [137].

The antenna system is assembled from several related LHC proteins (discussed above), the composition of which is dependent upon light intensity [113,137,138]. The significance of these changes in antennae composition is not clear. Although no specific LHC seems to be required for photosynthesis and the exact function for many is unknown, they are all likely are involved in the fine-tuning of light harvesting and are responsible for tolerance to the dynamic light environment [139].

4.3.4.1 LHC Regulation

Antennae size regulation is an integral component of acclimation and efficient photosynthesis, and examination of the molecular basis of these changes has been investigated in a number of studies. Plants and algae both experience downregulation of LHC transcript levels immediately upon a shift to high light and an increase when transferred to low light [90,96,101,134,135,140–148]. In *Chlamydomonas*, the LHC mRNA changes can be detected in as little as 15 minutes following a shift to high light and typically remain low for at least two to four hours [101,134].

Transcriptional control is responsible for much of this change in mRNA abundance as determined either directly using run-on assays [149,150] or inferred using a reporter gene [101]. With the exception of the CP29 transcript (*Lhcb4*), which undergoes a high-light-induced destabilization in *Chlamydomonas* [101], the stability of the primary LHCIIs is not significantly different under different light intensities [101,150,151]. After the initial change in gene expression following a shift in light intensity, transcript levels return to their pre-acclimated levels and the relationship to light intensity is lost [101,134]. Similar transient LHC expression was observed in *Arabidopsis* when shifted to low light where there was a burst of expression 24 hours after transfer that returned to pre-stress levels by day 2 [90].

Thus, under the acclimated state, there is a poor correlation with transcript abundance and protein levels, suggesting additional post-transcriptional mechanisms are major contributors to LHC regulation under steady-state conditions. Because of this, one must be careful to ensure the transcript changes are captured in the time course of the experiment.

Following a shift to low light, the increase in LHC transcription normally leads to greater LHC biosynthesis, but this needs to be coordinated with carotenoid and chlorophyll biosynthesis before a functional antennae protein can be assembled. The downregulation of antennae size in HL, however, not only requires that protein synthesis be reduced, but that pre-existing antennae proteins be removed. However, the strategies associated with this response are variable.

In plants, there is an active mechanism involving the proteolytic degradation of LHCs under excess light conditions [152]. The trimerized state of LHCII is unstable in high light and disassociates to form monomers, which can then be degraded [132]. Indeed the observation of smaller LHCII antennae and more PSII in high light can be explained by an increase in degradation of the LHCIIs with a concomitant synthesis of PSII [153]. The size of the LHCII component per PSII can be decreased without changing the number of LHCII complexes per chloroplast using this strategy. In microalgae, due to the faster generation time, they have a more passive strategy that depends on the dilution of LHC levels per cell through cell division, coupled with altered synthesis rates [101,135,141,154]. An increase in cell division rates is commonly seen following elevations of light intensity, serving to function as an electron sink and a means to dilute out photosynthetic components without proteolysis [99,101,135,143,154].

These observations suggest two levels of regulation: the initial light-density-dependent stressed condition that results in transcriptional regulation leading to less (or more) transcript levels, and an acclimated condition where post-transcriptional regulation predominates. For the latter case, it was proposed that LHC protein levels are responding to metabolic cues rather than light intensity per se [155,156]. This begs the question of what exactly senses light intensity and how does this information initiate the photoacclimatory changes in the expression of photosynthetic genes.

4.4 CHLOROPLAST TO NUCLEUS SIGNAL TRANSDUCTION

Changes in the concentration of photosynthetic proteins require the coordination of both nuclear and plastid gene expression because their genes are encoded by the two genomes. This necessitates some form of intracellular communication between the nucleus and the plastid. Clearly, the nucleus controls much of what occurs within the plastid through the regulated expression of structural and regulatory proteins. However, retrograde signals emanating from the plastid have been hypothesized to communicate its metabolic status to the nucleus. In plants, blocking carotenoid biosynthesis with norflurazon eliminated nuclear photosynthetic gene expression, indicating that a plastid specific event was responsible for the downregulation of specific nuclear genes (reviewed by Taylor [157]).

The hypothesized presence of such signals makes sense in terms of the apparent need to fine-tune the production of plastid proteins through feedback-type mechanisms that would allow the plant to respond to the ever-shifting environmental conditions. The identity of the signal, although more likely signals, remains elusive. There are several excellent reviews that cover the topic of plastid retrograde signaling that can be consulted for a more detailed discussion [4,158–163]. Recent work has begun to unravel the retrograde signal pathways in plants and algae, and should greatly assist our understanding of the molecular basis of environmental acclimation and photoacclimation in particular.

4.4.1 FEEDBACK FROM THE TETRAPYRROLE BIOSYNTHETIC PATHWAY

The concept that intermediates of tetrapyrrole biosynthesis have a role in retrograde signaling in plants has been around for two decades, although conclusive proof of a clear physiological role has not materialized. Some of the initial work involved feeding *Chlamydomonas* cells different intermediates of the chlorophyll biosynthetic pathway that led to the repression of LHC gene expression [164,165], suggesting that these intermediates have a role in the retrograde pathway. Certainly, this would seem a logical extension of mitochondrial to nucleus signaling in yeast where heme acts as a retrograde signal that can leave the organelle and interact with a transcription factor, HAP1 (heme-activating protein), to regulate the expression of mitochondrial genes [166,167].

A series of genomes uncoupled (GUN1-5) mutants that fail to downregulate LHC expression in the presence of norflurazon, and thus presumably are defective in chloroplast to nucleus signaling, have been isolated [168,169]. Four of these mutations have been identified and, interestingly, they all fall within the tetrapyrrole biosynthetic pathway. GUN2 and 3 are defective in heme oxygenase and phytochromobilin synthase, respectively, required for phytochromobilin synthesis [169]. GUN5 is defective in the H-subunit of Mg-chelatase [169], the first committed step of chlorophyll biosynthesis. GUN4, on the other hand, appears to be a plastid protein that can bind the chlorophyll intermediates protoporphyrin IX and Mg-protophyrin IX and likely has a role in regulating Mg-chelatase activity [170]. This clearly indicates that the perturbations in the tetrapyrrole biosynthetic pathway can influence the expression of nuclear photosynthetic genes under the conditions examined. Strand et al. [171] demonstrated that the active intermediate in the plastid to nucleus signaling pathway was Mg-protoporphyrin IX. Significantly, it was the same intermediate that was able to simulate the light induction of HSP70 genes when added to dark incubated *Chlamydomonas* cells [172,173]. Strand et al. [171] proposed that Mg-protoporphyrin IX accumulation leads to its movement out of the plastid where it is able to interact with regulatory proteins that lead to the transcriptional regulation of photosynthetic genes, although such regulatory components are unidentified.

The physiological basis of alterations in Mg-protoporphyrin IX concentration needs to be further examined (see discussion in [174]) to determine if this signaling pathway has a significant role in regulating such processes as photoacclimation. In the green alga *Chlorella*, Wilson et al. [175] measured the levels of Mg-protoporphyrin IX in response to acclimation to low temperature, and suggested that accumulation of this precursor may be a signal leading to the downregulation of photosynthetic genes.

4.4.2 PLASTID REDOX STATUS

Chloroplast-derived signals communicating the "redox status," or balance between the reduced and oxidized components of photosynthesis, is a particularly logical hypothesis as it offers a clear link between the physiology of the plastid and the expression of photosynthetic genes. This mechanism potentially represents a feedback loop that can optimize light harvesting and carbon fixation. Redox status can refer to many different components within the plastid including, but not limited to, the plastoquinone pool, ATP:ADP ratio, NADP:NADPH ratio, ferredoxin, thioredoxin, glutathione, ascorbate, and the *trans*-thylakoid pH.

In terms of photoacclimation, there is a clear relationship between shifts in light intensity and the redox state of the plastid: a shift to low light will cause the plastid to become more oxidized whereas a shift to high light will cause the components to become more reduced. Distinguishing between these different potential redox signals is not simple as these light changes will affect many different components concurrently.

4.4.2.1 Plastoquinone (PQ) Pool

Among the various potential signaling components, the redox state of the PQ pool has been viewed as an ideal sensor of light intensity. Given its intermediate position between photosystems II and I, it is potentially able to sense any imbalance in excitation energy between the two photosystems and the rates of carbon fixation [4]. The role of the PQ pool in regulating the synthesis of photosynthetic complexes has been proposed in cyanobacteria [176]. In eukaryotes, the role of the PQ pool in sensing light and regulating nuclear gene expression came through examining the expression of photosynthetic genes following the use of inhibitors that block the reduction [DCMU, 3-(3,4-dichlorophenyl)-1, 1-dimethylurea] or oxidation (DBMIB, 2,5-dibromo-3-methyl-6-isopropylbenzoquinone) of the PQ pool [150]. Additional work in the green alga, *Dunaliella*, indicates that in terms of nuclear photosynthetic gene regulation, the PQ redox state may be more significant in modulating long-term acclimation rather than the immediate responses to light intensity [151,177].

These studies provide the most direct evidence for the role of PQ in sensing light intensity, although one must interpret inhibitor studies with caution as they can have additional unknown consequences on cellular function, as is suspected for DBMIB [178]. DBMIB addition will also lead to increases in reactive oxygen species production [179], making the interpretation of data using DBMIB complex. However, noninhibitor studies have also suggested Q_a or the PQ pool as sensors in green algae [96], although the potential of other plastid redox signals, such as the *trans*-thylakoid pH, are recognized [83,177].

In plants, the role of the PQ pool in photoacclimation has been suggested for the expression of plastid-localized genes using a combination of light quality and photosynthetic inhibitors [161,180,181]. In *Arabidopsis*, the redox state of PQ also has a proposed role in the expression of a variety of nuclear photosynthetic genes, including the high-light-induction of ascorbate peroxidase [109] and PetF transcriptional regulation [158]. The suspected involvement of the PQ pool in sensing was again based on studies using inhibitors to oxidize (DCMU) and reduce (DBMIB) the PQ pool.

However, other plant studies suggest that the PQ pool alone is insufficient to initiate photoacclimation [24,161,182–185] and that it may be only one of the signals emanating from the plastid [186]. Thus, the role of the PQ pool redox state in light sensing remains poorly understood, and the significance of its role in higher plants in terms of retrograde signaling needs to be re-evaluated in the context of the involvement of other potential signals.

For the majority of studies on photoacclimation and retrograde signaling, only a few transcripts have been studied, the LHCs being the primary target. Fey et al. [163], however, used *Arabidopsis* microarrays to examine global patterns of gene expression in response to light shifts and a photosynthetic electron transport inhibitor (DCMU) to generate various plastid redox states. Of the 2000 light-regulated genes detected, only 15% showed redox regulation under the conditions tested. Only a small proportion of these were photosynthetic proteins, indicating that they are regulated at levels other than transcription or that there are additional plastid signals or, more likely, both.

Other studies have indicated that redox regulation is not limited to transcriptional control [187,188]. For instance, ferredoxin mRNA stability, and hence expression, is dependent upon photosynthesis [187] and the expression of the LHCII in *Lemna* is responsive to PQ redox state but post-translationally [189]. Transcript abundance-based studies such as microarrays would clearly underestimate the regulatory potential of plastid redox status.

Additional redox-active components have been implicated in different aspects of light acclimation. Of these, there is evidence that thioredoxin, which is downstream of PSI, is involved in the light regulation of D1 translation in the plastid [190] and LHCII kinase inactivation during HL exposure [75]. Glutathione has also been proposed as a mediator of light-induced changes in Rubisco expression [179]. With the number of sensor/signal candidates increasing, it is likely best to describe the plastid redox signal as simply arising from photosynthetic electron transport until components within the signaling pathway are identified with greater certainty.

4.4.2.2 Sugars

The alteration of sugar concentrations, the products of photosynthesis, in and outside the chloroplast during changes in light intensity could also provide the necessary metabolic cues to regulate nuclear, photosynthetic gene expression. Sugars are known to have a role in controlling a number of vital processes, including development, photosynthesis, germination, and growth in addition to an involvement in integrating hormone and light signals (see reviews [191] and [192]). Sugars, particularly glucose, are able to repress the expression of nuclear photosynthetic genes. Glucose levels are sensed via hexokinase, the first enzyme in glycolysis, through which sugar repression of nuclear photosynthetic genes is mediated [193]. As glycolysis is a metabolic link between plastids and mitochondria, such a sensor is likely to have a significant role in environmental acclimation and be in a position to integrate multiple metabolic cues.

With respect to photoacclimation, the phenotypes of hexokinase mutations in *Arabidopsis* were enhanced by increases in light intensity that included reduced growth, delayed senescence, and reduced cell expansion [194], indicating that hexokinase may have a role in the physiological acclimation to light intensity. Examining these mutants for photoacclimation responses in mature, light-shifted plants would help to characterize the role of hexokinase in this process. This same group also showed that hexokinase integrates with other signaling pathways to respond appropriately to light, nutrient status, and hormones, similar to the finding by Oswald et al. [183] who noted an interaction between plastid redox state and sugar signals.

4.4.2.3 Reactive Oxygen Species (ROS)

Reactive forms of oxygen are receiving considerable attention for their involvement in the initiation and transduction of signals from a variety of pathways, including pathogen defense, stress responses, programmed cell death, and response to high light (see reviews [195] and [196]). Cells can produce a host of reactive oxygen species (ROS) such as singlet oxygen ($^1O_2$), superoxide (O_2^-), hydrogen peroxide (H_2O_2), and hydroxyl radicals ($\cdot OH$), all of which are toxic to cells due to the oxidation of lipids and proteins. If detoxification/scavenging pathways are unable to cope with ROS production, the cell may die. There are multiple sites where significant ROS are generated that include plastids, mitochondria, and peroxisomes.

H_2O_2 production in the chloroplast as a result of exposure to high light (or any number of other stresses) is not only involved in the induction of detoxification genes (such as ascorbate peroxidase) in stressed tissues [197], but is also able to induce secondary responses in nonstressed tissues, as seen in "systemic acquired acclimation" [109]. It seems plausible at least that ROS, such as H_2O_2, are acting as plastid retrograde signals that can modulate or initiate photoacclimation and work in combination with other signals. The rate of H_2O_2 production is certainly fast, with H_2O_2 levels increasing within at least 30 minutes of an increase in light intensity [179], and it is clearly diffusible across the chloroplast envelope, meeting another requirement of communication [198]. In green algae, the control of LHC gene expression in response to light intensity is also predicted to be regulated by ROS [299]. How ROS such as interact with cellular components to affect the regulatory control is unknown and likely to be a difficult process to dissect.

4.4.3 Photoreceptors

Photoreceptors appear well suited to sensing the light environment, particularly light quality. In plants, phytochrome is involved with red light sensing, and cryptochrome and phototropins are able to sense blue light. It is known that these receptors have important roles in processes such as germination, development, growth, flowering, and senescence [200]. The role of phytochrome and cryptochrome in photoacclimation appears limited as judged by light responses in photoreceptor mutants [163,201]. However, because the role of these receptors is connected to developmental/photomorphogenic responses, they likely have some effects on long-term, whole-plant acclimation. Blue light is known to stimulate some avoidance mechanisms such as chloroplast movement through the action of phototropins [202], but the role of this novel class of photoreceptors in photoacclimation has not been thoroughly examined to our knowledge.

In microalgae, such as *Chlamydomonas*, avoidance and phototaxis responses are sensed through rhodopsins [203]. Other than persistent accounts of blue-light mediated alterations in transcription of LHC genes [164,204], a role of rhodopsin or other photoreceptors in long-term photoacclimation in algae has also not been presented. As with plant studies, examining the role of specific photoreceptors is difficult as exposing cells to different light qualities will surely alter the plastid redox state and potentially confuse interpretation. The use of photoreceptor mutants is an excellent approach to dissecting the involvement of the receptors in light acclimation.

4.5 FINAL THOUGHTS

The most exciting developments in the field of photoacclimation are in the area of regulation and the identity of the signals that initiate and communicate chloroplast metabolic status. Identifying the molecular components of this signal transduction pathway(s) is vital to understanding the dynamic equilibrium that plants maintain in responding to fluctuating light intensity and many other types of environmental stresses. Deciphering these signals and differentiating between developmental and acclimation-specific events are future challenges.

Further challenges also include integrating whole-plant metabolism for a more holistic view of acclimation in the context of nonphotosynthetic pathways. As the products of photosynthesis are metabolized to support growth and development, changes in the light environment will have many effects on the downstream side of immediate chloroplast functions. The importance of mitochondria and chloroplast interactions, for instance, have been recently illustrated by the identification of a nuclear, mitochondrial targeted protein that is required for the proper control of state transitions [205]. Progress will also come in understanding the physiological role context of alternate electron sinks, such as chlororespiration and plastid alternative oxidases [206,207], and their role in development and acclimation. It is in the context of all the energy generating and utilizing pathways that photoacclimation must eventually be viewed to understand what triggers, regulates, and modulates plant acclimation to changes in the light environment.

REFERENCES

1. Falkowski, P.G. and Laroche, J., Acclimation to spectral irradiance in algae. *J. Phycol.*, 27:8, 1991.
2. Horton, P., Ruban, A., and Walters R., Inhibition of delta-Ph-dependent control of chloroplast light harvesting by binding of DCCD to LHCII. *Photosyn. Res.*, 34:121, 1992.
3. Anderson, J.M., Chow, W.S., and Park, Y.I., The grand design of photosynthesis: acclimation of the photosynthetic apparatus to environmental cues. *Photosyn. Res.*, 46:129, 1995.
4. Durnford, D. and Falkowski, P., Chloroplast redox regulation of nuclear gene transcription during photoacclimation. *Photosyn. Res.*, 53:229, 1997.

5. Huner, N.P.A., Oquist, G., and Sarhan, F., Energy balance and acclimation to light and cold. *Trends Plant Sci.*, 3:224, 1998.
6. Niyogi, K.K., Safety valves for photosynthesis. *Curr. Opin. Plant Biol.*, 3:455, 2000.
7. Murchie, E.H. and Horton, P., Acclimation of photosynthesis to irradiance and spectral quality in British plant species: Chlorophyll content, photosynthetic capacity and habitat preference. *Plant Cell Environ.*, 20:438, 1997.
8. Kasahara, M. et al., Chloroplast avoidance movement reduces photodamage in plants. *Nature*, 420:829, 2002.
9. Foster, K.W. et al., A rhodopsin is the functional photoreceptor for phototaxis in the unicellular eukaryote *Chlamydomonas*. *Nature*, 311:756, 1984.
10. Iseki, M. et al., A blue-light-activated adenylyl cyclase mediates photoavoidance in *Euglena gracilis*. *Nature*, 415:1047, 2002.
11. Wada, M., Kagawa, T., and Sato, Y., Chloroplast movement. *Ann. Rev. Plant Biol.*, 54:455, 2003.
12. Valladares, F. and Pugnaire, F.I., Tradeoffs between irradiance capture and avoidance in semi-arid environments assessed with a crown architecture model. *Annals of Botany*, 83:459, 1999.
13. Kagawa, T. and Wada, M., Velocity of chloroplast avoidance movement is fluence rate dependent. *Photoch Photobio. Sci.*, 3:592, 2004.
14. Park, Y.I., Chow, W.S., and Anderson, J.M., Chloroplast movement in the shade plant *Tradescantia albiflora* helps protect photosystem II against light stress. *Plant Physiol.*, 111:867, 1996.
15. Williams, W.E., Gorton, H.L., and Witiak, S.M., Chloroplast movements in the field. *Plant Cell Environ.*, 26:2005, 2003.
16. Pilon, J.J. et al., Leaf waxes of slow-growing alpine and fast-growing lowland *Poa* species: Inherent differences and responses to UV-B radiation. *Phytochemistry*, 50:571, 1999.
17. Pietrini, F., Iannelli, M., and Massacci, A., Anthocyanin accumulation in the illuminated surface of maize leaves enhances protection from photo-inhibitory risks at low temperature, without further limitation to photosynthesis. *Plant Cell Environ.*, 25:1251, 2002.
18. Neill, S.O. and Gould, K.S., Anthocyanins in leaves. Light attenuators or antioxidants? *Funct. Plant Biol.*, 30:865, 2003.
19. Yong, Y.Y.R. and Lee, Y.K., Do carotenoids play a photoprotective role in the cytoplasm of *Haematococcus-Lacustris* (*Chlorophyta*). *Phycologia*, 30:257, 1991.
20. Hagen, C., Braune, W., and Greulich, F., Functional-aspects of secondary carotenoids in *Haematococcus-Lacustris* [Girod] Rostafinski (Volvocales). 4. Protection from photodynamic damage. *J. Photochem. Photobiol. B.*, 20:153, 1993.
21. Demmig-Adams, B. and Adams, W.W., The role of xanthophyll cycle carotenoids in the protection of photosynthesis. *Trends Plant Sci.*, 1:21, 1996.
22. Bonaventura, C. and Myers, J., Fluorescence and oxygen evolution from *Chlorella pyrenoidosa*. *Biochim. Biophys. Acta*, 189:366, 1969.
23. Murata, N., Control of excitation transfer in photosynthesis. 1. Light-Induced change of chlorophyll fluorescence in *Porphyridium cruentum*. *Biochim. Biophys. Acta*, 172;242, 1969.
24. Anderson, J.M., Park, Y.I., and Chow, W.S., Photoinactivation and photoprotection of photosystem II in nature. *Physiol. Plant.*, 100:214, 1997.
25. Butler, W.L., Energy-distribution in photo-chemical apparatus of photosynthesis. *Ann. Rev. Plant Physiol.*, 29:345, 1978.

26. Lazar, D., Chlorophyll a fluorescence induction. *Biochim. Biophys. Acta - Bioenergetics*, 1412;1, 1999.
27. Lazar, D., Chlorophyll a fluorescence rise induced by high light illumination of dark-adapted plant tissue studied by means of a model of photosystem II and considering photosystem II heterogeneity. *J. Theor. Biol.*, 220:469, 2003.
28. Maxwell, K. and Johnson, GN., Chlorophyll fluorescence—a practical guide. *J. Exp. Bot.*, 51:659, 2000.
29. Schreiber, U., Schliwa, U., and Bilger, W., Continuous recording of photochemical and nonphotochemical chlorophyll fluorescence quenching with a new type of modulation fluorometer. *Photosynth Res.*, 10:51, 1986.
30. Genty, B., Briantais, J.M., and Baker, N.R., The relationship between the quantum yield of photosynthetic electron-transport and quenching of chlorophyll fluorescence. *Biochim Biophys Acta.*, 990:87, 1989.
31. Huner, N.P.A. et al., Sensing environmental temperature change through imbalances between energy supply and energy consumption: Redox state of photosystem 11. *Physiol. Plantarum*, 98:358, 1996.
32. Schreiber, U. et al., Assessment of photosystem-II photochemical quantum yield by chlorophyll fluorescence quenching analysis. *Aust. J. Plant Physiol.*, 22:209, 1995.
33. Aro, E.M., McCaffery, S., and Anderson, J.M., Photoinhibition and D1 protein-degradation in peas acclimated to different growth irradiances. *Plant Physiol.*, 103:835, 1993.
34. Bruce, D., Samson, G., and Carpenter, C., The origins of nonphotochemical quenching of chlorophyll fluorescence in photosynthesis. Direct quenching by P680(+) in photosystem II enriched membranes at low pH. *Biochemistry-US.*, 36:749, 1997.
35. Rohacek, K., Chlorophyll fluorescence parameters: The definitions, photosynthetic meaning, and mutual relationships. *Photosynthetica*, 40:13, 2002.
36. Müller, P., Li, X.P., and Niyogi, K.K., Non-photochemical quenching. A response to excess light energy. *Plant Physiol.*, 125:1558,2001.
37. Beddard, G.S., Davidson, R.S., and Trethewey, K.R., Quenching of chlorophyll fluorescence by beta-carotene. *Nature*, 267:373, 1977.
38. Young, A.J., The photoprotective role of carotenoids in higher-plants. *Physiol. Plantarum*, 83:702, 1991.
39. Gilmore, A.M., Mechanistic aspects of xanthophyll cycle-dependent photoprotection in higher plant chloroplasts and leaves. *Physiol. Plant.*, 99:197, 1997.
40. Crofts, A.R. and Yerkes, C.T., A molecular mechanism for Q(E)-quenching. *FEBS Lett.*, 352:265, 1994.
41. Buchel, C., Naqvi, K.R., and Melo, T.B., Pigment-pigment interactions in thylakoids and LHCII of chlorophyll a/c containing alga *Pleurochloris Meiringensis*: Analysis of fluorescence-excitation and triplet-minus-singlet spectra. *Spectrochimica Acta: Part A-Molecular Biomolecular Spectroscopy*, 54:719, 1998.
42. Telfer, A., What is beta-carotene doing in the photosystem II reaction centre? *Philos T. Roy. Soc. B.*, 357:1431, 2002.
43. Gruszecki, W. et al., Increased heat emission in photosynthetic apparatus of rye subjected to light stress. *J. Photochem. Photobiol. B,* 32:67, 1996.
44. Bilger, W. and Bjorkman, O., Role of the xanthophyll cycle in photoprotection elucidated by measurements of light-induced absorbency changes, fluorescence and photosynthesis in leaves of *Hedera canariensis*. *Photosynth. Res.*, 25:173, 1990.

45. Demmig-Adams, B. and Adams, W.W., Photoprotection and other responses of plants to high light stress. *Ann. Rev. Plant Physiol.*, 43:599, 1992.
46. Kramer, D.M., Sacksteder, C.A., and Cruz, J.A., How acidic is the lumen? *Photosyn. Res.*, 60:151, 1999.
47. Nield, J., Redding, K., and Hippler, M., Remodeling of light-harvesting protein complexes in *Chlamydomonas* in response to environmental changes. *Eukaryotic Cell*, 3:1370, 2004.
48. Avenson, T.J., Cruz, J.A., and Kramer, D.M., Modulation of energy-dependent quenching of excitons in antennae of higher plants. *Proc. Natl. Acad. Sci. USA*, 101:5530, 2004.
49. Aspinall-O'Dea, M. et al., *In vitro* reconstitution of the activated zeaxanthin state associated with energy dissipation in plants. *Proc. Natl. Acad. Sci. USA*, 99:16331, 2002.
50. Li X. et al., Regulation of photosynthetic light harvesting involves intrathylakoid lumen pH sensing by the PsbS protein. *J. Biol. Chem.*, 279:22866,2004.
51. Liu, Z.F. et al., Crystal structure of spinach major light-harvesting complex at 2.72 Angstrom resolution. *Nature*, 428:287, 2004.
52. Ivanov, B., Influence of ascorbate and the mehler peroxidase reaction on non-photochemical quenching of chlorophyll fluorescence in maize mesophyll chloroplasts. *Planta*, 210:765, 2000.
53. Niyogi, K.K. et al., Is PsbS the site of non-photochemical quenching in photosynthesis? *J. Exp. Bot.*, 56:375, 2005.
54. Bassi, R., Sandona, D., and Croce, R., Novel aspects of chlorophyll a/b-binding proteins. *Physiol. Plant.*, 100:769, 1997.
55. Dall'Osto, L., Caffarri, S., and Bassi, R., A mechanism of nonphotochemical energy dissipation, independent from PsbS, revealed by a conformational change in the antenna protein CP26. *Plant Cell*, 17:1217, 2005.
56. Andersson, J. et al., Antisense inhibition of the photosynthetic antenna proteins CP29 and CP26: Implications for the mechanism of protective energy dissipation. *Plant Cell*, 13:1193, 2001.
57. Andersson, J. et al., Absence of the Lhcb1 and Lhcb2 proteins of the light-harvesting complex of photosystem II– Effects on photosynthesis, grana stacking and fitness. *Plant J*, 35:350, 2003.
58. Li, X.P. et al., A pigment-binding protein essential for regulation of photosynthetic light harvesting. *Nature*, 403:391, 2000.
59. Li, X.P., Gilmore, A.M., and Niyogi, K.K., Molecular and global time-resolved analysis of a PsbS gene dosage effect on pH- and xanthophyll cycle-dependent nonphotochemical quenching in photosystem II. *J Biol Chem.*, 277:33590, 2002.
60. Peterson, R. and Havir, E., Contrasting modes of regulation of PSII light utilization with changing irradiance in normal and PsbS mutant leaves of *Arabidopsis thaliana*. *Photosyn. Res.*, 75:57, 2003.
61. Anwaruzzaman, M. et al., Genomic analysis of mutants affecting xanthophyll biosynthesis and regulation of photosynthetic light harvesting in *Chlamydomonas reinhardtii*. *Photosyn. Res.*, 82:265, 2004.
62. Elrad, D., Niyogi, K.K., and Grossman, A.R., A major light-harvesting polypeptide of photosystem II functions in thermal dissipation. *Plant Cell*, 14, 1801, 2002.
63. Polle, J.E.W. et al., Photosynthetic apparatus organization and function in the wild type and a chlorophyll b-less mutant of *Chlamydomonas reinhardtii*. Dependence on carbon source. *Planta*, 211:335, 2000.

64. Finazzi, G., The central role of the green alga *Chlamydomonas reinhardtii* in revealing the mechanism of state transitions. *J. Exp. Bot.*, 56:383, 2005.
65. Finazzi, G. et al., State transitions, cyclic and linear electron transport and photophosphorylation in *Chlamydomonas reinhardtii*. *Biochem. Biophys. Acta*, 1413:117, 1999.
66. Allen, J.F., Protein-phosphorylation in regulation of photosynthesis, *Biochim. Biophys. Acta*, 1098:275, 1992.
67. Zer, H. et al., Regulation of thylakoid protein phosphorylation at the substrate level: Reversible light-induced conformational changes expose the phosphorylation site of the light-harvesting complex II. *Proc. Natl. Acad. Sci. USA*, 96:8277, 1999.
68. Haldrup, A. et al., Balance of power: a view of the mechanism of photosynthetic state transitions. *Trends Plant Sci.*, 6:301, 2001.
69. Finazzi, G. et al. Photoinhibition of *Chlamydomonas reinhardtii* in State 1 and State 2—Damages to the photosynthetic apparatus under linear and cyclic electron flow. *J Biol Chem.*, 276:22251, 2001.
70. Zito, F., The Qo site of cytochrome B(6)F complexes controls the activation of the LHCII kinase. *EMBO J.*, 18:2961, 1999.
71. Rintamaki, E. et al., Phosphorylation of light-harvesting complex II and photosystem II core proteins shows different irradiance-dependent regulation in vivo—Application of phosphothreonine antibodies to analysis of thylakoid phosphoproteins. *J. Biol. Chem.*, 272:30476, 1997.
72. Snyders, S. and Kohorn, B.D., TAKs, thylakoid membrane protein kinases associated with energy transduction. *J. Biol. Chem.*, 274:9137, 1999.
73. Snyders, S. and Kohorn, B.D., Disruption of thylakoid-associated kinase 1 leads to alteration of light harvesting in *Arabidopsis*. *J. Biol. Chem.*, 276:32169, 2001.
74. Depege, N., Bellafiore, S., and Rochaix, J.D., Role of chloroplast protein kinase Stt7 in LHCII phosphorylation and state transition in *Chlamydomonas*. *Science*, 299:1572, 2003.
75. Rintamaki, E., Martinsuo, P., Pursiheimo, S., and Aro, E.M., Cooperative regulation of light-harvesting complex II phosphorylation via the plastoquinol and ferredoxin-thioredoxin system in chloroplasts. *Proc. Natl. Acad. Sci. USA.*, 97:11644, 2000.
76. Zer, H. et al., Light affects the accessibility of the thylakoid light harvesting complex II (LHCII) phosphorylation site to the membrane protein kinase(S). *Biochemistry-US*, 42:728, 2003.
77. Mullineaux, C.W. and Emlyn-Jones, D., State transitions: An example of acclimation to low-light stress. *J. Exp Bot.*, 56:389, 2005.
78. Lunde, C. et al., Plants impaired in state transitions can to a large degree compensate for their defect. *Plant Cell Physiol.*, 44:44, 2003.
79. Osmond, B., What is photoinhibition? "Some insights from comparisons of shade and sun plants." In *Photoinhibition of Photosynthesis: From Molecular Mechanisms to the Field*, Baker, N., Bowyer, J., Eds., Bios Science, Herdon, VA, 1994, p. 24.
80. Kudoh, H., and Sonoike, K., Irreversible damage to photosystem I by chilling in the light: cause of the degradation of chlorophyll after returning to normal growth temperature. *Planta*, 215:541, 2002.
81. Melis, A., Photosystem-II damage and repair cycle in chloroplasts: What modulates the rate of photodamage *in vivo*? *Trends Plant Sci.*, 4:130, 1999.
82. Oquist, G., Chow, W.S., and Anderson, J.M., Photoinhibition of photosynthesis represents a mechanism for the long-term regulation of photosystem-II. *Planta*, 1186:450, 1992.

83. Wilson, K. and Huner, N., The role of growth rate, redox-state of the plastoquinone pool and the *trans*-thylakoid delta Ph in photoacclimation of *Chlorella vulgaris* to growth irradiance and temperature. *Planta*, 212:93, 2000.
84. Gray, G.R. et al., photosystem II excitation pressure and development of resistance to photoinhibition. *Plant Physiol.*, 110:61, 1996.
85. Demeter, S. et al., Effects of *in-vivo* CO2-depletion on electron-transport and photoinhibition in the green-algae; *Chlamydobotrys stellata* and *Chlamydomonas reinhardtii*. *Biochem. Biophys. Acta*, 1229:166, 1995.
86. Chow, W.S., Adamson, H.Y., and Anderson, J.M., Photosynthetic acclimation of *Tradescantia-Albiflora* to growth irradiance — Lack of adjustment of light-harvesting components and its consequences. *Physiol. Plant*, 81:175, 1991.
87. Morin, F., Andre, M., and Betsche, T., Growth-kinetics, carbohydrate, and leaf phosphate content of clover (*Trifolium-Subterraneum* L) after transfer to a high CO2 atmosphere or to high light and ambient air. *Plant Physiol.*, 99:89, 1992.
88. Amancio, S., Rebordao, J.P., and Chaves, M.M., Improvement of acclimatization of micropropagated grapevine: Photosynthetic competence and carbon allocation. *Plant Cell Tiss. Org.*, 58:31, 1999.
89. Evans, J.R., The relationship between electron transport components and photosynthetic capacity in pea leaves grown at different irradiances. *Aust. J. Plant Physiol.*, 14:157, 1987.
90. Walters, R. and Horton, P., Acclimation of *Arabidopsis thaliana* to the light environment — Changes in composition of the photosynthetic apparatus. *Planta*, 195:248, 1994.
91. Walters, R.G. et al., Identification of mutants of *Arabidopsis* defective in acclimation of photosynthesis to the light environment. *Plant Physiol.*, 131:472, 2003.
92. Walters, R.G., A mutant of *Arabidopsis* lacking the triose-phosphate/phosphate translocator reveals metabolic regulation of starch breakdown in the light. *Plant Physiol.*, 135:891, 2004.
93. Krapp, A., Quick, W.P., and Stitt, M., Ribulose-1,5-Bisphosphate carboxylase-oxygenase, other calvin-cycle enzymes, and chlorophyll decrease when glucose is supplied to mature spinach leaves via the transpiration stream. *Planta*, 186:58, 1991.
94. Henley, W. et al., Photoacclimation and photoinhibition in Ulva-Rotundata as influenced by nitrogen availability. *Planta*, 184:235, 1991.
95. Ramalho, J. et al., Photosynthetic acclimation to high light conditions in mature leaves of *Coffea arabica* L.: Role of xanthophylls, quenching mechanisms and nitrogen nutrition. *Aust. J. Plant Physiol.*, 27:43, 2000.
96. Maxwell, D.P., Laudenbach, D.E., and Huner, N.P.A., Redox regulation of light-harvesting complex 11 and *cab* mKNA abundance in *Dunaliella salina*. *Plant Physiol.*, 109:787, 1995.
97. Wykoff, D.D. et al., The regulation of photosynthetic electron transport during nutrient deprivation in *Chlamydomonas reinhardtii*. *Plant Physiol.*, 117:129, 1998.
98. Lu, C.M., PSII photochemistry, thermal energy dissipation, and the xanthophyll cycle in *Kalanchoe daigremontiana* exposed to a combination of water stress and high light. *Physiol. Plant*, 118:173, 2003.
99. Shapira, M. et al., Differential regulation of chloroplast gene expression in *Chlamydomonas reinhardtii* during photoacclimation: Light stress transiently suppresses synthesis of the Rubisco Lsu protein while enhancing synthesis of the Ps II D1 protein. *Plant Mol. Biol.*, 33:1001, 1997.

100. Shapira, M. et al., Differential regulation of chloroplast gene expression in *Chlamydomonas reinhardtii* during photoacclimation: Light stress transiently suppresses synthesis of the Rubisco Lsu protein while enhancing synthesis of the Ps II D1 protein. *Plant Mol. Biol.*, 33:1001, 1997.
101. Durnford, D. et al., Light-harvesting complex gene expression is controlled by both transcriptional and post-transcriptional mechanisms during photoacclimation in *Chlamydomonas reinhardtii*. *Physiol. Plant*, 118:193, 2003.
102. Sukenik, A., Bennett, J., and Falkowski, P., Light-saturated photosynthesis — Limitation by electron-transport or carbon fixation. *Biochim. Biophys. Acta*, 891:205, 1987.
103. Anderson, J.M., Chow, W.S., and Goodchild, D.J., Thylakoid membrane organization in sun shade acclimation. *Aust. J. Plant Physiol.* 15:11, 1988.
104. Murchie, E.H. and Horton, P., Contrasting patterns of photosynthetic acclimation to the light environment are dependent on the differential expression of the responses to altered irradiance and spectral quality. *Plant Cell Environ.*, 21:139, 1998.
105. Kursar, T.A. and Coley, P.D., Contrasting modes of light acclimation in two species of the rainforest understory. *Oecologia*, 121:489, 1999.
106. Awada, T. et al., Ecophysiology of seedlings of three Mediterranean pine species in contrasting light regimes. *Tree Physiol.*, 23:33, 2003.
107. Feng, Y.L., Cao, K.F., and Zhang, J.L., Photosynthetic characteristics, dark respiration, and leaf mass per unit area in seedlings of four tropical tree species grown under three irradiances. *Photosynthetica*, 42:431, 2004.
108. Hjelm, U. and Ogren, E., Photosynthetic responses to short-term and long-term light variation in *Pinus sylvestris* and *Salix dasyclados*. *Trees-Struct. Funct.*, 18:622, 2004.
109. Karpinski, S. et al., Systemic signaling and acclimation in response to excess excitation Energy in *Arabidopsis*. *Science*, 284:654, 1999.
110. Naramoto, M., Han, Q., and Kakubari, Y., The influence of previous irradiance on photosynthetic induction in three species grown in the gap and understory of a *Fagus crenata* forest. *Photosynthetica*, 39:545, 2001.
111. Yano, S. and Terashima, I., Separate localization of light signal perception for sun or shade type chloroplast and palisade tissue differentiation in *Chenopodium album*. *Plant Cell Phys.*, 42:1303, 2001.
112. Terashima, I. and Saeki, T., A new model for leaf photosynthesis incorporating the gradients of light environment and of photosynthetic properties of chloroplasts within a leaf. *Ann. Bot.-(London)*, 56:489, 1985.
113. Bailey, S., Horton, P., and Walters, R., Acclimation of *Arabidopsis thaliana* to the light environment: The relationship between photosynthetic function and chloroplast composition. *Planta*, 218:793, 2004.
114. Vogelmann, T.C. and Evans, J.R., Profiles of light absorption and chlorophyll within spinach leaves from chlorophyll fluorescence. *Plant Cell Environ.*, 25:1313, 2002.
115. Evans, J.R. and Vogelmann, T.C., Profiles of C-14 fixation through spinach leaves in relation to light absorption and photosynthetic capacity. *Plant Cell Environ.*, 26:547, 2003.
116. Murchie, E.H. et al., Acclimation of rice photosynthesis to irradiance under field conditions. *Plant Physiol.*, 120:1999, 2002.
117. Walters, R.G., Towards an understanding of photosynthetic acclimation. *J. Exp. Bot.*, 56:435, 2005.

118. Yin, Z.H. and Johnson, G.N., Photosynthetic acclimation of higher plants to growth in fluctuating light environments. *Photosyn. Res.*, 63:97, 2000.
119. Oguchi, R., Hikosaka, K., and Hirose, T., Does the photosynthetic light-acclimation need change in leaf anatomy? *Plant Cell Environ.*, 26:505, 2003.
120. Melis, A., Dynamics of photosynthetic membrane-composition and function. *Biochim. Biophys. Acta*, 1058:87, 1991.
121. Chow, W.S., Melis, A., and Anderson, J.M., Adjustments of photosystem stoichiometry in chloroplasts improve the quantum efficiency of photosynthesis. *Proc. Natl. Acad. Sci. USA*, 87:7502, 1990.
122. Sonoike, K., Hihara, Y., and Ikeuchi, M., Physiological significance of the regulation of photosystem stoichiometry upon high light acclimation of *Synechocystis* Sp Pcc 6803. *Plant Cell Physiol.*, 42:379, 2001.
123. Murakami, A. and Fujita, Y., Regulation of photosystem stoichiometry in the photosynthetic system of the cyanophyte *Synechocystis* Pcc 6714 in response to light-intensity. *Plant Cell Physiol.*, 32:223, 1991.
124. Bailey, S. et al., Acclimation of *Arabidopsis thaliana* to the light environment: the existence of separate low light and high light responses. *Planta*, 213:794, 2001.
125. Fujita, Y., Ohki, K., and Murakami, A., Chromatic regulation of photosystem composition in the photosynthetic system of red and blue-green-algae. *Plant Cell Physiol.*, 26:1541, 1985.
126. Melis, A., Chromatic regulation in *Chlamydomonas reinhardtii* alters photosystem stoichiometry and improves the quantum efficiency of photosynthesis. *Photosyn. Res.*, 47:253, 1996.
127. Niyogi, K.K., Bjorkman, O., and Grossman, A.R., The roles of specific xanthophylls in photoprotection. *Proc. Natl. Acad. Sci. USA*, 94:14162, 1997.
128. Ben-Shem, A., Frolow, F., and Nelson, N., Crystal structure of plant photosystem I. *Nature*, 426:630, 2003.
129. Merchant, S. and Sawaya, M., The light reactions: A guide to recent aquitions for the picture gallery. *Plant Cell*, 17:648, 2005.
130. Wentworth, M., Ruban, A.V., and Horton, P., The functional significance of the monomeric and trimeric states of the photosystem II light harvesting complexes, *Biochemistry- (US)*, 43:501, 2004.
131. Garab, G. et al., Light-induced trimer to monomer transition in the main light-harvesting antenna complex of plants: Thermo-optic mechanism. *Biochemistry-(US)*, 41:15121, 2002.
132. Yang, D.H., Paulsen, H., and Andersson, B., The N-terminal domain of the light-harvesting chlorophyll a/b-binding protein complex (LHCII) is essential for its acclimative proteolysis. *FEBS Lett.*, 466:385, 2000.
133. Ruban, A.V. et al., Plants lacking the main light-harvesting complex retain photosystem II macro-organization. *Nature*, 421:648, 2003.
134. Elrad, D. and Grossman, AR., A genome's-eye view of the light-harvesting polypeptides of *Chlamydomonas reinhardtii*. *Curr. Genet.*, 45:61, 2004.
135. Fujita, Y. et al., Regulation of the size of light-harvesting antennae in response to light-intensity in the green-Alga *Chlorella pyrenoidosa*. *Plant Cell Physiol.*, 30:1029, 1989.
136. Smith, B., Response of the photosynthetic apparatus in *Dunaliella salina* (green algae) to irradiance stress. *Plant Physiol.*, 93:1433, 1990.
137. Sukenik, A., Bennett, J., and Falkowski, P.G., The molecular-basis of photoadaptation in the marine chlorophyte *Dunaliella tertiolecta*. *Isr. J. Bot.*, 36:50, 1987.

138. Sukenik, A., Bennett, J., and Falkowski, P., Changes in the abundance of individual apoproteins of light-harvesting chlorophyll-a/b protein complexes of photosystem-I and photosystem-II with growth irradiance in the marine chlorophyte *Dunaliella tertiolecta*. *Biochim. Biophys. Acta*, 932:206, 1988.
139. Ganeteg, U. et al., Is each light-harvesting complex protein important for plant fitness? *Plant Physiol.*, 134:502, 2004.
140. Sukenik, A. et al., Adaptation of the photosynthetic apparatus to irradiance in *Dunaliella tertiolecta* — A kinetic-study. *Plant Physiol.*, 92:891, 1990.
141. Laroche, J., Mortainbertrand, A., and Falkowski, PG., Light intensity-induced changes in cab messenger-RNA and light harvesting complex-II apoprotein levels in the unicellular chlorophyte *Dunaliella tertiolecta*. *Plant Physiol.*, 97:147, 1991.
142. Maxwell, D.P. et al., Growth at low-temperature mimics high-light acclimation in *Chlorella vulgaris*. *Plant Physiol.*, 105:535, 1994.
143. Webb, M.R. and Melis, A., Chloroplast response in *Dunaliella salina* to irradiance stress — Effect on thylakoid membrane-protein assembly and function. *Plant Physiol.*, 107:885, 1995.
144. Hahn, D. and Kuck, U., Identification of DNA sequences controlling light- and chloroplast-dependent expression of the *Lhcb1* gene from *Chlamydomonas reinhardtii*. *Curr. Genet.*, 34:459, 1999.
145. Masuda, T., Polle, J.E.W., and Melis, A., Biosynthesis and distribution of chlorophyll among the photosystems during recovery of the green alga *Dunaliella salina* from irradiance stress. *Plant Physiol.*, 128:603, 2002.
146. Rossel, J.B., Wilson, I.W., and Pogson, B.J., Global changes in gene expression in response to high light in *Arabidopsis*. *Plant Physiol.*, 130:1109, 2002.
147. Teramoto, H. et al., Light-intensity-dependent expression of LHC gene family encoding light-harvesting chlorophyll-a/b proteins of photosystem II in *Chlamydomonas reinhardtii*. *Plant Physiol.*, 130:325, 2002.
148. Murchie, E.H. et al., Acclimation of photosynthesis to high irradiance in rice: gene expression and interactions with leaf development. *J. Exp. Bot.*, 56:449, 2005.
149. Jasper, F. et al., Control of cab gene-expression in synchronized *Chlamydomonas reinhardtii* cells. *J. Photochem. Photobiol.*, 11:139, 1991.
150. Escoubas, J. et al., Light-intensity regulation of cab gene-transcription is signaled by the redox state of the plastoquinone pool. *Proc. Natl. Acad. Sci. USA*, 92:10237, 1995.
151. Masuda, T., Tanaka, A., and Melis, A., Chlorophyll antenna size adjustments by irradiance in *Dunaliella salina* involve coordinate regulation of chlorophyll a oxygenase (CAO) and *Lhcb* gene expression. *Plant Mol. Biol.*, 51:757, 2003.
152. Yang, D.H. et al., Induction of acclimative proteolysis of the light-harvesting chlorophyll a/b protein of photosystem II in response to elevated light intensities. *Plant Physiol.*, 118:827, 1998.
153. Walters, R. et al., Identification of mutants of *Arabidopsis* defective in acclimation of photosynthesis to the light environment. *Plant Physiol.*, 131:472, 2003.
154. Murakami, A. et al., Chromatic regulation in *Chlamydomonas reinhardtii*: time course of photosystem stoichiometry adjustment following a shift in growth light quality. *Plant Cell Physiol.*, 38:188, 1997.
155. Melis, A., The mechanism of photosynthetic membrane adaptation to environmental stress conditions: A hypothesis on the role of electron transport and of ATP/NADPH pool in the regulation of the thylakoid membrane organization and function. *Physiol. Veg.*, 23:757, 1985.

156. Jang, J.C. et al., Hexokinase as a sugar sensor in higher plants. *Plant Cell*, 9:5, 1997.
157. Taylor, W.C., Regulatory interactions between nuclear and plastid genomes. *Ann. Rev. Plant Physiol.*, 40:211, 1989.
158. Pfannschmidt, T. et al., A novel mechanism of nuclear photosynthesis gene regulation by redox signals from the chloroplast during photosystem stoichiometry adjustment. *J. Biol. Chem.*, 276:36125, 2001.
159. Rodermel, S., Pathways of plastid-to-nucleus signaling. *Trends Plant Sci.*, 6:471, 2001.
160. Surpin, M., Larkin, R.M., and Chory, J., Signal transduction between the chloroplast and the nucleus. *Plant Cell*, 14:S327, 2002.
161. Pfannschmidt, T., Chloroplast redox signals: How photosynthesis controls its own genes. *Trends Plant Sci.*, 8:33, 2003.
162. Strand, A., Plastid-to-nucleus signalling. *Curr. Opin. Plant Biol.*, 7:621, 2004.
163. Fey, V. et al., Photosynthetic redox control of nuclear gene expression. *J. Exp. Bot.*, 56:1491, 2005.
164. Johanningmeier, U. and Howell, SH., Regulation of light-harvesting chlorophyll-binding protein messenger-RNA accumulation in *Chlamydomonas reinhardi* — Possible involvement of chlorophyll synthesis precursors. *J. Biol. Chem.*, 259:3541, 1984.
165. Johanningmeier, U., Possible control of transcript levels by chlorophyll precursors in *Chlamydomonas*. *Eur. J. Biochem.*, 177:417, 1988.
166. Parikh, V.S. et al., The mitochondrial genotype can influence nuclear gene-expression in yeast. *Science*, 235:576, 1987.
167. Forsburg, S.L. and Guarente, L., Communication between mitochondria and the nucleus in regulation of cytochrome genes in the yeast *Saccharomyces-cerevisiae*. *Ann. Rev. Cell Biol.*, 5:153, 1989.
168. Susek, R.E., Ausubel, F.M., and Chory, J., Signal-transduction mutants of *Arabidopsis* uncouple nuclear cab and rbcs gene-expression from chloroplast development. *Cell*, 74:787, 1993.
169. Mochizuki, N. et al., *Arabidopsis* genomes uncoupled 5 (Gun5) mutant reveals the involvement of Mg-Chelatase H subunit in plastid-to-nucleus signal transduction. *Proc. Natl. Acad. Sci. USA*, 98:2053, 2001.
170. Larkin, R.M., GUN4, a regulator of chlorophyll synthesis and intracellular signaling, *Science*, 299:902, 2003.
171. Strand, A. et al., Chloroplast to nucleus communication triggered by accumulation of Mg-Protoporphyrin IX. *Nature*, 421:79, 2003.
172. Kropat, J. et al., Chlorophyll precursors are signals of chloroplast origin involved in light induction of nuclear heat-shock genes. *Proc. Natl. Acad. Sci. USA*, 94:14168, 1997.
173. Kropat, J. et al., Chloroplast signalling in the light induction of nuclear Hsp70 genes requires the accumulation of chlorophyll precursors and their accessibility to cytoplasm/nucleus. *Plant J.*, 24:523, 2000.
174. Gray, G. et al., The characterization of photoinhibition and recovery during cold acclimation in *Arabidopsis thaliana* using chlorophyll fluorescence imaging. *Physiol. Plant*, 119:365, 2003.
175. Wilson, K., Sieger, S., and Huner, N., The temperature-dependent accumulation of Mg-Protoporphyrin IX and reactive oxygen species in *Chlorella vulgaris*. *Physiol. Plant*, 119:126, 2003.

176. Fujita, Y., Murakami, A., and Ohki, K., Regulation of photosystem composition in the cyanobacterial photosynthetic system — The regulation occurs in response to the redox state of the electron pool located between the 2 photosystems. *Plant Cell Physiol.*, 28:283, 1987.
177. Chen, Y-B. et al., Plastid regulation of Lhcbl transcription in the chlorophyte alga *Dunaliella tertiolecta*. *Plant Physiol.*, 136:3737, 2004.
178. Im, C.S. and Grossman, A.R., Identification and regulation of high light-induced genes in *Chlamydomonas reinhardtii*. *Plant J.*, 30:301, 2002.
179. Irihimovitch, V. and Shapira, M., Glutathione redox potential modulated by reactive oxygen species regulates translation of Rubisco large subunit in the chloroplast. *J. Biol. Chem.*, 275:16289, 2000.
180. Pfannschmidt, T., Nilsson, A., and Allen, J.F., Photosynthetic control of chloroplast gene expression. *Nature*, 397:625, 1999.
181. Tullberg, A. et al., Photosynthetic electron flow regulates transcription of the psab gene in pea (*Pisum sativum* L.) chloroplasts through the redox state of the plastoquinone pool. *Plant Cell Physiol.*, 41:1045. 2000.
182. Anderson, J.M. et al., Reduced levels of cytochrome Bf complex in transgenic tobacco leads to marked photochemical reduction of the plastoquinone pool, without significant change in acclimation to irradiance. *Photosyn. Res.*, 53:215, 1997.
183. Oswald, O. et al., Plastid redox state and sugars: interactive regulators of nuclear-encoded photosynthetic gene expression. *Proc. Natl. Acad. Sci. USA*, 98:2047, 2001.
184. Pursiheimo, S. et al., Coregulation of light-harvesting complex II phosphorylation and *Lhcb* mRNA accumulation in winter rye. *Plant J.*, 3:317, 2001.
185. Walters, R. et al., A mutant of *Arabidopsis* lacking the triose-phosphate/phosphate translocator reveals metabolic regulation of starch breakdown in the light. *Plant Physiol.*, 135:891, 2004.
186. Sullivan, J.A. and Gray, J.C., Multiple plastid signals regulate the expression of the pea plastocyanin gene in pea and transgenic tobacco plants. *Plant J.*, 32:763, 2002.
187. Petracek, M.E. et al., Ferredoxin-1 mRNA is destabilized by changes in photosynthetic electron transport. *Proc. Natl. Acad. Sci. USA*, 95:9009, 1998.
188. Yang, D. et al., The redox state of the plastoquinone pool controls the level of the light-harvesting chlorophyll a/b binding protein complex II (Lhc II) during photoacclimation — Cytochrome B(6)F deficient *Lemna perpusilla* plants are locked in a dtate of high-light scclimation. *Photosyn. Res.*, 68:163, 2001.
189. Yang, D.H. et al., The redox state of the plastoquinone pool controls the level of the light-harvesting chlorophyll a/B binding protein complex II (LHC II) during photoacclimation — Cytochrome B(6)F deficient *Lemna perpusilla* plants are locked in a state of high-light acclimation. *Photosyn. Res.*, 68:163, 2001.
190. Danon, A. and Mayfield, S.P., Light-regulated translation of chloroplast messenger-Rnas through redox potential. *Science*, 266:1717, 1994.
191. Koch, K., Sucrose metabolism: Regulatory mechanisms and pivotal roles in sugar sensing and plant development. *Curr. Opin. Plant Biol.*, 7:235, 2004.
192. Rolland, F. and Sheen, J., Sugar sensing and signalling networks in plants. *Biochem. Soc. Trans.*, 33:269, 2005.
193. Sheen, J., Metabolic repression of transcription in higher-plants. *Plant Cell*, 2:1027, 1990.

194. Moore, B. et al., Role of the *Arabidopsis* glucose sensor Hxk1 in nutrient, light, and hormonal signaling. *Science*, 300:332, 2003.
195. Mahalingam, R. and Fedoroff, N., Stress response, cell death and signalling: The many faces of reactive oxygen species. *Physiol. Plant*, 119:56, 2003.
196. Mittler, R. et al., Reactive oxygen gene network of plants. *Trends Plant Sci.*, 9:490, 2004.
197. Karpinski, S. et al., Photosynthetic electron transport regulates the expression of cytosolic ascorbate peroxidase genes in *Arabidopsis* during excess light stress. *Plant Cell*, 9:627, 1997.
198. Mullineaux, P. and Karpinski, S., Signal transduction in response to excess light: Getting out of the chloroplast. *Curr. Opin. Plant Biol.*, 5:43, 2002.
199. Minagawa, J. et al., Molecular characterization and gene expression of Lhcb5 gene encoding CP26 in the light-harvesting complex II of *Chlamydomonas reinhardtii*. *Plant Mol. Biol.*, 46:277, 2001.
200. Chen, M., Chory, J., and Fankhauser, C., Light signal transduction in higher plants. *Ann. Rev. Gen.*, 38:87, 2004.
201. Walters, R. et al., Acclimation of *Arabidopsis thaliana* to the light environment: The role of photoreceptors. *Planta*, 209:517, 1999.
202. Jarillo, J.A. et al., Phototropin-related Npl1 controls chloroplast relocation induced by blue light. *Nature*, 410:952, 2001.
203. Sineshchekov, O.A., Jung, K.H., and Spudich, J.L., Two rhodopsins mediate phototaxis to low- and high-intensity light in *Chlamydomonas reinhardtii*. *Proc. Natl. Acad. Sci. USA*, 99:8689, 2002.
204. Kindle, K.L., Expression of a gene for a light-harvesting chlorophyll-a-binding chlorophyll-b-binding protein in *Chlamydomonas reinhardtii* — Effect of light and acetate. *Plant Mol. Biol.*, 9:547, 1987.
205. Schonfeld, C. et al., The nucleus-encoded protein Moc1 is essential for mitochondrial light acclimation in *Chlamydomonas reinhardtii*. *J. Biol. Chem.*, 279:50366, 2004.
206. Wu, D.Y. et al., The immutans variegation locus of *Arabidopsis* defines a mitochondrial alternative oxidase homolog that functions during early chloroplast biogenesis. *Plant Cell*, 11:43, 1999.
207. Peltier, G. and Cournac, L., Chlororespiration. *Ann. Rev. Plant Biol.*, 53:523, 2002.

5 Molecular Mechanisms in Hormonal Regulation of Drought Tolerance

Yajun Wu, Jenny Ballif, and Bingru Huang

CONTENTS

5.1 Introduction 101
5.2 The Involvement of Hormones in Plant Detection
     of Water Stress 102
5.3 ABA as the Primary Chemical Signal from the Roots
     under Drought 103
5.4 Regulation of ABA Accumulation under Drought 103
5.5 ABA and Stomatal Closure during Drought 106
5.6 ABA Signaling in Nonguard Cells during Drought 109
5.7 ABA-Induced Gene Expression 110
5.8 ABA in Relation to Plant Growth under Drought 111
5.9 Concluding Remarks 113
Acknowledgment 114
References 114

5.1 INTRODUCTION

Drought stress is one of the major abiotic stresses limiting plant growth and productivity [1,2]. It will continue to be a serious problem in agriculture because water is becoming more limited due to increased use by the expanding human population, decreasing precipitation, and lower potable water availability. Extensive research efforts have been taken to investigate mechanisms that impart drought tolerance of various plant species. During the process of plant adaptation to drought stress, plants have developed sophisticated systems to avoid, tolerate, or escape drought stress. Various physiological, biochemical, and molecular processes change in response to drought stress, which determine the survivability and persistence of plants in water-limiting environments. One of the important strategies of plant adaptation to drought stress is water conservation. This could be achieved by rapid stomatal closure, resulting in the reduction of water loss through transpiration. The question of how plants detect drought stress and close stomata during drought stress has received increasing attention during the last few decades [3–5].

Ample evidence suggests that some phytohormones such as abscisic acid and cytokinin are involved in plant response to drought stress by serving as signal molecules and key mediators for regulating specific pathways. The involvement of these hormones in root-to-shoot signaling, particularly through regulation of stomatal behavior and leaf growth, has been implicated in plant resistance to drought [3,6,7]. The molecular mechanisms of how these hormones function in plants during drought stress have only recently started to unfold. This has been accomplished by using mutants and other molecular tools in combination with physiological and biochemical approaches. The abscisic acid (ABA) signaling pathway has been extensively studied and well documented. Increasing evidence has demonstrated a complex hormonal regulation pathway for ABA alone or in interaction with other hormones. In this chapter, we review recent advances in the area of plant hormonal regulation of plant tolerance to drought stress, with an emphasis on ABA regulation of stomatal closure.

5.2 THE INVOLVEMENT OF HORMONES IN PLANT DETECTION OF WATER STRESS

Roots are capable of sensing a moisture gradient through a mechanism called hydrotropism [8]. Hydrotropism is clearly distinct from gravitropism, evidenced by mutants that lack gravitropism but are able to respond to a water potential gradient [9]. Further studies have shown that roots are able to work against gravitropism in order to grow toward a water source, probably through degradation of starch particles in root columella cells [10]. ABA and auxin seem to be involved in hydrotropism. Roots of an ABA-deficient mutant, *aba1–1*, and an ABA-insensitive mutant, *abi2–1*, were less responsive to water potential gradients than wild-type plants. When ABA was added back to the *aba1–1* mutant, the normal hydrotropic response was restored. Auxin-insensitive mutants, *axr1–3* and *axr2–1*, showed stronger hydrotropic responses than wild-type plants [11].

Because ABA is accumulated in water-stressed roots and required for maintenance of root elongation under drought [12], it was hypothesized that ABA may be a key player in regulating hydrotropism to obtain water under moderate stress conditions [13]. A *no hydrotropic response* mutant (*nhr1*) that lacks hydrotropic response was recently identified [14]. This mutant may lead to identification of other important components in hydrotropism and provide insights to understanding how plants sense and respond to a water potential gradient.

It was recently found that cytokinin receptor Cre1, a histidine kinase (ATHK4), could be activated by a reduction in turgor pressure when the gene was expressed in yeast cells [15], suggesting that Cre1 might serve as a direct sensor for dehydration stress. Cre1 activation upon a change in turgor pressure implies a potential role of cytokinin in drought stress response. However, whether this mechanism also applies to plant responses to water stress has not been determined. Another histidine kinase, ATHK1, in *Arabidopsis* was also believed to be an osmosensor when examined in yeast cells [16]. It will be interesting to see if the cytokinin receptor mutant, *cre1*, which shows a reduced response to cytokinin [17], will also show a decreased drought tolerance in plants.

5.3 ABA AS THE PRIMARY CHEMICAL SIGNAL FROM THE ROOTS UNDER DROUGHT

Water conservation through closing stomata during dry periods is a critical factor for plant survival of drought conditions. It was commonly believed that stomatal closure was induced by water loss and thus a reduction in turgor pressure under drought. However, increasing evidence suggests that a chemical signal moving from roots to shoots in response to soil drying may lead to reduction in guard cell turgor, induce stomatal closure, and subsequently reduce water loss [18–20]. This mechanism was confirmed by using split-root systems in which part of the root system was exposed to drying soil while the remaining roots were maintained in well-watered soil. In split root studies, it was found that stomata can be closed even when leaves are fully hydrated with the supply of water through the well-watered roots [20,21].

Abscisic acid has been identified as the primary chemical signal controlling stomatal response to drying soil [22]. ABA may be synthesized in roots exposed to drying soil and transported to shoots where it triggers a signal transduction cascade eventually leading to a reduction in guard cell turgor and stomatal closure [22–24]. However, not all plant species exhibit chemical regulation under soil drying, and variability in how plants sense and respond to decreases in water availability may be related to specific strategies for drought avoidance or drought tolerance. For example, Puliga et al. [25] reported that decreases in leaf turgor of *Eragrostis curvula* and *Sporobolus stapfianus* occurred before any significant increases in ABA were detected.

5.4 REGULATION OF ABA ACCUMULATION UNDER DROUGHT

ABA accumulates in plant tissues in response to drought stress, which has been found to play important roles in plant tolerance to drought stress [26–28]. The accumulation of ABA in drought-stressed plants is the result of interaction of synthesis, catabolism, and mobilization (see Figure 5.1). Regulatory mechanisms of ABA synthesis during drought stress have recently been reviewed [29–31]. However, the understanding of ABA synthesis pathways is far from being complete.

Zeaxanthin epoxidase (ZEP) is required for conversion of zeaxanthin to violaxanthin. This epoxidation is one of the earliest steps in the ABA synthesis pathway. The transcript level of the ZEP gene was found to be upregulated, which is associated with ABA accumulation in roots of tobacco (*Nicotiana tabacum*) and *Arabidopsis* when exposed to drought stress. However, the transcript level was

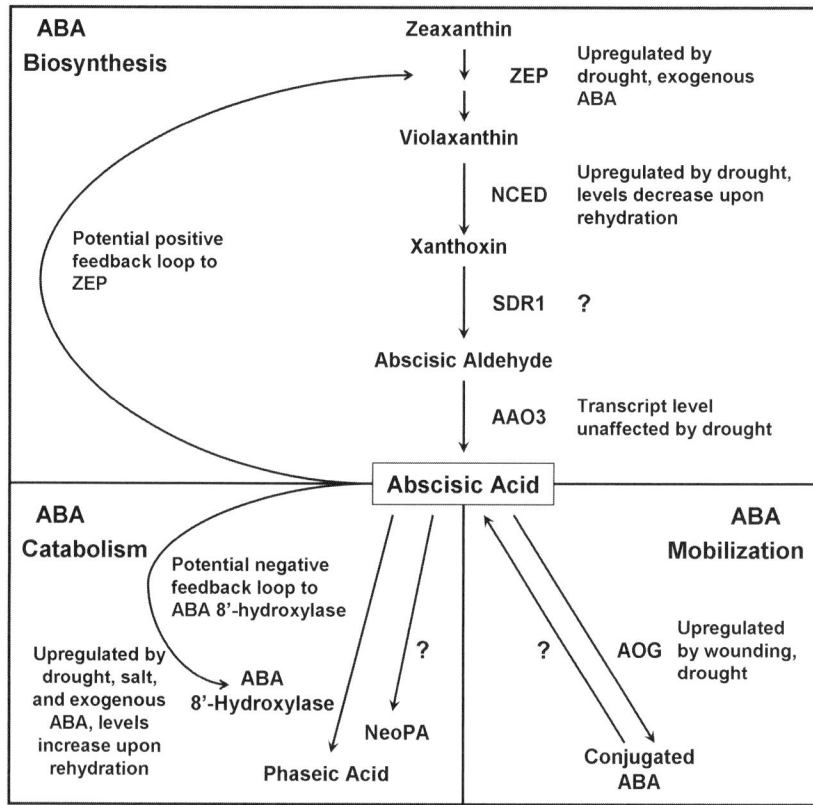

FIGURE 5.1 Regulation of ABA accumulation. ABA is produced through a series of reactions involving carotenoids. Zeaxanthin epoxidase (ZEP) catalyzes the synthesis of violaxanthin. Cleavage of the C_{40} violaxanthin (or neoxanthin, not shown) produces the C_{15} fragment xanthoxin. This reaction is catalyzed by a family of 9-*cis*-epoxycarotenoid dioxygenases (NCED). Xanthoxin is then converted to abscisic aldehyde by a short-chain dehydrogenase-reductase (SDR1 or ABA2). The final step in the synthesis of ABA is catalyzed by abscisic aldehyde oxidase (AAO3). ABA catabolism is primarily carried out by a group of ABA 8′-hydroxylases. These hydroxylases are coded for by genes CYP707A1–A4, members of the *Arabidipsis* P450 family. ABA can also be hydroxylated at its 9′ carbon to produce neophaseic acid. The enzyme responsible for this reaction has not yet been identified. An ABA-specific glucosyltransferase (AOG) catalyzes the conjugation of ABA to glucose. Glycosylated ABA has less biological activity, but it may serve as a reservoir for ABA because the molecule can be cleaved to release ABA. This may allow the plant to quickly mobilize ABA when needed, rather than synthesizing it *de novo*. The effect of drought stress and exogenous ABA, as described in the text, are shown in the figure.

unchanged despite ABA accumulation in tobacco leaves under drought, suggesting that ABA synthesis in leaves is not limited by substrate supply or is perhaps regulated by other factors [32,33]. A similar response was observed in tomato (*Lycopersicon esculentum* Mill.) where the ZEP transcript level was increased in roots but not in leaves [34]. However, in *Arabidopsis*, the transcript level of ZEP was upregulated in leaves in response to environmental stresses, including drought stress, indicating that different species may regulate ABA synthesis differently [35]. In addition, exogenous application of ABA elevated the ZEP transcript level, which suggests a positive feedback by ABA may exist [35].

9-*cis*-epoxycarotenoid dioxygenases (NCEDs) are enzymes working downstream of ZEP in the ABA synthesis pathway using violaxanthin or neoxanthin as a substrate [30,31,36]. Drought stress induced increases in the transcript level of NCED genes in many species, including maize (*Zea mays*), tomato, bean (*Phaseolus vulgaris*), *Arabidopsis*, cowpea (*Vigna unguiculata*), and avocado (*Persea americana*) (reviewed in [30,31]). In bean, both mRNA and protein levels were elevated by drought stress [37]. Moreover, overexpression of NCED in tobacco using *LeNCED1* from tomato, in *Arabidopsis* using *AtNCED3* from *Arabidopsis*, and in tobacco using *PvNCED1* from bean resulted in an accumulation of ABA [38–40]. Transgenic plants overexpressing the NCED gene showed greatly improved drought tolerance by at least partially reducing transpirational water loss [39,40] and by partially inducing or increasing mRNA levels of ABA responsive genes [39]. These results point to the possibility of improving plants' stress tolerance by manipulating the endogenous level of ABA. Using an inducible promoter would be beneficial because constitutive accumulation of ABA in seeds delayed seed germination [38,40].

It was recently identified that *ABA2* encodes a short-chain dehydrogenase-reductase (SDR1) in *Arabidopsis*. *ABA2* is an ABA-deficient mutant that blocks the steps of converting xanthoxin to ABA-aldehyde [41,42]. SDR1 transcript level was enhanced by glucose treatment [41] but not by salt stress [42]. Therefore, it is not clear whether the SDR1 transcript level is affected by drought stress.

The final step of ABA synthesis is catalyzed by ABA-aldehyde oxidase (AAO), which converts ABA-aldehyde to ABA. In *Arabidopsis*, *AtAAO3* was only expressed in leaves and its transcript levels were greatly increased when leaves were detached and dehydrated [43,44]. It is unknown whether overexpression of AAO3 in *Arabidopsis* could lead to an accumulation of ABA in plants. However, inasmuch as dehydration stress did not induce a change in protein level in *Arabidopsis*, it is believed that the last step of ABA synthesis is not rate-limiting [29,44]. Thus, overexpression of AAO3 may not be expected to have a major impact on ABA production in plants.

While ABA is accumulated during drought stress, phaseic acid (PA), a product during ABA catabolism, also increases at the same time [45–47]. It is believed that the balance between synthesis and catabolism of ABA is critical to maintaining ABA homeostasis in plants. One of the major ABA catabolic steps involves oxidation of ABA mediated by ABA 8′-hydroxylase. The genes that encode the hydroxylase have been recently identified. Using a combined approach of gene expression profiling and *in vitro* protein expression and activity assay, Kushiro et al. [47] identified four members of the *Arabidopsis* P450 family that were able to convert ABA into PA. More interestingly, the transcript levels of all four hydroxylase genes, *CYP707A1-A4*, were upregulated by dehydration stress. A rehydration treatment prompted an even greater increase in transcript levels, which was associated with a decrease in ABA content. However, the transcript level of *NCED3* decreased immediately upon rehydration [47].

In addition to the upregulation by salt, osmotic stress, and dehydration stress, the transcript levels of *CYP707As* were also found to be upregulated by ABA [48], suggesting that there is a negative feedback to control ABA content in plants. These results indicate that plants have developed a self-regulatory system so that ABA is maintained at physiologically optimum concentrations. ABA can also be hydroxylated at the 9′ carbon atom, yielding neophaseic acid (neoPA). NeoPA was found in dehydrated 10-d-old *B. napus* (cv. Quantum) seedlings and barley (*Hordeum vulgare* cv. Himalaya) seedlings, but not in well-watered seedlings [49], suggesting that a regulatory mechanism similar to the one for CYP707A exists. It would be interesting to find out which genes are involved in producing neoPA.

Glycosylation of ABA inactivates or decreases ABA activity, presenting another mechanism for the regulation of ABA. An ABA-specific glucosyltransferase gene (*AOG*) was recently identified from Adzuki bean (*Vigna angularis*). The *AOG* transcript level was induced by ABA, but not by other hormones tested. Dehydration and wounding also greatly induced the transcript level of *AOG* [50]. In certain species, conjugated ABA may serve as a storage form and can be reactivated by cleaving the sugar molecule. In water-stressed *Pseudotsuga menziesii* needles, it was found that the level of conjugated ABA decreased and the level of free ABA increased [51]. However, the significance of conjugated ABA in ABA accumulation is unclear.

Most tissues in plants are able to synthesize ABA in response to drought. However, little is know about specific cell types involved in ABA synthesis. Using the AAO3–GFP fusion protein as a marker, the GFP-fluorescence was localized in root tips, vascular bundles of roots, hypocotyls, inflorescence stems, and leaf veins. The AAO3 protein was also found in phloem companion cells, xylem parenchyma cells, and guard cells [52]. This specific expression pattern could facilitate ABA transport through phloem and xylem for distribution and action [53,54].

5.5 ABA AND STOMATAL CLOSURE DURING DROUGHT

ABA serves as a signal molecule to trigger a wide array of plant responses to drought stress. One of the major roles of ABA accumulation in drought-stressed plants is the induction of stomatal closure, minimizing water loss through transpiration. Although a receptor for ABA has not been revealed, many components in the pathway after the event of ABA perception have been identified mostly by using genetic approaches [4,5,55,56]. The signaling pathway of ABA-induced stomatal closure has been frequently reviewed [4,5]. In this complex pathway, Ca^{2+} plays a central role in ABA-induced stomatal closure as a secondary messenger. ABA induces an increase in Ca^{2+} concentration in the cytosol by taking up Ca^{2+} from the apoplast or releasing Ca^{2+} from storage inside the cell. The increased Ca^{2+} activates different anion efflux channels to release anions into cell walls. This in turn causes a membrane depolarization and activates outward-rectifying K^+ channels. At the same time, the increased Ca^{2+} inhibits K^+ uptake and plasma membrane proton pump activity, preventing stomatal opening. The consequence of this series of responses is a decrease in solute concentration or an increase in water potential inside the guard cell. A higher water potential in guard cells compared with adjacent leaf cells leads to water efflux, which reduces turgor pressure inside guard cells and causes stomatal closure [5].

Several components have been shown to be involved in regulating ABA-induced stomatal closure and could potentially improve plant water use efficiency and tolerance to drought (Figure 5.2). *ERA1* encodes a β-subunit of farnesyltransferase in *Arabidopsis*. A mutation of *ERA1* made plants hypersensitive to ABA, suggesting that *Arabidopsis* farnesyltransferase normally acts as a negative regulator in the ABA-induced stomatal closure pathway. Phenotypes include prolonged seed dormancy and greater stomatal closure when treated with ABA compared with wild-type plants [57]. Further studies demonstrated that the mutation of *ERA1* resulted in a greater cytosolic Ca^{2+} increase, which then induced stomatal closure [58]. In a dehydration stress experiment, the stomata of *era1* mutant plants closed faster and tighter than wild-type plants, resulting in slower water loss and delayed wilting [58].

ABH1 encodes a nuclear mRNA cap binding protein and is another negative regulator of the ABA-induced stomatal closure pathway. Mutants of *ABH1* (*abh1*) are hypersensitive to exogenous ABA during seed germination and stomatal closure [59]. The hypersensitivity to ABA in mutant guard cells was also associated with an enhanced level of cytosolic Ca^{2+}. Mutants of *ABH1* showed

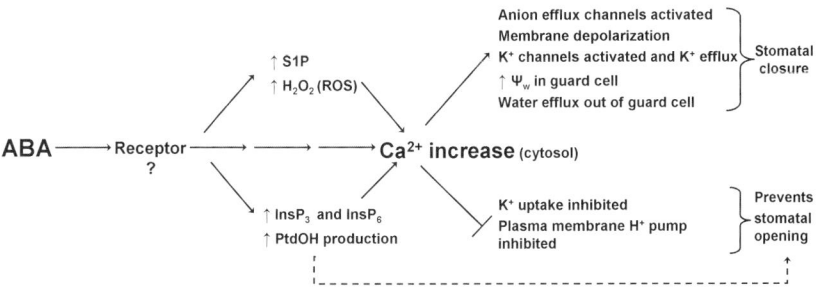

FIGURE 5.2 Simplified diagram of ABA-mediated stomatal closure. The diagram indicates the central role of Ca^{2+} in ABA-mediated stomatal closure. Many genes/proteins and signal molecules that have been recently identified and discussed in this chapter such as ERA1, ABH1, AtrbohD/AtrbohF, S1P, PtdOH, InsP$_3$, and InsP$_6$ close stomata through affecting cytosol Ca^{2+}.

improved tolerance to dehydration stress, which was associated with enhanced stomatal closure [59]. This study indicates that mRNA metabolism is a critical part of ABA signaling in *Arabidopsis* guard cells. Also, unlike *era1*, *abh1* has no pleiotropic phenotype and can be a great target for genetically engineering drought-tolerant plants.

Using the *faba* bean guard cell as a model system, Li et al. [60] identified a guard-cell-specific ABA-activated serine-threonine protein kinase (AAPK). Guard cells expressing a mutated AAPK failed to close stomata, indicating that AAPK is required for ABA signaling in *faba* bean stomatal closure. The *faba* bean guard cells that were overexpressing AAPK showed the same degree of stomatal closure as wild-type guard cells at 50 μM ABA. Further studies demonstrated that AAPK can directly activate AKIP1, a heterogeneous nuclear RNA-binding protein, only in the presence of ABA. The activated AKIP1 binds specifically to mRNA of dehydrin and a few others, suggesting AKIP may regulate stability or turnover of certain mRNAs [61]. These results, together with the *abh1* mutant study described above, provide further evidence that ABA plays an important role in post-transcriptional regulation of other genes. It would be interesting to see whether counterparts of AAPK and AKIP also exist in *Arabidopsis*.

Reactive oxygen species (ROS) have also been found to be involved in the ABA-induced stomatal closure process. Early work showed that H_2O_2 could induce stomatal closure by increasing Ca^{2+} levels in the cytosol. Moreover, ABA could induce H_2O_2 production in guard cells [62]. A recent study provided genetic evidence for ROS production and function *in vivo*. A double mutant of *atrbohD/F*, which has T-DNA insertions in genes for the catalytic subunit of the NADPH oxidase of AtrbohD and AtrbohF, showed no ABA-induced H_2O_2 production and reduced ABA-induced stomatal closure. However, stomata did close when guard cells were treated with H_2O_2 directly [63]. This study provides genetic evidence for the importance of ROS production in ABA-induced stomatal closing.

Although ABA induces stomatal closing, ABA also inhibits light-activated stomatal opening. By so doing, plants can ensure that stomata are tightly closed during drought conditions, even when they are exposed to light. During ABA inhibition of stomatal opening, it is known that ABA inhibits K^+ influx and activates certain types of anion efflux channels through Ca^{2+} signals [5,64]. In a recent study, the GTP-binding protein (G protein) was found to be involved in ABA inhibition of stomatal opening. Mutants of G protein lacking a functional α-subunit (GPA1) showed no ABA inhibition of inward K^+ channels or pH-independent ABA activation of anion channels in guard cells. The stomatal opening of mutant plants was less sensitive to ABA inhibition and plants lost more water during dehydration stress [64]. Whether genetic manipulation of G-proteins with a drought-inducible promoter could lead to a plant with higher sensitivity to ABA and improved drought tolerance deserves investigation.

ABA stimulates production of phosphatidic acid (PtdOH) in guard cells. Exogenous application of PtdOH to leaf epidermal peels induced stomatal closure and inhibited stomatal opening partially by inhibiting inward K^+ channels [65]. Moreover, it was found that overexpression of a phospholipase (AtPLDα) in *Arabidopsis*, which probably increases the level of PtdOH, enhanced stomatal sensitivity to ABA and thus reduced water loss. An antisense suppression of AtPLDα, however, reduced stomatal sensitivity to ABA and thus resulted in increased water loss [66]. Further studies revealed that PtdOH produced by PLDα inhibited the function of the negative regulator ABI1 or protein phosphatase 2C in the ABA signaling pathway, thus promoting ABA response [67]. Other phospholipid-related products have also been found to be involved in the ABA signaling pathway in guard cells. Both myo-inositol-hexakisphosphate ($InsP_6$) and inositol 1,4,5-triphosphate ($InsP_3$) production were stimulated by ABA in guard cells [68]. $InsP_3$ and $InsP_6$ were found to mobilize Ca^{2+} stored inside the cell and inhibit K^+ inward-rectifying channels [68,69]. Drought also increased the level of sphigosine-1-phosphate (S1P) in *Commelina communis* leaves [70]. S1P was able to increase cytosolic Ca^{2+} and promote stomatal closure. DL-threo-dihydrosphingosine, a competitive inhibitor of sphingosine kinase which decreases S1P levels, greatly reduced ABA-induced stomatal closure [70]. These results suggest that lipid molecules may serve as secondary messengers to mediate ABA-induced stomatal closure.

5.6 ABA SIGNALING IN NONGUARD CELLS DURING DROUGHT

ABA-induced stomatal closure is vital for plants to conserve water and survive under drought stress. However, plant responses to drought extend beyond the guard cells. Most tissues in plants are able to accumulate ABA during drought stress; this accumulation of ABA is required for activating or suppressing certain pathways for stress adaptation. The question is whether cells in other tissues use a pathway similar to that of guard cells and how the specificity is determined in each tissue. To address how guard cells respond to ABA differently from mesophyll cells, Leonhardt et al. [71] studied gene expression in isolated guard cells and mesophyll cells using a transcript profiling method. It was found that only 21 genes were induced in both guard cells and mesophyll cells among a total of 190 genes induced by ABA in both cell types. Among 118 genes suppressed by ABA in guard cells and mesophyll cells, only 3 genes were common to both cell types [71]. These results suggest that guard cells and mesophyll cells respond to ABA rather differently. However, due to genetic and functional redundancy in the plant genome, it is possible that two cell types would still share a similar mechanism of ABA response even if they use different genes.

A generalized signaling pathway for drought response in both ABA-dependent and ABA-independent cases has been proposed based on studies with different species and tissues using various approaches [26,55,56]. Many components found in ABA-induced stomatal closure in guard cells are also present in ABA signaling pathways in other tissues. For example, Ca^{2+} and phospholipids are proposed as secondary messengers (reviewed in [26,55]). FRY1 encodes an inositol polyphosphate 1-phosphatase, which reduces the amount of InsPs in plants. It was found that the *fry1* mutant accumulated more $InsP_3$ than wild-type plants and became hypersensitive to ABA as well as dehydration and salt stresses, based on leaf senescence and membrane leakage analysis [72]. These results provide genetic evidence for InsPs' role in ABA-mediated stress responses including dehydration stress response. It was hypothesized that FRY1 may be involved in the attenuation of ABA and stress response by breaking down IP_3 that was accumulated initially after stress treatment to serve as a secondary messenger [72]. An accumulation of IP_3 in *fry1* made plants more sensitive to ABA and stress.

Accumulation of PtdOH in water-stressed plants was associated with an increase in the transcript level of phospholipase D (AtPLDδ) in *Arabidopsis* [73]. A similar response was also observed in a desiccation-tolerant resurrection plant species (*Craterostigma plantagineum*) [74]. However, a transgenic plant that had suppressed expression of AtPLDδ did not confer any drought-sensitive phenotype, although the PtdOH production was significantly reduced during drought. Overexpression of AtPLDδ did not change PtdOH levels or the phenotype [73]. Thus, genetic evidence for involvement of PLD in drought response is still missing. Because *Arabidopsis* has three PLD genes, it is possible that genetic redundancy has masked a potential phenotype. Phosphatidylinositol 3,5-bisphosphate, phosphatidylinostol 4,5-bisphosphate, and $InsP_3$ were reported to accumulate upon hyperosmotic stress treatment in plants [75,76]. It is possible that these molecules are also involved in drought-induced ABA signaling.

Even though the major components of the ABA signaling pathway are similar between guard cells and nonguard cells, it does not necessarily mean that the signaling pathways operate in the same way. PLD was found to be a common component in ABA signaling pathways in leaf cells, barley aleurone cells, and guard cells of *Vicia faba* [77]. Depending on the cell type, PLD can be regulated at the transcript level or post-translational level and can be activated by G protein or by other signals. The downstream outputs from activation of PLD are different; in guard cells cytosolic Ca^{2+} increases in response to PLD action, but in aleurone cells it decreases [77]. Further research could reveal the differences in ABA signaling pathways for different tissues and cell types.

5.7 ABA-INDUCED GENE EXPRESSION

From transcript profiling studies we know that drought or dehydration stress changes the expression of a large number of genes in plants [78–82]. These genes cover almost every functional category, from transcription factors to cell structure, from energy metabolism to damage repair [83]. However, a large portion of drought-affected genes is classified as an "unknown function group." By comparing genes that are affected by dehydration stress and ABA treatment, Seki et al. [84] found that many genes were regulated by dehydration stress or ABA alone, whereas others responded to both ABA and drought. This demonstrates again that ABA-dependent and independent pathways work in a parallel manner in response to drought stress. Specific *cis*-acting elements in gene promoters are responsible for the specificity of gene regulation by ABA or by other signals. For instance, ABRE is an ABA-responsive element [85,86] and DRE is a dehydration-responsive element [87]. Among ABA-induced genes, ABRE was found in many promoters [71,84]. However, many genes that are inducible by ABA did not have an ABRE, implying that some novel ABREs may exist.

Because specific proteins such as transcription factors recognize *cis*-acting elements, transcription factors may play an important role in controlling gene expression and responding to drought and other stresses. In a transcript profiling study, it was found that about 11% of stress-induced genes (including dehydration stress) were transcription factors [81]. A few transcription factors have been identified to bind specifically to certain *cis*-acting elements of ABA-induced genes. ABA-responsive element-binding proteins (AREB1-3) are basic leucine zipper transcription factors and can bind specifically to ABREs in the promoter of the RD29B gene. AREB-mediated activation of the RD29B promoter requires ABA and protein modification, indicating that there are other components upstream of AREB in this regulatory pathway. In addition, the transcript levels of AREB1 and 2 were enhanced by ABA treatment [88]. These results demonstrate that these transcription factors could be regulated in a complex fashion and may play an important role in drought responses.

Other ABA-inducible genes, such as RD22, do not have an ABRE and may use other *cis*-acting elements for regulation. AtMYC2 and AtMYB2 are two transcription factors that can specifically bind to MYC and MYB *cis*-acting elements in the promoter of the RD22 and activate RD22 gene expression *in vitro* using a GUS reporter gene in a transient protoplast expression system. The transcript level of *AtMYBC2* itself was enhanced by dehydration and ABA treatment [89]. Recently, it was reported that transgenic plants overexpressing AtMYC2 or AtMYB2 are hypersensitive to ABA during seed germination and vegetative growth, which is associated with upregulation of several known ABA- and drought-induced genes. Transgenic plants that were overexpressing AtMYB2/AtMYC2 were smaller than wild-type plants. Based on electrolyte leakage assay, the transgenic plants were more tolerant to high concentrations of mannitol treatment [90]. These studies suggest that AtMYB2 and AtMYC2 are involved in regulation of gene expression in ABA signaling and drought response.

5.8 ABA IN RELATION TO PLANT GROWTH UNDER DROUGHT

Beside stomatal closure, ABA also causes changes in other physiological processes when ABA content rises during drought. It has been long believed that elevated ABA concentrations in drought-stressed plants cause inhibition of shoot growth [91,92]. Some recent studies found that ABA may be required for maintaining plant growth under drought by suppressing ethylene evolution [2,93].

Many ABA-deficient and insensitive mutants show pleiotropic phenotypes, including inhibition of growth. The reduction of growth has often been attributed to stomatal malfunction. *flc*, an ABA-deficient mutant in tomato, exhibits a wilt phenotype due to defective stomatal closure. Growing *flc* plants in a high humidity environment to achieve the same water status as in the wild-type plants did not significantly improve growth, suggesting that other factors besides water status were limiting plant growth. When ABA was added by foliar spray, *flc* mutants showed a dramatic recovery of plant growth in all examined parameters, such as leaf number, size, and fresh weight [94]. These results demonstrate that ABA is required for normal shoot growth in tomato. The same conclusion was drawn from another study using ABA-deficient *Arabidopsis* plants [95].

ABA is required for maintaining root elongation in maize under drought conditions. An ABA-deficient mutant, *vp5*, showed greater inhibition of root elongation than wild-type plants under drought. Root elongation was also impaired under drought when an ABA synthesis inhibitor blocked ABA accumulation. However, when ABA was added back to the ABA-deficient plants by supplying ABA in growth medium, root elongation was greatly recovered [12]. Drought-stressed roots of *vp5* mutants morphologically resembled ethylene-treated roots, being short and thick. Direct measurement of ethylene evolution from stressed *vp5* mutants confirmed large production of ethylene from roots. Exogenous application of ABA back to *vp5* roots recovered the wild-type phenotype, which was associated with a reduction of ethylene evolution. Using ethylene synthesis or perception inhibitors in stressed roots achieved a similar effect to that of ABA [96].

It is inferred from those studies that ABA accumulation in primary roots of drought-stressed maize may inhibit ethylene production, and thus help maintain root growth. ABA has also been shown to promote shoot growth in maize seedlings under drought at certain developmental stages [93]. However, it is not clear whether ABA is required for suppressing ethylene evolution as it is in roots.

Plant growth and development depend on a fine balance of phytohormones in plants. ABA-deficient plants are found to produce more ethylene in both tomato leaves (*flc*) and maize roots (*vp5*) [95,96]. However, ethylene-insensitive mutants, *era3* (enhanced response to ABA)/*ein2* (ethylene-insensitive) double mutant, and *etr1* (ethylene-insensitive mutant) had overaccumulated ABA [95,97]. These results suggest an antagonistic relationship between ABA and ethylene synthesis. Ethylene evolution has been found to increase during drought stress in some species. In jack pine (*Pinus banksiana* Lamb.) seedlings, drought stress greatly increased the transcript level of S-adenosylmethionine synthetase, one of the enzymes involved in ethylene synthesis. The increase in gene expression was associated with increased ethylene evolution [98]. A large increase in ethylene evolution was also observed in some cultivars of French bean (*Phaseolus vulgaris* L.) but not in other cultivars after drought stress [99].

It is not clear what caused the differences in ethylene evolution among different cultivars. If an accumulation of ABA in plants under drought is required for inhibiting ethylene evolution, it is possible that a difference in the level of accumulated ABA may result in a difference in ethylene evolution among cultivars during drought. Of course, ABA and ethylene may interact at other levels such as between signaling pathways as observed in nonstressed plants [97,100]. An increase in ethylene evolution was also observed in drought-stressed tomato leaves. The increase of ethylene evolution was associated with a reduction in leaf growth.

To examine whether ethylene was the cause of growth inhibition, an antisense transgenic plant for ACC oxidase to suppress ethylene synthesis was used. The transgenic plants showed little stress-induced ethylene evolution and greatly improved leaf elongation under drought [101]. Thus, a genetic approach can be a promising way to manipulate ethylene levels in plants under drought when endogenous ABA is not sufficient to suppress ethylene's inhibitory effect on growth.

5.9 CONCLUDING REMARKS

Plant hormones play important roles in plant responses to drought. A scheme of the hypothetic pathway involving hormones in drought response is presented in Figure 5.3. ABA, as an example, is probably involved in water gradient sensing by modulating hydrotropism. The accumulation of ABA under drought is a result of complex regulation of ABA synthesis, catabolism, and mobilization (as shown in Figure 5.1). The ABA accumulation activates a complex cascade involving many components, which either positively or negatively modulate the ABA signal, achieving an optimal final output. The molecular mechanisms of ABA action in guard cells during drought stress have begun to be unfolded (Figure 5.2) and will serve as a model system for studying ABA action in other cell types. With the advance in high-throughput technology, our knowledge of ABA action during drought response will expand rapidly.

Several areas need to be mentioned and deserve more attention in future studies. First, efforts should be made to address upstream signaling components

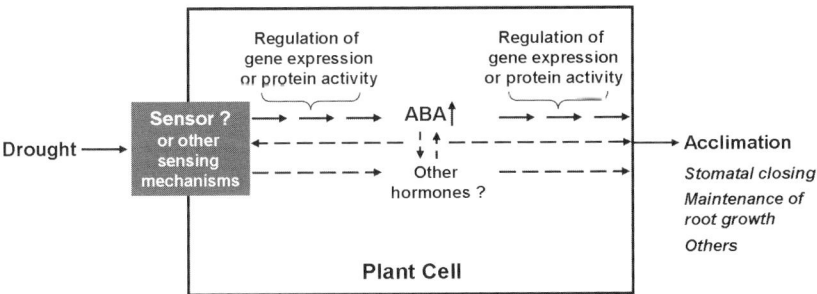

FIGURE 5.3 Scheme of hypothetic drought sensing and response involving ABA and other hormones. Plants need to sense drought. The sensor could be a protein or a complex in the membrane or inside the cell that can detect a change in water content probably by sensing (or being activated by) solute concentration or turgor change due to water loss. The signal promotes the ABA synthesis pathway (see Figure 5.1), although it is not clear how, resulting in an accumulation of ABA. The enhanced ABA level triggers versatile responses, leading to an acclimation of plants to drought stress. Very limited information has implied roles of other hormones in drought responses, such as CRE1 activation by turgor pressure [15]. Also, an interaction of different hormones could be important for a plant response to drought [11,93].

in the ABA signaling pathway. Little is known about how plants sense drought stress and what mechanisms activate ABA synthesis. Identification of mutants using a forward genetics approach that blocks the expression of drought-inducible genes in the ABA synthesis pathway can potentially lead to a discovery of upstream genes in the ABA signaling pathway [102]. Second, very little is known about how other hormones in plants are involved in drought response at the molecular level. Because mutants defective in synthesis or response for other hormones are available, a comprehensive test of these mutants on drought response and examination of expression of a set of known drought-responsive genes using transcript profiling will provide valuable insight into whether other hormones play roles in drought response and how different hormonal pathways interact with each other. A better understanding of cellular and molecular regulation of hormones in plant responses to drought stress will help design a strategy in generating more drought-tolerant plants through genetic engineering or traditional breeding.

ACKNOWLEDGMENT

We thank Elizabeth Davis for reading this manuscript and the Utah Agricultural Experiment Station for support.

REFERENCES

1. Boyer, J.S., Plant productivity and environment. *Science*, 218:443, 1982.
2. Bray, E.A., Bailey-Serres, J., and Weretilnyk, E., Response to abiotic stress. In: *Biochemistry and Molecular Biology of Plants*. Buchanan, B.B., Gruissem, W., and Russell, L.J., Eds., American Society of Plant Physiologist, MD, 2000.
3. Davies, W.J. and Zhang, J.H., Root signals and the regulation of growth and development of plants in drying soil. *Ann. Rev. Plant Physiol.*, 42:55, 1991.
4. Assmann, S.M. and Wang, X.Q., From milliseconds to millions of years: guard cells and environmental responses. *Curr. Opin. Plant Biol.*, 4:421, 2001.
5. Schroeder, J.I., Kwak, J.M., and Allen, G.J. Guard cell abscisic acid signalling and engineering drought hardiness in plants. *Nature*, 410:327, 2001.
6. Pospisilova, J., Synkova, H., and Rulcova, J. Cytokinins and water stress. *Biol. Plant*, 43:321, 2000.
7. Quarrie, S.A. Understanding plant responses to stress and breeding for improved stress resistance — The generation gap. In: *Plant Responses to Cellular Dehydration during Environmental Stress*, Close, T.J. and Bray E.A., Eds., *Proceedings of the16th Annual Riverside Symposium in Plant Physiology*, Riverside, CA, 1993, pp. 224–245.
8. Takahashi, H., Hydrotropism: The current state of our knowledge. *J. Plant Res.*, 110:163, 1997.
9. Takahashi, H. and Suge, H., Root hydrotropism of an agravitropic pea mutant, *ageotropum*. *Physiol. Plant*, 82:24, 1991.
10. Takahashi, N. et al., Hydrotropism interacts with gravitropism by degrading amyloplasts in seedling roots of *Arabidopsis* and radish. *Plant Physiol.*, 132:805, 2003.

11. Takahashi, N. et al., Hydrotropism in abscisic acid, wavy, and gravitropic mutants of *Arabidopsis thaliana*. *Planta*, 216:203, 2002.
12. Sharp, R.E. et al., Confirmation that abscisic-acid accumulation is required for maize primary root elongation at low water potentials. *J. Exp. Bot.*, 45:1743, 1994.
13. Eapen, D. et al., Hydrotropism: Root growth responses to water. *Trends Plant Sci.*, 10:44, 2005.
14. Eapen, D. et al., A *no hydrotropic response* root mutant that responds positively to gravitropism in *Arabidopsis*. *Plant Physiol.*, 131:536, 2003.
15. Reiser, V., Raitt, D.C., and Saito, H., Yeast osmosensor Sln1 and plant cytokinin receptor Cre1 respond to changes in turgor pressure. *J. Cell Biol.*, 161:1035, 2003.
16. Urao, T. et al., A transmembrane hybrid-type histidine kinase in *Arabidopsis* functions as an osmosensor. *Plant Cell*, 11:1743, 1999.
17. Inoue, T. et al., Identification of CRE1 as a cytokinin receptor from *Arabidopsis*. *Nature*, 409:1060, 2001.
18. Blackman, P.G. and Davies, W.J., Root to shoot communication in maize plants of the effects of soil drying. *J. Exp. Bot.*, 36:39, 1985.
19. Davies, W.J., Wilkinson, S., and Loveys, B., Stomatal control by chemical signaling and the exploitation of this mechanism to increase water use efficiency in agriculture. *New Phytol.*, 153:449, 2002.
20. Zhang, J.H. and Davies, W.J. Sequential response of whole plant water relations to prolonged soil drying and the involvement of xylem sap ABA in the regulation of stomatal behavior of sunflower plants. *New Phytol.*, 113:167, 1989.
21. Gowing, D.J.G., Davies, W.J., and Jones, H.G., A positive root-sourced signal as an indicator of soil drying in apple, *Malus x domestica* Borkh. *J. Exp. Bot.*, 41:1535–1540, 1990.
22. Wilkinson, S. and Davies, W.J., ABA-based chemical signaling: the co-ordination of responses to stress in plants. *Plant Cell Environ.*, 25:195, 2002.
23. Assmann, S.M. and Shimazaki, K., The multisensory guard cell. Stomatal responses to blue light and abscisic acid. *Plant Physiol.*, 119:809, 1999.
24. McAinsh, M.R., Brownlee, C., and Hetherington, A.M. Calcium ions as second messengers in guard cell signal transduction. *Physiol. Plant*, 100:16, 1997.
25. Puliga, S., Vazzana, C., and Davies, W.J. Control of crops leaf growth by chemical and hydraulic influences. *J. Exp. Bot.*, 47:529, 1996.
26. Finkelstein, R.R., Gampala, S.S.L., and Rock, C.D., Abscisic acid signaling in seeds and seedlings. *Plant Cell*, 14:S15, 2002.
27. Leung, J. and Giraudat, J., Abscisic acid signal transduction. *Ann. Rev. Plant Physiol.*, 49:199, 1998.
28. Koornneef, M. et al., The genetic and molecular dissection of abscisic acid biosynthesis and signal transduction in *Arabidopsis*. *Plant Physiol. Biochem.*, 36:83, 1998.
29. Bray, E.A., Abscisic acid regulation of gene expression during water-deficit stress in the era of the *Arabidopsis* genome. *Plant Cell Environ.*, 25:153, 2002.
30. Schwartz, S.H., Qin, X.Q., and Zeevaart, J.A.D., Elucidation of the indirect pathway of abscisic acid biosynthesis by mutants, genes, and enzymes. *Plant Physiol.*, 131:1591, 2003.
31. Xiong, L.M. and Zhu, J.K. Regulation of abscisic acid biosynthesis. *Plant Physiol.*, 133:29, 2003.
32. Audran, C. et al., Expression studies of the zeaxanthin epoxidase gene in *Nicotiana plumbaginifolia*. *Plant Physiol.*, 118:1021, 1998.

33. Audran, C., Liotenberg, S., Gonneau, M., et al., Localisation and expression of zeaxanthin epoxidase mRNA in *Arabidopsis* in response to drought stress and during seed development. *Aust. J. Plant Physiol.*, 28:1161, 2001.
34. Thompson, A.J., Jackson, A.C., Parker, R.A., et al., Abscisic acid biosynthesis in tomato: Regulation of zeaxanthin epoxidase and 9-cis-epoxycarotenoid dioxygenase mRNAs by light/dark cycles, water stress and abscisic acid. *Plant Mol. Biol.*, 42:833, 2000.
35. Xiong, L.M. et al., Regulation of osmotic stress-responsive gene expression by the *LOS6/ABA1* locus in *Arabidopsis*. *J. Biol. Chem.*, 277:8588, 2002.
36. Tan, B.C., Schwartz, S.H., Zeevaart, J.A.D., et al., Genetic control of abscisic acid biosynthesis in maize. *Proc. Natl. Acad. Sci. USA*, 94:12235, 1997.
37. Qin, X.Q. and Zeevaart, J.A.D., The 9-*cis*-epoxycarotenoid cleavage reaction is the key regulatory step of abscisic acid biosynthesis in water-stressed bean. *Proc. Natl. Acad. Sci. USA*, 96:15354, 1999.
38. Thompson, A.J. et al., Ectopic expression of a tomato 9-*cis*-epoxycarotenoid dioxygenase gene causes over-production of abscisic acid. *Plant J.*, 23:363, 2000.
39. Iuchi, S. et al., Regulation of drought tolerance by gene manipulation of 9-*cis*-epoxycarotenoid dioxygenase, a key enzyme in abscisic acid biosynthesis in *Arabidopsis*. *Plant J.*, 27:325, 2001.
40. Qin, X.Q. and Zeevaart, J.A.D., Overexpression of a 9-*cis*-epoxycarotenoid dioxygenase gene in *Nicotiana plumbaginifolia* increases abscisic acid and phaseic acid levels and enhances drought tolerance. *Plant Physiol.*, 128:544, 2002.
41. Cheng, W.H., Endo, A., Zhou, L., et al., A unique short-chain dehydrogenase/reductase in *Arabidopsis* glucose signaling and abscisic acid biosynthesis and functions. *Plant Cell*, 14:2723, 2002.
42. Gonzalez-Guzman, M., Apostolova, N., Belles, J.M., et al., The short-chain alcohol dehydrogenase ABA2 catalyzes the conversion of xanthoxin to abscisic aldehyde. *Plant Cell*, 14:1833, 2002.
43. Seo, M. et al., The *Arabidopsis* aldehyde oxidase 3 (AAO3) gene product catalyzes the final step in abscisic acid biosynthesis in leaves. *Proc. Natl. Acad. Sci. USA*, 97:12908, 2000.
44. Seo, M. et al., Abscisic aldehyde oxidase in leaves of *Arabidopsis thaliana*. *Plant J.*, 23:481, 2000.
45. Zeevaart, J.A.D., Changes in the levels of abscisic acid and its metabolites in excised leaf blades of *Xanthium strumarium* during and after water stress. *Plant Physiol.*, 66:672, 1980.
46. Cutler, A.J. and Krochko, J.E., Formation and breakdown of ABA. *Trends Plant Sci.*, 4:472, 1999.
47. Kushiro, T. et al., The *Arabidopsis* cytochrome P450CYP707A encodes ABA 8'-hydroxylases: Key enzymes in ABA catabolism. *Embo J.*, 23:1647, 2004.
48. Saito, S. et al., *Arabidopsis* CYP707As encode (+)-abscisic acid 8'-hydroxylase, a key enzyme in the oxidative catabolism of abscisic acid. *Plant Physiol.*, 134:1439, 2004.
49. Zhou, R. et al., A new abscisic acid catabolic pathway. *Plant Physiol.*, 134:361, 2004.
50. Xu, Z.J. et al., Cloning and characterization of the abscisic acid-specific glucosyltransferase gene from adzuki bean seedlings. *Plant Physiol.*, 129:1285, 2002.
51. Johnson, J.D. and Ferrell, W.K., The relationship of abscisic acid metabolism to stomatal conductance in Douglas fir during water stress. *Physiol. Plant.*, 55:431, 1982.

52. Koiwai, H. et al., Tissue-specific localization of an abscisic acid biosynthetic enzyme, AAO3, in *Arabidopsis*. *Plant Physiol.*, 134:1697, 2004.
53. Sauter, A., Davies, W.J., and Hartung, W., The long-distance abscisic acid signal in the droughted plant: The fate of the hormone on its way from root to shoot. *J. Exp. Bot.*, 52:1991, 2001.
54. Hartung, W., Sauter, A., and Hose, E., Abscisic acid in the xylem: Where does it come from, where does it go to? *J. Exp. Bot.*, 53:27, 2002.
55. Xiong, L.M., Schumaker, K.S., and Zhu, J.K., Cell signaling during cold, drought, and salt stress. *Plant Cell*, 14:S165, 2002.
56. Zhu, J.K., Salt and drought stress signal transduction in plants. *Ann. Rev. Plant Biol.*, 53:247, 2002.
57. Pei, Z.M. et al., Role of farnesyltransferase in ABA regulation of guard cell anion channels and plant water loss. *Science*, 282:287, 1998.
58. Allen, G.J., Murata, Y., Chu, S.P., et al., Hypersensitivity of abscisic acid-induced cytosolic calcium increases in the *Arabidopsis* farnesyltransferase mutant *era1-2*. *Plant Cell*, 14:1649, 2002.
59. Hugouvieux, V., Kwak, J.M., and Schroeder, J.I., An mRNA cap binding protein, ABH1, modulates early abscisic acid signal transduction in *Arabidopsis*. *Cell*, 106:477, 2001.
60. Li, J.X. et al., Regulation of abscisic acid-induced stomatal closure and anion channels by guard cell AAPK kinase. *Science*, 287:300, 2000.
61. Li, J.X., Kinoshita, T., Pandey, S., et al. Modulation of an RNA-binding protein by abscisic-acid-activated protein kinase. *Nature*, 418:793, 2002.
62. Pei, Z.M. et al., Calcium channels activated by hydrogen peroxide mediate abscisic acid signalling in guard cells. *Nature*, 406:731, 2000.
63. Kwak, J.M. et al., NADPH oxidase *AtrbohD* and *AtrbohF* genes function in ROS-dependent ABA signaling in *Arabidopsis*. *Embo J.*, 22:2623, 2003.
64. Wang, X.Q. et al., G protein regulation of ion channels and abscisic acid signaling in *Arabidopsis* guard cells. *Science*, 292:2070, 2001.
65. Jacob, T., Ritchie, S., Assmann, S.M., et al., Abscisic acid signal transduction in guard cells is mediated by phospholipase D activity. *Proc. Natl. Acad. Sci. USA*, 96:12192, 1999.
66. Sang, Y.M. et al., Regulation of plant water loss by manipulating the expression of phospholipase D alpha. *Plant J.*, 28:135, 2001.
67. Zhang, W.H. et al., Phospholipase D alpha 1-derived phosphatidic acid interacts with ABI1 phosphatase 2C and regulates abscisic acid signaling. *Proc. Natl. Acad. Sci. USA*, 101:9508, 2004.
68. Lemtiri-Chlieh, F., MacRobbie, E.A.C., and Brearley, C.A., Inositol hexakisphosphate is a physiological signal regulating the K^+-inward rectifying conductance in guard cells. *Proc. Natl. Acad. Sci. USA*, 97:8687, 2000.
69. Lemtiri-Chlieh, F. et al., Inositol hexakisphosphate mobilizes an endomembrane store of calcium in guard cells. *Proc. Natl. Acad. Sci. USA*, 100:10091, 2003.
70. Ng, C.K.Y. et al., Drought-induced guard cell signal transduction involves sphingosine-1-phosphate. *Nature*, 410:596, 2001.
71. Leonhardt, N. et al., Microarray expression analyses of *Arabidopsis* guard cells and isolation of a recessive abscisic acid hypersensitive protein phosphatase 2C mutant. *Plant Cell*, 16:596, 2004.

72. Xiong, L.M. et al., FIERY1 encoding an inositol polyphosphate 1-phosphatase is a negative regulator of abscisic acid and stress signaling in *Arabidopsis*. *Gene Dev.*, 15:1971, 2001.
73. Katagiri, T., Takahashi, S., and Shinozaki, K., Involvement of a novel *Arabidopsis* phospholipase D, AtPLD delta, in dehydration-inducible accumulation of phosphatidic acid in stress signalling. *Plant J.*, 26:595, 2001.
74. Frank, W. et al., Water deficit triggers phospholipase D activity in the resurrection plant *Craterostigma plantagineum*. *Plant Cell*, 12:111, 2000.
75. DeWald, D.B. et al., Rapid accumulation of phosphatidylinositol 4,5-bisphosphate and inositol 1,4,5-trisphosphate correlates with calcium mobilization in salt-stressed *Arabidopsis*. *Plant Physiol.*, 126:759, 2001.
76. Meijer, H.J.G. et al., Hyperosmotic stress rapidly generates lyso-phosphatidic acid in *Chlamydomonas*. *Plant J.*, 25, 541, 2001.
77. Ritchie, S.M., Swanson, S.J., and Gilroy, S. From common signalling components to cell specific responses: insights from the cereal aleurone. *Physiol. Plant.*, 115:342, 2002.
78. Seki, M. et al., Monitoring the expression pattern of 1300 *Arabidopsis* genes under drought and cold stresses by using a full-length cDNA microarray. *Plant Cell*, 13:61, 2001.
79. Kreps, J.A. et al., Transcriptome changes for *Arabidopsis* in response to salt, osmotic, and cold stress. *Plant Physiol.*, 130:2129, 2002.
80. Ozturk, Z.N. et al., Monitoring large-scale changes in transcript abundance in drought- and salt-stressed barley. *Plant Mol. Biol.*, 48:551, 2002.
81. Seki, M. et al., Monitoring the expression profiles of 7000 *Arabidopsis* genes under drought, cold and high-salinity stresses using a full-length cDNA microarray. *Plant J.*, 31:279, 2002.
82. Rabbani, M.A. et al., Monitoring expression profiles of rice genes under cold, drought, and high-salinity stresses and abscisic acid application using cDNA microarray and RNA get-blot analyses. *Plant Physiol.*, 133:1755, 2003.
83. Bray, E.A., Classification of genes differentially expressed during water-deficit stress in *Arabidopsis thaliana*: An analysis using microarray and differential expression data. *Ann. Bot. (London)*, 89:803, 2002.
84. Seki, M. et al., Monitoring the expression pattern of around 7,000 *Arabidopsis* genes under ABA treatments using a full-length cDNA microarray. *Funct. Integ. Genom.*, 2:282, 2002.
85. Marcotte, W.D., Russell, H.D., and Quatrano, R.S., Abscisic acid-responsive sequences from the *Em* gene of wheat. *Plant Cell*, 1:969, 1989.
86. Mundy, J., Yamaguchi-Shinozaki, K., and Chua, N.H., Nuclear proteins bind conserved elements in the abscisic acid-responsive promoter of a rice RAB gene. *Proc. Natl. Acad. Sci. USA*, 87:1406, 1990.
87. Yamaguchi-Shinozaki, K., Urao, T., and Shinozaki, K., Regulation of genes that are induced by drought stress in *Arabidopsis thaliana*. *J. Plant Res.*, 108:127, 1995.
88. Uno, Y. et al., *Arabidopsis* basic leucine zipper transcription factors involved in an abscisic acid-dependent signal transduction pathway under drought and high-salinity conditions. *Proc. Natl. Acad. Sci. USA*, 97:11632, 2000.
89. Abe, H. et al., Role of *Arabidopsis* MYC and MYB homologs in drought- and abscisic acid-regulated gene expression. *Plant Cell*, 9:1859, 1997.

90. Abe, H. et al., AtMYC2 and ATMYB2 act as transcriptional activators in ABA signaling. *Plant Cell Physiol.*, 44:S152, 2003.
91. Quarrie, S.A. and Jones, H.G., Effects of abscisic acid and water stress on development and morphology of wheat. *J. Exp. Bot.*, 28:192, 1977.
92. Trewavas, A.J. and Jones, H.G., An assessment of the role of ABA in plant development. In: *Abscisic Acid: Physiology and Biochemistry*, Davies, W.J. and Jones, H.G., Eds., BIOS Scientific, Oxford, 1991, pp. 169.
93. Sharp, R.E., Interaction with ethylene: Changing views on the role of abscisic acid in root and shoot growth responses to water stress. *Plant Cell Environ.*, 25:211, 2002.
94. Sharp, R.E. et al., Endogenous ABA maintains shoot growth in tomato independently of effects on plant water balance: Evidence for an interaction with ethylene. *J. Exp. Bot.*, 51:1575, 2000.
95. LeNoble, M.E., Spollen, W.G., and Sharp, R.E., Maintenance of shoot growth by endogenous ABA: Genetic assessment of the involvement of ethylene suppression. *J. Exp. Bot.*, 55:237, 2004.
96. Spollen, W.G. et al., Abscisic acid accumulation maintains maize primary root elongation at low water potentials by restricting ethylene production. *Plant Physiol.*, 122:967, 2000.
97. Ghassemian, M. et al., Regulation of abscisic acid signaling by the ethylene response pathway in *Arabidopsis*. *Plant Cell*, 12:1117, 2000.
98. Mayne, M.B., Coleman, J.R., and Blumwald, E., Differential expression during drought conditioning of a root-specific S-adenosylmethionine synthetase from jack pine (*Pinus banksiana* Lamb) seedlings. *Plant Cell Environ.*, 19:958, 1996.
99. Upreti, K.K., Murti, G.S.R., and Bhatt, R.M., Response of French bean cultivars to water deficits: Changes in endogenous hormones, proline and chlorophyll. *Biol. Plant*, 40:381, 1998.
100. Beaudoin, N. et al., Interactions between abscisic acid and ethylene signaling cascades. *Plant Cell*, 12:1103, 2000.
101. Sobeih, W.Y., Dodd, I.C., Bacon, M.A., et al., Long-distance signals regulating stomatal conductance and leaf growth in tomato (*Lycopersicon esculentum*) plants subjected to partial root-zone drying. *J. Exp. Bot.*, 55:2353, 2004.
102. Verslues, P.E. and Zhu, J.K., Before and beyond ABA: Upstream sensing and internal signals that determine ABA accumulation and response under abiotic stress. *Biochem. Soc. Trans.*, 33:375, 2005.

6 Salinity Tolerance: Cellular Mechanisms and Gene Regulation

R.K. Sairam, Aruna Tyagi, and Vishwanathan Chinnusamy

CONTENTS

6.1 Introduction 122
6.2 Physiological Effects of Salinity 123
    6.2.1 Plant Development 123
    6.2.2 Photosynthesis 125
        6.2.2.1 Effects of Salinity on Photosynthetic Pigments and Proteins 125
        6.2.2.2 Effects of Salinity on Proteins 125
        6.2.2.3 Mechanisms of Salinity Effects on Photosynthesis 126
6.3 Salt Response: A Multigenic Trait 128
    6.3.1 Perception of Salt Stress 129
    6.3.2 Second Messengers 130
    6.3.3 Ion Homeostasis 131
    6.3.4 Sodium Uptake 132
    6.3.5 Sodium Efflux 133
    6.3.6 Sodium Compartmentation 135
    6.3.7 Sodium Transport from Shoot to Root 136
6.4 Salinity Damage and Repair 137
    6.4.1 Polyamines 138
    6.4.2 Osmoprotectants/Compatible Solutes 139
    6.4.3 Metabolic Engineering of Osmoprotectants 142
    6.4.4 Regulation of Osmoprotectant Metabolism 145
6.5 Salt-Induced Oxidative Stress 146
6.6 Genes Encoding Proteins Involved in Cellular Protection 149
    6.6.1 LEA-Type Proteins 149
    6.6.2 Transcriptional Regulation of Stress Genes 151
        6.6.2.1 LEA/COR Genes 151
        6.6.2.2 Calcium Sensor Proteins 151

	6.6.2.3 Basic Lucine-Zipper Family Transcription Factors	153
	6.6.2.4 MYB/MYC-Type Transcription Factors	153
	6.6.2.5 C-Repeat Binding Proteins	154
6.7	Stress Signaling Pathways	155
6.8	Strategies to Improve Stress Tolerance	157
6.9	Conclusions and Prospects	158
References		159

6.1 INTRODUCTION

An excess amount of salts in soil adversely affects plant growth and development. Nearly 20% of the world's cultivated area and nearly half of the world's irrigated lands are affected by salinity [1]. Salinity stress threatens agricultural productivity in 77 million hectares of agricultural land, 45 million hectares of which (20% of irrigated area) are irrigated and 32 million hectares (2.1% of dry land) are unirrigated land [2]. Salinization is further spreading in irrigated land due to improper management of irrigation and drainage. Rain, tornadoes, and wind also add salts to coastal agricultural land. Soil salinity often leads to development of other problem soils such as soil sodicity and alkalinity. Soil sodicity is the result of binding of Na^+ to negatively charged clay particles leading to clay swelling and clay dispersal. Hydrolysis of the Na–clay complex results in soil alkalinity. The USDA Salinity Laboratory standards define a saline soil as a soil having the electrical conductivity of a saturated paste extract (EC_e) of 4 deci Siemens per meter (dS m^{-1}) or more. High concentrations of soluble salts such as chlorides of sodium, calcium, and magnesium contribute to the high electrical conductivity of saline soils. NaCl contributes most of the soluble salts in saline soils.

Plants are classified as glycophytes or halophytes according to their capacity to grow on a high salt medium. Most plants are glycophytes and cannot tolerate salt stress. The effects of salt stress are detrimental in several ways. First, high salt concentration decreases the osmotic potential of the soil solution, thus creating a water stress in the plant. Second, salts cause severe ion toxicity because Na^+ is not readily sequestered into vacuoles as it is in halophytes. In addition, the interactions of salts with mineral nutrition may result in nutrient imbalances and deficiencies. These factors interact and lead to cellular damage, cessation of growth, and ultimately, plant demise [3]. Processes such as seed germination, seedling growth and vigor, vegetative growth, flowering, and fruit set are adversely affected by high salt concentration, which can ultimately cause diminished agricultural quality and economic yield.

Development of salinity-tolerant crops is essential for sustaining agricultural production. To achieve salt tolerance, plants must either prevent or alleviate the damage associated with salt stress, and tolerate the high salt environment. However, barring a few exceptions, conventional breeding techniques have been unsuccessful in transferring salt tolerance to target species [4]. Slow progress in breeding for salt-tolerant crops can be attributed to the poor understanding of the molecular mechanisms associated with salt tolerance. Understanding the molecular basis of plant salt tolerance will also help in improving drought-, heat-, and cold-stress tolerance, inasmuch as osmotic and oxidative stresses are common to all of these abiotic stresses. A host of genes encoding different structural and regulatory proteins have recently been utilized over the past five to six years for the development of a range of abiotic stress-tolerant plants. An appreciation is growing for the usage of regulatory genes as an effective approach for developing stress-tolerant plants to increase productivity in saline areas. Thus, understanding the molecular basis of salt tolerance will be helpful in developing selection strategies for improving the growth and survival of plants in high salt environments. Identification of molecular markers linked to salinity and salinity tolerance traits has provided plant breeders a new tool for selecting cultivars with improved drought tolerance. The present review deals with changes in proteins/genes in crop plants in response to salinity stress.

6.2 PHYSIOLOGICAL EFFECTS OF SALINITY

6.2.1 PLANT DEVELOPMENT

Salinity affects almost all aspects of plant development such as germination, vegetative growth, and reproduction. Soil salinity imposes ion toxicity, osmotic stress, nutrient (N, Ca, K, P, Fe, Zn) deficiency, and oxidative stress on plants. Salinity also indirectly limits plant productivity through its adverse effects on the growth of beneficial and symbiotic microbes. High soil salt concentrations impose osmotic stress and thus limit water uptake from soil. Sodium accumulation in cell walls can rapidly lead to osmotic stress and cell death [2]. Ion toxicity is the result of replacement of K^+ by Na^+ in biochemical reactions, and Na^+ and Cl^- induced conformational changes in proteins. For several enzymes K^+ acts as co-factor, which cannot be substituted by Na^+. High K^+ concentration is also required for the binding of tRNA to the ribosomes and thus protein synthesis is affected [5,6]. Ion toxicity causes metabolic imbalances, which in turn leads to oxidative stress [7].

Based on the tolerance of plants to salinity, plants are classified as either halophytes, which can grow and reproduce under high salinity (>400 mM NaCl), or glycophytes, which cannot survive high salinity. Most of the grain crops and vegetables are glycophytes and are highly susceptible to soil salinity even when the soil EC_e is <4 dS m^{-1}. Crops such as bean (*Phaseolus vulgaris* L.), eggplant (*Solanum melongena* L.), corn (*Zea mays* L.), potato (*Solanum tuberosum* L.), and sugarcane (*Saccharum officinarum* L.) are highly susceptible to salinity with a threshold EC_e of <2 dS m^{-1}, whereas sugar beet (*Beta vulgaris* L.) and barley (*Hordeum vulgare* L.), although highly sensitive to salinity during germination, can tolerate soil EC_e up to 7 dS m^{-1} at later phases of crop development [8]. Soil type (particularly Ca^{2+} and clay content), rate of transpiration (which determines the amount of salt transported to the shoot for any given rate of salt uptake by the roots), and radiation may further alter salt tolerance of crops.

The adverse effects of salinity on plant development are more profound during reproductive growth. Figure 6.1 shows the adverse effects of salinity on vegetative

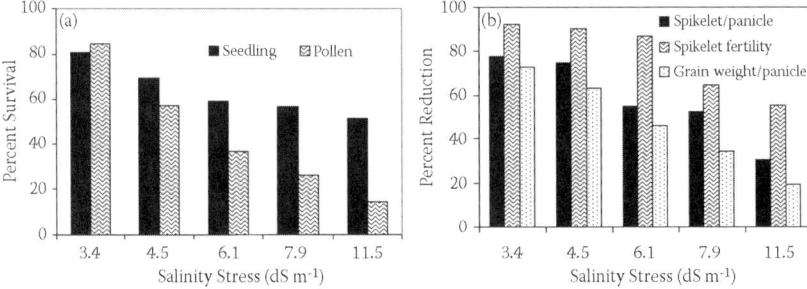

FIGURE 6.1 Salinity stress differentially affects yield and its components in rice. (a) Seedling stage is more tolerant to salinity than reproductive stage. The seedling survival rate was calculated as the percentage of live seedlings under different stress levels when compared to those under control. Pollen grains collected from plants grown under different stress levels were stained with 0.9% (w/v) thiazolyl blue to calculate the viable pollens and compared with those of control to calculate survival; (b) spikelet number and grain yield are more sensitive to salinity than spikelet fertility. (This graph was drawn using the data from Zeng and Shannon [9].)

and reproductive development and differential sensitivity of yield components to different intensities of salt stress in rice (*Oryza sativa* L.) [9]. Wheat (*Triticum aestivum* L.) plants stressed at 100 to 175 mM NaCl exhibited a significant reduction in spikelets per spike, delay in spike emergence, and reduction in fertility, leading to poor grain yield. However, the shoot apex of these wheat plants contained Na^+ and Cl^- concentrations below 50 and 30 mM, respectively, which are too low to limit metabolic reactions [10]. Hence, the adverse effects of salinity may have been attributed to salt stress effects on cell division and differentiation.

Salinity transiently arrests the cell cycle, resulting in fewer cells in the meristem and thus limits growth [11]. Salinity also arrests the cell cycle by reducing the expression and activity of cyclins and cyclin-dependent kinases. In *Arabidopsis*, reduction in root meristem size and root growth during salt stress is correlated with the downregulation of *CDC2a* (cyclin-dependent kinase), *CycA2,1* and *CycB1,1* (mitotic cyclins) expression [12]. The activity of cyclin-dependent kinase is also diminished by post-translational inhibition during salt stress [11]. Salt-stress-induced ABA may also affect cell cycle regulation. ABA upregulates the expression of an inhibitor of cyclin-dependent kinase (*ICK1*), which is a negative regulator of CDC2a [13]. Salinity adversely affects reproductive development due to inhibition of microsporogenesis, stamen filament elongation, enhancement of programmed cell death in some tissue types, ovule abortion, and senescence of fertilized embryos. In *Arabidopsis*, 200 mM NaCl stress caused up to 90% ovule abortion [14].

Plant responses to salt stress are controlled by multiple genes and hence salt tolerance is a complex phenomenon. Transcriptome analysis of salt-stressed (100 mM NaCl) *Arabidopsis* plants using a GeneChip microarray showed that about 424 genes were upregulated (>2-fold increase) in roots, and about 278 genes were upregulated in leaves [15]. In another study using *Arabidopsis*, cDNA microarray analysis also showed about 194 genes are upregulated and about 89 genes were downregulated by salt stress [16]. In rice, of the 1700 cDNAs analyzed, about 57 genes were upregulated by NaCl stress [17]. Many of the NaCl upregulated genes were also upregulated by dehydration, cold, and ABA [15–17], which suggested that some of the stress responses are common to all these abiotic stresses.

6.2.2 PHOTOSYNTHESIS

Analysis of the effect of salt stress on photosynthesis has revealed that salinity affects photosynthesis mainly through reduction in leaf area, chlorophyll content, and stomatal conductance, and to a smaller extent through a decrease in photosystem II efficiency [18].

6.2.2.1 Effects of Salinity on Photosynthetic Pigments and Proteins

Chlorophyll and total carotenoid content of leaves decreases under salt stress. The oldest leaves are often the first to develop chlorosis and to abscise under prolonged periods of salt stress [19–22]. However, Wang and Nil [23] reported that chlorophyll content increased under conditions of salinity in *Amaranthus*. In *Grevilea*, protochlorophyll, chlorophylls, and carotenoids were significantly reduced under NaCl stress, but the rate of decline of protochlorophyll and chlorophyll is greater than that of Chl a and carotenoids. In leaves of tomato (*Lycopersicon esculentum* Mill.), the contents of total chlorophyll (Chl a + b) and Chl a and β-carotene decreased under NaCl stress [24]. Under salinity stress, leaf pigments studied in nine genotypes of rice were reduced in general, but relatively high pigment levels were found in six genotypes [25]. In the cyanobacterium *Spirulina platens*, a decrease in the phycocyanin/chlorophyll and no significant change in the carotenoid/chlorophyll ratio were observed under salt stress [26]. Salinity caused a significant decrease in Chl a, Chl b, and carotenoid in leaves of *B. parviflora* [27].

6.2.2.2 Effects of Salinity on Proteins

Soluble protein contents of leaves decreased in response to salinity [21,23,25,27,28]. Agastian et al. [22] reported that soluble protein increases at low salinity and decreases at high salinity in mulberry (*Morus sp.*). SDS-PAGE analysis of proteins in peanut (*Arachis hypogaea* L) revealed that plants grown under NaCl exhibited induction (127 and 52 kDa) or repression (260 and 38 kDa) in the synthesis of some polypeptides [29]. Salinity induced six new proteins in roots of barley, which were of low molecular weight (24 to 27 kDa) with an isoelectric point of 6.1 to 7.6. In contrast to protein expression in roots, five new shoot proteins were induced with molecular weights and isoelectric points falling within the range of 20 to 24 kDa and 6.3 to 7.2, respectively. In addition, salinity also inhibited the synthesis of a majority of shoot proteins [30].

In radish (*Raphanus sativus*) salt stress caused accumulation of a 22 kDa protein and its mRNA in leaves [31]. A significant increase in the amount of protein, ranging from a molecular weight of 23–25 kDa, is observed under NaCl stress in cultured cells derived from Shamuti orange (*Citrus sinensis* L. Osbeck) ovular callus cells [32]. Yen et al. [33] reported accumulation of five polypeptides with estimated molecular masses of 40, 34, 32, 29, and 14 kDa by SDS and 2D-PAGE in callus of *Mesembryanthemum crystallinum* under NaCl stress, and these polypeptides were classified into two distinct groups according to their course of induction: early-responsive (40, 34, 29 kDa) and late-responsive (32, 14 kDa) proteins. The disappearance of a 6-kDa polypeptide in response to salinity has been reported in *Prosopsis* [34]. In wheat, content of a 26 kDa protein increased in NaCl-treated plants, and the stimulation was more pronounced in roots than in shoots, in contrast; the contents of 13 and 20 kDa proteins decreased and the 24 kDa protein disappeared with NaCl treatment [35]. NaCl induced accumulation of four polypeptides with molecular mass of 61, 51, 39, and 29 kDa in maize roots [36]. Fractionation of cells followed by SDS-PAGE and 2D-PAGE revealed an increase in the levels of membrane proteins of 39 and 50 kDa and a decrease in the level of a membrane protein of 52 kDa with increasing levels of external NaCl in *Rhodobacter sphaeroides*. After sequence analysis the polypeptide of 50 kDa was assigned to an inner membrane ATP synthase beta chain and a 52 kDa polypeptide in the outer membrane to a flagellar filament protein. Because the N-terminal of the 39 kDa protein in the outer membrane was blocked, partial proteolysis was carried out and four peptides were sequenced. Each sequence exhibited no significant homology with those available in databases, suggesting that the polypeptide of 39 kDa (named SspA) is a novel salt-stress-induced protein [37].

6.2.2.3 Mechanisms of Salinity Effects on Photosynthesis

Plant growth is the result of integrated and regulated physiological processes. Physiological processes are affected by a number of environmental factors, and they determine the response of plants to stress. Limitation of plant growth by environmental factors cannot be assigned to a single physiological process. The dominant physiological process is photosynthesis. Plant growth as biomass production is a measure of net photosynthesis and, therefore, environmental stress affecting growth also affects photosynthesis.

Photosynthesis involves membrane-bound electron transfer mechanisms, enzymes, and intermediate products and is regulated by number of external and internal factors. Photosynthetic efficiency depends on the sequence of metabolic events such as photochemical reactions, enzymes involved in carbon assimilation, the structure of the photosynthetic apparatus, and transport of photosynthetic intermediates between subcellular compartments. Photosynthetic rate is lower in salt-treated plants, but the photosynthetic potential is not greatly affected when rates are expressed in terms of chlorophyll or leaf area basis. Decrease in photosynthetic rate is due to several reasons: (1) dehydration of cell membranes that reduces their permeability to CO_2; (2) salt toxicity; (3) reduction of CO_2 supply because of hydroactive closure of stomata; (4) enhanced senescence induced by salinity; (5) changes of enzyme activity induced by changes in cytoplasmic structure; and (6) negative feedback by reduced sink activity.

Salt stress causes either short-term or long-term effects on photosynthesis. Short-term effects are mainly within hours or days of the onset of exposure. Long-term effects occur after several days of exposure to salt and result in reduced carbon assimilation due to salt accumulation in developing leaves. Although there are reports of suppression of photosynthesis by salt stress [38–40], there are also reports that photosynthesis was not affected by salinity and was even stimulated by low salt concentration in some cases [41,42]. In *Alhagi pseudoalhagi*, a leguminous plant, leaf CO_2 assimilation rate (Pn) increased at low salinity (50 mM NaCl), but it was not affected significantly at 100 mM NaCl, whereas it was reduced to about 60% of that of controls in 200 mM NaCl. Stomatal conductance (g_s) was consistent with the CO_2 assimilation rate regardless of the treatments, and intercellular CO_2 concentration (Ci) was lower in the NaCl-treated plants than in the controls [42].

In mulberry, transpiration rate (E) declined under salt stress, whereas Ci increased [21]. NaCl stress resulted in decreased chlorophyll content and net Pn rate and increased the rate of respiration and CO_2 compensation point concentration, with no significant change in carotenoid content in leaves of alfalfa plants (*Medicago sativa*) [43]. In *Atriplex lentiform*, net Pn decreased with salinity and the ratio of ribulose bisphosphate carboxylase oxygenase (Rubisco) activity to that of phosphoenol pyruvate carboxylase (PEPC) also decreased. Under salinity, PEPC activity on a leaf area basis increased linearly with salinity, whereas Rubisco activity remained relatively constant [44]. In the cyanobacterium *Spirulina platens*, salt stress reduced the apparent quantum efficiency of photosynthesis and PSII activity and stimulated PSI activity and dark respiration significantly.

Salt stress also results in a decrease in overall activity of the electron transport chain, which could not be restored by diphenyl carbazide, an artificial electron donor to PSII [26]. A significant reduction in g_s by salt treatment was also observed in horse gram *(Macrotyloma uniflorum)*, and salinity also inhibited the photosynthetic electron transport and the activity of Calvin cycle enzymes ribulose-I, 5-biphosphate carboxylase, ribulose-5-phosphate kinase, ribulose-5-phosphate isomerase, and NADP glyceraldehyde-3 phosphate dehydrogenase [45]. In four cultivars of rice, gradual decreases in the activity of PS I and PS II, as well as chlorophyll fluorescence transients and emission at 688 nm, were observed; an increase in NaCl concentration and a drastic decrease in net photosynthetic rate were found [46]. Mishra et al. [47] have reported that in wheat salt stress had no direct effect on electron transport activity and the ratio of variable fluorescence to maximum fluorescence (Fv/Fm ratio), suggesting that the efficiency of the photochemistry of PSII was not affected and the decrease in Fm due to salt stress may have influenced reduction of plastoquinone-A (Q_A). These results on fluorescence indicate that salt stress predisposes plants to photoinhibition and also reduces the ability to recover from photoinhibition.

Light stress and salt stress are major environmental factors that limit the efficiency of photosynthesis. However, Allakhverdiev et al. [48] have reported in the cyanobacterium *Synecchocystis sp.*, strain PCC 6803, that light and salt stress affected photosystem II (PS II) in different manners. Strong light induces photodamage to PS II, whereas salt stress inhibits the repair of the photodamaged PS II and does not accelerate damage to PS II directly. The combination of light and salt stress appears to inactivate PS II very rapidly as a consequence of their synergistic effects. Radioactive labeling of cells revealed that salt stress inhibited the synthesis of proteins *de novo* and, in particular, the synthesis of the D1 protein of PS II. Northern- and western-blot analyses demonstrated that salt stress inhibited the transcription and the translation of *psbA* genes, which encode D_1 protein [48]. DNA microarray analysis indicated that the light-induced expression of various genes is suppressed by salt stress. It has been suggested that salt stress inhibits the repair of PS II via suppression of the activities of the transcriptional and translational machinery [48].

Pn activity decreases as the water potential of leaves decreases. The reduction in Pn activity depends on two aspects of salinization, that is, the total concentration of salts and their ionic composition. High salt concentrations in soil and water create low osmotic potentials, which reduce the availability of water to plants. Decrease in water potential causes osmotic stress, which reversibly inactivates photosynthetic electron transport via shrinkage of intercellular spaces, due to efflux of water through water channels in the plasma membrane [49]. Decrease in osmotic potential under high salt conditions causes Na^+ ions to leak into the cytosol [50] and inactivates both photosynthetic and respiratory electron transport [51]. High salt (NaCl) uptake competes with the uptake of other nutrient ions, especially K^+, leading to K^+ deficiency. Under conditions of high salinity and K^+ deficiency, there is a reduction in the quantum yield of oxygen evolution due to malfunctioning of photosystem II.

Reduction in photosynthetic rates is associated with restricted availability of CO_2 for carboxylation reactions due to reduced g_s [52]. The extent to which stomatal closure affects photosynthetic capacity depends on the magnitude of the partial pressure of CO_2 inside the leaf. There are also reports of nonstomatal inhibition of photosynthesis under salt stress. Nonstomatal inhibition is due to increased resistance to CO_2 diffusion in the liquid phase from the mesophyll wall to the site of CO_2 reduction in the chloroplast and also reduced Pn efficiency of ribulose 1,5-bisphosphate carboxylase (RuBPCase).

Additionally, reduction in capacity may be the consequence of inhibition of certain carbon metabolism processes by feedback from other salt-induced reactions [53]. Under reduced water potential, stromal levels of the substrate fructose 1,6-bisphosphate accumulate and fructose-1,6-bisphosphatase (FBPase) product fructose-6-phosphate is reduced so that FBPase becomes rate limiting to Pn [54].

6.3 SALT RESPONSE: A MULTIGENIC TRAIT

Salt and dehydration stresses show a high degree of similarity with respect to physiological, biochemical, molecular, and genetic effects [55]. This is potentially due to the osmotic effect in salt-affected plants, which are apparently similar to the

Salinity Tolerance: Cellular Mechanisms and Gene Regulation

effects of water deficit and, to an extent, of cold and heat stresses [56]. The halophyte *Mesembryanthemum crystallinum*, the common ice plant, has emerged as a model system for understanding the molecular response to salt stress. This plant switches from C_3 photosynthesis to crassulacean acid metabolism (CAM) in response to salt and drought stress. Organic acids oxalate and malate are important osmolytes in CAM plants. A cDNA clone encoding NADP-malic enzyme has been isolated from *M. crystallinum* [57,58]. A large number of genes are concomitantly up and down regulated for this switch to be operational [59]. More than 100 genes are induced and up to three times as many transcripts repressed in response to salt stress in *M. crystallinum* [60]. Employing two-dimensional protein gel electrophoresis, it has been noted that application of salt to plants brought about major changes in the protein profile [61]. Therefore, not only it is imperative to ask how many and which genes, but also in what hierarchical order they are expressed. Salinity is a quantitative trait, and arrays of salt-induced genes have been isolated [62].

6.3.1 Perception of Salt Stress

The ability of plants to resist environmental stresses is determined by the efficiency of the plant to sense the environmental stress and activate its defense response. Plants perceive salt stress as ionic and osmotic stress. Excess Na^+- and Cl^--induced conformational changes in protein structure and membrane depolarization can lead to the perception of ion toxicity. Plasma membrane proteins, ion transporters, or Na^+-sensitive enzymes have been hypothesized as sensors of toxic Na^+ concentrations in extracellular and intracellular sites. Regulation of ion (Na and K) homeostasis involving salt overly sensitive (SOS) genes has recently been suggested by the SOS pathway. The input of the SOS pathway is due to excessive intracellular or extracellular Na^+, which triggers a cytoplasmic Ca^{2+} signal [63]. The outputs are expression and activity changes of transporters for ions such as Na^+, K^+, and H^+. The input for osmotic stress signaling is likely to be a change in turgor. A myristoylated calcium binding protein (myristic acid residue attached to the protein) encoded by SOS3 presumably senses the salt-elicited calcium signal and translates it to downstream responses [64]. SOS3 interacts with and activates SOS2, a serine/threonine protein kinase [65,66]. SOS2 and SOS3 regulate the expression level of SOS1, a salt-tolerance affector gene encoding a plasma membrane Na^+/H^+ antiporter [67]. SOS1 by itself can slightly increase the salt tolerance of a mutant yeast strain lacking all endogenous Na^+-ATPase and Na^+/H^+ antiporter [68]. Expression of a constitutively activated SOS2 mutant also increased the salt tolerance of SOS1 in the yeast mutant, implying that SOS2 kinase activity is sufficient for SOS1 activation.

In a complementary study, Qiu et al. [69] showed that constitutively active SOS2 kinase could enhance a Na^+/H^+ exchange activity in purified plasma membrane vesicles from the wild-type but not in SOS1-1 mutants [70]. In SOS2-2 and SOS3-1 mutants [71], the plasma membrane Na^+/H^+ exchange activity was much lower, but could be recovered to near wild-type levels by addition of activated SOS2 in vitro to the membrane vesicle preparations [69]. The plasma membrane Na$^+$/H$^+$ antiporter SOS1 has a very long tail that protrudes on the cytoplasmic side [67], which has been proposed to function as a sensor for different solutes. The possibility of SOS1 being both a transporter and a sensor cannot be dismissed. In addition to regulation by SOS2, SOS1 activity may also be regulated by SOS4. SOS4 catalyzes the formation of pyridoxal-5-phosphate, a co-factor that may serve as a legend ligand for SOS1, because the latter contains a putative binding sequence for this co-factor [72].

Apse et al. [73] have engineered a single endogenous gene (AtNHX1) encoding a Na$^+$/H$^+$ antiporter protein. In *Arabidopsis*, a Na$^+$/H$^+$ antiporter gene has been identified and characterized based on its similarity to bacterial, fungal, and mammalian homologues [73,74]. *Arabidopsis* plants engineered to express greater levels of AtNHX1 exhibited increased vacuolar Na$^+$ compared to the wild-type. Transgenic plants were significantly more salt tolerant and were able to thrive in soil irrigated with 200 mM NaCl.

6.3.2 Second Messengers

The ameliorative effect of Ca^{2+} in maintaining plant growth under salinity [75] and Ca^{2+}-induced ion channel discrimination against Na$^+$ [76] are well known. It has been established that in addition to its effect on preventing Na$^+$ entry into cells, Ca^{2+} acts as a signaling molecule in salt stress [77,78]. Cytosolic Ca^{2+} oscillations during salt stress are regulated through activities of mechanosensitive and ligand-gated Ca^{2+} channels on the plasma membrane, endoplasmic reticulum, and vacuole [5,79]. Excess Na$^+$-induced membrane depolarization may activate stretch and mechanosensitive Ca^{2+} channels to generate a Ca^{2+} signature under salt stress [6,78].

Pharmacological studies and genetic analysis have shown the involvement of inositol 1,4,5-tris phosphate (IP$_3$)-gated Ca^{2+} channels in the regulation of Ca^{2+} signatures during salt stress [80–82]. The *FRY1* locus of *Arabidopsis* encodes an inositol polyphosphate 1-phosphatase, which catabolizes IP$_3$. The *Arabidopsis fry1* mutant is impaired in inositol polyphosphate 1-phosphatase and thus lacks abscisic acid (ABA)–induced IP$_3$ transients. The *fry1* mutation leads to sustained accumulation of IP$_3$ and hypersensitivity to ABA, cold, and salt stresses. Thus IP$_3$ plays a crucial role in cytosolic Ca^{2+} oscillations during ABA, salt, and cold stress signaling [82]. Salinity stress also leads to synthesis of ABA [83–85] and accumulation of reactive oxygen species (ROS) [7]. Calcium and hydrogen peroxide (H$_2$O$_2$) each act as second messengers for ABA-induced stomatal closure and gene expression under abiotic stresses [86,87]. Transient expression analysis revealed that IP$_3$ and cyclic ADP ribose (cADPR)-gated calcium channels are involved in ABA-induced cytosolic Ca^{2+} oscillations [88]. ABA induces the expression and activity of ADP-ribosyl (ADPR) cyclase that synthesizes cADPR [89].

Involvement of a heterotrimeric GTP-binding G-protein has been demonstrated in ABA signal transduction during guard cell regulation [90]. Because ABA synthesis is induced under salinity, the G-protein-associated receptors may also elicit Ca^{2+} signatures during salinity stress. Salt-stress-induced Ca^{2+} signatures are then sensed and transduced by calcium sensor proteins, namely, SOS3,

SOS3-like calcium binding proteins (SCaBPs), and calcium-dependent protein kinases (CDPKs), as well as calmodulins (CaMs).

6.3.3 Ion Homeostasis

Plants achieve ion homeostasis by restricting the uptake of toxic ions, maintaining the uptake of essential ions, and compartmentation of toxic ions into the vacuole of specific tissue types. In most crop plants, Na^+ is the primary cause of ion toxicity, and hence management of cellular Na^+ concentration is critical for salt tolerance [6]. Sodium ions can be kept below toxic levels in the cytosol by (1) restriction of Na^+ entry at the root cortex cells, (2) excretion of Na^+ from root cells into the soil solution, (3) retrieval of Na^+ from the transpirational xylem stream to recirculate it to the roots, (4) storage of Na^+ in the vacuole of matured cells, and (5) Na^+ excretion from the leaves [5]. Among these mechanisms Na^+ excretion through salt glands is significant only in halophytes. Biochemical, electrophysiological, and molecular genetic evidence show that the SOS pathway plays a crucial role in the regulation of cellular and whole-plant ion homeostasis (Figure 6.2) [79].

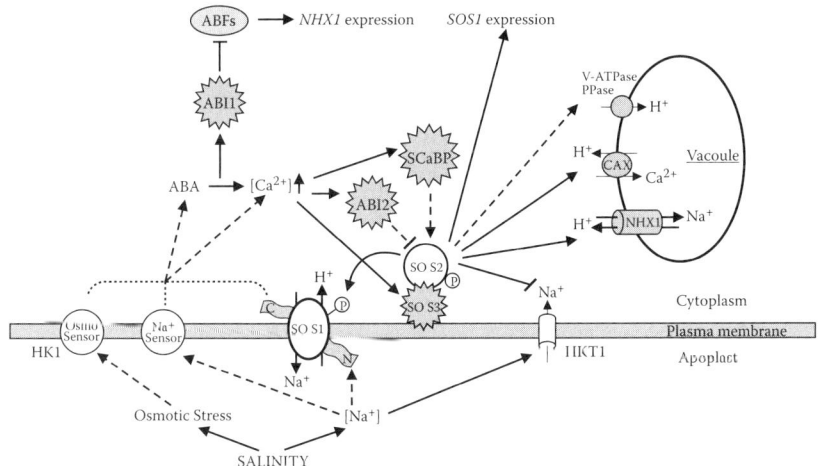

FIGURE 6.2 Regulation of ion homeostasis by SOS signaling a pathway during salt stress. Salt stress is probably sensed by putative salt-stress sensors such as salt overly sensitive 1 (SOS1) or two-component histidine kinase (HK1). These sensors induce elevation of cystolic Ca^{2+} concentration. A calcium sensor protein, SOS3, perceives this Ca^{2+} signal and activates SOS2 protein kinase. Activated SOS2 phosphorylates SOS1, a plasma membrane Na^+/H antiporter, which then transports Na^+ out of the cytosol. The SOS3-dependent SOS2 kinase pathway positively regulates SOS1 transcript level and negatively regulates the Na^+ transporter HKT1. The SOS2 kinase, probably through SOS3-like Ca^{2+} binding proteins (SCaBPs)-dependent pathway, activates the tonoplast Na^+/H antiporter (NHX1) and vacuolar H^+/Ca^{2+} antiporter (VCX1). Expression of NHX1 is regulated through ABFs (ABA-responsive element binding factors), which are under the negative control of ABI1 protein phosphatase 2C, whereas ABI2 negatively regulates ion homeostasis either by inhibiting SOS2 kinase activity or the activities of SOS2 targets.

6.3.4 Sodium Uptake

Restriction of Na^+ entry into root cells and the transpirational stream is critical for prevention of toxic levels of salt in the shoot. Both glycophytes and halophytes must exclude about 97% of the Na^+ present in the soil solution at the root surface in order to prevent toxic levels of Na^+ accumulation in the shoots [91]. Sodium entry into the transpirational stream depends upon the amount of Na^+ uptake by Na^+-specific and nonspecific cation transporters, and the percentage of water entering the apoplastic pathway in the xylem tissue. Na^+ from the soil solution initially enters into the cells through the root epidermis and cortex. The casparian strip in the endodermis plays a crucial role in preventing apoplastic Na^+ influx into the root stele. Halophytes such as salt cress (*Thellungiella halophila*) developed both an extra endodermis and cortex cell layer in roots compared to *Arabidopsis* [92]. Maize seedlings stressed at 200 mM NaCl showed an increase in the radial width of the casparian strip by 47% compared to that of control seedlings [93]. This may help to reduce Na^+ entry into the transpirational stream. In crops like rice, water entry into the xylem through the apoplastic pathway accounts for all the Na^+ buildup in the shoots, whereas in crops like wheat, transport protein-mediated Na^+ uptake accounts for most of the Na^+ buildup in shoots [94]. Silica deposition and polymerization of silicate in the endodermis and rhizodermis block Na^+ influx through the apoplastic pathway in the roots of rice [95]. Regulation of these anatomical and morphological changes in root development during salt stress needs further investigation.

Both voltage-dependent and voltage-independent cation channels mediate Na^+ uptake; however, the role of voltage-independent cation channels in Na^+ uptake is poorly understood. Voltage-dependent cation channels, such as K^+ inward rectifiers (HKT, HAK, and KUP), mediate Na^+ uptake into root cells. Sodium competes with K^+ uptake through Na^+-K^+ co-transporters and may also block the K^+-specific transporters of root cells [79]. Expression studies in yeast cells revealed that high-affinity K^+-uptake activity of both *Arabidopsis* AtKUP1 and barley HvHAK1 was inhibited by millimolar concentration of Na^+ [96,97]. Cellular K^+ concentration can be maintained by activity/expression of inward rectifying K^+-specific transporters under salinity. Salt stress, as well as K^+ starvation, upregulate the expression of *Mesembryanthemum crystallinum* high-affinity K^+ transporter genes (*McHAK*). McHAKs specifically mediate K^+ uptake and show high discrimination for Na^+ at high salinity [98]. In contrast, high affinity K^+ transporters (HKT) of wheat [99,100], *Arabidopsis* [101], and *Eucalyptus* [102] also act as low affinity Na^+ transporters when expressed in *Xenopus* oocytes. HKT transporters of *Eucalyptus camaldulensis* and wheat possess Na^+:K^+ symport activity but mediate mainly Na^+ transport under salinity [99,102]. The expression of *OsHKT1* was significantly downregulated in the salt-tolerant rice (cv. Pokkali) when compared to that of the salt-sensitive rice (cv. IR29) during 150 mM NaCl stress [103]. Transgenic wheat plants expressing wheat *HKT1* in antisense showed significantly less $^{22}Na^+$ uptake and enhanced growth under salinity stress when compared with the control plants [104]. These results suggest that HKT1 homologues contribute to Na^+ influx during salt stress, and downregulation of *HKT1* expression may help in limiting the Na^+ influx into roots.

In yeast, *HAL1* and *HAL3* regulate the expression of P-type ATPase, Na^+ efflux, and K^+ uptake. Transgenic overexpression of the yeast *HAL1* gene enhanced salt tolerance of melon *(Cucumis melo* L.) shoots in vitro [105], tomato [106,107], and watermelons (*Citrullus lanatus*) [108]. Transgenic tomato plants overexpressing yeast *HAL1* showed increased K^+ accumulation. Irrigation with 35 mM NaCl until maturity decreased the fruit yield by 57.5% in control plants, whereas transgenic tomato plants showed 24 to 42% decreases in fruit yield. However, under nonsaline growing conditions the transgenic lines were less productive than the wild-type [107]. Overexpression of the *Arabidopsis HAL3a* gene also enhanced the salt tolerance of transgenic *Arabidopsis* [109].

Electrophysiological evidence suggests that cyclic nucleotides (cAMP and cGMP) may minimize Na^+ influx into the cell by downregulating voltage-independent cation channels in *Arabidopsis* [110]. Exposure of *Arabidopsis* plants to salt and osmotic stress resulted in an increase in the cytosolic cGMP concentration within five seconds [111]. Pyrridoxal-5-phosphate is a co-factor involved in the biosynthesis of amino acids that are precursors for nucleotide biosynthesis. The *Arabidopsis sos4* mutant defective in a pyridoxal kinase gene showed hypersensitive root growth in NaCl and KCl stresses and accumulated more Na^+ but less K^+. Pyridoxal-5-phosphate and its derivatives act as ligands for P2X receptor ion channels in animals [72]. Pyridoxal-5-phosphate may regulate Na^+ efflux by SOS1, because SOS1 contains a putative pyridoxal-5-phosphate binding domain [5]. Thus, regulation of K^+ and Na^+ uptake by pyridoxal-5-phosphate and cyclic nucleotides may contribute to plant salt tolerance. Signaling pathways that regulate Na^+ and K^+ uptake by higher plants during salinity need further study.

6.3.5 SODIUM EFFLUX

Sodium efflux from root cells is a frontline defense that prevents accumulation of toxic levels of Na^+ in the cytosol and Na^+ transport to the shoot. Plasma membrane Na^+/H^+ antiporters pump out Na^+ from root cells. In *Arabidopsis*, the plasma membrane Na^+/H^+ antiporter, SOS1, mediates Na^+ efflux, and its activity during salt stress is regulated by a SOS3-SOS2 kinase complex [79] (Figure 6.2). Salt-stress-induced Ca^{2+} signatures are sensed by SOS3. SOS3 has three calcium binding EF hands, an N-myristoylation motif, and shows sequence similarity to the calcineurin B subunit of yeast and neuronal Ca^{2+} sensors of animals [64,112]. Calcineurin is a protein phosphatase (PP2B) that regulates salt tolerance in yeast. SOS3 and SOS3-like calcium-binding proteins (SCaBPs) identified in *Arabidopsis* differ from yeast calcineurin structurally and functionally. SCaBPs do not have a calcineurin A subunit catalytic domain. Unlike calcineurin, which activates protein phosphatases, SOS3 activates Ser/Thr protein kinase during salt stress. Thus SOS3 and SCaBPs are a new class of Ca^{2+} sensor proteins in higher plants.

Ishitani et al. [64] reported that mutations which disrupt either Ca^{2+} binding (*sos3-1*) or myristoylation (G2A) of SOS3 cause salt stress hypersensitivity in *Arabidopsis*. SOS3 binds Ca^{2+} with low affinity when compared to other Ca^{2+}-binding proteins such as caltractin and calmodulin [64]. The differences in the affinity of these Ca^{2+} sensors may be employed by cells to distinguish various Ca^{2+} signals. The SOS3 transduces the salt-stress signal by activating SOS2, a Ser/Thr protein kinase with an N-terminal kinase catalytic domain that is similar to that of yeast sucrose nonfermenting 1 (SNF1) and animal AMP-activated kinase (AMPK), and a unique C-terminal regulatory domain. The C-terminal regulatory domain of SOS2 consists of an auto-inhibitory FISL motif [66]. Under normal cellular conditions, the catalytic and regulatory domains of SOS2 interact with each other, likely preventing substrate phosphorylation by blocking substrate access. Yeast two-hybrid and in vitro binding assays have shown that in the presence of Ca^{2+} SOS3 binds to and activates the SOS2 kinase [65]. The FISL-motif in the regulatory domain of SOS2 is necessary and sufficient for interaction with SOS3 and deletion of this FISL-motif constitutively activates the SOS2. Replacement of Thr^{168} in the kinase domain by Asp also results in a constitutively active SOS2 kinase [113].

Molecular genetic analyses led to the identification of targets of the SOS3–SOS2 regulatory pathway. One of the targets of the SOS pathway is SOS1. The SOS1 has significant protein sequence homology and conserved domains similar to that of the plasma membrane Na^+/H^+ antiporter from bacteria, fungi, and animals. Expression of *SOS1* is ubiquitous but stronger in epidermal cells surrounding the root tip and in parenchyma cells bordering the xylem.

Expression of SOS1-GFP fusion protein and anti-SOS1 antibody has confirmed that SOS1 is localized in the plasma membrane of root and leaf cells [67,72,114]. The *sos1* mutant plants showed hypersensitivity to salt stress (100 mM NaCl) and accumulated more Na^+ in shoots than the wild-type plants [67]. Isolated plasma membrane vesicles from *sos1* mutants showed significantly less inherent as well as salt-stress-induced Na^+/H^+ antiporter activity than did vesicles from the wild-type [115]. These results showed that SOS1 functioned as a Na^+/H^+ antiporter on the plasma membrane and played a crucial role in Na+ efflux from the root cells. Indeed, transgenic *Arabidopsis* plants overexpressing *SOS1* exhibited less Na^+ in the xylem transpirational stream and in the shoot compared to the wild-type plants, and showed enhanced salt tolerance. Transgenic plants grew, bolted, and flowered at increasing concentrations of salt stress (50 mM to 200 mM NaCl), whereas control plants became necrotic and did not bolt [116]. The expression level of *SOS1* was also significantly higher in salt cress than that of *Arabidopsis* even in the absence of salt stress [117].

The Na^+/H^+ exchange activity of SOS1 is regulated by SOS3–SOS2 complex under salt stress. Isolated plasma membrane vesicles from *sos3* and *sos2* mutants showed significantly less Na^+/H^+ exchange activity than that of wild-type plants. Consistent with this finding, the mutants also accumulate higher levels of Na^+, similar to those accumulated by the *sos1* mutant. However, addition of activated SOS2 was sufficient to rescue the Na^+/H^+ exchange activity of plasma membrane vesicles from *sos3* and *sos2* mutants [115,118]. The SOS3–SOS2 kinase complex phosphorylates SOS1 proteins and activates SOS1-Na^+/H^+ antiporter activity [118]. *SOS1* upregulation during salt stress is also under the regulatory control of the SOS pathway, as evident from impaired expression of *SOS1* in salt-stressed *sos2* and *sos3* mutants [67]. Overexpression of the active form (Thr168 to Asp mutation) of *SOS2*, under the control of the CaMV 35S promoter (*35S::T/D SOS2*), rescued the *sos2* and *sos3* mutants under salinity. Transgenic *Arabidopsis* expressing *35S::T/DSOS2* showed enhanced SOS1 transporter activity as well as better vegetative and reproductive growth than that of wild-type plants, when grown in soils irrigated with 200 mM NaCl [119]. Co-expression of *SOS1*, *SOS2*, and *SOS3* rescued the yeast cells deficient in Na^+ exchangers. Co-expression of *SOS2* and *SOS3* significantly increased SOS1-dependent Na^+ tolerance of the yeast mutant [118]. This evidence demonstrated that SOS3 senses the salt-stress-induced Ca^{2+} signals and activates SOS2 kinase, which in turn regulates the Na^+/H^+ exchange activity and expression of *SOS1* [79] (Figure 6.2).

6.3.6 Sodium Compartmentation

To maintain water uptake during osmotic stress, plants have evolved a mechanism known as osmotic adjustment. Osmotic adjustment is the active accumulation of solutes such as inorganic ions (Na^+ and K^+) and organic solutes (proline, betaine, polyols, and soluble sugars). Vacuolar sequestration of Na^+ is an important and cost-effective strategy for osmotic adjustment, and at the same time, it also reduces the cytosolic Na^+ concentration during salinity. Vacuolar Na^+/H^+ antiporters use the proton gradient generated by vacuolar H^+-adenosine triphosphatase (H^+-ATPase) and H^+-inorganic pyrophosphatase (H^+-PPase) for Na^+ sequestration into the vacuole. Hence coordinated regulation of Na^+/H^+ antiporters, H^+-ATPase, and H^+-PPase are crucial for salt tolerance. Salt stress induces tonoplast H^+-ATPase and H^+-PPase activities [120]. Transgenic *Arabidopsis* plants overexpressing the *AVP1* (H^+-PPase) showed enhanced sequestration of Na^+ into the vacuole, maintained higher relative leaf water content and enhanced salt and drought stress tolerance [121]. Nitric oxide (NO) mediated signaling is implicated in the activation of plasma membrane H^+-ATPase [122] but the regulators of tonoplast H^+-ATPases and H^+-PPase are yet to be identified.

Vacuolar Na^+ sequestration is further regulated at the level of expression and activity of tonoplast Na^+/H^+ antiporters (NHX). Expression of *NHX1* is induced by salinity and ABA in *Arabidopsis* [74,123], rice [124], and cotton [125]. The expression level of *NHX1* is correlated with genotypic differences in salt tolerance in cotton (*Gossipium hirsutum*) [125]. Complementation studies showed that *AtNHX1* [74] and *OsNHX1* [124] could complement the yeast *nhx1* mutant. Transgenic *Arabidopsis* plants overexpressing *AtNHX1* showed significantly higher salt (200 mM NaCl) tolerance than wild-type plants [73]. Transgenic tomato plants overexpressing *AtNHX1* were able to grow and produce fruits in the presence of very high salt concentrations (200 mM NaCl) at which wild-type plants did not survive. The yield and fruit quality of transgenic tomato plants under salinity were equivalent to that of control plants under nonstress conditions [126]. Similar results were obtained with transgenic canola (*Brassica napus*) overexpressing *AtNHX1* [127]. These tomato and canola plants accumulated high concentrations of Na^+ in older leaves but not in reproductive parts [126,127]. Similar results were found for transgenic rice plants overexpressing *Atriplex gmelini NHX1* [128], transgenic rice overexpressing *OsNHX1* [124], and transgenic tobacco overexpressing cotton *NHX1* [125]. These transgenic plants exhibited better salt tolerance than that of control plants in the vegetative stage. However, transgenic rice plants overexpressing *OsNHX1* did not show a K^+/Na^+ ratio significantly different from that of control plants [124]. Analysis of the salt tolerance of the *osnhx1* null mutant of rice [124] may shed further light on the role of NHX1 in salt tolerance.

The SOS pathway and ABA regulate *AtNHX1* gene expression and its antiporter activity under salt stress. The promoter of *AtNHX1* contains putative ABA-responsive elements (ABRE) between base pair -736 to -728 from the initiation codon. *AtNHX1* expression under salt stress is partially dependent on ABA biosynthesis and ABA signaling through ABI1. This is because the salt-stress-induced upregulation was reduced in ABA-deficient mutants (*aba2-1* and *aba3-1*) and the ABA-insensitive mutant, *abi1-1* [123]. In cotton, *GhHNX1* expression was induced by ABA and appeared to be regulated by MYB/MYC-type transcription factors [125]. Analysis of the tonoplast Na^+/H^+-exchange activity in wild-type and *sos* mutants (*sos1*, *sos2*, and *sos3*) revealed that SOS2 regulates tonoplast Na^+/H^+-exchange activity. The impaired tonoplast Na^+/H^+-exchange activity from the isolated *sos2* tonoplast could be restored to the wild-type level by the addition of activated SOS2 protein. Because the Na^+/H^+-exchange activity was unaffected in the *sos3* mutant, regulation of tonoplast Na^+/H^+-exchange activity by SOS2 was SOS3-independent [129]. SOS2 has been found to interact with plant calcium sensor proteins, such as SOS3, SCaBP1 (SOS3-like calcium-binding proteins 1), SCaBP3, SCaBP5, and SCaBP6 [113]. One of these SCaBPs might signal SOS2 to regulate tonoplast Na^+/H^+-exchange activity (Figure 6.2) [129]. SOS2 also has an additional SOS3-independent role in regulating vacuolar H^+/Ca^{2+} antiporter VCX1, which plays a crucial role in regulating the duration and amplitude of cytosolic Ca^{2+} oscillations [130].

6.3.7 Sodium Transport from Shoot to Root

Many glycophytes have limited ability to sequester Na^+ in leaf vacuoles, but instead recirculate excess Na^+ from leaves to roots. Sodium transport from shoots to roots is probably mediated by SOS1 and HKT1 in *Arabidopsis*. Under salt stress (100 mM NaCl), Na^+ accumulation in shoots of *sos1* mutant plants was more than that of the wild-type plants. Strong expression of *SOS1* in cells bordering the xylem suggested that SOS1 mediated either Na^+ release into the xylem or retrieval of Na^+ from the xylem stream; which of these two phenomena occurs depends on salt-stress intensity and thus is critical for controlling long-distance Na^+ transport from root to shoot [72].

Comparison of the expression pattern of *HKT1* in wheat and *Arabidopsis* revealed that AtHKT and wheat HKT1 might have different functions. AtHKT1 was mainly expressed in phloem tissue but not in root peripheral cells, whereas that of wheat HKT1 was localized to root epidermis and leaf vascular tissue. The sodium overaccumulation in the *sas2-1* mutant of *Arabidopsis* showed significantly higher shoot Na^+ content, but lower root Na^+ content and lower Na^+ concentration in the phloem sap exuded from leaves. The *sas2-1* mutation impaired *AtHKT1* and thus its Na^+ transport activity in *Xenopus* oocytes [131]. T-DNA mutation in the *AtHKT1* gene also resulted in higher shoot Na^+ content and lower root Na^+ content [132]. Moreover, AtHKT1 did not show significant K^+ transport activity in *Xenopus* oocytes. A single-point mutation, Ser-68 to glycine, was sufficient to restore K^+ permeability to AtHKT1 [132]. These results show that AtHKT1 probably mediated Na^+ loading into the phloem sap in shoots and unloading in roots, and thus helped to maintain a low Na^+ concentration in the shoot [131]. The *Arabidopsis athkt1* mutation suppressed the salt hypersensitivity and K^+-deficient phenotype of *sos3* [133]. Hence, the SOS pathway may regulate and coordinate the activities of AtHKT1 and SOS1 to control Na^+ transport from shoots to roots.

Salt-stress-induced ABA accumulation, in addition to cytosolic Ca^{2+}, may also regulate the SOS pathway through the ABI2 protein phosphatase 2C. ABI2 interacts with the protein phosphatase interaction (PPI) motif of SOS2. This interaction is abolished by the *abi2-1* mutation, which enhances tolerance of seedlings to salt shock (150 mM NaCl) and causes ABA insensitivity. Hence, the wild-type ABI2 may negatively regulate salt tolerance by inactivating SOS2 or the SOS2 regulated ion channels such as HKT1, Na^+/H^+ antiporters, SOS1, and NHX1 (Figure 6.2) [134].

Transgenic manipulations of ion homeostasis have demonstrated the possibilities of genetic engineering of salt-tolerant crop plants. Although multiple genes govern salt-stress tolerance, significant increases in salt tolerance have been achieved by single gene manipulations, as revealed from the *SOS1* [116] and *NHX1* [73,126,127] overexpressing mutants. These plants were able to grow and produce flowers at a salt concentration of 200 mM NaCl, which was lethal to wild-type plants. In addition, the transgenic plants did not have any obvious growth abnormalities or change in the quality of the fruit [126,127]. Hence, genetic engineering for ion homeostasis by tissue-specific overexpression of *SOS1*, *NHX1*, and the active form of SOS2, may provide improvement in the salt tolerance of crop plants.

6.4 SALINITY DAMAGE AND REPAIR

Nature has provided various mechanisms for controlling salinity-induced damage, including polyamines, osmotic adjustment, osmoprotectant accumulation, oxidative stress management, induction of stress proteins, and modifications in root and shoot growth and transpiration.

6.4.1 POLYAMINES

Polyamines are low molecular weight polycationic molecules that are thought to play important roles in a number of physiological and developmental processes [135]. Diamine putrescine, spermidine, and spermine have been detected in most plant species, whereas diamine cadavarine only in *Leguminosae*. A number of stress factors such as potassium deficiency, osmotic stress, low pH, nutrient deficiency, or light have been shown to stimulate the accumulation of polyamines. Plant production of putrescence is catalyzed by arginine decarboxylase (ADC) and additional reactions convert putrescine into spermidine and spermine. These steps are catalyzed by spermidine and spermine synthases, which add polyamine groups generated from S-adenosyl methionine by S-adenosyl methionine decarboxylase (SAMDC).

In plants, polyamines accumulate under several abiotic stress stimuli, including salt and drought. It has been suggested that this increase in polyamine concentration could be considered as an indicator of plant stress. The first demonstration of involvement of polyamines in the stress response in plants was documented by accumulation of putrescine in response to suboptimal K^+ levels [136]. Since then, the link between increased putrescine levels and several abiotic stresses has been established. For example, Krishnamurthy and Bhagwat [137] reported accumulation of spermidine and spermine in salt-tolerant rice cultivars and accumulation of putrescine in cultivars sensitive to salinity stress. Polyamines have also recently been implicated in the protection of seedlings from the adverse effects of salinity. Suppression of polyamine biosynthesis by cyclohexylamine resulted in increased ethylene synthesis and seed germination [138]. This suggested a crosslinking of the pathways for polyamine and ethylene biosynthesis. Lin and Kao [139] reported an increase in the level of spermidine under salinity, but low levels of putrescine in the shoots and roots of rice seedlings. Accumulation of spermidine and spermine along with the activity of arginine decarboxylase in rice seedlings plays a significant role in salt tolerance.

The physiological role of an increase in putrescence accumulation following abiotic stresses is still unclear and is a matter of considerable debate. It has been very difficult to establish directly a cause-and-effect relationship between increased polyamine levels in plants and abiotic stress. The increase in putrescine levels in plants under stress might be the cause of stress-induced injury or alternatively a means of protection against stress. Shevyakova et al. [140] reported that stress-induced cadaverine in *Mesembryanthemum crystallinum* acted as a heat shock signal. They further reported that ethylene-induced production of cadaverine was mediated by protein phosphorylation and dephosphorylation [141]. Earlier experiments with rice plants expressing oat *ADC*-cDNA, under the control of an ABA-inducible promoter, resulted in transgenic rice plants with increased biomass when grown under salt stress [142,143]. The same experimenters expressed *Tritordeum-SMACD*-cDNA in rice under control of the same promoter. Under salt stress these plants showed increased seedling growth compared to the wild-type plants. Under salinity and drought conditions, polyamines and their corresponding enzyme activities were substantially enhanced [144]. DNA in plant mitochondria and chloroplasts is regulated and stabilized by putrescine and polyamines, whereas nuclear DNA is stabilized by histones in eukaryotic organisms. In addition, polyamines can stabilize biomembranes and stimulate protein biosynthesis, perhaps through interactions with nucleic acids. Results from a number of studies suggest that polyamines, particularly spermidine and spermine, are involved in regulation of gene expression by enhancing DNA binding activities to particular transcription factors. Polyamines are believed to have an osmoprotectant function in plant cells under water deficit.

Putrescine accumulation during environmental stress is correlated with increased arginine decarboxylase (ADC) activity in oats (*Avena sativa* L.). Recent studies with transgenic carrot (*Daucus carota* L.) cells overexpressing ornithine decarboxylase (ODC) cDNA showed that these cells were significantly more tolerant to both salt and water stress [59]. Transgenic rice with *ADC* cDNA showed an increase in biomass under salinity stress conditions compared to the control [142]. Ten-day-old *Zea mays* plants exposed to salt stress for eight days had increased the content of putrescine and spermidine in the roots and leaves, and the increase in leaves was higher than in roots [145].

Mutations that impair arginine decarboxylase (ADC, catalyzes the first committed step in polyamine biosynthesis) resulted in salt hypersensitivity [146,147]. Transgenic rice plants expressing the *Datura ADC* gene accumulated up to twofold more putrescine in leaf tissues compared to the wild-type. All wild-type plants wilted and showed drought-induced rolling of leaves following 20% PEG treatment. Such symptoms were completely absent from *ADC*-transgenic plants. After the six-day drought-stress period, the phenotype of the transgenic plant was indistinguishable from the nonstressed wild-type. Based on these observations, a model consistent with a mechanism linking polyamine metabolism to drought tolerance was developed.

Expression of the *ADC*-transgene driven by the strong maize Ubi-1 promoter augmented the putrescine pool to levels that extended beyond the critical threshold required to initiate the conversion of excess putrescine to spermidine and spermine [148]. Spermidine and spermine *de novo* synthesis in transgenic plants was corroborated by the activation of the rice *SAMDC* gene under drought stress. Transcript levels for rice *SAMDC* reached their maximum levels at six days after stress induction. Such increases in the endogenous spermidine and spermine pools of transgenic plants not only regulated the putrescence response, but also exerted an antisenescence effect at the whole-plant level, resulting in phenotypically normal plants. Wild-type plants, however, were not able to raise their spermidine and spermine levels after six days of drought stress, and consequently, exhibited the classical drought-stress response [149].

6.4.2 OSMOPROTECTANTS/COMPATIBLE SOLUTES

Molecules such as glycerol and sucrose were discovered by empirical methods to protect biological macromolecules against the damaging effects of salinity.

Later, more systematic examination of the molecules that accumulate in halophytes and halotolerant organisms led to the identification of a variety of compounds that were also able to provide protection [150,151]. Characteristically, these types of molecules are not highly charged, but are polar, highly water soluble, and have a large hydration shell. Such molecules are preferentially solublized in the bulk water of the cell where they could interact directly with macromolecules. Under osmotic stress an important protective mechanism is the accumulation of osmotically active compounds (osmolytes) to lower the osmotic potential. These are referred to as compatible metabolites, because they do not apparently interfere with normal cellular metabolism.

Some of the biochemical pathways involved in the production of compatible solutes are now better known. For example, genes involved in rate limiting steps have been cloned and transferred into crop plants to raise the level of osmolytes. Tobacco plants have been modified by the introduction of the *E. coli mtlD* gene, which encodes mannitol-1-phosphate dehydrogenase [152]. The enzyme is not normally produced by the wild-type tobacco; however, many organisms, including some plants, synthesize and accumulate mannitol. In the absence of salt stress, wild-type and transformed plants had similar height and fresh weight gains, but in the presence of 250 mol m^{-3} salt, the *mtlD*-transformed plants had better growth compared to wild-type plants, with faster growth, and more new leaf and root production.

Binzel et al. [153] found that tobacco cells adapted to 428 mM NaCl could maintain cytosolic Na$^+$ and Cl$^-$ levels at less than 100 mM. Although mannitol only partially decreased the amount of inorganic ion accumulation in the cytosol, its protective effect as a compatible solute may have been sufficient to give the marginal growth advantage observed in transformed plants. Su et al. [154] obtained three transgenic rice lines with bacterial *mtlD* and demonstrated that biosynthesis and accumulation of mannitol in plants are correlated with the salt-stress tolerance of plants. *Arabidopsis* plants transformed with bacterial *mtlD* encoding mannitol-1-phosphate dehydrogenase had a higher mannitol content and were able to withstand NaCl salinity up to 400 mol m^{-3}, whereas the wild-type seeds did not germinate at 100 mol m^{-3} NaCl [155].

Cyclic sugar alcohols (e.g., pinnitol and ononitol) are stored in a variety of species that are consistently exposed to saline conditions or accumulated in tolerant species when exposed to saline environments [156]. Facultative halophytes such as *M. crystallinum* accumulate these compounds only when subjected to water and salinity stresses. The proposed synthetic pathway consists of methylation of myo-inositol to the intermediate ononitol followed by epimerization to pinnitol [157]. An inositol methyl transferase (*Imt*) cDNA was isolated from transcripts induced in *M. crystallinum* plants by NaCl [158]. Tobacco transformed with inositol methyl transferase (*Imt*) has been obtained [159]. Similar to plants transformed with mannitol-1-phosphate dehydrogenase, growth of wild- and *Imt*-transformed plants was not distinguishable in the absence of stress, but the latter had a growth advantage over control plants in the presence of salt.

Sorbitol, the sugar alcohol of glucose, is found in a variety of plant species, usually as a constituent of seeds. Sorbitol accumulation has been reported in seeds of many crop plants [160]. In *Rosaceae* species, it functions as a translocated carbohydrate and also is reported in vegetative parts in the halotolerant *Plantago maritime* [161]. Increasing salinity from 0 to 400 mol m^{-3} resulted in an 8-fold increase of sorbitol concentration in shoot tissues and a 100-fold increase in root tissues. Accumulation in *P. maritima* serves an osmoregulatory function and its accumulation in plant seeds suggests it may contribute to the desiccation tolerance of the mature embryo [162]. The conversion of glucose to its sugar alcohol is catalyzed by aldose reductase. An aldose reductase-like protein accumulated during the period of embryo maturation in barley (*Hordeum vulgare* L.) when desiccation tolerance was obtained [163].

In organisms ranging from bacteria to higher plants, there is a strong correlation between increased cellular proline levels and the capacity to survive both water deficit and salinity. It may also serve as an organic nitrogen reserve that can be utilized during recovery. Sairam and Dube [164] and Sairam et al. [165] reported an increase in proline content in wheat genotypes under water and salinity stresses. Higher proline accumulation during drought and salinity was associated with retention of higher RWC; tolerant genotype Kharchia 65 showed the highest proline accumulation under salt stress. In *Lathyrus sativus*, a grain legume that can withstand drought, high proline accumulation was observed in leaves and roots under water stress [166]. Although proline can be synthesized from either glutamate or ornithine, glutamate is the primary precursor in osmotically stressed cells. The biosynthetic pathway consists of two important enzymes, pyrroline carboxylic acid synthetase and pyrroline carboxylic acid reductase. Transcripts corresponding to both cDNAs accumulate in response to NaCl treatment. Both of these regulatory steps are keys to developing strategies for overproducing proline in a selected plant species.

There is also evidence that degradation of proline in mitochondria is directly coupled to respiratory electron transport and ATP production. A pyrroline-5-carboxylate synthetase (P_5CS) cDNA from moth-bean (*Phaseolus aconitifolius*) was introduced into rice, and led to stress-induced overproduction of the P_5CS enzyme and proline accumulation in transgenic rice plants. Second-generation transgenic plants showed an increase in biomass under salt- and water-stress conditions [167].

Glycine-betaine levels in *Poaceae* species are correlated with salt tolerance. Highly tolerant *Spartina* and *Distichlis* accumulated the highest levels, moderately tolerant species accumulated intermediate levels, and sensitive species accumulated low levels or no glycine-betaine [168]. Glycine-betaine is synthesized from choline in two steps, first being catalyzed by choline mono-oxygenase leading to synthesis of betaine-aldehyde, which is further oxidized by betaine-aldehyde dehydrogenase. Salinity stress induces both enzyme activities [58,169]. Genetic evidence that glycine-betaine improves salinity tolerance has been obtained for barley and maize [168,170]. Isogenic barley lines containing different levels of glycine-betaine have different abilities to adjust osmotically. Transgenic rice plants expressing betaine-aldehyde dehydrogenase converted higher levels of exogenously applied betaine-aldehyde to glycine-betaine than did wild-type plants. The elevated level of glycine-betaine in transgenic plants conferred significant tolerance to salt, cold, and heat stresses.

6.4.3 Metabolic Engineering of Osmoprotectants

Huang et al. [171] reported metabolic limitation in betaine production in transgenic plants. *Arabidopsis*, canola, and tobacco were transformed with bacterial choline oxidase cDNA. The levels of glycine-betaine were 18.6, 12.8, and 13.0 μmol g^{-1} dry weights in *A. thaliana*, canola, and tobacco, respectively, which were 10- to 20-fold lower than the levels found in natural betaine producers. A moderate stress tolerance was noted in some transgenic lines based on relative shoot growth in response to salinity, drought, and freezing. However, choline-fed transgenic plants synthesized substantially more glycine-betaine, suggesting that there was need to enhance the endogenous supply of choline to support accumulation of a physiologically relevant amount of betaine.

These organic solutes may protect plants from abiotic stresses by osmotic adjustment that helps in turgor maintenance, detoxification of reactive oxygen species, and stabilization of the quaternary structure of proteins [172]. Polyols and proline act as antioxidants [173]. Glycine-betaine and trehalose stabilize the quaternary structures of proteins and the highly ordered state of membranes. Mannitol serves as a free radical scavenger. Proline also stabilizes subcellular structures (membranes and proteins) and buffers cellular redox potential under stress. Glycine-betaine also reduces lipid peroxidation during salinity stress. Hence, these organic osmolytes are known as osmoprotectants [5,172,174]. Genes involved in osmoprotectant biosynthesis are upregulated under salt stress, and the concentrations of accumulated osmoprotectants correlate with osmotic stress tolerance [5,174]. Halophytes such as *T. halophila* accumulated a significantly higher concentration of proline than that of *Arabidopsis* even under nonstress conditions [117]. Genetic analysis of the *Arabidopsis t365* mutant with an impaired S-adenosyl-L-methionine: phosphoethanolamine N-methyltransferase (*PEAMT*) gene involved in glycine-betaine biosynthesis (Figure 6.3) exhibited hypersensitivity to salt stress [175]. Thus, glycine-betaine accumulation is critical for salt tolerance. Several efforts have been made to engineer salt and abiotic stress resistance in plants through genetic manipulation of the osmoprotectant metabolism in plants. The pathways of various osmoprotectant biosynthesis are shown in Figure 6.3. Genes of these pathways that are employed in genetic engineering for salt tolerance and results are briefly reviewed in Table 6.1.

Genetically engineered overproduction of compatible osmolytes in transgenic plants such as *Arabidopsis*, tobacco, rice, wheat, and *Brassica* have also been shown to enhance stress tolerance at the vegetative stage, as measured by germination, seedling growth, survival, recovery, and photosystem II yield (Table 6.1) [152,177,183,190,191]. Abiotic stress tolerance of these transgenics

TABLE 6.1
Metabolic Engineering of Osmoprotectant Accumulation for Salt Stress Tolerance in Plants

Gene/Gene Product	Gene Source/Type	Plant	Reference/Nos.
Glycine betaine	*Arthrobacter globiformis* choline oxidase (*CodA*)	*Arabidopsis thaliana*	Hayashi et al. [176]
	A. globiformis CodA under CaMV 35S promoter	*Arabidopsis thaliana*	Sulpice et al. [177]
	A. globiformis CodA	Rice	Sakamoto et al. [178]
	A. globiformis CodA	*Brassica juncea*	Prasad et al. [179]
	Arthrobacter pascens choline oxidase (*COX*)	*Arabidopsis thaliana*, *Brassica napus*, and tobacco	Huang et al. [171]
	E. coli choline dehydrogenase (*betA*) and betaine aldehyde dehydrogenase (*betB*) genes	Tobacco	Holmstrom et al. [180]
	Atriplex hortensis BADH driven by maize ubiquitin promoter	*Triticum aestivum*	Guo et al. [181]
	Peroxisomal *BADH* of barley	Rice	Kishitani et al. [182]
Proline	*Vigna aconitifolia* L. *P5CS* (Δ^1-pyrroline-5-carboxylate synthetase) gene	Tobacco	Kishor et al. [183]
	V. aconitifolia L. *P5CS* that lacks end product (proline) inhibition	Tobacco	Hong et al. [184]
	V. aconitifolia L. *P5CS* gene under barley HVA22 promoter	Rice	Zhu et al. [78]
	Antisense proline dehydrogenase gene	*Arabidopsis thaliana*	Nanjo et al. [185]
	Antisense Δ^1-pyrroline-5-carboxylate reductase gene under heat stress inducible promoter	Soybean	De Ronde et al. [186]

(continued)

TABLE 6.1 (Continued)
Metabolic Engineering of Osmoprotectant Accumulation for Salt Stress Tolerance in Plants

Gene/Gene Product	Gene Source/Type	Plant	Reference/Nos.
Trehalose	*E. coli OstA* (trehalose 6P synthase) & *OstB* (trehalose 6P phosphatase) driven by ABA responsive promoter	Rice	Garg et al. [187]
	E. coli OstA & *OstB* driven by maize ubiquitin promoter	Rice	Jang et al. [188]
Mannitol	*E. coli mt1D* (mannitol-1-phosphate dehydrogenase) driven by CaMV 35S promoter	Tobacco	Tarczynski et al. [152]
	E. coli mt1D	*Arabidopsis thaliana*	Thomas et al. [155]
	E. coli mt1D	Tobacco	Karakas et al. [189]
	E. coli mt1D	*Triticum aestivum* L.	Abebe et al. [190]
	Celery mannose 6-P reductase driven by CaMV 35S promoter	*Arabidopsis thaliana*	Zhifang and Loescher [191]
D-Ononitol	Ice plant myo-inositol O-methyl transferase (*IMT1*)	Tobacco	Sheveleva et al. [192]
Sorbitol	Apple Stpd1 (sorbitol-6-phosphate dehydrogenase) driven by CaMV 35S promoter	Japanese persimmon, *Diospyros kaki* Thunb.	Gao et al. [113]

was mainly attributed to the osmoprotectant effect of the compounds (Table 6.1). In most cases, the contribution of engineered osmoprotectant concentration to osmotic adjustment was not measured, or the contribution to osmotic adjustment was much less. Furthermore, compartmentation of these osmoprotectants may also be required for enhanced tolerance. For example, transgenic rice plants that overexpressed choline oxidase targeted to chloroplasts showed better tolerance to photoinhibition under salt and low-temperature stresses than did plants overexpressing choline oxidase targeted to the cytosol [178]. Often, engineered osmoprotectant overaccumulation resulted in impaired plant growth and development even under nonstress conditions. Transgenic tobacco

Salinity Tolerance: Cellular Mechanisms and Gene Regulation

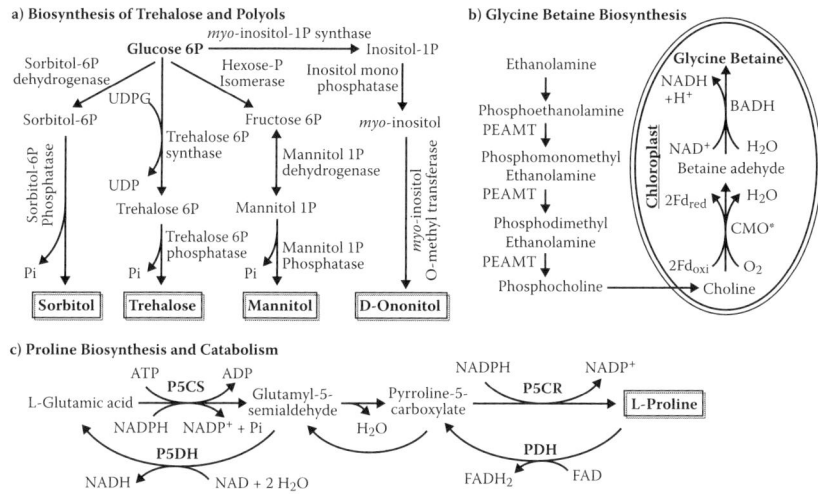

FIGURE 6.3 Compatible osmolyte metabolism. The role of osmolytes in abiotic stress tolerance has been examined by genetic engineering in plants (PEAMT: phosphoethanolamine N-methyltransferase; CMO; choline monooxygenase; BADH: betaine aldehyde dehydrogenase; * = E. coli, CDH (choline dehydrogenase) catalyzes this reaction. In Arthrobacter, a single enzyme choline oxidase (COD) converts choline into glycine betaine; P5CS: Δ^1pyrroline-5 carboxylate synthetase; P5CR: Δ^1pyrroline-5 carboxylate reductase; PDH: proline dehydrogenase; P5DH: pyrroline-5-carboxylate dehydrogenase).

plants overaccumulating sorbitol [193] or mannitol [189] exhibited stunted growth. Often, engineered alterations in osmoprotectant accumulation resulted in infertility depending upon the concentration of osmoprotectant [190,193]. Use of a stress-inducible promoter to overexpress osmoprotectant biosynthesis may help overcome the growth defects while protecting the plants during osmotic stresses [187]. Although transgenic tobacco overexpressing the myo-inositol O-methyl transferase gene accumulated D-ononitol in the cytosol up to 600 mM during salt stress, D-ononitol did not enter the vacuole [192]. Hence, further understanding of the metabolic flux and compartmentation of the osmoprotectant will help in precise engineering of osmoprotectant metabolism in plants for salt-stress tolerance.

6.4.4 Regulation of Osmoprotectant Metabolism

Evidence from genetic analysis, gene expression, and transgenic studies suggested that osmoprotectant biosynthesis and accumulation in the appropriate cellular organelle is critical for plant salt tolerance. However, the signaling cascades that regulate osmoprotectant biosynthesis and catabolism during salt and other osmotic stresses in higher plants are poorly understood. A signaling cascade similar to that of the yeast mitogen-activated protein kinase-high osmotic glycerol 1 (MAPK-HOG1) pathway may be involved in the regulation of osmoprotectant metabolism [5,194]. *Arabidopsis AtHK1*, a putative osmosensory two-component hybrid histidine kinase, is implicated in osmosensing during salt stress. The *AtHK1* expression is induced by salt stress, and it complements the yeast double mutant *sln1Δ sho1Δ* that lacks osmosensors. Similar to the SLN1 osmosensor of yeast, the AtHK1 is also probably active at low osmolarity and may inactivate a response regulator by phosphorylation. High osmolarity caused by salt stress may inactivate AtHK1, which results in accumulation of the active form of the nonphosphorylated response regulator, and in turn activates osmolyte biosynthesis in plants by activating MAPK pathway(s) [194]. Moreover, constitutive overexpression of a dominant negative mutant form of AtHK1 in transgenic *Arabidopsis* resulted in enhanced tolerance to salt and drought stress [195]. Results from the complementation analysis in yeast and transgenic *Arabidopsis* suggest that AtHK1 may act as an osmosensor in *Arabidopsis*. Determination of the in vivo role of higher plant putative sensory kinases in higher plants and the identification of signaling intermediates and targets will shed more light on salt-stress signaling.

Genetic analysis of ABA-deficient mutants *los6/aba1* and *los5/aba3* of *Arabidopsis* revealed that proline biosynthesis during osmotic stress was regulated by ABA, because salt and other abiotic stress induction of the P_5CS gene expression was either diminished or blocked in these mutants [196,197]. In *Arabidopsis* and *Medicago truncatula*, of the two P_5CS genes, only the expression of one gene is NaCl and osmotic stress regulated [198,199], which suggests that the promoters of these genes were regulated by developmental and osmotic stress cues. Biochemical analysis implicates phospholipase D (PLD) as a negative regulator of proline biosynthesis in *Arabidopsis* [200]. Recent studies have shown that proline can act as a signaling molecule to autoregulate the proline concentration and induce salt-stress-responsive proteins. In a desert plant *Pancratium maritimum*, severe salt stress resulted in an inhibition of antioxidative enzymes such as catalase and peroxidase. Exogenous application of proline helped to maintain the activities of these enzymes and also upregulated several salt-stress-responsive dehydrin proteins [201]. Microarray and RNA gel blot analyses have shown that 21 proline-inducible genes have a proline- or hypoosmolarity-responsive element (PRE, ACTCAT) in their promoter [202,203]. Transient activation analysis of a PRE-containing promoter led to the identification of four bZIP transcription factors that may regulate proline dehydrogenase and other proline- or hypoosmolarity-responsive genes in *Arabidopsis* [204]. Greater insight into the signaling events that regulate osmoprotectant metabolism during stress and recovery will be useful in improving salt and osmotic stress tolerance of crop plants.

6.5 SALT-INDUCED OXIDATIVE STRESS

Exposure of plants to unfavorable environmental conditions, including salinity, can increase the production of reactive oxygen species (ROS) such as singlet oxygen ($^1O_2$), superoxide radical ($O_2^{·-}$), hydrogen peroxide (H_2O_2), and the hydroxyl radical (OH·). Plants possess both enzymatic and nonenzymatic

Salinity Tolerance: Cellular Mechanisms and Gene Regulation

mechanisms for scavenging of ROS. The enzymic mechanisms are designated to minimize the concentration of O_2^- and H_2O_2. Antioxidant enzymes include superoxide dismutase (SOD), ascorbate peroxidase (APX), catalase (CAT), glutathione reductase (GR), and glutathione synthesizing enzymes.

The superoxide radical is regularly synthesized in chloroplasts [205] and mitochondria [206], although some quantity is also reportedly produced in microbodies [207]. Scavenging of O_2^- by SOD results in production of H_2O_2, which is removed by ascorbate peroxidase [208] or catalase [209]. However, both O_2^- and H_2O_2 are not as toxic as OH·, but can damage chlorophyll, protein, DNA, lipids, and other important macromolecules [210–214], ultimately affecting plant growth and yield [215,216]. A schematic presentation of production and scavenging of O_2^-, H_2O_2, and OH is presented in Figure 6.4.

Increases in activities of SOD, APX, CAT, and GR under drought, high temperature, and salinity, have been observed in genotypes of many species [20,217–219,220–223]. Sairam and Srivastava [224] reported comparatively higher Cu/Zn-SOD, Fe-SOD, APO, and GR activity in the chloroplastic fraction,

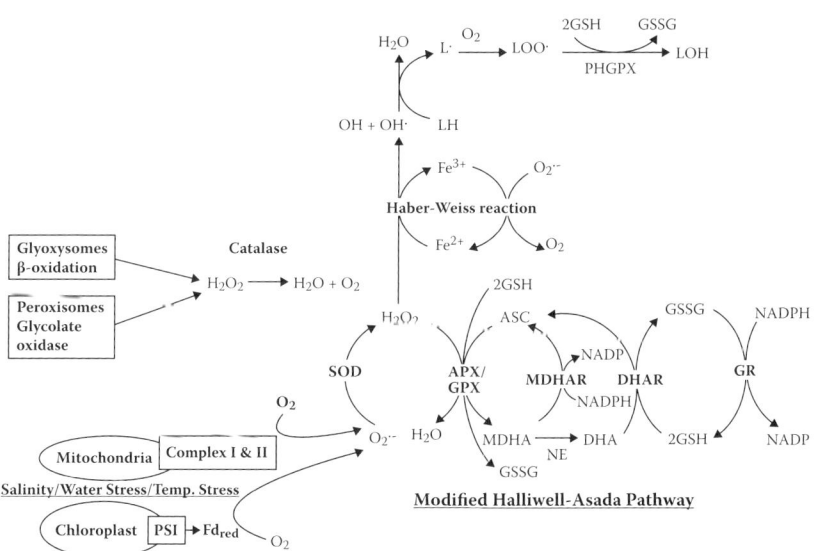

FIGURE 6.4 Generation and scavenging of superoxide radical and hydrogen peroxide, and hydroxyl radical (OH) induced lipid peroxidation and glutathione peroxidase mediated lipid (fatty acid) stabilization. (APX: ascorbate peroxidase, ASC: ascorbate, DHA: dehydroascorbate, DHA: dehydroascorbate, DHAR: dehydroascorbate reductase, Fd_{red}: reduced ferredoxin, GR: glutathione reducase, GSH: red-glutathione, GPX: glutathione peroxidase, GSSG: oxi-glutathione, HO: hydroxyl radical, LH: lipid, L; LOO: unstable lipid peroxy radical, LOH: stable lipid (fatty acid), MDHA: monodehydroascorbate, MDHAR: monodehydroascorbate reductase, NE: nonenzymatic reaction, PHGPX: phospholipid-hydroperoxide glutathione peroxidase, SOD: superoxide dismutase.

and Mn-SOD in the mitochondrial fraction in tolerant wheat genotypes in response to salt stress. Hernandez et al. [223] reported that NaCl enhanced mRNA expression and activity of Mn-SOD, APX, GR, and monodehydro-ascorbate reductase in tolerant pea (*Pisum sativum*) cv. Granada, whereas in salinity sensitive cv. Chillis no significant changes in activity and mRNA levels of those enzymes were observed.

Little work has been done on the development of transgenic plants overexpressing antioxidant enzyme activities under salt stress. Roxas et al. [225] reported overexpression of a tobacco glutathione-S-transferase (GST) and glutathione peroxidase (GPX) in transgenic tobacco seedlings under a variety of stresses. Salt-stress treatment inhibited the growth of the wild-type and caused increased lipid peroxidation, whereas GST-transformed seedlings did not exhibit any increased levels of lipid peroxidation. GST/GPX overexpression provides increased glutathione-dependent peroxidase scavenging and alterations in glutathione and ascorbate metabolism leading to reduced oxidative damage. Transgenic rice overexpressing yeast Mn-SOD exhibited increased levels of APX and chloroplastic SOD in transformed rice compared to the wild-type. The transformed rice plants also showed more salinity tolerance than the wild-type [226].

Transgenic plants overexpressing various antioxidant enzymes and showing tolerance to drought and chilling have been reported by many workers. Poplar (*Populus tremula* x *Populus alba*) transformed with *Arabidopsis thaliana* Fe-superoxide dismutase cDNA targeted to chloroplasts resulted in substantial overexpression of foliar SOD and dehydro ascorbate reductase, whereas no changes were observed in the activities of ascorbate peroxidase and monodehydro ascorbate reductase [227]. Reduced glutathione plays an important role as a plant antioxidant, especially in the recharging/reduction of ascorbic acid and α-tocopherol. Noctor et al. [228] examined the role of enzymes associated with glutathione synthesis in poplar plants transformed with *E. coli* γ-glutamyl-cysteine synthetase (γ-GCS) or glutathione synthetase in the chloroplasts. The transformed plants overexpressed the introduced genes (mRNA) and enzyme activity. Enhanced γ-GCS activity increased γ-glutamyl-cysteine and GSH levels.

Wu et al. [88] isolated cDNA encoding chloroplastic Cu/Zn-SOD and mitochondrial Mn-SOD from wheat. Northern blot analysis showed Mn-SOD genes were stress inducible. Although the Cu/Zn-SOD gene did not increase under drought, there was an increase in expression after rewatering. The results show that both Mn-SOD and Cu/Zn-SOD play definitive roles in stress tolerance, although at different phases. To test the hypothesis that enhanced tolerance to oxidative stress would improve winter survival, two clones of *Medicago sativa* were transformed with Mn-SOD cDNA targeted to the mitochondria or to the chloroplasts. Transformed plants have higher Mn-SOD activity and up to 100% more winter survival and herbage yield [229]. Expression of certain genes in an antisense mode has resulted in increased antioxidant activity in transformed plants [230]. *Arabidopsis thaliana* lines transformed with barley 2-cysteine peroxiredoxin cDNA in an antisense

Salinity Tolerance: Cellular Mechanisms and Gene Regulation   149

mode showed high ascorbate peroxidase and monodehydro-ascorbate reductase, although the mRNA levels and activity of glutathione peroxidase were lowered.

The glyoxylase system is ubiquitous in nature and consists of two enzymes, glyoxylase I and glyoxylase II, which act coordinately to convert 2-oxo-aldehydes into 2-hydroxy acids using reduced glutathione. Their primary function seems to be to remove methyl-glyoxal, the primary substrate for glyoxylase I, a cytotoxic compound known to arrest growth and react with DNA and proteins. Transgenic tobacco plants overexpressing glyoxylase I from *Brassica juncea* showed significant tolerance to salt stress, which was correlated with the degree of *Gly I* expression [231]. Post-transcriptional regulation of GR has been studied in maize [232]. GR may act as a central determinant of the overall cellular redox state, involving redox signaling for the expression of specific genes in optimal and stressed conditions [232]. The limitations on the regulation of such signaling pathways by the absence of GR may render tissues, such as those in maize leaves, sensitive to low temperature. Lipoxygenases are nonheme iron containing enzymes that catalyze the hydroperoxidation of fatty acids. *Lathyrus sativus* exhibited two- to threefold induction of lipoxygenase expression under water stress [233].

Transgenic tobacco plants overexpressing *AtAPX* targeted to the chloroplasts exhibited enhanced tolerance to salinity and oxidative stress [234]. The *Arabidopsis pst1* (*photoautotrophic salt tolerance 1*) mutant was more tolerant to salt stress than the wild-type, which was attributed to higher activities of SOD and APX activites than that of wild-type *Arabidopsis* [235]. These results showed that ROS detoxification may be an important trait of plant salt-stress tolerance. Pyramiding of chloroplastic and mitochondrial Mn-SOD in alfalfa resulted in lower biomass production compared to transgenic plants expressing either of the Mn-SOD genes [236]. Engineering alterations in antioxidant systems may alter the pool size of ROS, which are involved in developmental, biotic, and abiotic stress signaling [237,238]. In field environments, crop plants often experience more than one biotic and abiotic stress. Critical evaluation of the engineered alterations in antioxidant systems on crop productivity under multiple stress environments under field conditions, and understanding the signaling components that regulate ROS detoxification during salinity will be needed to use this trait for genetic engineering of plant salt tolerance.

6.6 GENES ENCODING PROTEINS INVOLVED IN CELLULAR PROTECTION

6.6.1 LEA-Type Proteins

In higher plants, osmotic stresses and ABA induce several late embryonic abundant (LEA) proteins in vegetative tissues. LEA protein expression levels are correlated to the acquisition of desiccation tolerance of vegetative tissues, pollen, and seeds [239,240]. LEA proteins include: dehydrins, responsiveness to dehydration (RD), early responsiveness to dehydration (ERD), cold inducible (KIN), cold regulated (COR), and responsiveness to ABA (RAB) [241]. The proposed functions of LEA proteins under stress include (1) protection of the cellular structure by acting as a hydration buffer, (2) protection of proteins and membranes, and (3) renaturation of denatured proteins [239,240]. LEA proteins were induced to higher levels by salt or ABA in salt-tolerant rice varieties than those of salt-sensitive rice varieties [242]. Genetically engineered rice plants constitutively overexpressing a barley *LEA* gene (*HVA1*) driven by the rice actin 1 promoter showed better salt- (200 mM NaCl) and water-stress tolerance and faster recovery once the stress was relieved. Wilting, dying of old leaves, and necrosis of young leaves was delayed in transgenic rice, when compared to the control plants under both salt and water stresses [243].

LEA proteins have been classified into different groups. Group 1 LEA proteins contain a high proportion of glycine and charged amino acids, and thus are highly hydrophilic. Group 1 LEA proteins are related to the *Em* gene that encodes wheat EM protein, which is more hydrated than other globular proteins [244,245]. Consequently, the predicted role of group 1 LEA proteins is for water binding. This function is important because it provides a protective aqueous environment for cellular components. The expression of members of this group of genes in vegetative tissues is induced by salt [245], water deficit stress [246,247], and ABA [245].

The expression of group 2 LEA proteins in vegetative tissues is responsive to ABA and water deficit-related stresses, including salinity and cold. Group 2 LEA proteins are often referred to as dehydrin or RAB, the latter term pertaining to the responsiveness of their expression to ABA. There is considerable variability in size and structural properties among members of this family of proteins. However, they are all characterized by a conserved lysine-rich amino acid domain located in the C-terminus, and at least one more at an upstream position. Suggested roles for these proteins under water deficit include the exclusion of solutes from the surface of membranes and cytosolic proteins, thereby preventing denaturation and maintaining the solvation of structural surfaces [248,249].

Two groups of LEA proteins, LEA 3 and 5, are characterized by the presence of tandemly repeated 11-mer amino acid motifs repeated many times within the protein [250]. These proteins exist as dimers and the polar faces of these dimerized helices are exposed and capable of binding ions via the formation of salt bridges. As a result, the likely roles of these proteins in salt or water deficit-stressed cells are their ion sequestering actions. Genes encoding group 3 LEA proteins are expressed in response to salt, water deficit, and ABA in soybean (*Glycine max*) and barley [184,251], and salt and ABA in roots of rice [252]. Naot et al. [253] isolated a gene from a salt-tolerant line of Shamuti orange (*Citrus sinensis*), which encoded a group 5-*lea* homologue in response to salt and water deficit.

The group 4 LEA proteins contain conserved N-terminal domains that form an -helix and a less conserved C-terminus, rich in glycine and amino acids that contain hydroxyl groups and form an unstructured random coil [254]. These have been suggested to bind water molecules and may also act as reverse chaperones, stabilizing the surface of membranes and possibly proteins by binding water and functioning as a solvation film. Genes encoding group 4 LEA are expressed in vegetative tissues in response to salinity, drought, ABA, and low temperature [251,255].

Galau et al. [256] reported LEA D95 protein in response to water stress in cotton leaves. This protein is unusual because there are some hydrophobic characteristics of the protein. LEA D95 is homologous to a cDNA, pcC27-45 from *Craterostigma plantigineum*, where it is expressed in response to salt in callus tissues and in response to desiccation and ABA in both leaves and callus [257].

6.6.2 Transcriptional Regulation of Stress Genes

6.6.2.1 LEA/COR Genes

ABA regulates several aspects of plant development, including seed development, seed desiccation tolerance, and seed dormancy, and plays a crucial role in abiotic and biotic stress tolerance of plants. Genetic analysis of ABA-deficient mutants established the essential role of ABA signaling in stomatal control of transpiration [87]. As discussed earlier, the rate of transpiration determines the amount of salt transport into the shoot; therefore stomatal regulation by ABA is an important trait of plant salt tolerance. Salt and osmotic stress regulation of *LEA* genes expression is mediated by both ABA-dependent and independent signaling pathways. Both pathways appear to employ Ca^{2+} signaling, at least in part to induce *LEA* gene expression during salinity and osmotic stresses [5,239]. Northern analysis of *COR* gene expression in ABA-deficient mutants, namely *los5/aba3* and *los6/aba1* of *Arabidopsis*, showed that ABA played a pivotal role in salt and osmotic stress-regulated gene expression. Expression of *RD29A, RD22, COR15A,* and *COR47* was severely reduced or completely blocked in the *los5* mutant [196], whereas in *los6*, the expression of *RD29A, RD19, COR15A, COR47,* and *KIN1* was lower than in wild-type plants [197].

Promoters of *LEA/COR* genes contain dehydration-responsive elements/C-Repeat (DRE/CRT), ABA-responsive elements (ABREs), MYC recognition sequence (MYCRS), or MYB recognition sequence (MYBRS) *cis*-elements. Regulation of gene expression through DRE/CRT *cis*-elements appears to be mainly ABA-independent, whereas ABRE and MYB/MYC element-controlled gene expression are ABA-dependent [239,258]. However, recent studies have shown that crosstalk exists between ABA-dependent and independent pathways. For example, *RD29A* expression is interdependent on both DRE and ABRE elements [259], and ABA can also induce the expression of C-repeat binding proteins (CBF1, CBF2, and CBF) [260]. Salt-stress signaling through Ca^{2+} and ABA mediates expression of *LEA* genes by transcription factors that activate *CRT, ABRE,* and *MYC/MYB cis*-elements (Figure 6.5).

6.6.2.2 Calcium Sensor Proteins

In an earlier section, we discussed the role of ABA in regulating cytosolic Ca^{2+} signature during salinity. Genetic and biochemical evidences show that ABA-mediated *COR* gene expression is regulated by Ca^{2+} signaling. In addition to SOS3, salt-stress-induced Ca^{2+} oscillations may also be perceived by Ca^{2+}-dependent protein kinases (CDPKs) and calmodulins (CaMs). *Arabidopsis AtCDPK1* and *AtCDPK2* were induced by salt and drought stress [261]. In rice, salt, drought,

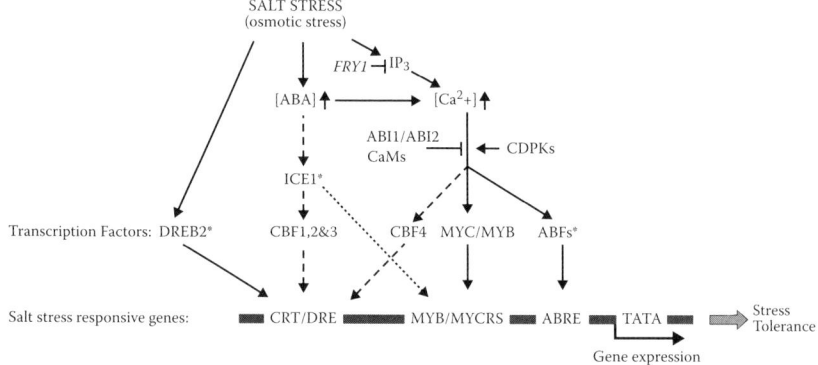

FIGURE 6.5 Salt-stress responsive genes are regulated through ABA-dependent and independent signaling pathways. Osmotic stress caused by salinity induces ABA accumulation and increases cytosolic Ca^{2+}. Ca^{2+} signaling is positively regulated by CDPKs and negatively regulated by ABI1/2 protein phosphatase 2C, SCaBP5-PKS3 complex, and CaMs. The ABA-dependent signaling pathway regulates the expression of salt-stress responsive genes through MYC/MYB and bZIP type transcription factors. Osmotic-stress-induced expression of salt-stress responsive genes is probably mediated by DREB 2 transcriptional factors through DRE/CRT cis-elements. ABA and salt stress induce the expression of ICE1, a MYC-like bHLH transcription factor. ABA has also been shown to induce the expression of DREB1 (CBF1-4) transcription factors, which may also regulate salt-stress-responsive genes through DRE/CRT cis-elements. (* = indicates post-translation activation requirement.)

and cold stresses induced the expression of *OsCDPK7* [262]. In *Mesembryanthemum crystallinum*, salinity and dehydration regulated myristoylation and localization of a CDPK (McCPK1) into the plasma membrane. When the humidity was reduced, McCPK1 changed in cellular localization from the plasma membrane to the nucleus, endoplasmic reticulum, and actin microfilaments [263]. McCDPK1 phosphorylated McCDPK1 substrate protein 1 (CSP1) in vitro in a Ca^{2+}-dependent manner, and salt stress induced co-localization of McCDPK1 and CSP1 in the nucleus of ice plants [264]. Salt stress and ABA-induced Ca^{2+} signals are perceived by CDPKs that regulate the expression of *LEA* type genes [262,265]. Transient expression analysis in maize protoplasts showed that increase in cytosolic Ca^{2+} concentration activated CDPKs that induce the stress-responsive *HVA1* promoter, and this gene expression was under the negative control of ABI1 protein phosphatase 2C [265]. Overexpression analysis also confirmed the regulatory role of CDPKs in salinity-induced LEA gene expression. Transgenic rice overexpressing *OsCDPK7* showed enhanced induction of a *LEA*-type gene (*RAB16A*) and salt and drought tolerance, whereas antisense transgenic plants were hypersensitive to salt and drought stress [262].

CaMs may act as negative regulators of salt-stress-induced Ca^{2+} signatures. Overexpression of *CaM3* in *Arabidopsis* repressed expression of *COR* genes (*RD29A* and *COR6.6*) [266], which was mediated by Ca^{2+} signals [267]. Ca-ATPases mediate Ca^{2+} efflux from the cytoplasm and thus regulate the magnitude and duration of cytosolic Ca^{2+} oscillations. Endoplasmic reticulum Ca-ATPase (ACA2) was activated by CaM and inhibited by CDPK [268]. Salinity-, dehydration-, and cold-stress-inducible *AtCaMBP25* (*Arabidopsis thaliana* calmodulin-binding protein of 25 kDa) binds to a canonical CaM in a Ca^{2+}-dependent manner. Transgenic plants overexpressing *AtCaMBP25* showed hypersensitivity to salt and osmotic stresses, whereas antisense *AtCaMBP25* transgenic plants were more tolerant to these stresses than wild-type plants.

These results suggested that the AtCaMBP25 may function as a negative effecter of salt and osmotic stress signaling [269]. The differences in affinity of SOS3, SCaBPs, CDPKs, and CaM for Ca^{2+} may determine the operation of a specific signaling cascade and interactions. Consequently, *LEA/COR* gene expression is regulated by the balance between the activities of CDPKs and CaMs. Ca–CaMs may also regulate cytoplasmic receptor-like kinases during salt and abiotic stress signaling. Salt-, cold-, and H_2O_2-inducible CaM binding cytoplasmic receptor-like kinase 1 (CRCK1) has been cloned from alfalfa [270]. Transcriptome analyses also showed induction of receptor-like kinase genes in *Arabidopsis* under salt stress [15,16]. However, the roles of these proteins in salt stress sensing and their targets are unknown.

6.6.2.3 Basic Lucine-Zipper Family Transcription Factors

ABA-dependent expression of *COR* genes under osmotic stress is regulated by basic lucine-zipper-(bZIP) [271] and MYB/MYC-type transcription factors [272] (Figure 6.5). Salt, drought, and ABA upregulate the expression of *Arabidopsis* bZIP transcription factors such as *ABREB1* (ABA-responsive element binding protein 1 = *ABF2*) and *ABREB2* (= *ABF4*) genes. These transcription factors have been shown to induce *RD29B* promoter-*GUS* in leaf protoplasts of wild-type *Arabidopsis* but not in *aba2* (ABA-deficient) and *abi1* (ABA-insensitive) mutants. Induction of *RD29B-GUS* by *ABREB*s is enhanced in an *era1* (enhanced response to ABA) mutant. This suggested that ABA was necessary for the expression and activation of ABREB1 and ABREB2, which in turn regulated *COR* gene expression [271]. Constitutive overexpression of *ABF3* and *ABREB2* (= *ABF4*) in *Arabidopsis* enhanced the expression level of target LEA genes (*RAB18* and *RD29B*). These transgenic plants showed hypersensitivity to ABA, sugar, and salt stresses during germination, but enhanced drought tolerance in the seedling stage [273].

6.6.2.4 MYB/MYC-Type Transcription Factors

MYB/MYC-type transcription factors, such as AtMYC2 (= RD22BP1) and AtMYB2, regulate *LEA* gene expression in *Arabidopsis* during osmotic stress (Figure 6.5). Transgenic *Arabidopsis* plants overexpressing *AtMYC2* and *AtMYB2* showed constitutive expression of *RD22* and *AtADH*, and the expression levels were further increased upon ABA treatment. Expression of *RD22* and *AtADH* genes is impaired in the *atmyc2* mutant. Transgenic *Arabidopsis* plants overexpressing *AtMYC2* and *AtMYB2* showed enhanced osmotic stress tolerance [272], although their salt-stress tolerance was not determined. In transgenic *Arabidopsis* plants, overexpression of ABA- and abiotic stress-inducible *Craterostigma plantagineum-MYB10* gene enhanced salinity and desiccation tolerance. These transgenics also showed ABA hypersensitivity and altered sugar sensing. An in vitro promoter-binding assay showed that CpMYB10 binds to a *LEA Cp11-24* promoter [274].

6.6.2.5 C-Repeat Binding Proteins

CBPs (C-repeat binding proteins) or DREBs (dehydration responsive element binding proteins) belong to the EREBP/AP2 domain transcription factor family. CBPs activate the expression of *LEA/COR* genes through *DRE/CRT cis*-elements in response to abiotic stresses. *Arabidopsis DREBs* are classified into two groups, *DREB1* (*DREB1A* = *CBF3*, *DREB1B* = *CBF1*, *DREB1C* = *CBF2* and *CBF4*) and *DREB2* (*DREB2A* and *DREB2B*). Expression of *CBF1*, *CBF2,* and *CBF3* is induced by cold stress, CBF4 expression is induced by drought stress, and *DREB2A* and *DREB2B* expression is induced by dehydration and salt stresses [239,258,275,276]. Similar to *Arabidopsis DREB2*, rice *OsDREB2A* is also induced by dehydration and salt stress [277]. Osmotic stress-induced expression of *CBF4* appears to be mainly mediated by ABA [276]. ABA has been shown to induce *CBF1*, *CBF2*, and *CBF3*, although their ABA-induced expression level is significantly lower than that of cold stress [260]. Transgenic plants overexpressing *CBF* (*CBF1-4*) genes showed constitutive activation of *DRE/CRT cis*-elements that were dependent on *COR* gene expression [275–280]. Transcriptional activation of *COR* genes by CBF transcription factors is conserved across plant species such as *Arabidopsis*, wheat, *Brassica napus* [280], barley, and rice [277]. Transcriptome analysis of *CBF* overexpressing transgenic *Arabidopsis* showed that about 13 *LEA/dehydrin* genes were under the transcriptional control of CBFs [281]. Recently, an ICE1 (Inducer of CBF Expression 1), a MYC-type basic helix–loop–helix transcription factor, as an upstream regulator of CBFs under cold stress was identified in *Arabidopsis* [282] (Figure 6.5). Upstream transcription factors that regulate the expression of *DREB2/CBFs* during salt stress have yet to be identified.

In tobacco *Tsi1* (tobacco-stress-induced-gene 1, a member EREBP/AP2 transcription factor family), gene expression was rapidly induced by salt stress but not by drought or ABA. Overexpression of *TSI1* enhanced retention of chlorophyll content when leaves were floated in 400 mM NaCl solution for 48 and 72 h [283]. Further detailed studies are needed to identify the targets of TSI1.

Transgenic *Arabidopsis* overexpressing *CBF1* or *CBF3* showed enhanced tolerance to salt, drought, and freezing stresses [275,278–280]. Transgenic wheat plants expressing *RD29A::CBF3* also showed enhanced osmotic stress tolerance [284]. Overexpression of the rice *OsDREB1A* gene in *Arabidopsis* resulted in activation of target *LEA* genes and conferred abiotic stress tolerance including salt stress [277]. Constitutive overexpression of *CBF1* or *CBF3* resulted in growth abnormalities of the transgenic plants [275,278–280,285,286]. This problem has been overcome by the use of a stress-responsive promoter to drive the expression of *CBFs* [279,284]. Salt and abiotic stress tolerances of CBF overexpressing transgenic plants were attributed to enhanced expression of *LEA* genes [279,280], accumulation of compatible osmolytes [287], and enhanced oxidative stress tolerance [285,286]. Genomewide expression analysis showed that *CBF* overexpression also induced transcription factors such as AP2 domain proteins (*RAP2.1* and *RAP2.6*) and putative zinc finger protein, R2R3-MYB73 [281], which might regulate genes involved in osmolyte biosynthesis and antioxidant defense. These results show that expression of several genes can be manipulated in transgenic plants engineered with a single CBF transcription factor, and enhanced expression of *LEA* genes is critical for salt and other abiotic stress tolerance.

6.7 STRESS SIGNALING PATHWAYS

Salt stress [7,288] and ABA [238,289] enhance production of H_2O_2. ABA-dependent ROS production is catalyzed by NADPH oxidase, as revealed from the analysis of the *atrbohD/F* double mutant of *Arabidopsis*, which is impaired in ABA-induced ROS production [290]. ABA-elicited H_2O_2 production is negatively regulated by the ABI2 protein [291]. H_2O_2 acts as a systemic molecule in regulating the expression of *GST* and *GPX* genes [292]. Accumulation of H_2O_2 in leaves of catalase-deficient tobacco plants was sufficient to induce the production of defense proteins (GPX, PR-1) locally as well as systemically [293]. Promoter analysis of the salt-stress-inducible *Citrus sinensis GPX1* (phospholipid hydroperoxide) gene suggests that *GPX1* upregulation under salinity is mediated by H_2O_2 but not by superoxide [234]. Promoters of genes that encode ROS detoxifying enzymes contain antioxidant-responsive elements (ARE), ABA-responsive elements (ABRE), NF-κB redox-regulated transcription factor recognition sequences, heat shock elements (HSE), and redox-regulated transcription factor Y-box *cis*-elements [294]. Hence, ABA and H_2O_2 may act as second messengers to regulate antioxidant defense genes during salinity stress.

ROS signaling in plants during various stresses is mediated by mitogen-activated protein kinase (MAPK) signaling pathways [295,296]. Salt stress triggers the activation and enhanced gene expression of MAPK signaling cascades, some components of which are common for both salt and ROS [297,298]. The *Arabidopsis* genome encodes about 60 MAPKKKs but only about 10 MAPKKs and 20 MAPKs [299]. Hence signals perceived by the 60 MAPKKKs have to be transduced through 10 MAPKKs to 20 MAPKs. Thus MAPK cascades offer potential modes for stress, hormonal, and developmental signal crosstalk. Salt stress activated *Arabidopsis* AtMEKK1 (a mitogen-activated protein kinase) [300], AtMKK2 (a mitogen-activated protein kinase) [301], and MAPKs (ATMPK3, ATMPK4, and ATMPK6) [302,303]. The active form of AtMEKK1 has been shown to activate AtMPK4 in vitro [304].

Yeast two-hybrid analysis, in vitro and in vivo protein kinase assays, and analysis of the *mkk2* null mutant have led to the identification of a MAPK signaling pathway consisting of AtMEKK1, AtMEK1/AtMKK2, and AtMPK4/AtMPK6 [300,301] involved in the transduction of salt and other abiotic stress signals in *Arabidopsis*. Transgenic *Arabidopsis* plants overexpressing *AtMKK2* exhibited constitutive AtMPK4 and AtMPK6 activity and enhanced salt and freezing tolerance, whereas *mkk2* mutant plants exhibited impaired activation of AtMPK4 and AtMPK6 and thus hypersensitivity to salt and cold stresses [301]. In addition to salinity, H_2O_2 also activated AtMPK3 and AtMPK6 [305], probably through H_2O_2-activated ANP1 (a MAPKKK) [306]. Transgenic tobacco plants overexpressing a constitutively active tobacco *ANP1* orthologue, *NPK1*, exhibited constitutive AtMPK3 and AtMPK6 activity and enhanced salt-, drought-, and cold-stress tolerance [306].

Gene expression analysis of *AtMKK2* and *ANP1* overexpressing transgenic *Arabidopsis* led to the identification of target genes of this MAPK pathway. Overexpression of the active form of ANP1 showed activation of the *GST6* and *HSP18.2* promoters but not the RD29A promoter. A single amino acid mutation in the ATP-binding site of ANP1 abolished the ANP1 effect on these promoters [306]. Microarray analysis of the transcriptome profile of *MKK2* overexpressing plants identified about 152 target genes. Upregulated genes included CBF2, RAV1, RAV2, MYB, and WRKY transcription factors, which may further regulate the expression of subregulons [301].

The *Arabidopsis* MAPK phosphatase 1 (*mkp1*) mutant exhibits salinity tolerance but is hypersensitive to genotoxic stress induced by UV-C. In a yeast two-hybrid system MKP1 interacted with AtMPK3, 4, and 6. Microarray analysis of *mkp1* revealed that AtMKP1 negatively regulates a putative Na^+/H^+-antiporter AT4G23700 [307]. Hence MKP1 may negatively regulate salt-stress signaling through AtMPK4. *Arabidopsis* nucleoside diphosphate kinase 2 (AtNDPK2) has been shown to interact with and activate AtMPK3 and AtMPK6 in a yeast two-hybrid assay and transgenic *Arabidopsis*. Furthermore, these transgenic plants accumulated lower levels of ROS and exhibited an enhanced tolerance to salinity and other abiotic stresses. Deletion mutation of *AtNDPK2* impaired AtMPK3 and AtMPK6 activities.

This evidence suggested that AtNDPK2 was a positive regulator of stress signaling through MAPK pathways [305]. In rice, gene expression and kinase activity of *OsMAPK5* were regulated by ABA, salt, drought, wounding, and cold. Transgenic rice overexpressing *OsMAPK5* also showed increased tolerance to several abiotic stresses including salt stress [308]. This evidence demonstrated that diverse abiotic stress signals converge at MAPK cascades to regulate stress tolerance. Thus in *Arabidopsis*, MAPK cascades consisting of AtMEKK1/ANP1, AtMEK1/AtMKK2, and AtMPK3/AtMPK4/AtMPK6 are involved in salt-stress signaling. These MAPK cascades are further finetuned by a negative regulator, AtMKP1, and a positive regulator, AtNDPK1 (Figure 6.6).

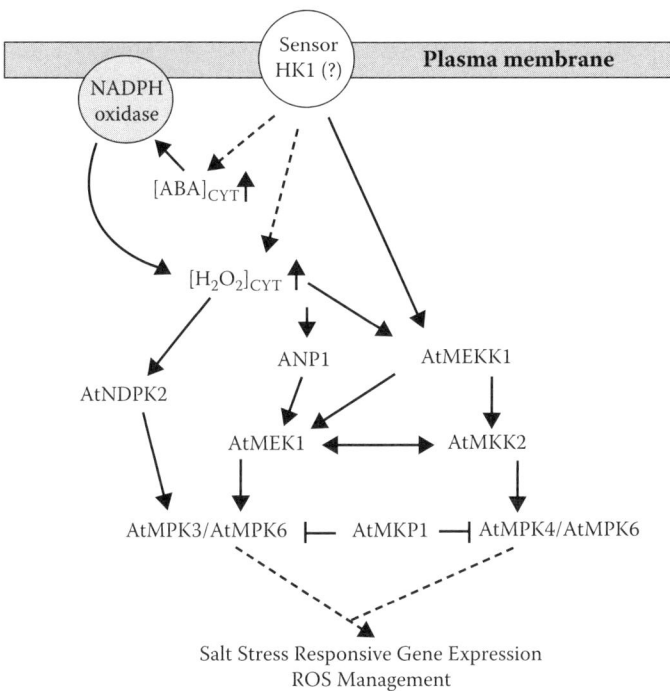

FIGURE 6.6 Regulation of salt-stress responsive genes by mitogen activated protein kinases (MAPKs) in *Arabidopsis*. Salt-stress-induced ABA and reactive oxygen species (ROS) activate MAPK cascades. Activated MAPK kinase kinases (ANP1 and AtMEKK1) activate species-regulated NDP kinase 2 (NDPK2) and are inactivated by MAPK phosphatase 1 (AtMKP1). Activated MAPKs induce gene expression by activating transcriptional activators/repressors.

6.8 STRATEGIES TO IMPROVE STRESS TOLERANCE

Recent advances in plant genome mapping and techniques in molecular biology offer new opportunities for understanding the genetics of stress-resistance genes and their contribution to plant performance under stress. These biotechnological advances will provide new tools for breeding for stress environments. Molecular genetic maps have been developed for major crop plants, including rice, wheat, maize, barley, sorghum, and potato, which make it possible for scientists to tag desirable traits using known DNA landmarks. Molecular genetic markers allow breeders to track genetic loci controlling stress resistance without having to evaluate the phenotype, thus reducing the need for extensive field testing over time and space. Moreover, gene pyramiding or introgression can be done more precisely using molecular tags. Together, molecular genetic markers offer a new strategy known as marker-assisted selection.

Gene cloning and plant transformation technology is another molecular strategy for genetic engineering of selected genes into elite breeding lines. The success of genetic engineering depends on three factors: (1) the isolation of the gene of interest, (2) an effective technique for transferring the desired gene from one species to another, and (3) the availability of promoter sequences for regulated expression of the gene. Among these, obtaining the relevant gene is considered a rate-limiting factor, even though many stress-induced genes have been isolated [309]. Stress-responsive genes can be analyzed following targeted or nontargeted strategies. The targeted approach relies upon the availability of relevant biochemical information (i.e., well-defined enzyme, protein, a biochemical reaction, or a physiological phenomenon). The nontargeted strategy to obtain a desired gene is indirect and includes differential hybridization and shotgun cloning. The list of genes whose transcription is upregulated in response to stress is rapidly increasing. An understanding of the mechanisms that regulate gene expression and the ability to transfer genes between different plants will expand the ways in which plants can be manipulated. To exploit the full potential of these approaches, it is essential that the knowledge is applied to agriculturally and ecologically important plant species.

6.9 CONCLUSIONS AND PROSPECTS

Constitutive overexpression of signaling components, osmoprotectants, and stress-responsive genes often results in a reduction in plant size and other growth abnormalities even under normal growth conditions. Kasuga et al. [279] demonstrated that the use of a stress-responsive promoter can overcome this problem. Hence, selection of stress-responsive and tissue-specific promoters for engineering the stress-tolerance trait is critical. Overexpression of osmoprotectant and antioxidant systems has been shown to protect transgenic plants from salt stress. Engineering for antioxidant systems may alter the pool size of H_2O_2, a signaling molecule involved in developmental and stress signaling. Hence, careful examination is needed to employ these traits to engineer salt-tolerant crops.

During the past decade, applications of molecular tools such as gene disruption and transgenic approaches have significantly enhanced our knowledge of salt-stress tolerance. Significant progress has been made in understanding salt-stress signaling and the factors that control ion homeostasis and salt tolerance. The salt overly sensitive (SOS) pathway regulates ion homeostasis during salt stress in *Arabidopsis*. Salt-stress-sensor-induced cytosolic Ca^{2+} signals are perceived by SOS3, which in turn activates SOS2 kinase. Activated SOS2 kinase regulates Na^+ efflux and sequestration of Na^+ into the vacuole by activating Na^+/H^+ antiporters of the plasma membrane and tonoplast, respectively. Osmotic-homeostasis and stress-damage control appear to be regulated by salt-stress-induced ABA, ROS, a putative osmosensory histidine kinase (AtHK1), and MAPK cascades. However, components and targets of these signaling pathways are not yet fully understood. CBFs, bZIP, MYB-, and MYC-type transcription factors induce LEA gene expression during osmotic stresses. Molecular, genetic, and cellular biological approaches to identify signaling components and biochemical characterization of signaling complexes will be required to further the understanding of salt-stress signaling pathways and how they may be used in crop improvement. The transgenic approach demonstrates the possibilities of gene transfer across organisms and engineering of salt tolerance by manipulation of a single or a few genes. Genetic engineering of ion transporters has been shown to significantly enhance salt tolerance [116,126,127]. Transgenic manipulation of signaling molecules and transcription factors will be advantageous, as engineering of a single gene can lead to change in the expression of several target genes involved in the stress response and provide multiple abiotic stress tolerances [119,275,278,279,300,307].

Most of the transgenic manipulations discussed here were conducted in model plants, and stress tolerance was assessed at the vegetative phase of growth under controlled conditions for short durations. Experiments in which transgenic plants are not evaluated under realistic stress conditions (e.g., low or no transpiration) are often criticized [4]. However, high salt-stress levels are given to clearly show the survival of transgenic plants and death of control plants, rather than comparing their productivity under long-term realistic salinity levels. Hence, the effect of stresses in relation to plant ontogeny should be assessed at realistic stress levels, and under combinations that occur in nature, by using transgenic crop plants in the field.

REFERENCES

1. Zhu, J.K., Over-expression of a delta-pyrroline-5-carboxylate synthetase gene and analysis of tolerance to water and salt stress in transgenic rice. *Trends Plant Sci.*, 6:66, 2001.
2. Munns, R., Comparative physiology of salt and water stress. *Plant Cell Environ.*, 25:239, 2002.
3. McCue, K.F. and Hanson, A.D., Salt-inducible betaine aldehyde dehydrogenase from sugar beet: cDNA cloning and expression. *Trends Biotech.*, 8:358, 1990.
4. Flowers, T.J., Improving crop salt tolerance. *J. Exp. Bot.*, 55:307, 2004.
5. Zhu, J.K., Salt and drought stress signal transduction in plants. *Ann. Rev. Plant Bol.*, 53:247, 2002.
6. Tester, M. and Davenport, R.A., Na^+ tolerance and Na^+ transport in higher plants. *Ann. Bot.*, 91:503, 2003.
7. Hernandez, J.A. et al., Antioxidant systems and O_2^-/H_2O_2 production in the apoplast of pea leaves, Its relation with salt-induced necrotic lesions in minor veins. *Plant Physiol.*, 127:817, 2001.
8. Maas, E.V., Crop salt tolerance. In *Agricultural Salinity Assessment and Management*, Tanji, K.K., Ed., ASCE Manuals and Reports on Engineering No. 71, American Society of Civil Engineers, New York, 1990, Chapter 13, p. 262.
9. Zeng, L. and Shannon, M.C., Salinity effects on seedling growth and yield components of rice. *Crop Sci.*, 40:996, 2000.
10. Munns, R. and Rawson, H.M., Effect of salinity on salt accumulation and reproductive development in the apical meristem of wheat and barley. *Aust. J. Plant Physiol.*, 26:459, 1999.

11. West, G., Inze, D., and Beemster, G.T., Cell cycle modulation in the response of the primary root of *Arabidopsis* to salt stress. *Plant Physiol.*, 135:1050, 2004.
12. Burssens, S. et al., Expression of cell cycle regulatory genes and morphological alterations in response to salt stress in *Arabidopsis thaliana. Planta*, 211:632, 2000.
13. Wang, H. et al., ICK1, a cyclin-dependent protein kinase inhibitor from *Arabidopsis thaliana* interacts with both Cdc2a and CycD3, and its expression is induced by abscisic acid. *Plant J.*, 15:501, 1998.
14. Sun, K., Hunt, K., and Hauser, B.A., Ovule abortion in Arabidopsis triggered by stress. *Plant Physiol.*, 135:2358, 2004.
15. Kreps, J.A. et al., Transcriptome changes for *Arabidopsis* in response to salt, osmotic, and cold stress. *Plant Physiol.*, 130:2129, 2002.
16. Seki, M. et al., Monitoring the expression profiles of 7000 *Arabidopsis* genes under drought, cold and high-salinity stresses using a full-length cDNA microarray. *Plant J.*, 31:279, 2002.
17. Rabbani M.A. et al., Monitoring expression profiles of rice genes under cold, drought, and high-salinity stresses and abscisic acid application using cDNA microarray and RNA gel-blot analyses. *Plant Physiol.*, 133:1755, 2003.
18. Netondo, G.W., Onyango, J.C., and Beck, E., Sorghum and salinity: II. Gas exchange and chlorophyll fluorescence of sorghum under salt stress. *Crop Sci.*, 44:806, 2004.
19. Hernandez, J.A. et al., Salt-induced oxidative stress in chloroplasts of pea plants. *Plant Sci.*, 105:151, 1995.
20. Hernandez, J.A. et al., Response of antioxidant systems and leaf water relations to NaCl stress in pea plants. *New Phytol.*, 141:241,1999.
21. Gadallah, M.A.A., Effects of proline and glycinebetaine on *Vicia faba* response to salt stress. *Biol. Plant.*, 42:249, 1999.
22. Agastian, P., Kingsley, S.I., and Vivekanandan, M., Effect of salinity on photosynthesis and biochemical characteristics in mulberry genotypes. *Photosynthetica*, 38:287, 2000.
23. Wang, Y. and Nil, N., Changes in chlorophyll, ribulose biphosphate carboxylase/oxygenase, glycine betaine content, photosynthesis and transpiration in Amaranthus tricolor leaves during salt stress. *J. Hort. Sci. Biotech.*, 75:623, 2000.
24. Khavarinejad, R.A. and Mostofi, Y., Effects of NaCl on photosynthetic pigments, saccharides, and chloroplast ultrastructure in leaves of tomato cultivars. *Photosynthetica*, 35:151, 1998.
25. Alamgir, A.N.M. and Ali, M.Y., Effect of salinity on leaf pigments, sugar and protein concentrations and chloroplast ATPAase activity of rice (*Oryza sativa* L). *Bangladesh J. Botany* 28:145, 1999.
26. Lu, C.M. and Vonshak, A., Characterization of PS II photochemistry in salt-adapted cells of cyanobacterium *Spirulina platensis. New Phytol.*, 141:231, 1999.
27. Panda, A., Das, A.B., and Das, P., NaCl stress causes changes in photosynthetic pigments, proteins and other metabolic components in the leaves of a true mangrove, *Bruguiera parviflora,* in hydroponic cultures. *J. Plant Biol.*, 45:28, 2002.
28. Muthukumarasamy, M., Gupta, S.D., and Pannerselvam, R., Enhancement of peroxidase, polyphenol oxidase and superoxide dismutage activities by triadimefon in NaCl stressed *Raphanus sativus* L. *Biol. Plant*, 43:317, 2000.
29. Hassanein, A.M., Alterations in protein and esterase patterns of peanut in response to salinity stress. *Biol. Plant*, 42:241, 1999.

30. Ramagopal, S., Salinity stress induced tissue-specific proteins in barley seedlings. *Plant Physiol.*, 84:324, 1987.
31. Lopez, F. et al., Accumulation of a 22-kDa protein in the leaves of Raphanus sativus in response to salt stress or water deficit, *Physiol. Plant*, 91:605, 1994.
32. Benhayyim, G. et al., Isolation and characterization of salt-associated protein in *Citrus. Plant Sci.*, 88:129, 1993.
33. Yen, H.E. et al., Salt-induced changes in protein composition in light-grown callus of Mesembryanthemurn crystallinum. *Physiol. Plant*, 101:526, 1997.
34. Munoz, G.E., Marin, K., and Gonzalez, C., Polypeptide profile in *Prosopsis* seedlings growing in saline conditions. *Phyton-Int. J. Exp. Bot.*, 61:17, 1997.
35. Elshintinawy, F. and Elshourbagy, M.N., Alleviation of changes in protein metabolism in NaCl-stressed wheat seedlings by thiamine. *Biol. Plant*, 44:541, 2001.
36. Tamas, L., Huttova, J., and Mistrik, I., Impact of aluminium, NaCl and growth retardant tetcyclacis on growth and protein composition of maize roots. *Biologia*, 56:441, 2001.
37. Xu, X.Y. et al., Salt-stress-responsive membrane proteins in *Rhodobacter sphaeroides* f. sp. denitrificans IL106. *J. Biosci. Bioeng.*, 91:228, 2001.
38. Soussi, M., Ocana, A., and Lluch, C., Effects of salt stress on growth, photosynthesis and nitrogen fixation in chick-pea *(Cicer arietinum* L.). *J. Exp. Bot.*, 49:1329, 1998.
39. AliDinar, H.M., Ebert, G., and Ludders, P., Growth, chlorophyll content, photosynthesis and water relations in guava *(Psidium guajava* L) under salinity and different nitrogen supply. *Gartenbauwissenschaft*, 64:54, 1999.
40. Romeroaranda, R., Soria, T., and Cuartero, L., Tomato plant-water uptake and plantwater relationships under saline growth conditions. *Plant Sci.*, 160:265, 2001.
41. Rajesh, A., Arumugam, R., and Venkatesalu, V., Growth and photosynthetic characterics of *Ceriops roxburghiana* under NaCl stress. *Photosynthetica*, 35:285, 1998.
42. Kurban, H. et al., Effect of salinity on growth, photosynthesis and mineral composition in leguminous plant *Alhagi pseudoalhagi* (Bieb.). *Soil Sci. Plant Nutr.*, 45:851, 1999.
43. Khavarinejad, R.A. and Chaparzadeh, N., The effects of NaCl and $CaCl_2$ on photosynthesis and growth of alfalfa plants. *Photosynthetica* 35:461, 1998.
44. Zhu, J. and Meinzer, F.C., Efficiency of C-4 photosynthesis in *Atriplex lentiformis* under salinity stress. *Aust. J. Plant Physiol.*, 26:79, 1999.
45. Reddy, M.P., Sanish, S., and Iyengar, E.R.R., Photosynthetic studies and under saline conditions, *Photosynthetica*, 26:173, 1988.
46. Tiwari, B.S., Bose, A., and Ghosh, B., Photosynthesis in rice under salt stress. *Photosynthetica*, 34:303–306.
47. Mishra, S.K., Subrahmanyam, D., and Singhal, G.S., Interrelationship between salt and light stress on primary processes of photosynthesis. *J. Plant Physiol.* 138:92, 1991.
48. Allakhverdiev, S.I. et al., Salt stress inhibits the repair of photo damaged photosystem II by suppressing the transcription and translation of psbA genes in *Synechocystis. Plant Physiol.*, 130:1443, 2002.
49. Allakhverdiev, S.I. et al., Inactivation of photo systems I and II in response to osmotic stress in *Synechococcus,* contribution of water channels, *Plant Physiol.*, 122:1201, 2000.
50. Papageorgiou, G.C. et al., A method to probe the cytoplasmic osmolarity and osmotic water and solute fluxes across the cell membrane of Cyanobacteria with Chl a florescence: Experiments with *Synechococcus* sp, PCC 7942. *Physiol. Plant*, 103:215, 1998.

51. Allakhverdiev, S.I. et al., Genetic engineering of the unsaturation of fatty acids in membrane lipids alters the tolerance of *Synechocystis* to salt stress, *Proc. Natl. Acad. Sci. USA.*, 96:5862, 1999.
52. Brugnoli, E. and Bjorkman, O., Growth of cotton under continuous salinity stress: Influence on allocation pattern, stomatal and non-stomatal components of photosynthesis and dissipation of excess light energy. *Planta*, 187:335, 1992.
53. Greenway, H. and Munns, R., Mechanisms of salt tolerance in non-halophytes. *Annu. Rev. Plant Physiol.*, 31:149, 1980.
54. Heuer, B., Photosynthetic carbon metabolism of crops under salt stress. In *Hand Book of Photosynthesis,* Pesserkali, M., Ed., Marcel Dekker, Boca Raton, FL, 1996, p. 887.
55. Cushman, J.C., De Rocher, E.J., and Bohnert, H.J., Gene expression during adaptation to salt stress. In *Environmental Injury of Plants*, Kalterman, F., Ed., Academic, San Diego, 1990, p. 173.
56. Almoguera, C., Coca, M.A., and Jouanin, L., Differential accumulation of sunflower tetraubiquitin mRNAs during zygotic embryogenesis and developmental regulation of their heat shock response. *Plant Physiol.,* 107:765, 1995.
57. Munns, R.A., Physiological processes limiting plant growth in saline soils: Some dogmas and hypothesis. *Plant Cell Environ.*, 16:15, 1993.
58. Weretilnyk, E.A. and Hanson, A.D., Molecular cloning of a plant betaine-aldehyde dehydrogenase, an enzyme implicated in adaptation to salinity and drought. *Proc. Natl. Acad. Sci. USA,* 87:2745, 1990.
59. Bohnert, H.J., Nebson, D.W., and Jensen, R.G., Adaptation to environmental stress. *Plant Cell*, 7:1099, 1995.
60. Meyer, G., Schmitt, J.M., and Bohnert, H.J., Direct screening of a small genome: Estimation of the magnitude of plant gene expression during adaptation to high salt. *Mol. Gen. Genet.*, 224:347, 1990.
61. Ramgopal, S. and Carr, J.B., Sugarcane proteins and messenger RNAs regulated by salt in suspension cells. *Plant Cell Environ.,* 1991:14, 47–46.
62. Bray, E.A. et al., Regulation of gene expression by endogenous ABA during drought stress. In *Plant Responses to Cellular Dehydration during Environmental Stress: Current Topics in Plant Physiology*, Close, T.J. and Bray, E.A., Eds., American Society of Plant Physiologist Series, Rockville, MD, 1993, p. 167.
63. Zhu, J.K., Genetic analysis of plant salt tolerance using *Arabidopsis thaliana*. *Plant Physiol.*, 124:941, 2000.
64. Ishitani, M. et al., SOS 3 function in plant salt tolerance requires myristoylation and calcium binding. *Plant Cell*, 12:1667, 2000.
65. Halfter, U., Ishitani, M., and Zhu, J.K., The Arabidopsis SOS 2 protein kinase physically interacts with and is activated by the calcium binding protein SOS 3. *Proc. Natl. Acad. Sci. USA*, 97:3730, 2000.
66. Liu, J. et al., The *Arabidopsis thaliana SOS 2* gene encodes a protein kinase that is required for salt tolerance. *Proc. Natl. Acad. Sci. USA*, 97:3735, 2000.
67. Shi, H. et al., The *Arabidopsis thaliana* salt tolerance gene *SOS 1* encode a putative Na^+/H^+ antiporter. *Proc. Natl. Acad. Sci. USA*, 97:6896, 2000.
68. Shi, H. et al., The putative plasma membrane Na^+/H^+ antiporter SOS1 controls long-distance Na^+ transport in plants. *Plant Cell*, 14:465, 2002a.
69. Qiu, Q.S. et al., Abstr. Characterization of plasma membrane Na^+/H^+ exchange in *Arabidopsis thaliana*. *12th International Workshop on Plant Membrane Biology*, Madison, WI, 2001, p. 235.

70. Wu, S.J., Lei, D., and Zhu, J.K., SOS 1, a genetic locus essential for salt tolerance and potassium acquisition. *Plant Cell*, 8:617, 1996.
71. Zhu, J.K., Liu, J., and Xiong, L., Genetic analysis of salt tolerance in *Arabidopsis thaliana*: Evidence of a critical rate for potassium nutrition. *Plant Cell*, 10:1181, 1998.
72. Shi, H. et al., The *Arabidopsis salt overly sensitive 4* mutants uncover a critical role for vitamin B6 in plant salt tolerance. *Plant Cell*, 14:575, 2002b.
73. Apse, M.P. et al., Salt tolerance conferred by overexpression of a vacuolar Na^+/H^+ antiport in *Arabidopsis*. *Science,* 285:1256, 1999.
74. Gaxiola, R.A. et al., The *Arabidopsis thaliana* proton transporters, AtNhx1 and Avp1, can function in cation detoxification in yeast. *Proc. Nat. Acad. Sci. USA*, 96:1480, 1999.
75. Kurth, E. et al., Effects of NaCl and $CaCl_2$ on cell enlargement and cell production in cotton roots. *Plant Physiol.*, 82:1102, 1986.
76. Schroeder, J.I. and Hagiwara, S., Cytosolic calcium regulates ion channels in the plasma membrane of *Vicia faba* guard cells. *Nature*, 338:427, 1989.
77. Knight, H., Trewavas, A.J., and Knight, M.R., Calcium signalling in *Arabidopsis thaliana* responding to drought and salinity. *Plant J.*, 12:1067, 1997.
78. Sanders, D., Brownlee, C., and Harper, J.F., Communicating with calcium. *Plant Cell*, 11:691, 1999.
79. Zhu, J.K., Regulation of ion homeostasis under salt stress. *Curr. Opin. Plant Biol.*, 6:441, 2003.
80. DeWald, D.B. et al., Rapid accumulation of phosphatidylinositol 4,5-bisphosphate and inositol 1,4,5-trisphosphate correlates with calcium mobilization in salt-stressed *Arabidopsis*. *Plant Physiol.*, 126:759, 2001.
81. Takahashi, S. et al., Hyperosmotic stress induces a rapid and transient increase in inositol 1,4,5-trisphosphate independent of abscisic acid in *Arabidopsis* cell culture. *Plant Cell Physiol.*, 42:214, 2001.
82. Xiong, L. et al., *FIERY1* encoding an inositol polyphosphate 1-phosphatase is a negative regulator of abscisic acid and stress signaling in *Arabidopsis*. *Genes Dev.*, 15:1971, 2001.
83. Jia, W. et al., Salt-stress-induced ABA accumulation is more sensitively triggered in roots than in shoots, *J. Exp. Bot.*, 53:2201, 2002.
84. Xiong, L., Schumaker, K.S., and Zhu, J.K., Cell signalling for. cold, drought, and salt stresses. *Plant Cell*, 14:165, 2002.
85. Xiong, L. and Zhu, J.K., Regulation of abscisic acid biosynthesis. *Plant Physiol., 133:29, 2003.*
86. Leung, J. and Giraudat, J., Abscisic acid signal transduction. *Ann. Rev. Plant Physiol. Plant Mol. Biol.*, 49:199, 1998.
87. Schroeder, J.I. et al., Guard cell signal transduction. *Ann. Rev. Plant Physiol. Plant Mol. Biol.*, 52:627, 2001.
88. Wu, R. et al., Isolation, chromosomal location and differential expression of mitochondrial manganese-superoxide dismutase and chloroplastic copper/zinc-superoxide dismutase genes in wheat. *Plant Physiol.*, 120:513, 1999.
89. Sanchez, J.P., Duque, P., and Chua, N.H., ABA activates ADPR cyclase and cADPR induces a subset of ABA-responsive genes in *Arabidopsis*. *Plant J.*, 38:381, 2004.
90. Wang, X.-Q. et al., G protein regulation of ion channels and abscisic acid signaling in *Arabidopsis* guard cells. *Science*, 292:2070, 2001.

91. Munns, R. et al., Genetic variation for improving the salt tolerance of durum wheat. *Aust. J. Agric. Res.*, 51:69, 2000.
92. Inan, G. et al., Salt cress: A halophyte and cryophyte *Arabidopsis* relative model system and its applicability to molecular genetic analyses of growth and development of extremophiles. *Plant Physiol.*, 135:1718, 2004.
93. Karahara, I. et al. Development of the Casparian strip in primary roots of maize under salt stress. *Planta*, 219:41, 2004.
94. Garcia, A. et al., Sodium and potassium transport to the xylem are inherited independently in rice and the mechanism of sodium:potassium selectivity differs from rice and wheat. *Plant Cell Environ.*, 20:1167, 1997.
95. Yeo, A.R. et al., Silicon reduces sodium uptake in rice (*Oryza sativa* L.) in saline conditions and this is accounted for by a reduction in the transpirational bypass flow. *Plant Cell Environ.*, 22:559, 1999.
96. Santa-Maria, G.E. et al., The *HAK1* gene of barley is a member of a large gene family and encodes a high-affinity potassium transporter. *Plant Cell*, 9:2281, 1997.
97. Fu, H.H. and Luan, S., AtKUP1: a dual-affinity K+ transporter in Arabidopsis. *Plant Cell*, 10:63, 1998.
98. Su, H. et al., The expression of HAK-type K+ transporters is regulated in response to salinity stress in common ice plant. *Plant Physiol.*, 129:1482, 2002.
99. Rubio, F., Gassmann, W., and Schroeder, J.I., Sodium driven potassium uptake by the plant potassium transporter HKT1 and mutations conferring salt tolerance. *Science*, 270:1660, 1995.
100. Gorham, J. et al., Analysis and physiology of a trait for enhanced K^+/Na^+ discrimination in wheat. *New Phytol.*, 137:109, 1997.
101. Uozumi, N. et al., The *Arabidopsis HKT1* gene homolog mediates inward Na^+ currents in *Xenopus laevis* oocytes and Na^+ uptake in *Saccharomyces cerevisiae*. *Plant Physiol.*, 122:1249, 2000.
102. Liu, W. et al., Characterization of two HKT1 homologues from Eucalyptus camaldulensis that display intrinsic osmosensing capability. *Plant Physiol.*, 127:283, 2001.
103. Golldack, D. et al., Characterization of a HKT-type transporter in rice as a general alkali cation transporter. *Plant J.*, 31:529, 2002.
104. Laurie, S. et al., A role for HKT1 in sodium uptake by wheat roots. *Plant J.*, 32:139, 2002.
105. Bordás, M. et al., Transfer of the yeast salt tolerance gene HAL1 to *Cucumis melo* L. cultivars and in vitro evaluation of salt tolerance. *Transgenic Res.*, 6:41, 1997.
106. Gisbert, C. et al., The yeast *HAL1* gene improves salt tolerance of transgenic tomato. *Plant Physiol.*, 123:393, 2000.
107. Rus, A.M., Expressing the yeast *HAL1* gene in tomato increases fruit yield and enhances K^+/Na^+ selectivity under salt stress. *Plant Cell Environ.*, 24:875, 2001.
108. Ellul, P. et al., The expression of the *Saccharomyces cerevisiae* HAL1 gene increases salt tolerance in transgenic watermelon [*Citrullus lanatus* (Thunb.) Matsun. & Nakai.]. *Thoer. Appl. Genet.*, 107:462, 2003.
109. Espinosa-Ruiz, A. et al., *Arabidopsis thaliana AtHAL3*: A flavoprotein related to salt and osmotic tolerance and plant growth. *Plant J.*, 20:529, 1999.
110. Maathuis, F.J.M. and Sanders, D., Sodium uptake in Arabidopsis roots is regulated by cyclic nucleotides. *Plant Physiol.*, 127:1617, 2001.
111. Donaldson, L. et al., Salt and osmotic stress cause rapid increases in *Arabidopsis thaliana* cGMP levels. *FEBS Lett.*, 569:317, 2004.

112. Liu, J. and Zhu J.K., A calcium sensor homolog required for plant salt tolerance. *Science,* 280:1943, 1998.
113. Guo, Y. et al., Molecular characterization of functional domains in the protein kinase SOS2 that is required for plant salt tolerance. *Plant Cell,* 13:1383, 2001.
114. Qiu, Q.S. et al., Na^+/H^+ exchange activity in the plasma membrane of *Arabidopsis,* *Plant Physiol.,* 132:1041, 2003.
115. Qiu, Q.S. et al., Regulation of SOS1, a plasma membrane Na^+/H^+ exchanger in *Arabidopsis thaliana,* by SOS2 and SOS3. *Proc. Natl. Acad. Sci. USA,* 99:8436, 2002.
116. Shi, H. et al., Overexpression of a plasma membrane Na^+/H^+ antiporter improves salt tolerance in *Arabidopsis. Nat. Biotechnol.,* 21:81, 2003.
117. Taji, T. et al., Comparative genomics in salt tolerance between *Arabidopsis* and *Arabidopsis*-related halophyte salt cress using *Arabidopsis* microarray. *Plant Physiol.* 135:1697, 2004.
118. Quintero, F.J. et al., Reconstitution in yeast of the *Arabidopsis* SOS signaling pathway for Na^+ homeostasis. *Proc. Nat. Acad. Sci. USA,* 99:9061, 2002.
119. Guo, Y. et al., Transgenic evaluation of activated mutant alleles of SOS2 reveals a critical requirement for its kinase activity and C-terminal regulatory domain for salt tolerance in *Arabidopsis thaliana. Plant Cell,* 16:435, 2004.
120. Fukuda, A. et al., Effect of salt and osmotic stresses on the expression of genes for the vacuolar H^+- pyrophosphatase, H^+-ATPase subunit A and Na^+/H^+ antiporter from barley. *J. Exp. Bot.,* 55:585, 2004.
121. Gaxiola, R.A. et al., Drought- and salt-tolerant plants result from overexpression of the *AVP1* H^+-pump, *Proc. Nat. Acad. Sci. USA,* 98:11444, 2001.
122. Zhao, L. et al., Nitric oxide functions as a signal in salt resistance in the calluses from two ecotypes of Reed. *Plant Physiol.,* 134:849, 2004.
123. Shi, H. and Zhu, J.K., Regulation of expression of the vacuolar Na^+/H^+ antiporter gene *AtNHX1* by salt stress and ABA. *Plant Mol. Biol.,* 50:543, 2002.
124. Fukuda, A., Nakamura, A., and Tanaka, Y., Molecular cloning and expression of the Na^+/H^+ exchanger gene in *Oryza sativa. Biochem. Biophys. Acta,* 1446:149, 1999.
125. Wu, C.A. et al., The cotton *GhNHX1* gene encoding a novel putative tonoplast Na(+)/H(+) antiporter plays an important role in salt stress. *Plant Cell Physiol.,* 45:600, 2004.
126. Zhang, H.X. and Blumwald, E., Transgenic salt-tolerant tomato plants accumulate salt in foliage but not in fruit. *Nat. Biotech.,* 19:765, 2001.
127. Zhang, H.X. et al., Engineering salt-tolerant *Brassica* plants: Characterization of yield and seed oil quality in transgenic plants with increased vacuolar sodium accumulation. *Proc. Nat. Acad. Sci. USA,* 98:12832, 2001.
128. Ohta, M. et al., Introduction of a Na^+/H^+ antiporter gene from *Atriplex gmelini* confers salt tolerance to rice. *FEBS Lett.,* 532:279, 2002.
129. Qiu, Q.S. et al., Regulation of vacuolar Na^+/H^+ exchange in *Arabidopsis thaliana* by the SOS pathway. *J. Biol. Chem.,* 279:207, 2004.
130. Cheng, N.H. et al., The protein kinase SOS2 activates the *Arabidopsis* H^+/Ca^{2+} antiporter CAX1 to integrate calcium transport and salt tolerance, *J. Biol. Chem.,* 279:2922, 2004.
131. Berthomieu, P. et al., Functional analysis of AtHKT1 in Arabidopsis shows that Na(+) recirculation by the phloem is crucial for salt tolerance. *Eur. Mol. Biol. Org. J.,* 22:2004, 2003.

132. Maser, P. et al. Altered shoot/root Na+ distribution and bifurcating salt sensitivity in *Arabidopsis* by genetic disruption of the Na+ transporter *AtHKT1*. *FEBS Lett.*, 531:157, 2002.
133. Rus, A.M. et al., AtHKT1 is a salt tolerance determinant that controls $Na^{(+)}$ entry into plant roots. *Proc. Nat. Acad. Sci. USA*, 98:14150, 2001.
134. Ohta, M. et al., A novel domain in the protein kinase SOS2 mediates interaction with the protein phosphatase 2C ABI2. *Proc. Natl. Acad. Sci. USA*, 100:11771, 2003.
135. Malmberg, R.L. et al., Molecular genetic analysis of plant polyamines. *Critical Rev. Plant Sci.*, 17:199, 1998.
136. Gallardo, M. et al., Inhibition of polyamine synthesis by cyclohexylamine stimulates the ethylene pathway and accelerates the germination of *Cicer arietinum* seeds. *Physiol. Plant.*, 91:9, 1995.
137. Lin, C.C. and Kao, C.H., Levels of endogenous polyamines and NaCl inhibited growth of rice seedlings. *Plant Growth Regul.*, 17:15, 1995.
138. Richards, F.J. and Coleman, R.G., Occurrence of putrescine in potassium deficient barley. *Nature,* 170:460, 1952.
139. Krishnamurthy, R. and Bhagwat, K.A., Polyamines as modulators of salt tolerance in rice cultivars. *Plant Physiol.,* 91:500, 1989.
140. Shevyakova, N.I. et al., Cadaverine as a signal of heat shock in plants, *Dokl. Biol. Sci.*, 375:657, 2000.
141. Shevyakova, N.I. et al., Ethylene-induced production of cadaverine is mediated by protein phosphorylation and dephosphorylation. *Dokl. Biol. Sci.*, 395:127, 2004.
142. Roy, M. and Wu, R. Arginine decarboxylase transgene expression and analysis of environmental stress tolerance in transgenic rice. *Plant Sci.,* 160:869, 2001.
143. Roy, M. and Wu, R. Over-expression of S-adenosyl methionine decarboxylase gene in rice increases and enhances sodium chloride stress tolerance. *Plant Sci.,* 163:987, 2002.
144. Lefevre, I. and Lutts, S., Effects of salt and osmotic stress on free polyamine accumulation in moderately salt resistant rice cultivar AIW 4. *Int. Rice Res. Note*, 25:36, 2000.
145. Caldevia, H.D.Q.M. and Caldevia, G., Free polyamine accumulation in unstressed and NaCl stressed maize plants. *Agronomia Lusitana*, 47:209, 1999.
146. Kasinathan, V. and Wingler, A., Effect of reduced arginine decarboxylase activity on salt tolerance and on polyamine formation during salt stress in *Arabidopsis thaliana*. *Physiol. Plant.* 121:101, 2004.
147. Urano, K. et al., *Arabidopsis* stress-inducible gene for arginine decarboxylase AtADC2 is required for accumulation of putrescine in salt tolerance. *Biochem. Biophys. Res. Commun.*, 313:369, 2004.
148. Bassie, L. et al., Promoter strength influences polyamine metabolism and morphogenic capacity in transgenic rice tissues expressing the oat *adc* cDNA constitutively. *Transgenic Res.*, 9:33, 2000.
149. Capell, T., Bassie, L., and Christou, P., Modulation of the polyamine biosynthesis pathway in transgenic rice confers tolerance to drought stress. *Proc. Natl. Acad, Sci. USA*, 101:9909, 2004.
150. Arabawa, T. and Timasheff, S.N., The stabilization of proteins by osmolytes. *Biophys. J.*, 47:411, 1985.
151. Wiggins, P.M., Role of water in some biological processes. *Microbial. Rev.*, 54:432, 1990.

152. Tarcynski, M.C., Jensen, R.G., and Bohnert, H.J., Stress protection of transgenic tobacco by production of the osmolyte mannitol. *Science*, 259:508, 1993.
153. Binzel, M.L. et al., Intracellular compost mention of ions in salt adapted tobacco cells. *Plant Physiol.*, 86:607, 1988.
154. Su, J., Chen, P.L., and Wu, R., Transgene expression of mannitol-1-phosphate dehydrogenase enhanced the salt stress tolerance of the transgenic rice seedlings. *Scientia Agricultura Sinica*, 32:101, 1999.
155. Thomas, J.C. et al., Enhancement of seed germination in high salinity by engineering mannitol expression in *Arabidopsis thaliana*. *Plant Cell Environ.*, 18:801, 1995.
156. Paul, M.J. and Cockburn, W., Pinnitol, a compatible solute in *Mesembryanthemum crystallinum* L, *J. Expt. Bot.*, 40:1093, 1989.
157. Loewus, F.A. and Dickinson, D.B., Cyclitols. In *Encyclopedia of Plant Physiology: Plant Carbohydrates 1. Intracellular Carbohydrates*, Vol. 13A, Loewus, F.A. and Tanner, W., Eds., Springer-Verlag, Berlin, 1982, p. 193.
158. Vernon, D.M. and Bohnert, H.J., A novel methyl transferase induced by osmotic stress in the facultative halophytes *Mesembryanthemum crystallinum*. *EMBO J.*, 11:2077, 1992.
159. Vernon, D.M. et al., Cyclitol production in transgenic tolerance. *Plant J.*, 4:199, 1993.
160. Kuo, T.M., Doehlert, D.C., and Crawford, C.G., Sugar metabolism in germinating soybean seeds. *Plant Physiol.*, 93:1514, 1990.
161. Ahmad, I., Larhar, F., and Stewart, G.R., Sorbitol a compatible osmotic solute in *Plantago maritime*. *New Physiol.*, 82:671, 1979.
162. Colaco, C. et al., Extra ordinary stability of enzymes dried in trehalose: Simplified molecular biology. *Biotechnology*, 10:1007, 1992.
163. Bartels, D. et al., An ABA and GA modulated gene expressed in the barley encodes an aldose reductase related protein. *Eur. Mol. Biol. Org. J.*, 10:1037, 1991.
164. Sairam, R.K. and Dube, S.D., Effect of moisture stress on proline accumulation in wheat in relation to drought tolerance. *Indian J. Agric. Sci.*, 54:146, 1984.
165. Sairam, R.K, Rao, K.V. and Srivastava, G.C., Differential response of wheat genotypes to long-term salinity stress in relation to oxidative stress, antioxidant activity and osmolyte concentration, *Plant Sci.*, 163:1037, 2002.
166. Tyagi, A., Santha, I.M., and Mehta, S.L., Effect of water stress on proline content and transcript levels in *Lathyrus sativus*. *Indian J. Biochem. Biophys.*, 36:207, 1999.
167. Zhu, B. et al., Overexpression of a pyrroline-5-carboxylate synthetase gene and analysis of tolerance to water and salt stress in transgenic rice. *Plant Sci.*, 139:41, 1998.
168. Rhodes, D. et al., Development of two isogenic sweet corn hybrids differing for glycine betaine content. *Plant Physiol.*, 91:1112, 1989.
169. Arakawa, K., Katayama, M., and Takabe, T., Levels of betaine and betaine aldehyde dehydrogenase activity in the green leaves and etiolated leaves and roots of barley. *Plant Cell Physiol.*, 31:797, 1990.
170. Grumet, R. and Hanson, A.D., Genetic betaine accumulation in barley. *Aust. J. Plant Physiol.*, 13:353, 1986.
171. Huang, J. et al., Genetic engineering of glycinebetaine production toward enhancing stress tolerance in plants: metabolic limitations. *Plant Physiol.*, 122:747, 2000.
172. Bohnert, H.J. and Jensen, R.G., Strategies for engineering water stress tolerance in plants. *Trends Biotech.*, 14:89, 1996.

173. Smirnoff, N. and Cumbes, Q.J., Hydroxyl radical scavenging activity of compatible solutes. *Phytochemistry*, 28:1057, 1989.
174. Chen, T.H.H. and Murata, N., Enhancement of tolerance of abiotic stress by metabolic engineering of betaines and other compatible solutes. *Curr. Opin. Plant Biol.*, 5:250, 2002.
175. Mou, Z. et al., Silencing of phosphoethanolamine *N*- Methyltransferase results in temperature-sensitive male sterility and salt hypersensitivity in *Arabidopsis*. *Plant Cell*, 14:2031, 2002.
176. Hayashi, H. et al., Transformation of *Arabidopsis thaliana* with the *codA* gene for choline oxidase; accumulation of glycinebetaine and enhanced tolerance to salt and cold stress. *Plant J.*, 12:133, 1997.
177. Sulpice, R. et al., Enhanced formation of flowers in salt-stressed Arabidopsis after genetic engineering of the synthesis of glycine betaine. *Plant J.*, 36:165, 2003.
178. Sakamoto, A., Alia, H., and Murata, N., Metabolic engineering of rice leading to biosynthesis of glycinebetaine and tolerance to salt and cold. *Plant Mol. Biol.* 38:1011, 1998.
179. Prasad, K.V.S.K. et al., Transformation of *Brassica juncea* (L.) Czern with bacterial *codA* gene enhances its tolerance to salt stress. *Mol. Breed.*, 6:489, 2000.
180. Holmstrom, K.O. et al., Improved tolerance to salinity and low temperature in transgenic tobacco producing glycine betaine. *J. Exp. Bot.*, 51:177, 2000.
181. Guo, B.H. et al., Transformation of wheat with a gene encoding for the betaine aldehyde dehydrogenase (BADH). *Acta Bot. Sinica*, 42:279, 2000.
182. Kishitani, S. et al., Compatibility of glycinebetaine in rice plants: Evaluation using transgenic rice plants with a gene for peroxisomal betaine aldehyde dehydrogenase from barley. *Plant Cell Environ.*, 23:107, 2000.
183. Kishor, P.B.K. et al., Overexpression of [delta]-pyrroline-5-carboxylate synthetase increases proline production and confers osmotolerance in transgenic plants. *Plant Physiol.*, 108:1387, 1995.
184. Hong, B., Barg, R., and Ho, T.H.D., Developmental and organ-specific expression of an ABA and stress-induced protein in barley. *Plant Mol. Biol.*, 18:663, 1992.
185. Nanjo, T. et al., Antisense suppression of proline degradation improves tolerance to freezing and salinity in *Arabidopsis thaliana. FEBS Lett.*, 461:205, 1999.
186. De Ronde, J.A., Spreeth, M.H., and Cress, W.A., Effect of antisense L-D1-pyrroline-5-carboxylate reductase transgenic soybean plants subjected to osmotic and drought stress. *Plant Growth Regul.*, 32:13, 2000.
187. Garg, A.K. et al., Trehalose accumulation in rice plants confers high tolerance levels to different abiotic stresses. *Proc. Nat. Acad. Sci. USA*, 99:15898, 2002.
188. Jang, I.C. et al., Expression of a bifunctional fusion of the *Escherichia coli* genes for trehalose-6-phosphate synthase and trehalose-6-phosphate phosphatase in transgenic rice plants increases trehalose accumulation and abiotic stress tolerance without stunting growth. *Plant Physiol.*, 131:516, 2003.
189. Karakas, B. et al., Salinity and drought tolerance of mannitol-accumulating transgenic tobacco,.*Plant Cell Environ.*, 20:609, 1997.
190. Abebe, T. et al., Tolerance of mannitol-accumulating transgenic wheat to water stress and salinity. *Plant Physiol.*, 131:1748, 2003.
191. Zhifang, G. and Loescher, W.H., Expression of a celery mannose 6-phosphate reductase in *Arabidopsis thaliana* enhances salt tolerance and induces biosynthesis of both mannitol and a glucosyl-mannitol dimer. *Plant Cell Environ.*, 26:275, 2003.

192. Sheveleva, E. et al., Increased salt and drought tolerance by D-ononitol production in transgenic *Nicotiana tabacum* L. *Plant Physiol.*, 115:1211, 1997.
193. Sheveleva, E.V. et al., Sorbitol-6-phosphate dehydrogenase expression in transgenic tobacco. High amounts of sorbitol lead to necrotic lesions. *Plant Physiol.*, 117:831, 1998.
194. Urao, T. et al., A transmembrane hybrid-type histidine kinase in *Arabidopsis* functions as an osmosensor. *Plant Cell*, 11:1743, 1999.
195. Urao, T. and Yamaguchi-Shinozaki, K., An osmosensor as a molecular tool for the genetic improvement of drought tolerant crops. In *JIRCAS Annual Report*, 2000, p. 47.
196. Xiong, L. et al., The *Arabidopsis LOS5/ABA3* locus encodes a molybdenum cofactor sulfurase and modulates cold stress- and osmotic stress-responsive gene expression. *Plant Cell*, 13:2063, 2001.
197. Xiong, L. et al., Regulation of osmotic stress responsive gene expression by *LOS6/ABA1* locus in *Arabidopsis*. *J. Biol. Chem.*, 277:8588, 2002.
198. Abraham, E. et al., Light-dependent induction of proline biosynthesis by abscisic acid and salt stress is inhibited by brassinosteroid in *Arabidopsis*. *Plant Mol Biol.*, 51:363, 2003.
199. Armengaud, P. et al., Transcriptional regulation of proline biosynthesis in *Medicago truncatula* reveals developmental and environmental specific features. *Physiol. Plant.*, 120:442, 2004.
200. Thiery, L. et al., Phospholipase-D is a negative regulator of proline biosynthesis in *Arabidopsis thaliana, J. Biol. Chem.*, 279:14812, 2004.
201. Khedr, A.H.A. et al., Proline induces the expression of salt-stress-responsive proteins and may improve the adaptation of *Pancratium maritimum* L. to salt-stress. *J. Exp. Bot.*, 54:2553, 2003.
202. Satoh, R. et al., ACTCAT, a novel cis-acting element for proline- and hypoosmolarity-responsive expression of the *ProDH* gene encoding proline dehydrogenase in *Arabidopsis*. *Plant Physiol.* 130:709, 2002.
203. Oono, Y. et al., Monitoring expression profiles of *Arabidopsis* gene expression during rehydration process after dehydration using *ca.* 7000 full-length cDNA microarray. *Plant J.*, 34:868, 2003.
204. Satoh, R. et al., A novel subgroup of bZIP proteins functions as transcriptional activators in hypoosmolarity-responsive expression of the *ProDH* gene in Arabidopsis. *Plant Cell Physiol.*, 45:309, 2004.
205. Elstner, E.F., Mechanisms of oxygen activation in different compartments of plant cells. In *Active Oxygen/Oxidative Stress and Plant Metabolism,* Pell, E.J. and Steffen, K.L., Eds., American Soc. Plant Physiol., Rockville, MD, 1991, p. 13.
206. Rich, P.R. and Bonner, W.D., Jr., The sites of superoxide anion generation in higher plant mitochondria. *Arch. Biochem. Biophys.*, 188:206, 1978.
207. Lindquist, Y. et al., Spinach glycolate oxidase and yeast flavocytochrome b_2 are structurally homologous and evolutionarily related enzymes with distinctly different function and flavin mononucleotide binding. *J. Biol. Chem.*, 266:235, 1991.
208. Asada, K., Ascorbate peroxidase — A hydrogen peroxide scavenging enzyme in plants. *Physiol. Plant.*, 85:235, 1992.
209. Scandalios, J.G., Responses of plant antioxidant defence genes to environmental stress. *Adv. Genet.*, 28:1–41, 1990.
210. Fenton, H.J.H., Oxidation of certain organic acids in the presence of ferrous salts. *Proc. Chem. Soc.,* 25:224, 1899.

211. Haber, F. and Weiss, J., The catalytic decomposition of hydrogen peroxide by iron salts. *Proc. Royal Soc. A.*, 147:332, 1934.
212. Frankel, E.N., Chemistry of free radical and singlet oxidation of lipids. *Prog. Lipid Res.*, 23:197, 1985.
213. Farr, S.B. and Kogoma, T., Oxidative stress responses in *Escherichia coli* and *Salmonella typhimurium*. *Microbiol. Rev.*, 55:561, 1991.
214. Imlay, J.A. and Linn, S., DNA damage and oxygen radical toxicity. *Science*, 240:1302, 1986.
215. Arora, A., Sairam, R.K., and Srivastava, G.C., Oxidative stress and antioxidative system in plants. *Curr. Sci.*, 82:1, 2002.
216. Sairam, R.K. and Tyagi, A., Physiology and molecular biology of salinity stress tolerance in plants. *Curr. Sci.*, 86:407, 2004.
217. Sairam, R.K., Deshmukh, P.S., and Saxena, D.C., Role of antioxidant systems in wheat genotypes tolerance to water stress. *Biol. Plant.*, 41:384, 1998.
218. Sairam, R.K., Srivastava, G.C., and Saxena, D.C., Increased antioxidant activity under elevated temperature: A mechanism of heat stress tolerance in wheat genotypes. *Biol. Plant.*, 43:245, 2000.
219. Sairam, R.K., Chandrasekhar, V., and Srivastava, G.C., Comparison of hexaploid and tetraploid wheat cultivars in their response to water stress. *Biol. Plant.*, 44:89, 2001.
220. Chen, Y.W., Shao, G.H., and Chang, R.Z., The effect of salt stress on superoxide dismutase in various organelles of cotyledons of soybean seedlings. *Acta Agronomica Sinica*, 23:214, 1997.
221. Gueta-Dahan, Y. et al., Salt and oxidative stress: Similar and specific responses and their relation to salt tolerance in Citrus. *Planta*, 203:460, 1997.
222. Sreenivasulu, N. et al., Differential response of antioxidant components to salinity stress in salt tolerant and salt sensitive seedlings of foxtail millet (*Setaria italica*). *Phyisol. Plant.*, 109:435, 2000.
223. Hernandez, J.A., Jimerez, A., Mullineaux, P.M., and Sevilla, P.F., Tolerance of pea (*Pisum sativum*) to long-term salt stress is associated with induction of antioxidant defenses. *Plant Cell Biol.*, 23:853, 2000.
224. Sairam, R.K. and Srivastava, G.C., Changes in antioxidant activity in sub-cellular fractions of tolerant and susceptible wheat genotypes in response to long-term salt stress. *Plant Sci.*, 162:897, 2002.
225. Roxas, V.P. et al., Stress tolerance in transgenic tobacco seedlings that overexpress glutathione S-transferase/glutathione peroxidase. *Plant Cell Physiol.*, 41:1229, 2000.
226. Tanaka, Y. et al., Salt tolerance of transgenic rice over-expressing yeast mitochondrial Mn-SOD in chloroplasts. *Plant Sci.*, 148:131, 1999.
227. Arisi, A.-C.M. et al., Over-expression of iron-superoxide dismutase in transformed poplar modifies at low CO_2 partial pressure or following exposure to the peroxidant herbicide methyl viologen. *Plant Physiol.*, 117:565, 1998.
228. Noctor, G. et al., Manipulation of glutathione and amino acid biosynthesis in the chloroplast, *Plant Physiol.*, 118:471, 1998.
229. McKersie, B.D., Bowley, S.R., and Jones, K.S., Winter survival of transgenic alfalfa overexpression superoxide dismutase. *Plant Physiol.*, 119:839, 1999.
230. Baier, M. et al., Antisense suppression of 2-cysteine peroxiredoxin in *Arabidopsis* specifically enhances the activities and expression of enzymes associated with ascorbate metabolism but not glutathione metabolism. *Plant Physiol.*, 124:823, 2000.

231. Veena, Reddy, S.V., and Sopory, S.K., Glycoxalase I from *Brassica juncea*: Molecular cloning, regulation and its over-expression confer tolerance in transgenic tobacco under stress. *Plant J.*, 17:385, 1999.
232. Pastori, G.M., Mullineaux, P.M., and Foyer, C.H., Post transcriptional regulation prevents accumulation of glutathione reductase protein and activity in the bundle sheath cells of maize. *Plant Physiol.*, 122:667, 2000.
233. Tyagi, A., Santha, I.M., and Mehta, S.L., Molecular response to water stress in *Lathyrus sativus*. *J. Plant Biochem. Biotech.*, 4:47, 1995.
234. Badawi, G.H. et al., Over-expression of ascorbate peroxidase in tobacco chloroplasts enhances the tolerance to salt stress and water deficit. *Physiol. Plant*, 121:231, 2004.
235. Tsugane, K. et al., A recessive *Arabidopsis* mutant that grows photoautotrophically under salt stress shows enhanced active oxygen detoxification. *Plant Cell*, 11:1195, 1999.
236. Samis, K., Bowley, S., and McKersie, B., Pyramiding Mn-superoxide dismutase transgenes to improve persistence and biomass production in alfalfa. *J. Exp. Bot.*, 53:1343, 2002.
237. Alvarez, M.E. et al., Reactive oxygen intermediates mediate a systemic signal network in the establishment of plant immunity. *Cell* 92:773, 1992.
238. Pei, Z.M. et al., Calcium channels activated by hydrogen peroxide mediate abscisic acid signaling in guard cells. *Nature*, 406:731, 2000.
239. Ingram, J. and Bartels, D., The molecular basis of dehydration tolerance in plants. *Ann. Rev. Plant Physiol. Plant Mol. Biol.* 47:377, 1996.
240. Wise, M.J. and Tunnacliffe, A., POPP the question: What do LEA proteins do? *Trends Plant Sci.*, 9:13, 2004.
241. Shinozaki, K. and Yamaguchi-Shinozaki, K., Molecular response to dehydration and low temperature: Differences and cross-talk between two stress signaling pathways. *Curr. Opin. Plant Biol.*, 3:217, 2000.
242. Moons, A. et al., Molecular and physiological responses to abscisic acid and salts in roots of salt-sensitive and salt-tolerant Indica rice varieties. *Plant Physiol.*, 107:177, 1995.
243. Xu, D. et al., Expression of a late embryogenesis abundant protein gene, *HVA1*, from barley confers tolerance to water deficit and salt stress in transgenic rice. *Plant Physiol.*, 110:249, 1996.
244. McCubbin, W.D., Kay, C.M., and Lane B.G., Hydrodynamic and optical properties of the wheat germ Em protein. *Can. J. Biochem. Cell Biol.*, 63:803, 1985.
245. Bostock, R.M. and Quairano, R.S., Regulation of Em gene expression in rice. *Plant Physiol.*, 98:1356, 1992.
246. Marcotte, W.R., Bayley, C.C., and Quatrano, R.S., Regulation of a wheat promoter by abscisic acid in rice protoplasts. *Nature*, 335:454, 1988.
247. Almoguera, C. and Jordeno, J., Developmental and environmental concurrent expression of sunflower dry seed stored low molecular weight heat-shock protein and Lea mRNAs. *J. Plant Mol. Biol.*, 19:781, 1992.
248. Mundy, J. and Chua, N.H., Developmental and environmental concurrent expression of sunflower dry seed stored low molecular weight heat-shock protein and Lea mRNAs. *EMBO J.*, 7:2279, 1988.
249. Gilmour, S.J., Artus, N.N., and Thomashaw, M.F., cDNA sequence analysis and expression of two cold-regulated genes of *Arabidopsis thaliana*. *Plant Mol. Biol.*, 18:13, 1992.

250. Dure, L.S., The Lea proteins of higher plants. In *Control of Plant Gene Expression*, Verma, D.P.S., Ed., CRC Press, Boca Raton, FL, 1993, p. 325.
251. Hsing, Y.C. et al. Unusual sequence of group 3 LEA mRNA inducible by maturation or drying in soybean seeds. *Plant Mol. Biol.*, 29:863, 1995.
252. Moons, A., De Keyser, A., and Van Montagu, M., A group 3 LEA cDNA of rice, responsive to abscisic acide, but not to jasmonic acid, shows variety-specific differences in salt stress response. *Gene*, 191:197, 1997.
253. Naot, D. et al., Drought, heat, and salt stress induce the expression of a citrus homologue of an atypical late embryogenesis Lea5 gene. *Plant Mol. Biol.*, 27:619, 1995.
254. Dure, L.S., A repeating 11-mer amino acid motif and plant desiccation in C_3 and C_4 plants. *Plant Physiol.*, 59:86, 1993.
255. Cohen, A. et al., Organ-specific and environmentally regulated expression of two abscisic acid induced genes of tomato. *Plant Physiol.*, 97:1367, 1991.
256. Galau, G.A., Wang, H.Y.C., and Hughes, D.W., Cotton Lea5 and Lea14 encode atypical late embryogenesis-abundant proteins. *Plant Physiol.*, 101:695, 1993.
257. Piatkowski, D. et al., Characterization of five abscisic acid responsive cDNA clones isolated from the desiccation tolerant planti *Craterostigma plantagineum* and their relationship to other water stress genes. *Plant Physiol.*, 94:447, 1990.
258. Thomashow, M.F., Plant cold acclimation: Freezing tolerance genes and regulatory mechanisms. *Ann. Rev. Plant Physiol. Plant Mol. Biol.*, 50:571, 1999.
259. Narusaka, Y. et al., Interaction between two *cis*-acting elements, *ABRE* and *DRE*, in ABA-dependent expression of *Arabidopsis rd29A* gene in response to dehydration and high-salinity stresses. *Plant J.*, 34:137, 2003.
260. Knight, H. et al., Abscisic acid induces CBF gene transcription and subsequent induction of cold-regulated genes via the CRT promoter element. *Plant Physiol.*, 135:1710, 2004.
261. Urao, T. et al., Two genes that encode Ca^{2+}-dependent protein kinases are induced by drought and high-salt stresses in *Arabidopsis thaliana*, *Mol. Gen. Genet.*, 244:331, 1994.
262. Saijo, Y. et al., Over-expression of a single Ca^{2+} dependent protein kinase confers both cold and salt/drought tolerance on rice plants, *Plant J.*, 23:319, 2000.
263. Chehab, E.W. et al., Autophosphorylation and subcellular localization dynamics of a salt- and water deficit-induced calcium-dependent protein kinase from ice plant, *Plant Physiol.*, 135:1430, 2004.
264. Patharkar, O.R. and Cushman, J.C., A stress-induced calcium-dependent protein kinase from Mesembryanthemum crystallinum phosphorylates a two-component response regulator, *Plant J.*, 24:679, 2000.
265. Sheen, J., Ca^{2+}-dependent protein kinases and stress signal transduction in plants, *Science*, 274:1900, 1996.
266. Townley, H.E. and Knight, M.R., Calmodulin as a potential negative regulator of *Arabidopsis COR* gene expression. *Plant Physiol.*, 128:1169, 1997.
267. Viswanathan, C. and Zhu, J.K., Molecular genetic analysis of cold regulated gene transcription. *Phil. Trans. R. Soc. London B. Biol. Sci.*, 357:877, 2002.
268. Hwang, I., Sze, H., and Harper, J.F., A calcium-dependent protein kinase can inhibit a calmodulin-stimulated Ca2+ pump (ACA2) located in the endoplasmic reticulum of *Arabidopsis. Proc. Natl. Acad. Sci. USA*, 97:6224, 2000.
269. Perruc, E., et al., A novel calmodulin-binding protein functions as a negative regulator of osmotic stress tolerance in *Arabidopsis thaliana* seedlings. *Plant J.*, 38:410, 2004.

270. Yang, T. et al., Calcium/calmodulin up-regulates a cytoplasmic receptor-like kinase in plants. *J. Biol. Chem.*, 279:42552, 2004.
271. Uno, Y. et al., Novel *Arabidopsis* bZIP transcription factors involved in an abscisic-acid-dependent signal transduction pathway under drought and high salinity conditions. *Proc. Natl. Acad. Sci. USA*, 97:11632, 2000.
272. Abe, H. et al., *Arabidopsis* AtMYC2 (bHLH) and AtMYB2 (MYB) function as transcriptional activators in abscisic acid signaling. *Plant Cell,* 15:63, 2003.
273. Kang, J.Y. et al., *Arabidopsis* basic leucine zipper proteins that mediate stress-responsive abscisic acid signaling. *Plant Cell*, 14:343, 2002.
274. Villalobos, M.A., Bartels, D., and Iturriaga, G., Stress tolerance and glucose insensitive phenotypes in *Arabidopsis* over-expressing the *CpMYB10* transcription factor gene. *Plant Physiol.*, 135:309, 2004.
275. Liu, Q. et al., Two transcription factors, DREB1 and DREB2, with an EREBP/AP2 DNA binding domain separate two cellular signal transduction pathways in drought- and low-temperature-responsive gene expression, respectively, in Arabidopsis. *Plant Cell*, 10:1391, 1998.
276. Haake, V. et al., Transcription factor CBF4 is a regulator of drought adaptation in *Arabidopsis. Plant Physiol.*, 130:639, 2002.
277. Dubouzet, J.G. et al., *OsDREB* genes in rice, *Oryza sativa* L., encode transcription activators that function in drought-, high-salt- and cold-responsive gene expression. *Plant J.*, 33:751, 2003.
278. Jaglo-Ottosen, K.R. et al., *Arabidopsis* CBF1 overexpression induces *cor* genes and enhances freezing tolerance. *Science*, 280:104, 1998.
279. Kasuga, M. et al., Improving plant drought, salt, and freezing tolerance by gene transfer of a single stress-inducible transcription factor. *Nat. Biotechnol.*, 17:287, 1999.
280. Jaglo, K.R. et al., Components of the *Arabidopsis* C-repeat/dehydration-responsive element binding factor cold-response pathway are conserved in *Brassica napus* and other plant species. *Plant Physiol.*, 127:910, 2001.
281. Fowler, S. and Thomashow, M.F., Arabidopsis transcriptome profiling indicates that multiple regulatory pathways are activated during cold acclimation in addition to the CBF cold response pathway. *Plant Cell*, 14:1675, 2002.
282. Chinnusamy, V. et al., ICE1: A regulator of cold-induced transcriptome and freezing tolerance in *Arabidopsis. Genes Dev.*, 17:1043, 2003.
283. Park, J.M. et al., Overexpression of the tobacco *Tsi1* gene encoding an EREBP/AP2–type transcription factor enhances resistance against pathogen attack and osmotic stress in tobacco. *Plant Cell*, 13:1035, 2001.
284. Pellegrineschi, A. et al., Progress in the genetic engineering of wheat for water-limited conditions. *JIRCAS Working Report*, 2002:55, 2002.
285. Hsieh, T.H. et al., Tomato plants ectopically expressing *Arabidopsis CBF1* show enhanced resistance to water deficit stress. *Plant Physiol.*, 130:618, 2002.
286. Hsieh, T.H. et al., Heterology expression of the Arabidopsis *C-Repeat/Dehydration Response Element Binding Factor 1* gene confers elevated tolerance to chilling and oxidative stresses in transgenic tomato. *Plant Physiol.*, 129:1086, 2002.
287. Gilmour, S.J. et al., Overexpression of the *Arabidopsis CBF3* transcriptional activator mimics multiple biochemical changes associated with cold acclimation. *Plant Physiol.*, 124:1854, 2000.

288. Gomez, J.M. et al., Differential response of antioxidative enzymes of chloroplast and mitochondria to long term NaCl stress of pea plants. *Free Radic. Res.*, 31(Suppl.):11, 1999.
289. Guan, L.Q.M., Zhao, J., and Scandalios, J.G., Cis-elements and trans-factors that regulate expression of maize *Cat1* antioxidant gene in response to ABA and osmotic stress: H_2O_2 is the likely intermediary signaling molecule for the response. *Plant J.*, 22:87, 2000.
290. Kwak, J.M. et al., NADPH oxidase *AtrbohD* and *AtrbohF* genes function in ROS-dependent ABA signaling in *Arabidopsis. EMBO J.*, 22:2623, 2003.
291. Murata, Y. et al., Abscisic acid activation of plasma membrane Ca^{2+} channels in guard cells requires cytosolic NAD(P)H and is differentially disrupted upstream and downstream of reactive oxygen species production in *abi1-1* and *abi2-1* protein phosphatase 2C mutants. *Plant Cell*, 12:2513, 2001.
292. Levine, A. et al., H_2O_2 from the oxidative burst orchestrates the plant hypersensitive disease resistance response. *Cell*, 79:583, 1994.
293. Chamnongpol, S. et al., Defense activation and enhanced pathogen tolerance induced by H_2O_2 in transgenic tobacco. *Proc. Natl. Acad. Sci. USA*, 95:5818, 1998.
294. Vranova, E., Inze, D., and Van Breusegem, F., Signal transduction during oxidative stress. *J. Exp. Bot.*, 53:1227, 2002.
295. Apel, K. and Hirt, H., Reactive oxygen species: Metabolism, oxidative stress and signal transduction. *Ann. Rev. Plant Biol.*, 55:373, 2004.
296. Laloi, C., Apel, K., and Danon, A., Reactive oxygen signalling: The latest news. *Curr. Opinion Plant Biol.*, 7:323, 2004.
297. Chinnusamy, V. and Zhu, J.K., Plant salt tolerance. *Topics. Curr. Genet.*, 4:241, 2003.
298. Chinnusamy, V., Schumaker, K., and Zhu, J.K., Molecular genetic perspectives on cross-talk and specificity in abiotic stress signalling in plants. *J. Exp. Bot.*, 55:225, 2004.
299. Arabidopsis Genome Initiative, Analysis of the genome sequence of the flowering plant *Arabidopsis thaliana. Nature*, 408:796, 2000.
300. Ichimura, K. et al., Isolation of ATMEKK1 (a MAP kinase kinase kinase)-interacting proteins and analysis of a MAP kinase cascade in *Arabidopsis. Biochem. Biophys. Res. Comm.*, 253:532, 1998.
301. Teige, M. et al., The MKK2 pathway mediates cold and salt stress signaling in *Arabidopsis. Mol. Cell,* 15:141, 2004.
302. Mizoguchi, T. et al., A gene encoding a mitogen-activated protein kinase kinase kinase is induced simultaneously with genes for a mitogen-activated protein kinase and an S6 ribosomal protein kinase by touch, cold, and water stress in *Arabidopsis thaliana. Proc. Nat. Acad. Sci. USA*, 93:765, 1996.
303. Ichimura, K. et al., Various Abiotic stresses rapidly activate *Arabidopsis* MAP kinases ATMPK4 and ATMPK6. *Plant J.*, 24:655, 2000.
304. Huang, Y. et al., ATMPK4, an *Arabidopsis* homolog of mitogen-activated protein kinase, is activated *in vitro* by AtMEK1 through threonine phosphorylation. *Plant Physiol.*, 122:1301, 2000.
305. Moon, H. et al., NDP kinase 2 interacts with two oxidative stress-activated MAPKs to regulate cellular redox state and enhances multiple stress tolerance in transgenic plants. *Proc. Nat. Acad. Sci. USA*, 100:358, 2003.

306. Kovtun, Y. et al., Functional analysis of oxidative stress-activated mitogen-activated protein kinase cascade in plants. *Proc. Nat. Acad. Sci. USA*, 97:2940, 2000.
307. Ulm, R. et al., Distinct regulation of salinity and genotoxic stress responses by *Arabidopsis* MAP kinase phosphatase 1. *EMBO J.*, 21:6483, 2002.
308. Xiong, L. and Yang, Y., Disease resistance and abiotic stress tolerance in rice are inversely modulated by an abscisic acid–inducible mitogen-activated protein kinase. *Plant Cell*, 15:745, 2003.
309. Bartels, D. et al., Desiccation related gene products analyzed in a resurrection plant and in barley embryos. In *Plant Response to Cellular Dehydration during Environmental Stress: Current Topics in Plant Physiology,* Close, T.J. and Bray, E.A., Eds., American Society of Plant Physiologist Series, Rockville, MD, 1993, p. 119.

7 Cellular and Molecular Mechanisms of Plant Tolerance to Waterlogging

B. Ricard, S. Aschi-Smiti, I. Gharbi, and R. Brouquisse

CONTENTS

7.1 Introduction 178
7.2 Stress Perception 178
    7.2.1 Potential Environmental Signals 178
    7.2.2 Possible Cellular Sensors of Oxygen 179
        7.2.2.1 Mitochondria and Other ROS Producers 179
        7.2.2.2 Phytohemoglobin 180
    7.2.3 ATP and Metabolic Flux Sensors 180
    7.2.4 pH Sensors 181
    7.2.5 Second Messengers 182
    7.2.6 Signal Transduction 182
7.3 Mechanisms of Waterlogging Tolerance 182
    7.3.1 Morphology and Anatomy 183
        7.3.1.1 Aerenchyma 183
        7.3.1.2 Nodal or Adventitious Roots 184
        7.3.1.3 Barriers to Radial Oxygen Loss 184
    7.3.2 Metabolism 184
        7.3.2.1 Growth Inhibition and Cell Death in O_2-Limited Organs 184
        7.3.2.2 Regulation of Anaerobic Cell Metabolism 186
        7.3.2.3 Post-Anoxic Damage and Recovery 193
7.4 Genetic Manipulation of Tolerance to Waterlogging 194
    7.4.1 Genetic Engineering 194
        7.4.1.1 Transcription Factors 194
        7.4.1.2 Fueling of Glycolysis 194
        7.4.1.3 Fermentation Pathways 195

	7.4.1.4	Hormone Levels	195
	7.4.1.5	Phytohemoglobin	196
	7.4.1.6	Protection against Post-Anoxic Injury	196
	7.4.2	Production of Stress-Tolerant Germplasm	196
7.5	Evaluation of Plant Tolerance to Waterlogging		197
7.6	Summary and Prospects		200
References			200

7.1 INTRODUCTION

Waterlogging can occur in diverse environments and vary in intensity and duration from temporary to intermittent or continuous, and in timing during different periods of the plant growth cycle. It is not only a major constraint for crop productivity but strongly affects species distribution. Waterlogging is distinguished from flooding by the limitation of excess water to the root system. The adverse effects of waterlogging have been attributed principally to lowered oxygen supply, which is exacerbated by soil microorganisms that consume oxygen and may also diminish the supply of nitrate. Highly toxic microelements accumulate, as do CO_2 and ethylene. Tolerance to waterlogging can be defined either in physiological terms as maintenance of high biomass production or minimal growth reduction due to waterlogging, or in agronomic terms as achieving high grain yields or less reduction in yields under waterlogged conditions. Tolerance is thus a complex trait, undoubtedly involving numerous mechanisms at cellular and molecular levels. The initial step must be the capacity to perceive and transmit a signal to its point of action. Some progress has been made in understanding mechanisms of waterlogging stress sensing and regulation, although it is still unclear which of many possible signals and which sensing mechanisms are involved in eliciting an appropriate plant response. The subject of perception and signaling of soil water saturation (also see [1]) is discussed in the first part of this chapter. The mechanisms of anoxic tolerance have previously been extensively reviewed [2,3]. This chapter succinctly presents and updates our understanding of plant tolerance to waterlogging. It also reviews studies with an aim to improve waterlogging tolerance by genetic manipulation and finally, addresses how the level of plant tolerance to waterlogging can be evaluated. Varietal differences in tolerance at the germination stage often differ from tolerance at later stages of development, in support of the hypothesis that different mechanisms are involved. This chapter discusses mechanisms of waterlogging tolerance of plants with developed root systems at the vegetative stage of their development.

7.2 STRESS PERCEPTION

7.2.1 POTENTIAL ENVIRONMENTAL SIGNALS

The first event in waterlogging is soil water saturation. A recent study has identified an osmosensor in *Arabidopsis* [4], which has led to speculations that plants could detect changes in water homeostasis [1]. However, excess soil water

Cellular and Molecular Mechanisms of Plant Tolerance to Waterlogging

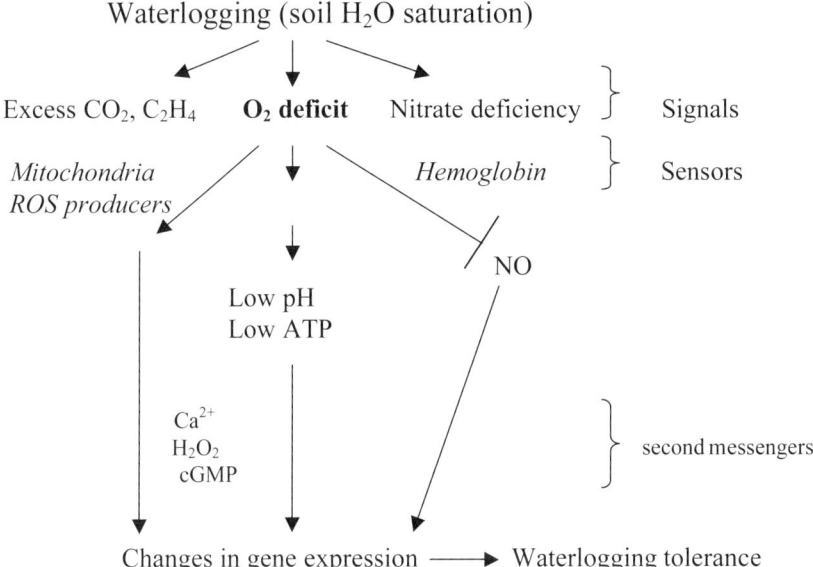

FIGURE 7.1 Schematic diagram of main potential signals resulting from soil water saturation and possible transduction pathways for oxygen deficit. Decreasing oxygen levels could be detected directly by mitochondria or other ROS producers. Phytohemoglobin could modulate the reponse by interference with NO signaling. Low oxygen could also trigger cellular and metabolic changes that then initiate signal cascades, which result in changes in gene expression patterns leading to adaptive responses that increase plant tolerance to waterlogging.

leads to numerous other soil physicochemical changes. The earliest sign of waterlogging is the depletion of oxygen, which is generally considered to constitute the primary signal. The diffusion rate of oxygen in water is 10,000-fold slower than in air, which results in increased levels of phytotoxic byproducts of plant metabolism, such as carbon dioxide, ethylene, and lactic acid. Soil pH and redox potential are reduced, thereby affecting the availability or restriction of nutrients. These soil signals could be detected by a number of potential sensors (Figure 7.1).

7.2.2 Possible Cellular Sensors of Oxygen

7.2.2.1 Mitochondria and Other ROS Producers

As major cellular oxygen consumers, mitochondria have been implicated in the hypoxic response of animals and constitute a prime suspect in plants, more probably in connection with the production of reactive oxygen species (ROS) as signaling molecules. Cytochrome oxidase appears unlikely to be directly involved because hypoxic responses occur at oxygen concentrations above its K_m (O_2). Mitochondria are not the only source of ROS. In *Arabidopsis*, another ROS generator, a flavin-binding oxidase, was found to be necessary for increased expression of alcohol dehydrogenase (ADH) but also of Rop GTPase activating protein (RopGAP) under low oxygen conditions [5]. Rop signaling thus involves a feedback mechanism that has been proposed to allow for flexibility in the regulation of anaerobic metabolism, the management of carbohydrate consumption, and the avoidance of oxidative stress [6].

Plants possess two classes of GTP-binding proteins (G-proteins). Rop (monomeric RHO of plant) G-proteins are plant-specific and known to regulate many cellular processes, such as Ca^{2+} gradients, hormonal responses, and hydrogen peroxide (H_2O_2) production [7]. Rop is signaling competent in its GTP-bound form and is inactive in its GDP-bound form. Hypoxia was shown to activate Rop signaling of H_2O_2 production via a flavin-binding oxidase. ROS production induced ADH expression under low oxygen but also that of RopGAP, resulting in the hydrolysis of Rop-GTP to Rop-GDP and the attenuation of Rop signaling. Moreover, a RopGAP loss-of-function mutant and one that constitutively bound GTP or GDP were all hypersensitive to hypoxia, further indicating that transient activation of Rop signaling and moderate ADH induction were both necessary for hypoxic tolerance [5,6].

7.2.2.2 Phytohemoglobin

Oxygen sensors have been described in many living aerobic organisms and usually involve oxygen as a ligand binding reversibly to a haem moiety. The binding or nonbinding of oxygen results in an allosteric modification thought to initiate a signaling cascade that regulates genes responsible for tolerance. Two types of hemoglobin have been identified in plants, symbiotic leghemoglobin and a nonsymbiotic phyto or class 1 hemoglobin. Several roles have been proposed for the nonsymbiotic phytohemoglobin (pHb), among which is that of an oxygen sensor. However, the high avidity of pHb for oxygen indicates that the protein is probably oxygenated at oxygen concentrations well below those at which anaerobic processes are activated, making a direct role as an oxygen sensor unlikely [8]. A more plausible possibility is that pHb plays a role in the modulation of the low oxygen response, possibly through the regulation of signaling by nitric oxide (NO).

Nitric oxide, probably produced by nitrate reductase and a plant-specific NO synthase, is known to be involved in H_2O_2-mediated signaling. The rise in NO levels during hypoxia appeared to be modulated by pHb, whose levels similarly rose during hypoxia. By converting NO to nitrate, pHb could not only contribute to the detoxification of NO, but could also regulate NO-mediated signaling [9–11].

7.2.3 ATP and Metabolic Flux Sensors

A decrease in ambient oxygen may not itself or alone be the perceived signal but could trigger metabolic responses, which then initiate the signaling cascade (Figure 7.1). The immediate consequence of the depletion of oxygen is a reduction in respiration, thus diminishing ATP generation and leading to a decrease in the ATP/ADP ratio and the adenylate energy charge. Changes in ATP levels itself could conceivably serve as a signal for subsequent adaptive responses, perhaps by regulating ion channel activity as in the case of anoxia-tolerant freshwater turtles and fish [12] or by modifying calcium fluxes [1].

Low ATP levels lead to an energy crisis to which plants can respond by increasing glycolytic flux (Pasteur effect) and switching to alcoholic fermentation. High glycolytic flux can be supported only if sufficient carbohydrate is available. Phloem transport of sucrose from leaves to roots, import of sucrose into cells via plasmodesmata or transporters, and sucrose cleavage by apoplastic or intracellular invertases or sucrose synthase (Susy) are therefore crucial for the supply of hexoses, but may also initiate sugar signals (for review, cf [13]). Hexokinase (HK) catalyzes the first irreversible step in glycolysis but also functions as a sugar sensor [14], and has recently been implicated in hormonal and light signaling as well [15]. The enzyme could be directly involved in sensing or mediating an increase in glycolytic flux. Its induction during hypoxia in maize (*Zea mays*), rice (*Oryza sativa*), and in *Echinochloa* lends support for this role.

An increase in glycolytic flux, however, can only be transient and permit the root system to survive short-term hypoxia or anoxia. Normoxic ATP levels are never attained. Long-term survival in anoxia-tolerant vertebrates is known to involve the matching of ATP production with ATP utilization, not by increasing ATP production but by decreasing the utilization of ATP to levels met by anaerobic metabolism. Thus, longer-term survival in plants, such as that shown by rhizomes, bulbs, and seeds, can probably be explained by the inhibition of ATP-consuming processes. Inhibition of ATP-consuming pathways involved in biosynthesis, growth, and wounding responses has also been shown to occur in potato (*Solanum tuberosum*) tuber discs in response to low oxygen [16,17]. A switch to pathways that consume less ATP is another means to conserve ATP and to improve survival. Such a strategy would explain the importance of the Susy over the invertase pathway for sucrose degradation during low oxygen conditions, as shown by the enhanced sensitivity of maize double mutant for Susy [18] as well as by the pattern of repression and induction of invertase and Susy genes in maize root tips [19,20] and in potato tubers [17].

7.2.4 pH Sensors

A decline in pH is another early event following root hypoxia, which might serve as a signal for adaptive responses (Figure 7.1). For instance, cytosolic pH is known to regulate gene expression in fungi [21] and recently was discovered to regulate water transport in roots during low-oxygen stress [22]. Moreover, protein synthesis under oxygen deprivation has been shown to be sensitive to pH, with many polyribosomes being stalled when translated at low pH [23]. This suggests that the translation machinery can sense intracellular pH, thus providing a means to modulate gene expression.

7.2.5 SECOND MESSENGERS

A number of molecules have been proposed to act as second messengers in the hypoxic signaling cascade. These include H_2O_2, Ca^{2+}, and cyclic GMP (Figure 7.1). Rop signaling leading to H_2O_2 production could involve a flavin-binding oxidase, as suggested by the blockage of hypoxia-induced increase in H_2O_2 production by a flavin analogue [5], but does not exclude a mitochondrial source. Disruption of the mitochondrial electron chain by low oxygen could also lead to H_2O_2 production.

Evidence has shown that calcium fluxes, probably at least in part from mitochondrial storages, are involved in the hypoxic signal transduction pathway. A transient rise in cytosolic Ca^{2+} levels occurred very early among hypoxic responses and was localized near mitochondria [24]. Interestingly, ROS produced by the Ca^{2+}-dependent plasma membrane NAD(P)H oxidase has been shown to activate plasma membrane Ca^{2+} channels [25]. Furthermore, changes in Ca^{2+} fluxes in hypoxic maize roots led to ethylene biosynthesis and subsequent aerenchyma formation [26,27]. Calcium is thought to act as a switch that activates calcium-dependent mediators, such as calcium-dependent protein kinases, calcinulin, and calmodulins. This is in accord with results showing that the synthesis of a Ca^{2+}/calmodulin-dependent glutamate decarboxylase is stimulated by cytosolic Ca^{2+} in response to flooding stress [28,29] and that a calmodulinlike calcium-binding protein is upregulated in hypoxic *Arabidopsis* root cultures [30]. A rapid (2 min) and transient increase in cyclic GMP content in roots and coleoptiles of rice seedlings occurred under anoxia [31]. This suggests that cyclic GMP may be involved in hypoxic signaling in plants and in the response of mammalian cells to ischemia and anoxia [32].

7.2.6 SIGNAL TRANSDUCTION

The transmission of these signals and the activation of an adaptive physiological response probably involve a battery of signaling cascades shared by other biotic and abiotic stresses, and further involving considerable crosstalks. Secondary messengers such as Ca^{2+} [33] regulated by calmodulin [34], H_2O_2, and NO may all play a role in stress-signal transduction. Plant growth regulators such as ethylene [35,36], abscisic acid (ABA) [37,38], gibberellic acid (GA) [39], auxin (IAA) [30], and cytokinins (CK) [40] may also be involved in stress signal transduction.

7.3 MECHANISMS OF WATERLOGGING TOLERANCE

The first adverse effect of waterlogging attributed to partial oxygen shortage is stunted root growth [41] followed by root tip death, necrosis, and eventually ablation of entire roots [42]. Root-to-shoot signaling occurring during the first hours of soil flooding [43] results in epinastic leaf curvature and stomatal closure. The former is mediated by an ethylene signal originated in roots [44,45] and the latter is in association with increases in ABA in leaves.

Stomatal closure could be due to increases in ABA supply from roots to shoots or water deficit due to the loss of hydraulic conductance [46]. The resultant reduction in photosynthesis is attributed to diffusion limitations due to stomatal closure and metabolic inhibition, which affects shoot growth and biomass production.

7.3.1 MORPHOLOGY AND ANATOMY

Tolerance to waterlogging can be achieved by the avoidance of low oxygen stress through morphological and anatomical adaptations that improve gas exchange and transport. These adaptations include the development of nodal or adventitious roots, increases in aerenchyma and root porosity, and possibly barriers to radial oxygen loss (ROL).

7.3.1.1 Aerenchyma

Aerenchyma are interconnected gas-filled intercellular spaces, which are categorized as schizogenous or lysigenous depending on whether formed by cell separation or selective death. Aerenchyma are constitutive in many plants inhabiting aquatic, wetland, or flood-prone environments, but can also occur in some dryland plants [47]. In dryland cereals such as maize, barley (*Hordeum vulgare*), and wheat (*Triticum aestivum*), aerenchyma are produced in response to flooding, whereas in many wetland species, aerenchyma constitutively formed during development become more extensive. Evidence to support the importance of internal aeration for growth in waterlogged soils includes mathematical models that show the correlation of internal aeration capacity and relative waterlogging tolerance [48].

The formation of aerenchyma in roots depends first on the perception of a decrease in oxygen levels, next on the expression of a set of genes involved in cell death and lysis, and finally on the breakdown of cell components and structure. How oxygen shortage is sensed and transduced is not well understood, but strong evidence implies that the accumulation of endogenous ethylene in submerged tissues may be involved in sensing and transduction of oxygen shortage [27]. Partial oxygen shortage initially promotes the biosynthesis of ethylene [49], which is then trapped within roots by water covering [27]. Finally, ethylene accumulation induces programmed cell death in target cells of the cortex, creating longitudinally interconnected gas spaces that replace the original files of cells [50]. The sensing of ethylene is relayed by a signal transduction cascade that involves Ca^{2+} [51], protein kinases [26], and cytoplasmic acidification [52]. Surprisingly rapid (<0.5 days [53]) cell wall changes closely accompany cytoplasmic changes reminiscent of animal cell apoptosis. De-esterification of pectins [53] are followed by debris cleanup mediated by hydrolytic enzymes such as cellulase [54], xylanases [55], and possibly xyloglucan *endo*transglycosylase (XET [56]). Additional cell wall and cytoplasmic degradation enzymes are also likely to be involved in this process.

7.3.1.2 Nodal or Adventitious Roots

A widespread adaptive response to waterlogging is the emergence of aerenchymatous adventitious roots from the shoot base [57]. Such roots may also develop from aerial parts, such as stem or hypocotyl nodes [42] or the upper part of the primary root [43]. The expression of cell-cycle gene *cdc2* differs in internodal stem or root meristems [35], suggesting that different mechanisms regulate the initiation of adventitious roots from a stem or root. Root and stem meristems are also activated by different hormones, gibberellic acid and ethylene, respectively.

The proposed role of ethylene in adventitious root formation has varied from major to minor, from direct to indirect, which remains unclear. Auxin has been shown to induce the formation of adventitious roots in *Rumex* [58] and in excised hypocotyls of *Pinus taeda* [59] but also increases the biosynthesis of ethylene [60].

7.3.1.3 Barriers to Radial Oxygen Loss

Roots of many (but not all) wetland species not only contain large volumes of aerenchyma (root porosity can reach 55%) but a barrier impermeable to radial oxygen loss (ROL) often occurs in basal zones [61]. Little is known of the anatomical and physiological basis of the barrier to ROL, which may consist of an exodermis originating from sclerenchymatous fibers, thickly packed cells in outer tissues, suberin deposits, and lignification. Such a barrier can enhance longitudinal oxygen diffusion in aerenchyma and possibly protect roots against soil-derived toxins [62] but a "tight" barrier might also impede water and nutrient uptake. The inducibility of the barrier to ROL in some wetland species and the absence of a tight barrier in all dryland species examined to date have led to the suggestion that the capacity for an inducible ROL could be an adaptive trait for transient waterlogging [63].

7.3.2 Metabolism

7.3.2.1 Growth Inhibition and Cell Death in O_2-Limited Organs

Transient flooding or waterlogging can rapidly lead to low oxygen levels within a root. Short exposure to complete absence of oxygen is sufficient to kill nonacclimated roots, that is, roots which were not previously exposed to low oxygen [64]. However, small amounts of external oxygen supply (6 to 10 µM O_2 in solution) are sufficient to prevent cell death and allow recovery after return to aerated conditions. Under anoxic conditions, when the activity of cytochrome oxidase (Km [O_2] is close to 140 nM [65]) ATP formation through oxidative phosphorylation is inhibited and ATP is produced by fermentation [64]. Because the efficiency of ATP formation in fermenting cells is low and because oxygen is required in various metabolic pathways, such as heme, sterol, or fatty acid biosynthesis, cellular metabolism is significantly reduced. Inhibition of metabolism and eventual cell death are generally attributed mainly to insufficient ATP regeneration to satisfy demand or to injury from products generated by anaerobic metabolism.

7.3.2.1.1 ATP Supply and Demand

The chemical energy generated by anaerobic roots derives mainly from glycolysis, which yields only 2 to 3 molecules of ATP per molecule of hexose equivalent, compared to 24 to 36 molecules by oxidative phosphorylation in aerated conditions. The fueling of glycolysis requires a continuous supply of glucose and NAD^+. Lack of ATP and sugars needed to support ATP synthesis is a major cause of cell injury and death. Root tips can survive anoxia temporarily for lengths of time that vary with species and the degree of O_2 limitation, with the duration of survival presumably being dependent on how long supply and demand can be kept in balance. Experimental evidence for energy starvation as a cause of growth inhibition and root death is considerable [3]. For instance, in *Arabidopsis*, labeling studies showed that falling oxygen concentrations were accompanied by a progressive decrease in biosynthetic fluxes to lipids, starch, proteins, and cell walls [65]. In potato tubers, low oxygen inhibited the energy-consuming wounding response and reduced phenylalanine ammonia lyase activity and phenylpropanoid synthesis [16]. Finally, in rice coleoptiles, long-term oxygen deprivation decreased the energy requirements for maintenance and ion fluxes [66].

Marked decreases in ATP after exposure to anoxia have been linked with root death. Treatments that improve root survival (i.e., hypoxic pre-treatment with 3% oxygen for several hours) raise ATP levels and promote glycolysis and fermentation. Similarly, mutations that inhibit glycolysis by strongly suppressing ADH activity depress fermentation and shorten root survival time. However, ATP supply may be less critical than is often thought. Only a small amount of ATP is required to prolong survival of hypoxically acclimated roots [67]. Similarly, experiments with hypoxically treated potato tuber slices showed that the Krebs cycle and glycolysis were coordinately downregulated at oxygen concentrations well above the K_m for the cytochrome oxidase, making it unlikely that oxygen was limiting for electron transport and ATP regeneration [16,17].

7.3.2.1.2 Self-Injury from Products of Anaerobic Metabolism

Cell death from anoxia has traditionally been attributed to the toxic effects of ethanol generated by fermentation. An excess of protons in the cytoplasm leading to a potentially fatal drop of 0.5 to 0.6 pH units [68] is now thought to be the more likely cause of anoxia-induced cell death. Thus, treatments that increase cytoplasmic acidity may enhance root damage from anoxia, whereas those limiting acidosis may improve survival. Early death from cytoplasmic acidosis could be prevented by roots themselves switching from lactate to ethanolic fermentation. Experimental support for the importance of this switch was obtained by the use of a weak base to offset acidification and dampen ethanol biosynthesis [69]. Similarly, in mutants incapable of ethanolic fermentation, lactate continues to be produced, resulting in earlier death in association with cytoplasmic acidification [70]. However, the complete picture concerning the role of proton poisoning is undoubtedly more complex.

Beside the perturbations in glycolysis and protein synthesis or ion transport, another deleterious effect of acidosis in anoxic tissues is the loss of membrane selectivity and integrity [3], as indicated by nonselective loss of solutes, such as ions, amino acids, and sugars. Decreased selectivity could be due to an increased permeability of the lipid bilayer or to leakage of molecules and ions through the various specific channels and carriers. In the absence of acclimation, loss of membrane selectivity is followed by the loss of membrane integrity, characterized by the release of proteases, lipases, and phosphatases from the vacuole, which ultimately leads to cell death [71].

Another contributing factor to seedling growth inhibition under low oxygen concentrations could be reduced phloem transport, which would limit carbon supply to the roots. One early response of plants to anoxia is reduced water uptake by roots due to decreased water permeability [72]. Root water uptake is mediated largely by aquaporins, which are water channel proteins [73]. Recently, the whole root and cell basis for inhibition of water uptake by anoxia was linked to cytosol acidosis [22]. The inhibition of aquaporin activity by acidic cytosolic pH could thus explain the inhibition of water permeability by anoxia and perhaps also the inhibition of sucrose transport and phloem unloading observed in hypoxic organs.

The various pathways and processes that may serve to regulate cytoplasmic pH include lactate excretion, excretion of protons, and maintenance of proton-consuming pathways [74]. However, the source of the damaging protons (e.g., the vacuole) and the mechanisms that regulate their concentration have not yet been made clear.

7.3.2.2 Regulation of Anaerobic Cell Metabolism

The processes postulated to have priority for energy produced under anoxia are (1) the maintenance of membrane selectivity and integrity, (2) the *de novo* synthesis of specific proteins and the maintenance of some energy-dependent solute transport, and (3) the regulation of cytosolic pH. The acclimative increase in glycolytic flux (the Pasteur effect) during anoxia partly compensates for the low energy yield in the absence of oxidative phosphorylation. However, even in plant tissues that experience a substantial Pasteur effect, the rate of ATP production is only one-third that in aerated tissues. Thus, mechanisms to increase energy production must be accompanied by a switch to pathways that consume less ATP and utilize O_2 more efficiently. For glycolysis and energy production to continue during anoxia, sugars must be available and oxidized nucleotides reduced during glycolysis must be regenerated. In addition, there must be some mechanisms to avoid accumulation of toxic endproducts. The main adaptive mechanisms available to plant cells in response to O_2-limitation are summarized below.

7.3.2.2.1 Regulation of Gene Expression and Enzyme Synthesis

Regulation of anaerobic gene expression involves proteinaceous *trans*-acting factors (transcription factors) that bind to sites present in the promoter regions (*cis*-elements) and constitute response elements required for gene activation or repression. The first anaerobic response element (ARE) described consisted of GC and GT motifs in the promoter region of the ADH1 gene [75]; similar ARE motifs have since been found in many anaerobically regulated genes. In addition to ARE, other *cis*-elements, such as half G-boxes are involved in regulating anaerobic gene expression. Both sites are common to the promoters of many stress-induced genes [30].

Several transcription factors have been identified, which bind to the ARE and the half G-boxes. MYB factors known to bind to ARE-like elements are involved in the induction of the *Arabidopsis* ADH1 gene [76]. Low oxygen affects the expression of several MYB-related genes [77,30] as well as of WRKY-type, ZAT12, AP2-domain, bZIP, and ERF-like transcription factors [30]. A G-box factor (GBF) has been found to bind to the maize ADH1 promoter in vitro [78], forming a complex associated with a 14-3-3 protein that is itself induced by hypoxia [79]. 14-3-3 proteins are brain regulatory proteins involved in signaling pathways and may be ubiquitous in eukaryotes. The finding that plant GBF can interact with a 14-3-3 protein that phosphorylates and binds Ca^{2+} [80] reinforces the proposed link between anoxic gene expression and Ca^{2+} signaling. These results indicate a complex interplay of several *cis* elements interacting with numerous proteins which may be further regulated by phosphorylation and changes in calcium levels.

A dramatic change in protein synthesis occurs in roots during anaerobiosis. In maize roots, a set of ~20 anaerobic proteins (ANPs) were synthesized selectively; most of these proteins have been identified as enzymes of glycolysis or sugar-phosphate metabolism [81]. ANPs that are part of other metabolic processes also have been reported [82], indicating that the low oxygen response involves more than a simple adaptation in energy metabolism. Careful control of oxygen supply has shown that maximal induction of ANP gene expression occurs in cells that are only partially oxygen-deficient rather than fully anoxic. Increased synthesis of the ANPs takes place only during the hypoxic period preceding anoxia, leading to the suggestion that ANPs be termed hypoxically induced proteins [83]. Hypoxically induced proteins comprise enzymes involved in energy metabolism, pH regulation, aerenchyma formation, protective functions, signal transduction, and several other activities. The synthesis of these enzymes and the arrest of synthesis of other proteins are known collectively as the "anaerobic response," which is regulated at transcriptional and post-transcriptional levels. Post-transcriptional regulations include changes in transcript stability and efficiency of ribosome loading [84]. Regulation of RNase activity could play a role in conserving nontranslating ribosomal and poorly translated mRNAs during low oxygen stress [85]. A variety of mechanisms for the selective translation of anaerobic rather than aerobic mRNA has been described. These include interactions within the sequence of ADH mRNA [86] and the phosphorylation of proteins involved in the translation machinery [87].

In recent years, high-throughput analysis of gene transcript and protein profiles has been applied to hypoxically treated rice [88], maize [82], and *Arabidopsis* [30].

Two-dimensional isoelectric focusing (IEF) SDS-PAGE linked to matrix-assisted laser desorption/ionization delayed-extraction reflectron time-of-flight (MALDI-DE-TOF) mass spectroscopy permitted over 200 proteins to be analyzed in maize root tips [82], of which 42 proteins were upregulated by hypoxia. These included the familiar metabolic enzymes, such as ADH, pyruvate decarboxylase (PDC), glyceraldehyde-3-phosphate dehydrogenase (GAPDH), enolase, malate dehydrogenase (MDH), NADP-isocitrate dehydrogenase, and aconitase. Serial analysis of gene expression (SAGE) on rice [88] and DNA microarray analysis on *Arabidopsis* [30] allowed the identification of more than 210 genes differentially regulated by hypoxia. Besides genes encoding fermentative enzymes (ADH, PDC, lactate dehydrogenase (LDH)), numerous genes encoding proteins involved in signal transduction, transcription, translation, detoxification of reactive oxygen species, proteolysis, amino acid metabolism, cell component synthesis, and defense response were regulated by hypoxia. These results suggest a complex network of interacting proteins that eventually may explain the acclimation process.

7.3.2.2.2 Downregulation of Biosynthetic Pathways

Acclimation to low oxygen includes general inhibition of biosynthetic processes, presumably to allow reduction of ATP consumption (Figure 7.2). Several experimental observations support this hypothesis. In potato tuber discs, decreasing oxygen concentrations to subambient concentrations between 21 and 0% resulted in a progressive and rapid (within a few hours) inhibition of sucrose, amino acid, protein, and lipid biosynthesis [16,17], and the energy-consuming defense response [89], as well as ion uptake [66]. Expression profile analyses carried out on *Arabidopsis* roots submitted to hypoxia [30] indicated that genes encoding for enzymes involved in cell wall, lipid, and flavonoid biosynthesis, and in the defense responses, were downregulated. This repression is consistent with the decrease in biosynthetic fluxes that occurs when internal oxygen concentrations fall.

7.3.2.2.3 Sugar Transport and Degradation

Oxygen shortage could affect energy metabolism by interfering with transport of fermentable carbohydrate from seeds or leaves to sink tissues via the phloem. All membrane-based sugar transporters so far isolated function as proton symporters and are thus energy-dependent. In tomato (*Solanum lycopersicum*) roots, the transport of sucrose and glucose was found to be upregulated after one week of hypoxic treatment (Gharbi, unpublished data). However, high sucrose concentrations within the phloem cells would be sufficient, in the absence of chemical energy, to move sucrose along the concentration gradient from cell to cell by diffusion through plasmodesmata. In roots, the available evidence supports a mostly symplasmic route for sugar entry [90]. Hypoxic pretreatment permits survival possibly by protecting plasmodesmata from the collapse that anoxia normally causes [83] or by increasing the size exclusion limit of plasmodesmata, in the case of wheat roots from <1 to 5 to10 kDa [91]. The latter would permit metabolites, including ATP and sugars, to flow from cells receiving sufficient O_2

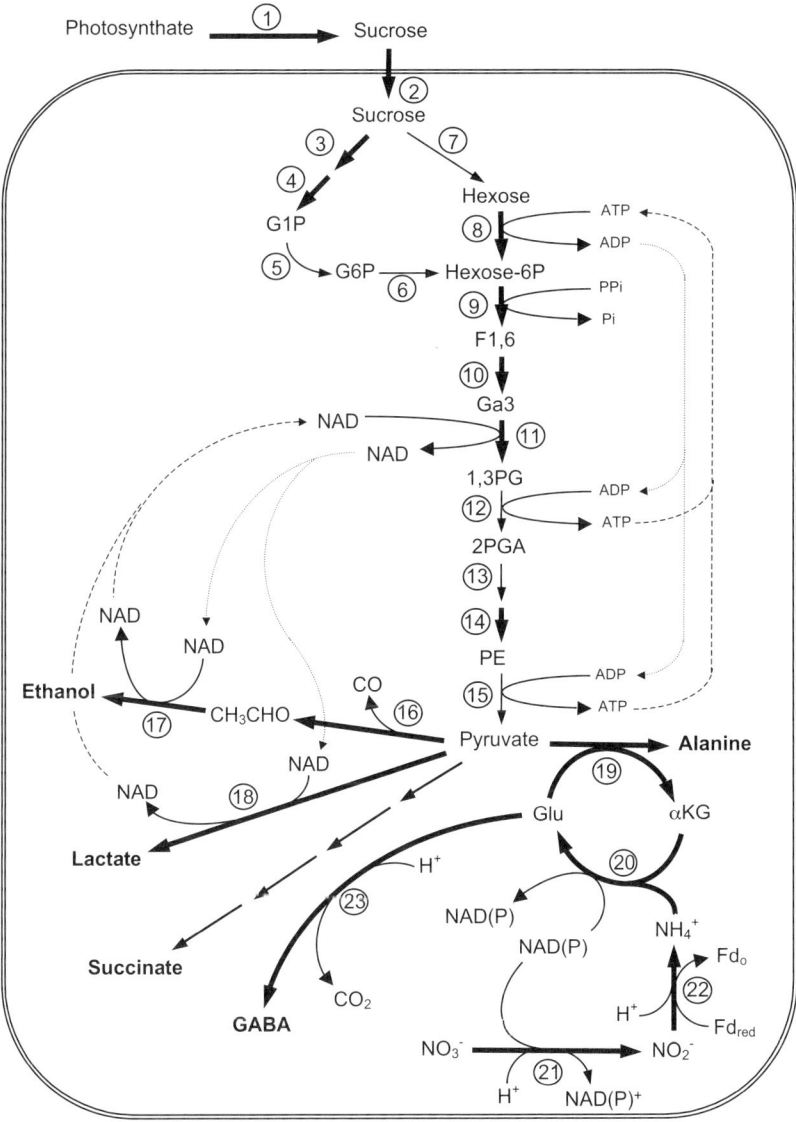

FIGURE 7.2 Schematic diagram of various pathways from sucrose involved in the tolerance to hypoxia. Major endproducts are typed in bold characters; bold arrows indicate upregulated steps in hypoxic cells. 1, photosynthate transport; 2, sugar translocation; 3, sucrose synthase; 4, UDPG pyrophosphorylase; 5, phosphoglucomutase; 6, phosphoglucose isomerase; 7, invertase; 8, hexokinase; 9, PPi-phosphofructokinase; 10, aldolase; 11, glyceraldehyde 3P dehydrogenase; 12, phosphoglycerate kinase; 13, phosphoglycerate mutase; 14, enolase; 15, pyruvate kinase; 16, pyruvate decarboxylase; 17, alcohol dehydrogenase; 18, lactate dehydrogenase; 19, alanine aminotransferase; 20, glutamate dehydrogenase; 21, nitrate reductase; 22, nitrite reductase; 23, glutamate decarboxylase.

for oxidative phosphorylation to anoxic cells. This might allow higher synthetic activity and possibly longer survival in tissues with anoxic zones [2].

Another strategy is to degrade sugars through alternative, energy-conserving pathways. The breakdown of a molecule of sucrose by invertase and HK requires two molecules of ATP, whereas breakdown by Susy and UDP-glucose pyrophosphorylase (UDPG-PPase) requires only one molecule of inorganic pyrophosphate (PPi) [92]. The importance of the latter pathway for hypoxic tolerance is consistent with the widely observed upregulation of Susy genes and activity and the repression of acid invertase genes and activity during low oxygen conditions (Figure 7.2).

Increasing sugar supply to allow high rates of glycolysis and fermentation in hypoxic roots may be another adaptive response and could be achieved by increased Susy expression and activity or by upregulation of HK [93,94]. However, the latter mechanism may not be ubiquitous since as overexpression of HK in tomato roots did not improve tolerance to hypoxia (Gharbi, unpublished data).

7.3.2.2.4 Glycolytic and Fermentative Enzymes

Ethanol and lactate, two main products of fermentation in plants submitted to flooding, both derive from pyruvate, the endproduct of glycolysis (Figure 7.2). Extensive literature and reviews deal with the induction of ethanol and lactate production in response to hypoxia and anoxia [2]. Ethanol is the major product of fermentation in higher plant tissues, whether they are tolerant to anoxia or not. In most cases where the time course of accumulation of fermentation products has been followed, the rate of ethanol production decreased during the hours or days following the transfer to anoxia [64]. Cell membranes are highly permeable to ethanol, and the ethanol synthesized during fermentation leaks from the tissues, thus preventing an excessive internal concentration. Thus, the catabolism of carbohydrates into ethanol through glycolysis and the PDC–ADH pathway appears to be a good way to regenerate ATP and NAD$^+$, without problematic endproduct accumulation and pH decrease. L-lactate is often produced prior to ethanol, within the first minutes after the transfer to anoxia and then during three to four hours. The importance of the pathway leading to lactate accumulation is unclear. However, circumstantial evidence suggests that recycling of lactate could not only prevent endproduct inhibition, but also contribute to carbon–nitrogen balance of a tissue [2].

The increase in glycolytic flux under low oxygen has been linked with the upregulation of a set of glycolytic and fermentative enzymes. In maize root tips, the activity of several glycolytic enzymes, together with the fermentative enzymes PDC, ADH, and LDH, increased during hypoxic pre-treatment and after anoxic shock [94]. Numerous genetic and transgenic approaches have been used to evaluate the genes that are involved in tolerance to anoxia. These approaches are often limited by the fact that many enzymes are encoded by multiple genes. However, ADH null mutants with much-reduced ADH activity were shown to be less tolerant to anoxia in a variety of species such as maize [95], rice [96], and *Arabidopsis* [38]. Likewise, maize double mutant for the two genes encoding Susy were also less tolerant to anoxia [18]. Antisense approaches have generally been disappointing, probably because enzyme activity was not sufficiently reduced. Overexpression has been more successful and is discussed later.

7.3.2.2.5 Cytoplasmic Acidosis

A second category of hypoxically induced processes that might be involved in acclimation includes those influencing cytoplasmic pH (Figure 7.2). Apart from ethanol, alanine is often the most prevalent endproduct of anaerobic fermentation in roots [2], in correlation with the hypoxic induction of alanine amino transferase [30,97]. Alanine synthesis produces as much ATP as ethanolic fermentation, provided that stored or imported glutamine is the source of the amino group and does not modify cytoplasmic pH. However, if the source of the amino group is NH_4^+, incorporated via the ATP-consuming glutamine synthetase/glutamate synthase (GS-GOGAT) pathway, the ATP yield would be null. The labeling pattern of glutamine and glutamate in anoxic rice coleoptiles supplied with exogenous $^{15}NO_3^-$ indicates that during NO_3^- assimilation, a distinct pool of glutamate is synthesized via glutamate dehydrogenase (GDH) which does not consume ATP and recycles $NAD(P)^+$ [98]. Moreover, the reduction pathway of NO_3^- into NH_4^+, via nitrate reductase and nitrite reductase, was shown to be upregulated during anoxia [98]. As the conversion consumes 4 mol of NAD(P)H for each mol of NO_3^- reduced, this pathway provides a mechanism to regenerate oxidized nucleotides and contributes to the pH stat [2].

Another pathway activated by hypoxia includes γ-amino butyric acid (GABA) synthesis. GABA increased in response to hypoxia or anoxia in several types of organs including roots [99]. During the decarboxylation of glutamate to GABA, via glutamate decarboxylase, protons are consumed and hence contribute to increase pH. In maize root tips, the kinetics of glutamate decarboxylase activation coincided with the time course of pH stabilization soon after the onset of anoxia [69]. Similarly, succinate accumulated in some tissues under anoxia [100]. The formation of succinate from oxaloacetate follows the reverse tricarboxylic acid cycle direction and most probably involves mitochondrial enzymes. During hypoxia, succinate synthesis might be a way to consume H^+ and to regenerate NAD^+ from NADH produced by the other branch of the tricarboxylic acid cycle from pyruvate to 2-oxoglutarate. However, in most cases, the significance of the reduction of nitrate and the production of alanine, GABA, and succinate during anoxia is still a matter of debate. The synthesis of these metabolites represents only a minor part of the ethanol and lactate flux in hypoxic or anoxic tissues, but could nevertheless constitute important transient mechanisms during hypoxia, for example, to buffer the decrease of cytoplasmic pH generated by lactate formation, nucleoside triphosphate breakdown, and inhibition of proton pumping.

7.3.2.2.6 Other Routes to Tolerance

In plants, nonsymbiotic hemoglobin has been shown to be expressed in tissues exposed to hypoxia or anoxia [30], but its function is still matter for debate. Several possible roles have been proposed. First, nonsymbiotic phytohemoglobin (pHb) could serve as an O_2 carrier to help preserve mitochondrial respiration under anoxic conditions. However, the affinity of pHb for oxygen is in the range of 0.5 to 3 nM, which is inconsistent with a role for hemoglobin as an O_2 carrier to cytochrome oxidase that has a Km[O_2] of 140 nM. Moreover, the concentrations of pHb in plants are too low to facilitate O_2 diffusion, and its size (18 kDa) makes its transport through plasmodesmata of the roots seem unlikely. Second, pHb could help to sustain glycolytic metabolism in anoxic tissues. The high affinity of pHb for O_2 means that the free protein would remain oxygenated at very low oxygen concentrations, making it a good candidate to sequester O_2 under anoxic environments [101]. In addition, the interaction of pHb with another protein could create a complex capable of oxidizing NAD(P)H. On the basis of this evidence, it was suggested that pHb could function under low oxygen conditions to recycle NADH, and thus replace PDC and ADH, facilitate glycolysis, and increase substrate-level phosphorylation [8]. However, this alternative to the classical fermentation pathway should result in an accumulation of alternative endproducts which have not been detected to date. Third, pHb has been proposed to be part of an O_2 sensor system capable of regulating gene expression under anoxia [102], as previously discussed.

Hypoxia, like other environmental stresses, causes protein denaturation. Adaptive mechanisms involve the induction of specific proteolytic processes [103]. More specifically, proteolysis contributes to the degradation of proteins in lysed cells during the formation of aerenchyma. In *Arabidopsis* roots, genes that are involved in protein degradation are mostly repressed under hypoxia [30]. Similarly, hypoxia leads to a decrease in global proteolytic activity in mature roots of clover (*Trifolium subterraneum*) [104]. Protein degradation is an ATP-consuming process, and the repression of protein degradation allows protein turnover to be decreased, thereby conserving ATP and decreasing oxygen consumption. However, increased proteolytic activity has been observed during anoxic stress in maize seedling roots [105]. Protease and protein synthesis inhibitor studies showed that the major part of the proteolytic activity was due to a newly synthesized Ca^{2+}-dependent cysteine protease, which was named anoxia-induced protease (AIP) [105]. The appearance of this AIP in the root tip was spatially and temporally associated with the initiation of root tissue death. An increase in cysteine protease activity was also detected in roots of young clover seedlings subjected to hypoxia [104]. To date, no data report the involvement of the ubiquitin and proteasome-dependent proteolysis in low oxygen treated plants. Nevertheless, the induction of two ubiquitin and one ubiquitin-conjugating enzyme genes in hypoxia-treated *Arabidopsis* roots [30], suggest that this proteolysis could play a role in acclimation to hypoxia.

The metabolic processes described above are by no means an exhaustive list of the mechanisms involved in acclimation to hypoxia. High-throughput analysis of gene transcripts and proteins overexpressed in hypoxic tissues [30,82,88] indicate that many other metabolic pathways are upregulated in response to low oxygen. These include water transport, carbon and methyl metabolism, the defense response, ion transport, and lipid biosynthesis.

7.3.2.3 Post-Anoxic Damage and Recovery

As no plant tissues are able to survive indefinitely in the absence of oxygen, plants must be able to deal with the consequences of re-exposure to O_2. Re-exposure to oxygen after anoxia causes severe injuries to plant tissues and organs [106], mainly due to the generation of reactive oxygen species (ROS) such as superoxide radicals (O_2^-), singlet oxygen ($^{\cdot}O_2$), hydroxyl radicals ($^{\cdot}OH$), and H_2O_2. These ROS damage lipids, DNA, and proteins [107]. Lipid peroxidation impairs membrane function, inactivates membrane-bound receptors and enzymes, and increases nonspecific permeability to ions [108]. There are indications that flooding can induce oxidative stress, leading to increased production of ROS [109]. Indicators of oxidative stress, such as decreased membrane integrity [110], changes in lipid content and composition [111], and increases in the content of malondialdehyde (MDA), an important product of lipid peroxidation [112,113], have been reported in plants subjected to hypoxia or anoxia.

To control the level of ROS and to protect cells, plants possess a number of low molecular mass antioxidants (ascorbate, glutathione, phenolic coumpounds, carotenoides, tochopherols) and enzymes that act as ROS scavengers and regenerate the active form of the antioxidants such as superoxide dismutase (SOD), catalase, ascorbate and glutathione peroxidases, and ascorbate and glutathione reductases. High levels of some antioxidant enzymes and metabolites have been shown to be important in a wide variety of plants for survival of oxidative stress incurred after waterlogging. This is consistent with experiments showing that short-term post-anoxic injury could be avoided in soybean (*Glycine max*) seedlings if root tissue were incubated with ascorbate [114].

In anoxia-tolerant species, the lipid peroxidation marker MDA occurs to a much smaller extent than in intolerant species. This is consistent with the accumulation of antioxidants and the induction of SOD activity in the roots which would theoretically help prevent damage from free radicals during recovery from anoxic stress. In the roots of tomato and eggplant (*Solanum melongena*), ascorbate peroxidase (and not SOD) was found to be the most important antioxidant enzyme for tolerance to hypoxia [115]. Besides classical antioxidant mechanisms, mitochondrial alternative oxidase has been also proposed to play a protective role against oxidative stress following periods of anoxia [116]. In tobacco (*Nicotiana tabacum*) cells, the overexpression of alternative oxidase reduced the formation of ROS in mitochondria [117]. Furthermore, an anoxic pretreatment rendered soybean cells resistant to oxidative stress after return to air [116] and was related to the action of peroxidases and mitochondrial alternative oxidase.

Foliar tissues of flooded plants are submitted to oxidative stress as well. Inhibition of photosynthetic activity and increased level of lipid peroxidation and electrolyte leakage were reported in leaves of barley plants (*Hordeum vulgare* L.) subjected to flooding [110,118]. Flooding-induced oxidative stress induces antioxidant enzymes such as ascorbate peroxidase and catalase [110]. Whether this induction improves oxidative stress tolerance upon return to air is not yet known.

7.4 GENETIC MANIPULATION OF TOLERANCE TO WATERLOGGING

7.4.1 Genetic Engineering

7.4.1.1 Transcription Factors

As mentioned earlier, waterlogging leads to a number of physicochemical changes in the soil environment. Genetic manipulation of tolerance to waterlogging is based on the postulate that one or more of these changes constitutes signal(s) perceived by receptor(s), initiating a signal transduction pathway, which results in metabolic and anatomical responses permitting improved survival of waterlogging. Tolerance to freezing and low temperature, for instance, has been successfully increased in *Arabidopsis* by overexpressing transcription factor ABI3 [119]. Manipulation of such an early regulatory event is presumed to amplify the entire response pathway. To date, only one transcription factor with a role in the anaerobic response has been isolated (AtMYB2 [76]), but attempts to introduce AtMYB2 under the control of the strong constitutive 35S promoter into *Arabidopsis* were unsuccessful. None of the transgenic plants overexpressed AtMYB2 and the transgene was found to be truncated or reorganized, leading to the hypothesis that strong overexpression of AtMYB2 might be lethal [120].

7.4.1.2 Fueling of Glycolysis

As mentioned earlier, HKs catalyze the first committed step of glycolysis and are usually upregulated during hypoxic acclimation. However, the probability that HKs are involved in sugar, light, and hormonal signaling means that overexpression of HK could have unexpected consequences, with possible conflicting or overlapping responses to sugar and low oxygen, for instance. A sugar/oxygen signal overlap has already been shown for maize genes encoding Sus1 and Sh1 sucrose synthases, Ivr1 and Ivr2 invertases, and alcohol dehydrogneasel [121]. In addition to *Arabidopsis*, potato and tomato plants have been engineered with increased HK activity. Considerable changes were seen in carbohydrate metabolism for tomato fruit in contrast to potato tubers [122,123]. These changes could be rationalized on the basis of increased hexose phosphorylation alone, without recourse to signaling effects [124,125]. The transgenic tomato plants which overexpressed a yeast and an *Arabidopsis* HK were further examined for waterlogging tolerance. Preliminary experiments showed that tomato which overexpressed the plant HK was hypersensitive to waterlogging compared to tomato with wild-type and yeast HK overexpression. This result is difficult to explain on the basis of increased hexose phosphorylation alone (Gharbi, unpublished). More work is needed to distinguish between catalytic and possible signaling effects.

7.4.1.3 Fermentation Pathways

Manipulation of later anaerobic responses could be beneficial, if these responses were sufficiently important in conferring tolerance. For instance, alcohol fermentation has been correlated with improved tolerance, whereas lactate fermentation has long been considered to contribute to cytoplasmic acidosis, and eventually cell death. A number of attempts have been made to manipulate these pathways. When we attempted to use the classic antisense approach, we were unable to obtain transgenic tomato roots with significantly lowered levels of LDH activity, suggesting that expression of the gene(s) might be necessary [126]. Two *ldh* genes are present in tomato; the LDH2 isozyme present in aerobic roots is responsible for the lactic acid produced immediately upon oxygen deprivation, but the LDH1 isozyme induced during hypoxia is more efficient in the lactate to pyruvate direction. Consequently, LDH1 could actually limit lactic acid accumulation during anoxia [83]. This idea is supported by work in potato tubers, where the antisense technique was successful in generating transgenic potato plants with a preferential decrease in two of the five LDH isozymes. Transgenic tubers under anoxic conditions contained twofold more lactate than control tubers, due to decreased oxidation of lactate to pyruvate [127]. However, in the case of potatoes, the accumulation of lactate during anoxia was not accompanied by an induction of LDH activity or a change in isozyme distribution. In *Arabidopsis*, strong overexpression of LDH1 did not affect survival under low oxygen stress [120].

Manipulation of the ethanol fermentation pathway has targeted the genes encoding ADH and PDC in order to increase or decrease the production of these enzymes. Overexpression of the *Arabidopsis* PDC1 and PDC2 resulted in improved survival of three-week-old *Arabidopsis* plants to hypoxic shock (direct transfer to a 0.1% O_2 environment), whereas overexpression of ADH1 had no effect. ADH1 null mutants had decreased hypoxic survival; antisense reduction of PDC levels was insufficient for survival to be affected [128]. Increasing PDC levels in rice also increased submergence tolerance [129], but overexpression of a bacterial PDC gene in tobacco did not lead to improved hypoxic tolerance [130]. These different results may be explained by the different plant species studied or by different regulation of the PDC isozymes that were overexpressed.

7.4.1.4 Hormone Levels

Improving waterlogging tolerance through manipulation of hormone levels has met with some success. A positive correlation between cytokinin levels and oxygen tension has been shown in seeds, leading to the postulate that decreased oxygen levels could reduce cytokinin production and contribute to the premature senescence observed during flooding stress. Indeed, the introduction of the *ipt* gene coding for isopentenyl transferase, a rate-limiting enzyme in the cytokinin biosynthesis pathway, under the control of the senescence-specific SAG12 promoter, increased flooding tolerance in *Arabidopsis* [40]. Increased synthesis of stress ethylene by flooding of tomato plants could stimulate the development of epinasty, leaf chlorosis, and necrosis and lead to reduced fruit yield. ACC, which is the immediate precursor of ethylene in plants, can be catabolized by the microbial enzyme ACC deaminase. By overexpressing a bacterial ACC deaminase under the control of three different promoters (two tandem 35S, pathogenesis related PRB-1*b*, anaerobically inducible and root-specific *rolD*), transgenic tomato plants were produced that had decreased ethylene production during flooding stress and increased flooding tolerance. Plants with an ACC deaminase gene under the control of the *rolD* promoter showed the greatest tolerance [131].

7.4.1.5 Phytohemoglobin

In spite of the uncertainty of the exact role of pHb in the hypoxic response, its overexpression has met with some degree of success in conferring anoxic tolerance. Overexpression of nonsymbiotic hemoglobin or the symbiotic hemoglobin from *Parasponia andersonii* in *Arabidopsis* led to increased survival under severe hypoxia [132]. Hypoxia tolerance was lost when the transgene was a mutated hemoglobin with reduced oxygen-binding affinity [132]. Furthermore, the overexpression of a symbiotic hemoglobin from *Lotus japonica* in potato tubers led to the formation of hypertrophic lenticels and to higher internal oxygen concentration in tubers [16].

7.4.1.6 Protection against Post-Anoxic Injury

As discussed above, tissue damage occurs when plants are re-exposed to air following anoxic treatment, probably due to the generation of extremely toxic ROS. One protective system involves SOD which converts $^{\bullet}O_2^-$ radicals to H_2O_2, in turn reduced to water by peroxidases or catalases. SOD activity increased 13-fold during 28 days of anoxia in tolerant rhizomes of *Iris pseudacorus*, but failed to increase in less tolerant *Iris germanica* and *Glyceria maxima* [133,134]. Accordingly, the growth of transgenic tobacco overproducing Mn-SOD and Fe-SOD was not affected after 2 days of waterlogging, whereas nontransgenic plants showed decreased growth at this stage [135]. To date, no transgenic crop plant has been produced and field tested to validate waterlogging tolerance.

7.4.2 Production of Stress-Tolerant Germplasm

Prospects for breeding of waterlogging tolerance are considered to be good, if (1) the waterlogging conditions (intensity, duration, intermittent or continuous, and type of soil) in target environments are known; (2) local and international germplasm have been evaluated under target conditions and good genetic diversity for high growth rates, high grain yields, or survival of both seeds and plants have been demonstrated; and finally (3) adaptive traits correlated with waterlogging tolerance have been shown to be highly heritable [136]. This appears to be the case for cereals, where genetic diversity for waterlogging tolerance at the vegetative stage based on grain yields is clearly demonstrated from a review of published information in wheat, barley, and oats (*Avena sativa*) [136]. Many studies indicate that plants generally show considerable variation in their tolerance to waterlogging. In addition to wheat, barley, and oats, genetic variation in flooding tolerance has been found in rice, maize [137], clover [138], and *Arabidopsis*. In certain cases, this was due to mutations, for instance, in ADH and Susy genes.

In the absence of molecular markers, alternative methods that offer a simple, rapid approach to screening have been described. An example is the use of leaf chlorosis at CIMMYT and in China for the selection of tolerant wheat lines. In certain but not all waterlogged environments, leaf chlorosis is strongly correlated to grain yield and is a highly heritable trait [136]. Other screening criteria could be based on root traits such as aerenchyma production, reduced radial O_2 loss, tolerance to mineral nutrient toxicities, and growth in stagnant agar [136].

The development of molecular markers for waterlogging tolerance would undoubtedly be valuable and would permit marker-assisted selection. Molecular markers could be based on sequences from genes of known function, that is, of candidate genes. Because one mechanism of anoxia tolerance is thought to be activation of the ethanol fermentation pathway, evaluation of polymorphisms for known sequences of PDC or ADH might lead to the identification of QTLs associated with waterlogging-tolerant genotypes. The identification of other candidate genes could take into consideration specific environmental factors and many other strategies that plants have evolved to survive.

For instance, in rice plants, the two main strategies are to elongate and escape low oxygen conditions or not to elongate and conserve resources. For rainfed lowland rice, selection for minimal elongation during submergence has been exploited to identify one prominent locus for tolerance on chromosome 9. Designated as SUB1(T), this locus accounted for about 70% of the phenotypic variation for submergence tolerance derived from the tolerant parent FR13A [139]. Several molecular markers map sufficiently close to be suitable for selection of the locus and should prove useful in the production of submergence tolerance breeding [140].

7.5 EVALUATION OF PLANT TOLERANCE TO WATERLOGGING

Waterlogging tolerance has often been evaluated under glasshouse conditions that attempt to simulate field conditions and take into consideration growth or survival. Two techniques are widely used: (1) roots of intact plants are grown in stagnant nutrient solution, sometimes containing agar to minimize convection [141], and (2) roots of intact plants are grown in flooded soil. Both techniques confer hypoxic pretreatment, with oxygen being gradually depleted. The use of agar mimics the lack of venting of gaseous byproducts of anaerobic metabolism, such as ethylene and CO_2. The first technique allows the possibility of evaluating root elongation and function, but does not reproduce field conditions as well as the second technique.

Root tolerance of low oxygen has been extensively studied [41]. The major effect of low oxygen due to soil flooding is reduction in root growth and biomass. The length of time that root tips survive has therefore been used as a criterion for tolerance and the positive effects of a hypoxic pretreatment (maize [142], tomato [93], rice [143]). Root tip survival can be improved by hypoxic pretreatment, but is also dependent on carbohydrate status and environmental factors, such as temperature. Moreover, expanded tissues are generally more tolerant than growing root tips [144], possibly due to differences in symplastic or apoplastic unloading of transport sucrose [83]. However, in spite of metabolic adaptations, roots of vascular plants cannot survive extended periods of low oxygen. The only exception is the roots of 2-day-old rice seedlings that survived 19 days of anoxia [145].

The relevance of the primary root tips to anoxia tolerance has been questioned by the observation that removal of the primary root tips of three-day-old intact maize plants prior to anoxic treatment actually improved shoot growth during recovery [105]. The authors suggest that natural "de-tipping" is an adaptive character, presumably encouraging the development of lateral roots after loss of apical dominance. This hypothesis is supported by the observation that *Trifolium* species highly tolerant of waterlogging develop extensive lateral roots at the same time as the necrosis of the seminal root [42].

Anoxic tolerance of root tips cannot be directly extrapolated to root tolerance to waterlogging. Root tolerance to waterlogging involves many other adaptive mechanisms, in particular those improving internal aeration. These can be evaluated by a number of methods either directly at the root level or indirectly by effects on shoots. The presence or induction of internal aeration systems can be evaluated by measurements of root porosity, aerenchyma, adventitious roots, and radial oxygen loss. The general health of the root system can be quantitated by measuring the rate of root respiration, the integrity of root cell membrane, or water potential in the xylem.

At the shoot level, visual methods have proven useful; for instance, Southern soybean germplasm was evaluated for waterlogging injury using a rating from 0 (no symptom) to 9 (>90% severely chlorotic or dead) [146]. Visual scoring methods for waterlogging injury upon three bean genotypes were found to agree well with more quantitative methods [147]. A comparison of effects on leaf photosynthesis, pigment composition, chlorophyll content, maximal quantum efficiency of photosystem II (Fv/Fm), shoot and root growth in a population of lucerne cultivars submitted to 16 d of waterlogging showed that most of the parameters changed significantly with the duration of waterlogging. All cultivars had reduced shoots and even more reduced root mass. Of the other parameters, Fv/Fm ratio was concluded to be the most practical parameter for screening purposes [148].

Glasshouse or laboratory conditions control a limited number of environmental factors. Field experiments allow for the evaluation of waterlogging tolerance in natural environmental conditions. However, many field experiments limit observations to waterlogged conditions. In the absence of comparisons with nonwaterlogged conditions, high yields, biomass, or survival could be related to high yield potential or high seed vigor and not to true waterlogging tolerance [136]. Only some plant species grow under severe waterlogging and in such cases, growth is limited to shoots. Evaluation of tolerance can be based on the agronomic or the physiological definition, that is, high yield or survival/growth. Using the physiological definition, waterlogging tolerance is evaluated by shoot growth following a limited period of waterlogging. An increase in shoot length and biomass of one species or variety in comparison to another indicates better tolerance. For the agronomic definition of tolerance, grain or fruit yield is the determining criterion. Table 7.1 gives examples of articles published in 2003 to 2004, which illustrate methods widely used to evaluate waterlogging tolerance.

7.6 SUMMARY AND PROSPECTS

Many detrimental factors have been identified in flooded soil, but the shortage of oxygen is without doubt the most important. Accumulating evidence suggests that the decline in plant growth and survival under low oxygen conditions could be caused by the imbalance between ATP supply and demand, resulting in energy and sugar starvation, or by the toxic effects of the products in anaerobic metabolism. Plants with developed root systems are capable of sensing declining levels of oxygen in the soil and responding in ways that permit survival and even growth. These responses include the management of carbohydrate transport and consumption, the adaptation of energy use to limited supply, and mechanisms to limit or circumvent the damaging consequences of anaerobic metabolism. Longer-term plant tolerance depends largely on morphological and anatomical modifications, such as aerenchyma formation and adventitious root production that enable plants to avoid anoxia.

The possibility of improving plant tolerance to waterlogging relies on genetic diversity and on our understanding of the molecular mechanisms involved. The former allowed rice breeders, for instance, to identify a pertinent locus and to envisage its introduction into more resistant cultivars. Molecular approaches have sought to enhance anaerobic responses by introducing individual genes involved in signaling (hormones or a transcription factor) or in the different metabolic pathways shown to be important in anoxic tolerance. The latter have generally

TABLE 7.1
Evaluation of Waterlogging Tolerance at the Vegetative Stage

Plant Species	Waterlogging Duration and Timing	Experiment Conditions	Evaluation Criterion	Reference
Zea mays	7 d, knee-height	Field experiment: 16 cultivars; India; 5 cm submergence	Plant and ear height; mean grain yield; 100-seed weight; cob length and diameter; kernel rows/ear	136
Hordeum vulgare		Stagnant nutrient solution 35 "wild" + cultivated	Adventitious root mass; relative growth rate	149
Gossypium hirsutum	Intermittent	Field exp: Australia; soil with low drainage rates; extended irrigation	Boll number; total plant dry weight	150
Medigo sativa	16 d, 2-mo-old	1/2-strength Hoagland solution; 4 cultivars	Shoot and root fresh, and dry weight; Fv/Fm ratio	148
Hordeum vulgare	3 wk + 2 wk recovery, 3 or 4 expanded leaf stages	Artificial potting mix, Vertisol; Australia	Chlorophyll content; CO_2 assimilation rate; Fv/Fm; shoot and root growth	151
Vigna mungo	3, 6, and 9 d, vegetative	Field experiment: India	Chlorophyll and N content; pod fresh weight	152

been less efficient than the former. This is probably due to our inadequate knowledge of what regulates the physiological basis of waterlogging tolerance. Anoxic tolerance, for instance, cannot be directly extrapolated to waterlogging tolerance. The global approaches of genomics, proteomics, and metabolomics have already allowed a more integrated understanding of what constitutes waterlogging tolerance. It is only a matter of time and continued work before this knowledge can be applied to the development of plants with increased tolerance to waterlogging.

REFERENCES

1. Dat, J.F. et al., Sensing and signalling during plant flooding. *Plant Physiol. Biochem.*, 42:273, 2004.
2. Gibbs, J. and Greenway, H., Mechanisms of anoxia tolerance in plants. I. Growth, survival and anaerobic catabolism. *Func. Plant Biol.*, 30:1, 2003.
3. Greenway, H. and Gibbs, J., Mechanisms of anoxia tolerance in plants. II. Energy requirements for maintenance and energy distribution to essential processes. *Func. Plant Biol.*, 30:999, 2003.
4. Urao, T. et al., A transmembrane hybrid-type histidine kinase in *Arabidopsis* functions as an osmosensor. *Plant Cell*, 11:1743, 1999.
5. Baxter-Burrell, A. et al., RopGAP4-dependent Rop GTPase rheostat control of *Arabidopsis* oxygen deprivation tolerance. *Science*, 296:2026, 2002.
6. Fukao, T. and Bailey-Serres, J., Plant responses to hypoxia — Is survival a balancing act? *Trends Plant Sci.*, 9:449, 2004.
7. Yang, Z., Small GTPases: Versatile signalling switches in plants. *Plant Cell*, 14: 375, 2002.
8. Hill, R.D., What are hemoglobins doing in plants? *Can. J. Bot.*, 76:707, 1998.
9. Dordas, C. et al., Expression of a stress-induced hemoglobin affects NO levels produced by alfalfa root cultures under hypoxic stress. *Plant J.*, 35:763, 2003.
10. Dordas, C., Rivoal, J., and Hill, R.D., Plant haemoglobins, nitric oxide and hypoxic stress. *Ann. Bot.*, 91:73, 2003.
11. Dordas, C. et al., Class-1 hemoglobins, nitrate and NO levels in anoxic maize cell-suspension cultures. *Planta*, 219:66, 2004.
12. Buck, L.T., Adenosine as a signal for ion channel arrest in anoxia-tolerant organisms. In *Comparative Biochemistry and Physiology. Part B: Biochemistry and Molecular Biology*. In press.
13. Koch, K., Sucrose metabolism: regulatory mechanisms and pivotal roles in sugar sensing and plant development. *Curr. Opin. Plant Biol.*, 7:235, 2004.
14. Jang, J.C. et al., Hexokinase as a sugar sensor in higher plants, *Plant Cell*, 9:5, 1997.
15. Moore, B. et al., Role of the *Arabidopsis* glucose sensor HXK1 in nutrient, light, and hormonal signalling. *Science*, 300:332, 2003.
16. Geigenberger, P. et al., Metabolic activity decreases as an adaptive response to low internal oxygen in growing potato tubers. *Biol. Chem.*, 381:723, 2000.
17. Geigenberger, P., Response of plant metabolism to too little oxygen. *Curr. Opin. Plant Biol.*, 6:247, 2003.
18. Ricard, B. et al., Evidence for the critical role of sucrose synthase for anoxic tolerance of maize roots using a double mutant. *Plant Physiol.*, 116:1323, 1998.
19. Zeng, Y. et al., Differential regulation of sugar-sensitive sucrose synthases by hypoxia and anoxia indicate complementary transcription and posttranscriptional regulation. *Plant Physiol.*, 116:1573, 1998.
20. Zeng, Y. et al., Rapid repression of maize invertases by low oxygen. Invertase/sucrose synthase balance, sugar signalling potential, and seedling survival. *Plant Physiol.*, 121:599, 1999.
21. Denison, S.H., pH regulation of gene expression in fungi. *Fungal Genet. Biol.*, 29:61, 2000.
22. Tournaire-Roux, C. et al., Cytosolic pH regulates root water transport during anoxic stress through gating of aquaporins. *Nature*, 425:393, 2003.

23. Webster, C., Kim, C.Y., and Roberts, J.K.M., Elongation and termination reactions of protein synthesis on maize root tip polyribosomes studied in a homologous cell-free system. *Plant Physiol.*, 96:418, 1991.
24. Subbaiah, C.C., Bush, D.S., and Sachs, M.M., Elevation of cytosolic calcium precedes anoxic gene expression in maize suspension-cultured cells. *Plant Cell*, 6:1747, 1994.
25. Foreman, J. et al., Reactive oxygen species produced by NADPH oxidase regulate plant cell growth. *Nature*, 422:442, 2003.
26. He, C.J. et al., Ethylene biosynthesis during aerenchyma formation in roots of maize subjected to mechanical impedance and hypoxia. *Plant Physiol.*, 112:1679, 1996.
27. Drew, M.C., He, C.J., and Morgan, P.W., Programmed cell death and aerenchyma formation in roots. *Trends Plant Sci.*, 53, 123, 2000.
28. Baum, G. et al., A plant glutamate-decarboxylase containing a calmodulin-binding domain — Cloning, sequence, and functional analysis. *J. Biol. Chem.*, 268:19610, 1993.
29. Snedden, W.A. et al., Activation of a recombinant Petunia glutamate decarboxylase by calcium/calmodulin or by a monoclonal antibody which recognizes the calmodulin binding domain. *J. Biol. Chem.*, 271:4148, 1996.
30. Klok, E.J. et al., Expression profile analysis of the low-oxygen response in *Arabidopsis* root cultures. *Plant Cell*, 14:2481, 2002.
31. Reggiani, R., Alteration of levels of cyclic nucleotides in response to anaerobiosis in rice seedlings. *Plant Cell Physiol.*, 38:740, 1997.
32. Depre, C. and Hue, L., Cyclic GMP in the perfused rat heart. Effect of ischaemia, anoxia and nitric oxide synthase inhibitor. *FEBS Lett.*, 345:241, 1994.
33. Tsuji, H. et al., Transcript levels of the nuclear-encoded respiratory genes in rice decrease by oxygen deprivation: Evidence for involvement of calcium in expression of the alternative oxidase 1a gene. *FEBS Lett.*, 471:201, 2000.
34. Subbaiah, C.C. and Sachs, M.M., Maize cap1 encodes a novel SERCA-type calcium-ATPase with a calmodulin-binding domain. *J. Biol. Chem.*, 275:21678, 2000.
35. Lorbiecke, R. and Sauter, M., Adventitious root growth and cell-cycle induction in deepwater rice. *Plant Physiol.*, 119:21, 1999.
36. Grichko, V.P. and Glick, B.R., Ethylene and flooding stress in plants. *Plant Physiol. Biochem.*, 39:1, 2001.
37. Hwang, S.Y. and VanToai, T.T., Abscisic acid induces anaerobiosis tolerance in corn. *Plant Physiol.*, 97:593, 1991.
38. Ellis, M.H., Dennis, E.S., and Peacock, W.J., *Arabidopsis* roots and shoots have different mechanisms for hypoxic stress tolerance. *Plant Physiol.*, 119:57, 1999.
39. Rijnders, J.G. et al., Ethylene enhances gibberelin levels and petiole sensitivity in flooding tolerant Rumex in contrast to intolerant species. *Planta*, 203:20, 1997.
40. Zhang, J. et al., Development of flooding-tolerant *Arabidopsis thaliana* by autoregulated cytokinin production. *Mol. Breeding*, 6:135, 2000.
41. Jackson, M.B. and Ricard, B., Physiology, biochemistry and molecular biology of plant root systems subjected to flooding of the soil. In *Ecological Studies: Root Ecology*, Vol. 168, de Kroon, H. and Visser, E.J.W., Eds., Springer-Verlag, Berlin/Heidelberg, 2003, p. 193.

42. Aschi-Smiti, S., Bizid, E., and Hamza, M., Effect of waterlogging on growth of four varieties of clover (*Trifolium subterraneum* L.). *Agronomie*, 23:97, 2003.
43. Jackson, M.B., Long-distance signalling from roots to shoots assessed: the flooding story. *J Exp. Bot.*, 53:175, 2002.
44. Jackson, M.B., Hormones from roots as signals for the shoots of stressed plants. *Trends Plant Sci.*, 2:22, 1997.
45. Else, M.A. and Jackson, M.B., Transport of 1-aminocyclopropane-1-carboxylic acid (ACC) in the transpiration stream of tomato (*Lycopersicon esculentum*) in relation to foliar ethylene production and petiole epinasty. *Aust. J. Plant Physiol.*, 25:453, 1998.
46. Else, M.A. et al., Decreased root hydraulic conductivity reduces leaf water potential, initiates stomatal closure and slows leaf expansion in flooded plants of castor oil (*Ricinus communis*) despite diminished delivery of ABA from the roots in xylem sap. *Physiol. Plant*, 111:46, 2001.
47. Jackson, M.B. and Armstrong, W., Formation of aerenchyma and the processes of plant ventilation in relation to soil flooding and submergence. *Plant Biol.*, 1:274, 1999.
48. Sorrell, B.K. et al., Ecophysiology of wetland plant roots: A modeling comparison of aeration in relation to species distribution. *Ann. Bot.*, 86:675, 2000.
49. Brailsford, R.W. et al., Enhanced ethylene production by primary roots of *Zea mays* L. in response to sub-ambient partial pressures of oxygen. *Plant Cell Environ.*, 16:1071, 1993.
50. Gunawardena, H.L.A.N. et al., Characterization of programmed cell death during aerenchyma formation induced by ethylene or hypoxia in roots of maize (*Zea mays* L.). *Planta*, 212:205, 2001.
51. Subbaiah, C.C. and Sachs, M.M., Molecular and cellular adaptation of maize to flooding stress. *Ann. Bot.*, 90:119, 2003.
52. Kawai, M. et al., Cellular dissection of the degradation pattern of cortical cell death during aerenchyma formation of rice roots. *Planta*, 204:277, 1998.
53. Gunawardena, H.L.A.N., Rapid changes in cell wall pectic polysaccharides are closely associated with early stages of aerenchyma formation, a spatially localised form of programmed cell death in roots of maize (*Zea mays* L.) promoted by ethylene. *Plant Cell Environ.*, 24:1369, 2001.
54. Grineva, G.M. and Bragina, T.V., Formation of adaptations to flooding in corn — structural and functional parameters. *Russ. J. Plant Physiol.*, 40:583, 1993.
55. Grineva, G.M., Bragina, T.V., and Platonov, A.V., Ethylene-induced activation of hydrolytic enzymes in adventitious maize roots under conditions of advancing flooding. *Dok Bot. Sci.*, 373:23, 2000.
56. Saab, I.N. and Sachs, M.M., A flooding induced xyloglucan endo-transglycosylase homolog in maize is responsive to ethylene and associated with aerenchyma. *Plant Physiol.*, 112:385, 1996.
57. Vignolio, O.R., Fernandez, O.N., and Macerira, N.O., Flooding tolerance in five populations of *Lotus glaber* Mill. (syn. *Lotus tenuis* Waldst. Et. Kit.). *Aust. J. Agric. Sci.*, 50:555, 1999.
58. Visser, E.J.W. et al., Regulatory role of auxin in adventitious root formation in two species of Rumex, differing in their sensitivity to waterlogging. *Physiol. Plant.*, 93;116, 1995.

59. Greenwood, M.S., Cui, X., and Xu, F., Response to auxin changes during maturation-related loss of adventitious rooting competence in lobolly pine (*Pinus taeda*) stem cuttings. *Physiol. Plant*, 111:373, 2001.
60. Kelly, M.O. and Bradford, K.J., Ethylene synthesis and growth of tomato hypocotyls: Induction by auxin and fusicoccin and inhibition by vanadate. *J. Plant Growth Reg.*, 9:43, 1990.
61. McDonald, M.P., Galwey, N.W., and Colmer, T.D., Waterlogging tolerance in the tribe Triticeae: the adventitious roots of *Critesion marinum* have a relatively high porosity and a barrier to radial oxygen loss. *Plant Cell Environ.*, 24:585, 2002.
62. Chabbi, A., McKee, K.L., and Mendelssohn, I.A., Fate of oxygen losses from *Typha domingensis* (Typhaceae) and *Cladium jamaicense* (Cyperaceae) and consequences for root metabolism. *Am. J. Bot.*, 87:1081, 2000.
63. Colmer, T.D., Long-distance transport of gases in plants: A perspective on internal aeration and radial oxygen loss from roots. *Plant Cell Environ.*, 26:17, 2003.
64. Ricard, B. et al., Plant metabolism under hypoxia and anoxia. *Plant Physiol. Biochem.*, 32:1, 1994.
65. Gibon, Y. et al., Sensitive and high throughput metabolite assays for inorganic pyrophosphate, ADPGlc, nucleotide phosphates and glycolytic intermediates based on a novel enzymatic cycling system. *Plant J.*, 30:221, 2002.
66. Colmer, T.D., Huang, S., and Greenway, H., Evidence for down-regulation of ethanolic fermentation and K+ effluxes in the coleoptile of rice seedlings during prolonged anoxia. *J. Exp. Bot.*, 52:1507, 2001.
67. Xia, J.H., Saglio, P., and Roberts, J.K.M., Nucleotide levels do not critically determine survival of maize root tips acclimated to a low-oxygen environment. *Plant Physiol.*, 108:589, 1995.
68. Saint-Gès, V. et al., Kinetic studies of the variations of cytoplasmic pH, nucleotide triphosphates (^{31}P-NMR) and lactate during normoxic and anoxic transitions in maize root tips. *Eur. J. Biochem.*, 200:477, 1991.
69. Fox, G.G., McCallan, N.R., and Ratcliffe, R.G., Manipulating cytoplasmic pH under anoxia: A critical test of the role of pH in the switch from aerobic to anaerobic metabolism. *Planta*, 195:324, 1995.
70. Roberts, J.K.M., Cytoplasmic acidosis as a determinant of flooding intolerance in plants. *Proc. Natl. Acad. Sci. USA*, 81:6029, 1984.
71. Zhang, W.H. and Tyerman, S.D., Effects of low O_2 concentration and azide on hydraulic conductivity and osmotic volume of the cortical cells of wheat roots. *Aust. J. Plant Physiol.*, 18:603, 1991.
72. Kamaluddin, M. and Zwiazek, J.J., Metabolic inhibition of root water flow in red-osier dogwood (*Cornus stolonifera*) seedlings. *J. Exp. Bot.*, 52:739, 2001.
73. Javot, H. et al., Role of a single aquaporin isoform in root water uptake. *Plant Cell*, 15:509, 2003.
74. Ratcliffe, R.G., Metabolic aspects of the anoxic response in plant tissues. In *Environment and Plant Metabolism: Flexibility and Acclimation*, Smirnoff, N., Ed., Bios Scientific, Oxford, 1995, p. 111.
75. Walker, J. et al., DNA sequences required for anaerobic expression of the maize alcohol dehydrogenase I gene. *Proc. Natl. Acad. Sci USA*, 84:6624, 1987.
76. Hoeren, F.U. et al., Evidence for a role of a MYB2 in the induction of the *Arabidopsis thaliana* alcohol dehydrogenase gene (ADH1) by low oxygen. *Genetics*, 149:479, 1998.

77. Magaraggia, F. et al., I. Maturation and translation mechanisms involved in the expression of a myb gene of rice. *Plant Mol. Biol.*, 35:1003, 1997.
78. de Vetten, N.C. and Ferl, R.J., Characterization of a G-box binding factor that is induced by hypoxia. *Plant J.*, 7:589, 1995.
79. de Vetten, N.C., Lu, G., and Ferl, R.J., A maize protein associated with the G-box binding complex has homology to brain regulatory proteins. *Plant Cell*, 4:1295, 1992.
80. Lu, G. et al., Brain proteins in plants: an *Arabidopsis* homolog to neurotransmitter pathway activators is part of a DNA binding complex. *Proc. Natl. Acad. Sci. USA*, 9:11390, 1992.
81. Sachs, M.M., Subbaiah, C.C., and Saab, I.N., Anaerobic gene expression and flooding tolerance in maize. *J. Exp. Bot.*, 47:1, 1996.
82. Chang, W.P.P., et al., Patterns of protein synthesis and tolerance of anoxia in root tips of maize seedling acclimated to low oxygen environment, and identification of protein by mass spectrometry. *Plant Physiol.*, 122:295, 2000.
83. Saglio, P., Germain, V., and Ricard, B., The response of plant to oxygen deprivation: Role of enzyme induction in the improvement of tolerance to anoxia. In *Plant Response to Environmental Stresses*, Lerner, H.R., Ed., Marcel Dekker, New York, 1999, p. 373.
84. Fennoy, S.L. and Bailey-Serres, J., Posttranscriptional regulation of gene expression in oxygen-deprived roots of maize. *Plant J.*, 7:287, 1995.
85. Fennoy, S.L., Jayachandra, S., and Bailey-Serres, J., RNAse activities are reduced concomitantly with total conservation of total cellular RNA and ribosomes in O_2-deprived seedling roots of maize. *Plant Physiol.*, 115:1109, 1997.
86. Bailey-Serres, J. and Dawe, K., Both 5' and 3' sequences of maize *adh1* mRNA are required for enhanced translation under low-oxygen conditions. *Plant Physiol.*, 112:685, 1996.
87. Manjunath, S., Williams, A.J., and Bailey-Serres, J., Oxygen deprivation stimulates Ca^{2+}-mediated phosphorylation of mRNA cap-binding protein eIF4E in maize roots. *Plant J.*, 9:21, 1999.
88. Matsumara, H., Nirasawa, S., and Terauchi, R., Transcript profiling in rice (*Oryza sativa* L.) seedlings using serial analysis of gene expression (SAGE). *Plant J.*, 20:719, 1999.
89. Butler, W., Cook, L., and Vayda, M.E., Hypoxic stress inhibits multiple aspects of the potato tuber wound response. *Plant Physiol.*, 93:264, 1990.
90. Kühn, C. et al., Update on sucrose transport in higher plants. *J. Exp. Bot.*, 50:935, 1999.
91. Cleland, R.E., Fujiwara, T., and Lucas, W., Plasmodesmal-mediated cell to cell transport in wheat root is modulated by anaerobic stress. *Protoplasma*, 178:81, 1994.
92. Stitt, M., Pyrophosphate as an alternative energy donor in the cytosol of plant cells: An enigmatic alternative to ATP. *Bot. Acta*, 111:167, 1998.
93. Germain, V. et al., The role of sugars, hexokinase, and sucrose synthase in the determination of hypoxically induced tolerance to anoxia in tomato roots. *Plant Physiol.*, 114:167, 1997.
94. Bouny, M. and Saglio, P., Glycolytic flux and hexokinase activities in anoxic maize root tips acclimated by hypoxic pre-treatment. *Plant Physiol.*, 111:187, 1996.
95. Johnson, J.R., Cobb, B.G., and Drew, M.C., Hypoxic induction of anoxia tolerance in roots of ADH1 null *Zea mays* L. *Plant Physiol.*, 105:61, 1994.

96. Matsumura, H. et al., *Adh1* is transcriptionally active but its translational product is reduced in a *rad* mutant of rice (*Oryza sativa* L.), which is vulnerable to submergence stress. *Theoret. Appl. Genetics*, 97:1197, 1998.
97. Good, A.G. and Crosby, W.L., Anaerobic induction of alanine aminotransferase in barley root tissues. *Plant Physiol.*, 90:1305, 1989.
98. Fan, T.W.M. et al., Anaerobic nitrate and ammonium metabolism in flood-tolerant rice coleoptiles. *J. Exp. Bot.*, 48:1655, 1997.
99. Ford, Y., Ratcliffe, R.G., and Robins, R.J., Phytochrome induced GABA production in transformed root cultures of *Datura stramonium*: An in vivo ^{15}N NMR study. *J. Exp. Bot.*, 47:811, 1996.
100. Menegus, F. et al., Differences in the anaerobic lactate-succinate production and the changes of cell sap pH for plants with high and low resistance to anoxia. *Plant Physiol.*, 90:29, 1989.
101. Sowa, A.W. et al., Altering hemoglobin levels changes energy status in maize cells under hypoxia. *Proc. Natl. Acad. Sci. USA*, 95:10317, 1998.
102. Appleby, C.A. et al., A role for hemoglobin in all plant roots. *Plant Cell Environ.*, 11:359, 1988.
103. Vierstra, R.D., Proteolysis in plants: Mechanisms and functions. *Plant Mol. Biol.*, 32:275, 1996.
104. Aschi-Smiti, S. et al., Assessment of enzymes induction and aerenchyma formation as mechanisms for flooding tolerance in *Trifolium subterraneum* "Park." *Ann. Bot.*, 91:195, 2003.
105. Subbaiah, C.C., Kollipara, K.P., and Sachs, M.M., A Ca^{2+}-dependent cysteine protease is associated with anoxia-induced root tip death in maize. *J. Exp. Bot.*, 51:721, 2000.
106. Crawford, R.M., Plant survival without oxygen. *Biologist*, 40:110, 1993.
107. Foyer, C.H., Lelandais, M., and Kunert, K.J., Photooxidative stress in plants. *Physiol. Plant*, 92:696, 1994.
108. Gutteridge, J.M.C. and Haliwell, B., The measurement and mechanisms of lipid peroxidation in biological systems. *Trends Biochem. Sci.*, 7:270, 1990.
109. Crawford, R.M. and Woolenweber-Ratzer, B., Influence of ascorbic acid on post-anoxic growth and survival of chickpea seedlings (*Cicer aeriatum* L). *J. Exp. Bot.*, 43:703, 1992.
110. Yordanova, R.Y., Alexieva, V.S., and Popova, L.P., Influence of root oxygen deficiency on photosynthesis and antioxidants in barley plants. *Russ. J. Plant Physiol.*, 50:163, 2003.
111. Hetherington, A.M., Hunter, M.J., and Crawford, R.M.M., Contrasting effects of anoxia on rhizome lipids of *Iris* species. *Phytochem.*, 21:1275, 1982.
112. Leul, M. and Zhou, W.J., Alleviation of waterlogging damage in winter rape by uniconazole applications: effects on enzyme activity, lipid peroxidation, and membrane integrity. *J. Plant Growth Regul.*, 18:9, 1999.
113. Hurng, W.P. and Kao, C.H., Lipid peroxidation and antioxidative enzymes in senescent tobacco leaves following flooding. *Plant Sci.*, 96:41, 1994.
114. Biemelt, S., Keetman, U., and Albrecht, G., Re-aeration following hypoxia or anoxia leads to activation of the antioxidative defense system in roots of wheat seedlings. *Plant Physiol.*, 116:651, 1998.
115. Lin, K.H., Study of the root antioxidative system of tomatoes and eggplants under waterlogged conditions. *Plant Sci.*, 167:355, 2004.

116. Amor, Y., Chevion, M., and Levine, A., Anoxia pretreatment protects soybean cells against H_2O_2-induced cell death: Possible involvement of peroxidase and of alternative oxidase. *FEBS Lett.*, 477:175, 2000.
117. Maxwell, D., Wang, Y., and McIntosh, L., The alternative oxidase lowers mitochondrial reactive oxygen production in plant cells. *Proc. Natl. Acad. Sci. USA*, 96:8271, 1999.
118. Yordanova, R.Y. and Popova, L.P., Photosynthetic response of barley plants to soil flooding. *Photosynthetica*, 39:515, 2001.
119. Tamminen, I. et al., Ectopic expression of *ABI3* gene enhances freezing tolerance in response to abscisic acid and low temperature in *Arabidopsis thaliana*. *Plant J.*, 25:1, 2001.
120. Dolferus, R. et al., Enhancing the anaerobic response. *Ann. Bot.*, 91:111, 2003.
121. Koch, K.E. et al., Multiple paths of sugar-sensing and a sugar/oxygen overlap for genes of sucrose and ethanol metabolism. *J. Exp. Bot.*, 51:417, 2000.
122. Dai, N. et al., Overexpression of *Arabidopsis* hexokinase in tomato plants inhibits growth, reduces photosynthesis, and induces rapid senescence. *Plant Cell*, 11:1253, 1999.
123. Veramendi, J. et al., Potato hexokinase 2 complements transgenic *Arabidopsis* plants deficient in hexokinase 1 but does not play a key role in tuber carbohydrate metabolism. *Plant Mol. Biol.*, 49:491, 2002.
124. Roessner-Tunali, U. et al., Metabolic profiling of transgenic tomato plants overexpressing hexokinase reveals that the influence of hexose phosphorylation diminishes during fruit development. *Plant Physiol.*, 133:84, 2003.
125. Menu, T. et al., High hexokinase activity in tomato fruit perturbs carbon and energy metabolism and reduces fruit and seed size. *Plant Cell Environ.*, 27:89, 2003.
126. Germain, V., Réponse au déficit en oxygène chez *Lycopersicon esculentum* M., Thèse de doctorat de l'Université de Bordeaux 2, 1997.
127. Sweetlove, L.J. et al., Lactate metabolism in potato tubers deficient in lactate dehydrogenase activity. *Plant Cell Environ.*, 23:873, 2000.
128. Ismond, K.P. et al., Enhanced low oxygen survival in *Arabidopsis* through increased metabolic flux in the fermentative pathway. *Plant Physiol.*, 132:1292, 2003.
129. Quimio, C.A. et al., Enhancement of submergence tolerance in transgenic rice overproducing pyruvate decarboxylase. *J. Plant Physiol.*, 156:516, 2000.
130. Tadege, M., Brandle, R., and Kuhlemeier, C., Anoxia tolerance in tobacco roots: Effect of overexpression of pyruvate decarboxylase. *Plant J.*, 14:327, 1998.
131. Grichko, V.P. and Glick, B.R., Flooding tolerance of transgenic plants expressing the bacterial enzyme ACC deaminase controlled by the 35S, *rolD* or PRB-1*b* promoter. *Plant Physiol. Biochem.*, 39:19, 2001.
132. Hunt, P.W. et al., Increased level of hemoglobin 1 enhances survival of hypoxic stress and promotes early growth in *Arabidpopsis thaliana*. *Proc. Natl. Acad. Sci. USA*, 99:17197, 2002.
133. Armstrong, W., Brandle, R., and Jackson, M.B., Mechanisms of flood tolerance in plants. *Acta Bot. Neerl.*, 43:307, 1994.
134. Monk, L.S., Fagerstedt, K.V., and Crawford, R.M.M., Oxygen toxicity and superoxide dismutase as an anti-oxidant in physiological stress. *Physiol. Plant*, 76:456, 1989.

135. Yu, Q. and Rengel, Z., Waterlogging influences plant growth and activities of superoxide dismutases in narrow-leafed lupine and transgenic tobacco plants. *J. Plant Physiol.*, 155:431, 1999.
136. Setter, T.L. and Waters, L., Review of prospects for germplasm improvement for waterlogging tolerance in wheat, barley and oats. *Plant Soil*, 253:1, 2003.
137. Shailesh, T., Warsi, M.Z.K., and Verma, S.S., Water logging tolerance in inbred lines of maize (*Zea mays* L.). *Cereal Res. Comm.*, 31:221, 2003.
138. Gibberd, M.R. et al., Waterlogging tolerance among a diverse range of Trifolium accessions is related to root porosity, lateral root formation and "aerotropic rooting." *Ann. Bot.*, 88:579, 2001.
139. Xu, K. and Mackill, D.J., A major locus for submergence tolerance mapped on rice chromosome 9. *Mol. Breeding*, 2:219, 1996.
140. Ram, P.C. et al., Submergence tolerance in rainfed lowland rice: Physiological basis and prospects for cultivar improvement through market-aided breeding. *Field Crops Res.*, 76:131, 2002.
141. Wiengweera, A., Greenway, H., and Thomson, C.J., The use of agar nutrient solution to simulate lack of convection in waterlogged soils. *Ann. Bot.*, 80:113, 1997.
142. Saglio, P.H., Drew, M.C., and Pradet, P., Metabolic acclimation to anoxia induced by low (2-4 kPa parial pressure) oxygen pretreatment (hypoxia) in root tips of *Zea mays*. *Plant Physiol.*, 86:61, 1988.
143. Ricard, B., Anoxic responses of seedling roots of rice cultivars varying in tolerance to submergence. *Russ. J. Plant Physiol.*, 50:799, 2003.
144. Andrews, D.L. et al., Hypoxic and anoxic induction of alcohol dehydrogenase in roots and shoots of seedlings of *Zea mays*. *Plant Physiol.*, 101:407, 1992.
145. Couée, I. et al., Effects of anoxia on mitochondrial biogenesis in rice shoots: Modification of *in organello* translation characteristics. *Plant Physiol.*, 98:411, 1992.
146. Reyna, N. et al., Evaluation of a QTL for waterlogging tolerance in southern soybean germplasm. *Crop Sci.*, 43:2077, 2003.
147. Nelson, R.B. et al., Measurement of soil waterlogging tolerance in *Phaseolus vulgaris* L.: A comparison of screening techniques. *Scientia horticulturae.*, 20:303, 1983.
148. Smethurst, C.F. and Shabala, S., Screening methods for waterlogging tolerance in lucerne: Comparative analysis of waterlogging effects on chlorophyll fluorescence, photosynthesis, biomass and chlorophyll content. *Func. Plant Biol.*, 30:335, 2003.
149. Garthwaite, A.J., von Bothmer, R., and Colmer, T.D., Diversity in root aeration traits associated with waterlogging tolerance in the genus Hordeum. *Func. Plant Biol.*, 30:875, 2003.
150. Bange, M.P., Milroy, S.P., and Thongbai, P., Growth and yield of cotton in response to waterlogging. *Field Crops Res.*, 88:129, 2004.
151. Pang, J.Y. et al., Growth and physiological responses of six barley genotypes to waterlogging and subsequent recovery. *Aust. J. Agric. Res.*, 55:895, 2004.
152. Band, P.E. et al., Effect of waterlogging on bio-chemical, yield and yield contributing parameters in black gram. *J. Soils Crops*, 14:76, 2004.

8 Whole-Plant Adaptations to Low Phosphorus Availability

Jonathan P. Lynch and Kathleen Brown

CONTENTS

8.1 Low Soil Phosphorus Availability is a Primary Constraint to Plant Productivity 210
8.2 Plant Adaptations to Low Phosphorus Availability 211
    8.2.1 Growth Regulation 212
    8.2.2 Soil Exploration 213
        8.2.2.1 Topsoil Foraging 214
        8.2.2.2 Reducing the Metabolic Costs of Soil Exploration 215
        8.2.2.3 Aerenchyma Reduces Metabolic Costs of Soil Exploration 218
        8.2.2.4 Root "Etiolation" 221
        8.2.2.5 Alternative Respiration 222
        8.2.2.6 Root Turnover 223
        8.2.2.7 Root Hairs 223
    8.2.3 Phosphorus Mobilization 225
    8.2.4 Cluster Roots 226
    8.2.5 Microbial Symbioses 226
    8.2.6 Phenology 227
    8.2.7 Trait Interactions 228
    8.2.8 Tradeoffs 229
    8.2.9 Responses to Localized Availability of Phosphorus 230
8.3 Competition 231
8.4 Ecosystem Issues 232
8.5 Conclusions 233
References 234

8.1 LOW SOIL PHOSPHORUS AVAILABILITY IS A PRIMARY CONSTRAINT TO PLANT PRODUCTIVITY

Soil infertility is a primary constraint to plant productivity over the majority of the Earth's land surface. Nitrogen is often limiting in young soils of the temperate zone, whereas phosphorus (P) is a primary limitation in most forests, weathered soils, and the humid tropics, which support the majority of terrestrial plant biomass (Figure 8.1) [1,2]. Low soil phosphorus availability is caused by several factors, including the reactivity of orthophosphate with common soil constituents such as Fe and Al oxides, resulting in compounds of limited bioavailability, especially as soil weathering progresses, and the fact that the phosphorus cycle is open-ended and tends toward depletion.

Human activity in many managed ecosystems has further reduced phosphorus bioavailability through topsoil erosion, acidification, and nutrient mining, especially in developing countries [3]. Approximately 40% of the agricultural land in the world has been significantly degraded by human activity, including 65% of the agricultural soils of Africa [4]. Replenishment of soil phosphorus reserves through fertilization is common in developed countries, but the economic sustainability of this practice is in question, because analysts estimate that economically recoverable phosphorus reserves will be depleted by 50% by the middle of this century [5,6]. In many developing countries, especially in Africa, fertilizer use is negligible [7], and the productivity of many of these agroecosystems is phosphorus-limited. The response of terrestrial ecosystems to global climate change will depend on interactions of climate change variables with edaphic limitations to plant productivity, including

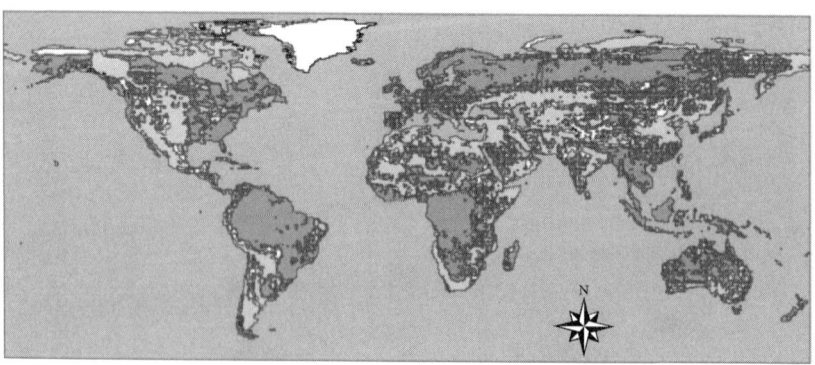

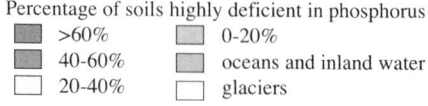

Percentage of soils highly deficient in phosphorus
- >60%
- 40-60%
- 20-40%
- 0-20%
- oceans and inland water
- glaciers

FIGURE 8.1 Map of soil phosphorus availability. (Modified from International Soil Reference and Information Centre, http://lime.isric.nl/. With permission.)

phosphorus [8]. The adaptation of plants to low phosphorus availability is therefore of considerable interest in both basic and applied plant biology.

8.2 PLANT ADAPTATIONS TO LOW PHOSPHORUS AVAILABILITY

Plants display a variety of adaptations to low phosphorus availability, including mycorrhizal symbioses [9], root-hair elongation, and proliferation [10–12], for example, rhizosphere modification through secretion of carboxylates [13,14], protons [15], and phosphatases [16], and modification of root architecture to maximize phosphorus acquisition efficiency (Figure 8.2) [17]. Phosphorus-deficient plants typically have higher root:shoot ratios than high-P plants, either because of allometric relationships [18] or because of increased biomass allocation to roots [19,20]. Increased root growth is obviously beneficial for phosphorus acquisition, because phosphorus is fairly immobile in soil, but can slow overall plant growth because of the increased respiratory burden of root tissue [20–24].

Several recent reviews summarize plant responses to low phosphorus availability, emphasizing cellular and biochemical or molecular processes [25–27] and specific adaptations to low phosphorus availability, such as cluster roots [28–30] and mycorrhizas [31–33]. The purpose of this review is to focus on adaptations

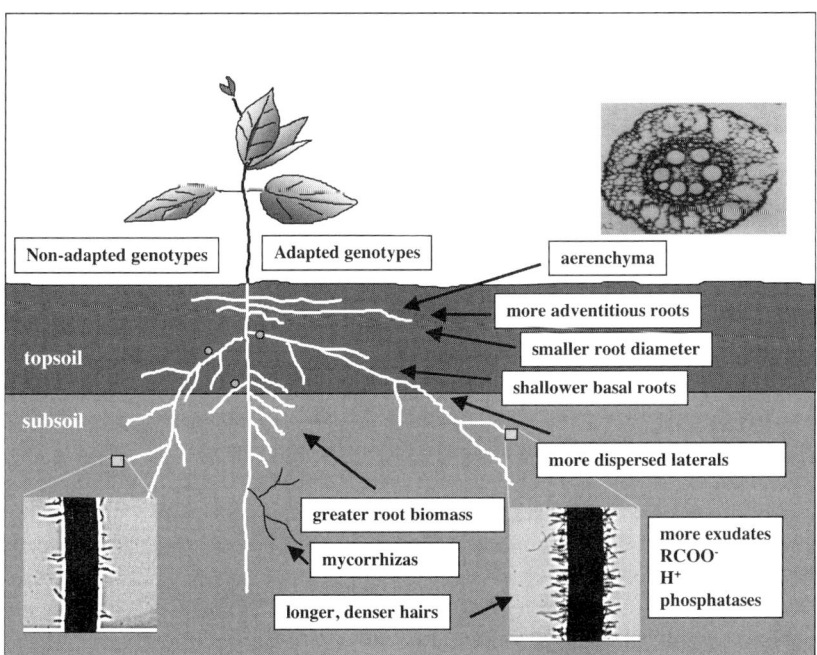

FIGURE 8.2 Some adaptive responses of root systems to low phosphorus availability.

to low phosphorus availability that manifest at the supracellular level, including tissue, organ, and organismic processes, which have not been the focus of recent reviews. We focus on the traits themselves rather than mechanisms by which those traits may be induced by low phosphorus availability. We also focus on traits that are adaptive, that is, that contribute to enhanced growth and productivity in low phosphorus environments, rather than those responses with uncertain function or that may be symptoms of stress rather than mechanisms of adaptation.

An example of an apparently nonadaptive response is the dramatic upregulation of phosphatases in leaf tissue of phosphorus-deprived plants, which for some years was assumed to be useful in enhancing phosphorus remobilization to growing tissue. However, a careful examination of nearly 100 recombinant inbred lines of common bean (*Phaseolus vulgaris* L.) that segregated for leaf phosphatase upregulation in response to phosphorus stress showed that there was no association of this trait with phosphorus remobilization or plant growth under phosphorus stress [34]. The discrimination of adaptive from nonadaptive responses is important in understanding plant responses to stress, inasmuch as severe or sustained stress will induce many plant reactions that are direct or indirect responses to injury and growth reduction rather than adaptive mechanisms. The identification of adaptive traits, and an understanding of the physiological and ecological tradeoffs associated with these traits, is especially important for breeding more phosphorus-efficient crops, an enterprise of great importance for developing countries [35,36].

8.2.1 Growth Regulation

A common response to phosphorus deficiency is an increase in root-to-shoot dry-weight ratio, resulting from a greater inhibition of shoot growth than root growth [37–39]. A portion of this apparent change is allometric; that is, root:shoot ratios normally decline with increase in plant mass, and because plants supplied with low phosphorus grow more slowly, their root:shoot ratios are higher at a given plant age. However, when this factor was eliminated by comparison of allometric coefficients among plants grown at different phosphorus levels, some genotypes showed a lower allometric coefficient (smaller increase in root DW relative to increases in shoot DW) with low phosphorus, whereas others did not [20]. In this study, genotypes of phosphorus-efficient (less yield depression under low phosphorus) common bean plants maintained a higher root:shoot ratio (higher allometric coefficient) with continued growth, supporting the idea that root growth is valuable for phosphorus acquisition. Low phosphorus availability reduces leaf appearance, leaf expansion, and shoot branching [38,40]. Among annuals, phosphorus stress decreased shoot growth in dicots more than in monocots, possibly because of differences in leaf morphology [41].

Root growth is a key trait for phosphorus efficiency, defined as little or no reduction in plant productivity under low phosphorus availability compared with high phosphorus availability [42,43]. Low phosphorus availability changes the distribution of growth among various root types. In common bean, growth of main root (primary and basal root) axes is maintained under low phosphorus, and initiation of lateral roots is reduced, so that lateral root density declines [44]. Mean lateral root length is unaffected.

In experiments with maize (*Zea mays* L.) subjected to phosphorus starvation, axial (seminal and nodal) root elongation and lateral root density were unaffected, but lateral root elongation was first promoted slightly, then severely retarded, as phosphorus starvation proceeded [39]. Initiation of new axial roots also ceased after six days of phosphorus starvation. The maintenance of elongation of main roots in maize and common bean could be interpreted as exploratory behavior, allowing these roots to grow maximally until they encounter localized patches of higher phosphorus availability. When the main root of a phosphorus-deficient plant encounters a patch of higher nutrient availability, lateral roots may proliferate within the patch (see Section 8.2.9) [45]. In *Arabidopsis thaliana* seedlings deprived of added phosphorus, lateral root number was also reduced, but in this case the remaining lateral roots elongated at the expense of the primary root [46–48]. One reason for the discrepancy may be that *Arabidopsis* plants lack main roots other than the primary root (i.e., they have no root type analogous to the basal roots of legumes or the seminal roots of grasses), so that a subset of lateral roots must take over the functions of the other main roots in species with more complex root systems. We have observed genetic variation for the effect of buffered low phosphorus on lateral root length and number in maize, with some genotypes showing an increase and others showing a decrease in these variables [49]. Genotypes with increased or sustained lateral root development under phosphorus deficiency had superior ability to acquire phosphorus and maintain growth.

In common bean, some genotypes preferentially increase growth of adventitious roots, which have the advantages of low construction cost and location within the topsoil (see Section 8.2.2.1) [50]. Adventitious rooting has long been associated with adaptation to waterlogging [51] and has recently been associated with root rot resistance [52] and responses to root herbivory [53]. In some crops, such as maize, a high proportion of the mature root systems consists of adventitious roots, so prevention of adventitious rooting reduces water uptake even in well-watered plants [54]. Under low phosphorus conditions, adventitious root development may be delayed or reduced, primarily as a result of overall growth inhibition [50,55]. In some genotypes of common bean, the maintenance of adventitious root formation when overall growth is inhibited by phosphorus deficiency results in an increased proportion of root length in the adventitious root system [50]. This characteristic is associated with phosphorus efficiency in soils with poor phosphorus availability.

8.2.2 Soil Exploration

Root growth and architecture are particularly relevant for phosphorus acquisition. Phosphorus is relatively immobile in soil, moving primarily by diffusion rather than mass flow [56]. Phosphorus distribution is highly heterogeneous in most soils, generally being greatest in surface horizons and decreasing with depth [57–59].

The availability of soil phosphorus is also highly heterogeneous because of spatial heterogeneity of pH, eH, microbial activity, temperature, and so on [60]. Phosphorus mobilization and uptake by the root itself creates zones of phosphorus depletion that vary sharply on the scale of millimeters [61]. As a result of the development of phosphorus depletion zones around existing roots, phosphorus acquisition is dependent on continued root growth and exploration of new soil domains that have not yet been depleted of phosphorus [60].

8.2.2.1 Topsoil Foraging

Because the topsoil is generally the soil stratum with greatest phosphorus bioavailability, adaptation to low soil phosphorus availability is associated with the variation in extent of topsoil foraging among genotypes of maize and common bean [62–66]. We recently reviewed the importance of topsoil exploration for phosphorus efficiency [17].

Architectural traits associated with enhanced topsoil foraging in common bean include shallower growth of basal roots, enhanced adventitious rooting, and greater dispersion of lateral roots (Figure 8.2). There are several lines of evidence that shallower basal root growth enhances topsoil foraging and thereby phosphorus acquisition efficiency. The geometric simulation model SimRoot was used to model the effect of changing basal root gravitropism on interroot competition for phosphorus [66]. This study showed that in soils with uniform phosphorus distribution, shallower root systems explored more soil per unit of root biomass than deeper systems, because shallower systems have more dispersed basal roots, and therefore less interroot competition, which occurs when neighboring roots have overlapping phosphorus depletion zones [66]. In stratified soils with more phosphorus in the topsoil, simulations showed that shallower root systems acquired more phosphorus than deeper ones, by concentrating root foraging in the topsoil [66]. These modeling results are supported by the significant correlation of basal root growth angle in young common bean plants in growth pouches with their yield in field trials in low phosphorus tropical soils [64].

In comparison of individual plants grown in pots of soil, genotypes with shallower basal roots had greater phosphorus uptake than those with deeper root systems (Figure 8.3) [65]. Common bean genotypes with shallower basal roots had superior growth in a low phosphorus field trial in Honduras [63]. Genetic analysis of common bean lines segregating for basal root shallowness showed co-segregation of QTL for root shallowness and phosphorus uptake in the field in Colombia [67]. In maize, genotypes with shallower seminal roots (analogous to basal roots in dicots) had superior growth in low phosphorus soils in the field and greenhouse [62]. Similar results have been observed with soybean (*Glycine max* Merr.) [68]. It therefore appears that basal root shallowness is an important trait for topsoil foraging and phosphorus acquisition efficiency in annual crops.

In crops such as common bean, adventitious roots emerge from the subterranean portion of the hypocotyl and grow horizontally through the topsoil. Adventitious rooting is therefore an important element of topsoil exploration by the root system.

Whole-Plant Adaptations to Low Phosphorus Availability

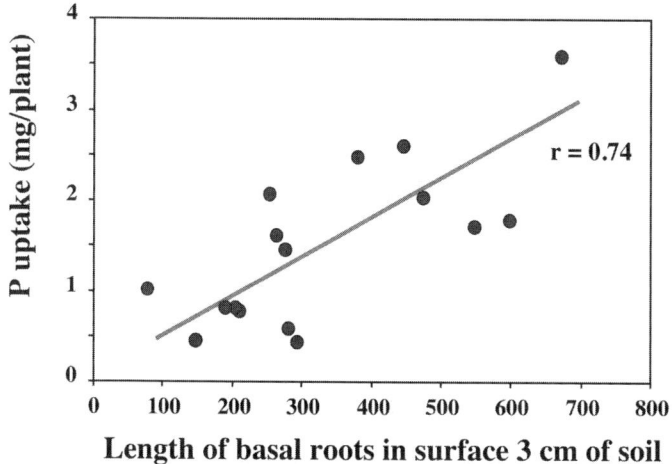

FIGURE 8.3 Phosphorus uptake among recombinant inbred lines of common bean (*Phaseolus vulgaris* L.) growing in low-P soil as a function of basal root shallowness.

Common bean genotypes differ in their extent of adventitious rooting and in the regulation of that trait by phosphorus [50]. As with basal root gravitropism, genotypic and phosphorus-induced adventitious rooting vary widely, from virtually no adventitious rooting in some conditions to dozens of adventitious roots in others [50]. A field study in a low phosphorus tropical soil showed that common bean genotypes with greater growth and phosphorus uptake had more adventitious rooting relative to basal root growth than did phosphorus-inefficient genotypes [50].

Adventitious roots may have several benefits for topsoil exploration. Obviously, their horizontal growth concentrates their foraging activity in the topsoil. Other advantages may relate to the anatomical and morphological differences between adventitious roots and basal roots. In common bean, adventitious roots have greater specific root length (root length per unit root mass) than other root types (Figure 8.4). This is advantageous for topsoil exploration because it enables the plant to explore a larger volume of soil per unit of metabolic investment in root tissue [24]. Adventitious roots may have a greater abundance of aerenchyma than other root types [51], which may be a mechanism of reducing the metabolic costs of soil exploration (see below). Finally, adventitious roots also have less lateral branching than basal roots, which would again serve to disperse root foraging over larger soil volumes for a given metabolic investment [50].

8.2.2.2 Reducing the Metabolic Costs of Soil Exploration

A number of studies have shown that the metabolic costs of soil exploration by root systems (which generally include mycorrhizal symbioses) are quite substantial, and can exceed 50% of daily photosynthesis [69]. Following the economic

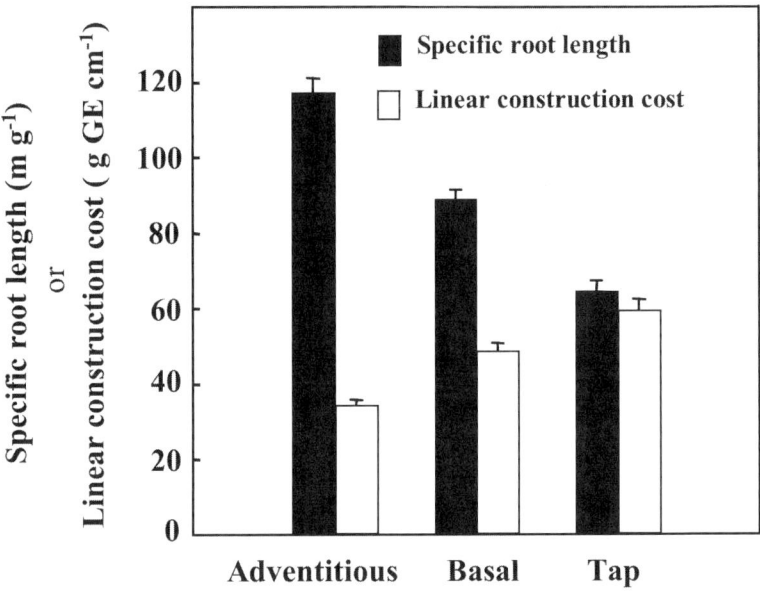

FIGURE 8.4 Specific root length and linear construction cost (in glucose equivalents per cm root length) of root classes of common bean (*Phaseolus vulgaris* L.). Each bar is the mean of four replicates, error bars = SEM. (From Lynch, J. and Ho, M., *Plant Soil*, 269:45–56, 2004. With permission.)

paradigm of plant resource allocation [70], we use the term "cost" to denote metabolic investment, including the production and maintenance of tissues, and measurable units of carbon (C is a convenient "currency" for our analysis; other "currencies," including phosphorus itself, may also be useful in some contexts [71–73]). Plant allocation to root growth typically increases under nutrient stress, and therefore the metabolic costs of root growth can be a significant component of plant fitness and adaptation under nutrient stress. All else being equal, a plant that is able to acquire a limiting soil resource at reduced metabolic cost will have superior fitness, because it will have more metabolic resources available for defense, growth, and reproduction.

The importance of root costs in plant adaptation to low phosphorus is illustrated by our work with the common bean. In common bean, low phosphorus availability increased the fraction of daily photosynthate respired by roots by 75% in both phosphorus-efficient and phosphorus-inefficient genotypes [20,22]. However, phosphorus-efficient genotypes had greater root growth per unit root respiration than did phosphorus-inefficient genotypes (Figure 8.5) [20], which enabled phosphorus-efficient genotypes to develop more than twice as much root biomass at low phosphorus than the phosphorus-inefficient genotypes. Phosphorus stress slightly increased the specific respiration rate (i.e., respiration per unit

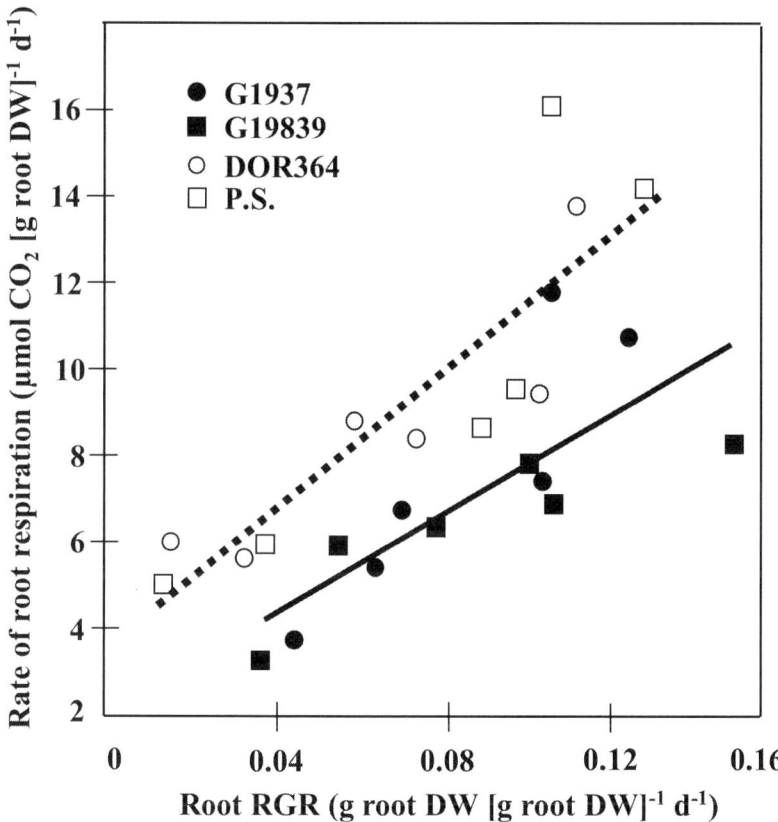

FIGURE 8.5 Relationship of root respiration and root relative growth rate in four genotypes of common bean (*Phaseolus vulgaris* L.). Open symbols represent genotypes that are P-inefficient (i.e., have poor growth in low-P media), closed symbols represent genotypes that are P-efficient. (From Nielsen, K.L., Eshel, A., and Lynch, J.P., *J. Exp. Bot.*, 52:329–339, 2001. With permission.)

biomass) of roots of the phosphorus-inefficient genotypes, but halved the respiration rate of roots of the phosphorus-efficient genotypes [24]. Thus, adaptation to low phosphorus availability in this species is associated with the ability to explore the soil at minimal metabolic cost. We refer to the metabolic cost of phosphorus acquisition as phosphorus acquisition efficiency, or PAE.

Several types of root traits could alter the relationship of root growth and root carbon cost. Geometric modeling suggests that root architecture can alter the carbon cost of soil exploration by regulating the extent of root competition within and among root systems [66,74]. The importance of root architecture for interplant competition for phosphorus was confirmed in field studies [75]. Morphological traits such as root hairs could enhance phosphorus acquisition at

TABLE 8.1
Maintenance Respiration Dominates Root Respiration Under Low Phosphorus in Common Bean[a]

Days after Planting	P-Level	Percent of Total Root Respiration		
		Rg	Riu	Rm
14	High	29	14	57
	Low	19	9	72
28	High	25	11	64
	Low	6	4	89

[a] Rg = growth respiration; Riu = ion uptake respiration; Rm = maintenance respiration.

minimal root carbon cost [11,76,77]. One mechanism of reducing root costs is to allocate more biomass to root classes that are less metabolically demanding per unit of phosphorus acquisition. We have shown that adventitious roots acquire phosphorus at less metabolic cost than basal and tap roots, and that phosphorus stress increases relative biomass allocation to adventitious roots, especially in phosphorus-efficient genotypes [50].

Root respiration can be divided into three components: growth of new tissue, maintenance of existing tissue, and ion uptake and assimilation [69,78,79]. As root systems mature and the proportion of nongrowing tissue increases, maintenance respiration becomes an increasing fraction of total respiration. Even in one-month-old common bean plants grown with low phosphorus, maintenance respiration comprises 90% of total root respiration (Table 8.1). In this context it is noteworthy that under phosphorus stress a phosphorus-efficient common bean genotype had 50% lower maintenance respiration than a phosphorus-inefficient genotype [24]. Reduced maintenance respiration of root tissue under phosphorus stress is an important adaptation to low phosphorus availability, by making more fixed carbon available for continued root growth.

8.2.2.3 Aerenchyma Reduces Metabolic Costs of Soil Exploration

Aerenchyma denotes tissue with large intercellular spaces [80]. Root aerenchyma is an adaptation to hypoxia (reviewed in [81]). In C3 plants, aerenchyma may also provide a photosynthetic benefit by channeling CO_2 from root respiration to leaf intercellular spaces [82,83]. Although the overwhelming majority of research on root aerenchyma has focused on its importance in hypoxia, root aerenchyma can also be induced by suboptimal nutrient availability. In aerated solution, aerenchyma was observed in maize roots when N or P was omitted from the medium [84–86]. This response was also observed in common bean [86,87] and rice (*Oryza sativa*) [88]. In maize, the induction of aerenchyma by low phosphorus may be related to increased ethylene sensitivity of phosphorus-stressed roots [89].

Anatomical traits affecting tissue composition could be an important means of reducing maintenance respiration. In this regard aerenchyma is particularly interesting, because it would dramatically reduce maintenance respiration by replacing living cortical cells with air space [36,86]. Besides reducing the ongoing carbon cost of root maintenance, lysis of cortical cells may contribute previously fixed carbon to root apices. An additional benefit from aerenchyma formation would be the reduced phosphorus requirement of root growth, which in conditions of phosphorus limitation can be as significant as C costs for metabolic efficiency [71,72]. Phosphorus released from cortical tissue by aerenchyma formation would be useful in meeting the phosphorus demands of new root elongation. A similar concept has been proposed for cortical senescence in grasses [90] (also see [91]).

Results from our lab support the hypothesis that aerenchyma formation is a useful adaptation to low phosphorus. In common bean and maize, we observed substantial genotypic variation in the induction of cortical aerenchyma by phosphorus stress [86]. Differences in aerenchyma formation induced by ethylene treatments and genotypic variation were correlated with proportionate reductions in root phosphorus concentration in low-P roots. Reduced phosphorus requirement for soil exploration would be advantageous in conditions of low phosphorus availability. Phosphorus liberated by senescing cortical cells could be used for continued apical growth. In low phosphorus conditions, most of the phosphorus taken up by roots is utilized first to meet local tissue demand [92].

Variation in aerenchyma formation was disproportionately correlated with root respiration (Figure 8.6) [86]. Root segments with 20% cross-sectional area as aerenchyma had half the respiration rate per unit root length of roots without aerenchyma. The disproportionate effect of aerenchyma on respiration may reflect the fact that the cortical cells lost during the formation of aerenchyma are metabolically active, whereas inactive tissues such as sclerenchyma and xylem vessels do not contribute to maintenance respiration. Results with isolated root segments were confirmed in intact plants; whole root systems of a maize genotype with abundant aerenchyma had less root respiration per unit of root length than did a genotype with less aerenchyma [86]. In greenhouse and field studies, the high-aerenchyma maize genotype Oh43 had better root growth in low phosphorus soil than the low-aerenchyma W64a [93,94]. In maize, root porosity was highly correlated with sustained root growth under low phosphorus (Figure 8.7).

Genetic variation for aerenchyma induction in response to waterlogging has been observed in many species, including banana (*Musa* spp.) [95], wheat (*Triticum aestivum* L.) [96], barley (*Hordeum vulgare* L.) [97], and maize [98]. Related species may also vary in constitutive (nonstressed) aerenchyma formation [99,100]. We observed large genotypic variation (200–300%) in aerenchyma formation in response to phosphorus stress in both maize and common bean [86]. Such variation raises interesting questions regarding the adaptive importance and functional tradeoffs for aerenchyma in diverse environments. The large intraspecific variation in important crop species also makes aerenchyma amenable to

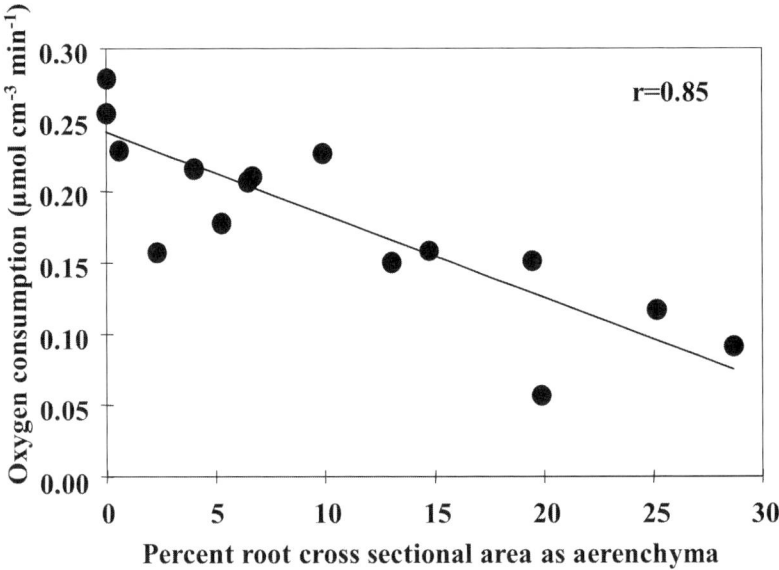

FIGURE 8.6 Correlation between aerenchyma area and respiration in maize (*Zea mays* L.) roots. Each data point is the mean of 6 measurements of respiration and 10–12 measurements of aerenchyma on comparable root segments. (From Fan, M.S., Zhu, J.M., Richards, C., Brown, K.M., and Lynch, J.P., Physiological roles for aerenchyma in phosphorus-stressed roots, *Func. Plant Biol.*, 30:493–506, 2003. With permission.)

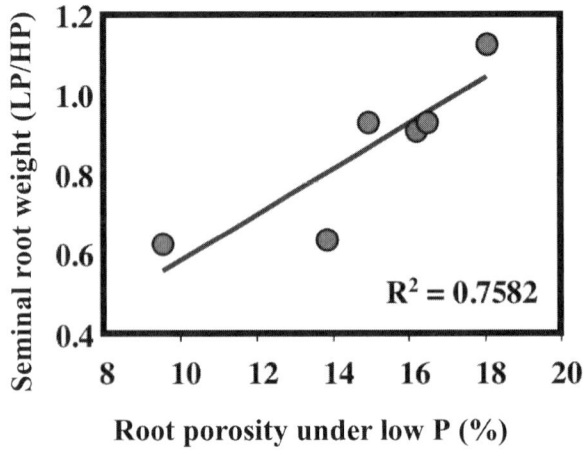

FIGURE 8.7 Maintenance of root growth in a low-P field as related to cortical aerenchyma formation in unrelated maize (*Zea mays* L.) genotypes. Root weights are expressed as the proportion of corresponding high-P roots. Each point is the mean of four replicates.

plant breeding, as is currently under way to enhance flooding tolerance in maize and other cereals [101,102].

8.2.2.4 Root "Etiolation"

Shoots respond to low light intensity by "etiolation," enhanced elongation at the expense of radial thickening and lateral shoot growth. This response is adaptive by increasing the likelihood of the shoot growing into better illumination and by increasing light capture in competitive situations. We hypothesize that an analogous process occurs in roots sensing low phosphorus availability.

There are many reports of increased fineness of roots under nutrient deficiency, usually described as reduced specific root length (SRL, root length per unit weight). However, increased SRL could result from the increased proportion of secondary roots, inasmuch as comparisons are not usually made within root classes [103,104]. Evidence for increased, reduced, or unchanged SRL can be found in the literature, but most reports do not consider variation in tissue density or variation in SRL within root classes, and are therefore not direct measurements of root diameter [105,106]. Careful studies of the effect of nutrient stress on root diameter within root classes and orders are needed to determine whether root etiolation could be an adaptive trait.

Under low phosphorus availability, root elongation is enhanced at the expense of lateral branching [44] and secondary growth [87]. There have been a few reports on increased diameter of specific root classes under high nutrient availability, including nitrate [105,107,108] and phosphorus [49,109]. Common bean basal roots show increased root diameter under high phosphorus, primarily in the older parts of the root (Figure 8.8). The larger diameter of the older parts of basal roots grown in high phosphorus was largely a result of a greater area of the stele, both in absolute area and relative to total root area [86]. Similarly, barley roots grown with high nitrate supply showed an increase in stele diameter [107], so this response is not restricted to dicots. In our study of maize genotypes with contrasting phosphorus efficiency, we found that lateral root SRL and diameter varied among genotypes, and that smaller diameter and greater SRL of lateral roots was associated with faster lateral root growth, which in turn was associated with higher shoot growth and phosphorus efficiency [49]. Furthermore, there was genetic variation in plasticity of this trait, that is, its response to phosphorus availability.

Particular root types may be more likely to alter their diameter in response to nutrient stress. In studies of barley, high nitrate increased the diameter of first- and second-order lateral roots, but not seminal roots [107]. In our study of elongation of the primary root of *Arabidopsis*, no difference in diameter could be discerned between high and low phosphorus treatments [110]. The timing and extent of etiolation may vary with root class, order, and extent of the nutrient stress. Root etiolation is presumably adaptive by reducing the metabolic costs of root extension into new soil domains that may have greater phosphorus availability. This phenomenon deserves further study.

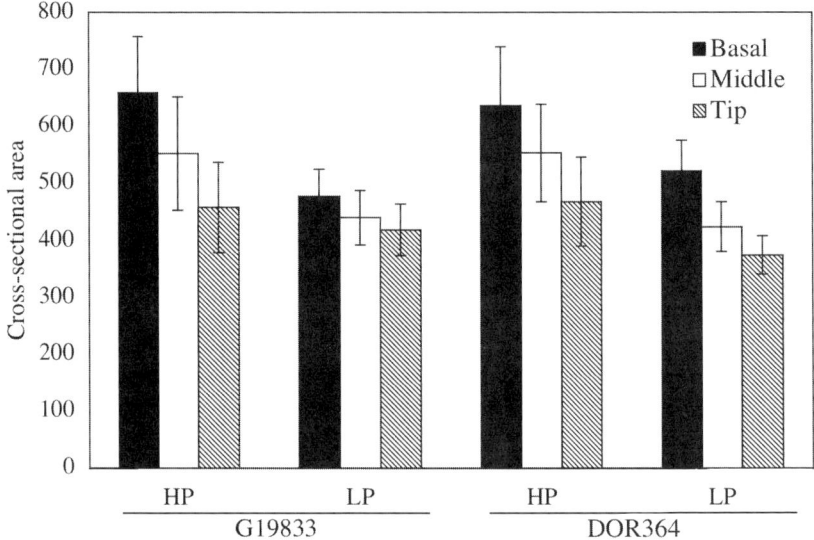

FIGURE 8.8 Cross-sectional area of common bean (*Phaseolus vulgaris* L.) basal roots grown for six weeks with high (1 mM, HP) or low (1 μM, LP) phosphorus. Total cross-sectional area was measured from segments of the most basal (2 cm from point of origin), central, and apical (2 cm from root tip) portions of one basal root from each of six plants per genotype and treatment, as described in Fan et. al., 2003. Values shown are means $\pm$ SEM.

Like shoot etiolation, root etiolation increases exploration at the expense of mechanical strength. Finer roots may be able to penetrate smaller pores in soil, but have less ability to push soil particles aside, so roots grown in soils with high bulk density tend to have a larger diameter and reduced branching [111]. In experiments on the effects of co-occurring soil compaction and phosphorus deficiency, roots increased their diameter with increasing bulk density only when supplied with phosphorus [112]. Root etiolation may also have tradeoffs in terms of turnover rates, desiccation tolerance, susceptibility to herbivory, and other characteristics [104].

8.2.2.5 Alternative Respiration

Mitochondrial respiration usually proceeds via the cytochrome-mediated pathway, which results in the phosphorylation of ADP to ATP [69]. In addition to the cytochrome pathway, two nonphosphorylating pathways exist in plant mitochondria, the cyanide-resistant and rotenone-insensitive pathways, which are induced under phosphorus stress conditions and allow respiration to proceed without depleting adenylate or phosphate pools [113–115]. Phosphorus stress induces other alterations in the respiration of common bean roots that appear to be related to oxidative stress [116]. It is not known if genotypic differences in root respiration under phosphorus stress are related to differential induction of alternative respiratory pathways.

8.2.2.6 Root Turnover

Root senescence or turnover through abiotic and biotic stress could have significant effects on the efficiency of phosphorus acquisition. Effects could be positive or negative from an adaptive perspective. Negative effects would result if roots are lost in fertile soil domains, resulting in loss of prior metabolic investment in those roots as well as the opportunity costs of phosphorus that is unexploited, or worse, exploited by a competitor. Positive effects could result from the pruning of roots in infertile soil domains, thereby avoiding ongoing maintenance costs of unproductive organs, which is important because maintenance costs rapidly overtake construction costs in most roots (e.g., Table 8.1) [22,117]. Regulated senescence of roots would permit the remobilization of root resources, including carbohydrates and nutrients, to other plant activities, notably to reproductive growth in annual plants. In common bean, there is no evidence that roots in infertile soil domains preferentially senesce [92], or that programmed root death occurs during reproductive development [118]. It appears that significant root turnover observed in the field is the result of biotic and abiotic stress rather than programmed plant responses [118,119]. Therefore, traits that affect root lifespan, such as defense chemistry or tissue composition, may have only indirect effects on low phosphorus adaptation.

8.2.2.7 Root Hairs

Root hairs are subcellular protrusions of root epidermal cells that are important for the acquisition of relatively immobile nutrients such as phosphorus [120–122]. Mathematical modeling indicates that root hairs substantially increase phosphorus acquisition from soil by expanding the soil volume subject to phosphorus depletion through diffusion to the root surface [123]. Indirect evidence from autoradiography demonstrated that root hairs increase the size of phosphorus depletion zones around roots [124,125]. The inclusion of root hairs in simulation models improved estimates of crop phosphorus uptake [126,127]. More recently, direct evidence for phosphorus uptake by root hairs was demonstrated [128].

Root hairs may also assist the dispersion of exudates such as carboxylates throughout the rhizosphere, which improves phosphorus bioavailability in many soils [14,15]. Mutants of *Arabidopsis* and barley lacking root hairs have severely impaired phosphorus uptake [76,77,129] and in the case of *Arabidopsis*, reduced competitiveness in low phosphorus soil [130]. Both root-hair length [10] and root-hair density [12] are highly regulated by phosphorus availability, which suggests that they have value to plants in low phosphorus soil. Geometric modeling indicated that responses of root hairs to phosphorus availability interact synergistically to improve phosphorus acquisition [11]. Variation among species in root-hair length is correlated with phosphorus acquisition [127,131,132], as is intraspecific variation among genotypes of white clover (*Trifolium repens* L.) [133], barley, wheat [134–136], and common bean [137,138]. In contrast to these comparisons, two maize root-hair mutants showed normal growth and development, which raises questions regarding the benefit of root hairs in this species [139].

Genotypic variation in root-hair length and density in maize and common bean is controlled by several major quantitative trait loci (QTL) [138,140], suggesting that this trait could be selected in breeding programs through marker-aided selection as well as through direct phenotypic screening. Root hairs are particularly important for phosphorus acquisition in nonmycorrhizal plants, inasmuch as mycorrhizal hyphae fulfill some of the same functions as root hairs. However, genotypic variation in root-hair length and density is important for PAE regardless of the mycorrhizal status of the plant (Figures 8.9 and 8.10) [137]. The large genotypic variation in root-hair traits, and the substantial effect this variation has on PAE, regardless of mycorrhizal status, together with the relatively simple genetic control of these traits and opportunities for direct phenotypic selection, makes them attractive targets for breeding programs [106].

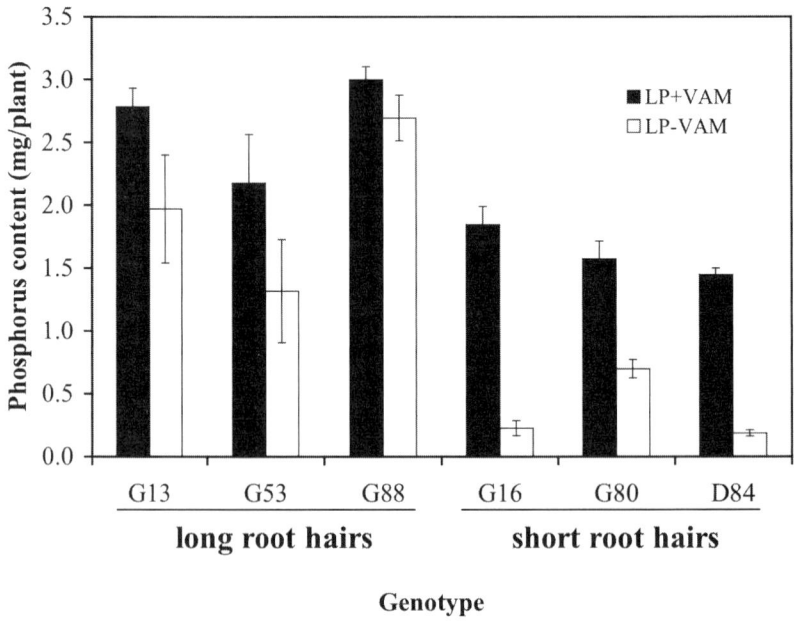

FIGURE 8.9 Effects of root-hair length and mycorrhizal inoculation on phosphorus content of common bean (*Phaseolus vulgaris* L.) genotypes. Plants were grown for 28 days in low-P soil with (+VAM) or without (–VAM) mycorrhizal inoculum. Genotypes are recombinant inbred lines having long or short root hairs. Each bar is the mean of four replicates; bars = SEM.

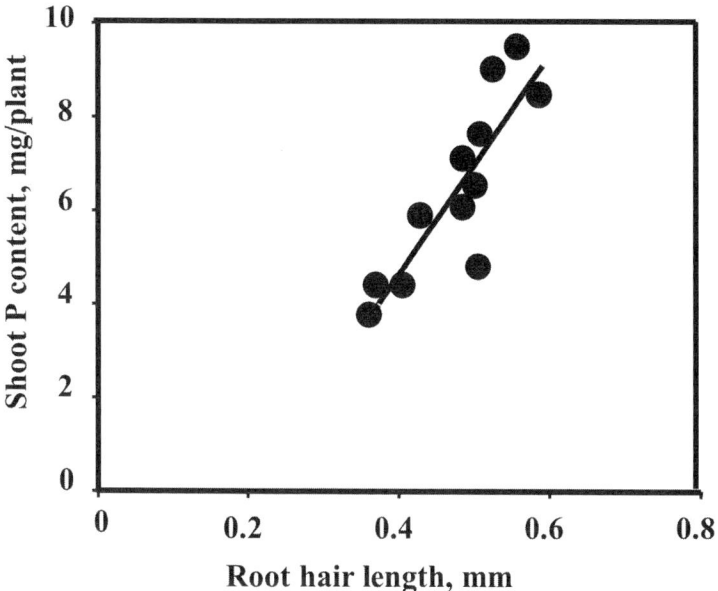

FIGURE 8.10 Effect of root-hair length on phosphorus content of common bean (*Phaseolus vulgaris* L.) genotypes. Plants were grown for 35 days in low-P soil in the field in Costa Rica. Each point is the mean of four replicates of one genotype; the set of genotypes are recombinant inbred lines having long or short root hairs.

8.2.3 Phosphorus Mobilization

In addition to solving the problem of access to soil phosphorus by producing roots in the most nutrient-rich soil domains, plants may also increase the availability of phosphorus at a particular site by altering the rhizosphere. Exudation of carboxylates, such as citrate and malate, is particularly important for phosphorus acquisition from phosphorus-fixing soils. Carboxylates chelate Al^{3+}, Fe^{3+}, and Ca^{2+}, which results in mobilization of phosphate from bound forms [141]. This activity is complemented in alkaline soils by rhizosphere acidification, which results in increased solubility of Ca-phosphates [15]. Although carboxylate excretion is accentuated under phosphorus-deficiency conditions in many species, recent evidence showed that this activity is constitutive in three genotypes of chickpea (*Cicer arietinum*) [142,143]. The subject of carboxylate excretion and its importance for release of phosphorus from inorganic forms has been discussed extensively in several recent reviews [14,15,144]. Carboxylate excretion is also important for aluminum resistance, which is related to phosphorus efficiency because excess aluminum availability coincides with phosphorus deficiency in many acid soils [145]. Overexpression of enzymes responsible for carboxylate production in roots improves plant growth in soils with excess aluminum or deficient in phosphorus [144,146,147].

Because a considerable proportion of phosphorus may occur in organic forms, plants may increase phosphorus availability in the rhizosphere by secreting phosphohydrolases to mineralize phosphate from organic compounds [25,27,141]. Secreted acid phosphatases are upregulated under phosphorus deficiency [25,141,148]. Recent work has demonstrated their significance for phosphorus nutrition under phosphorus-limiting conditions [148–150], although their importance seems to vary with species, cropping system, and forms of organic phosphorus in the soil [149–152].

8.2.4 Cluster Roots

Cluster roots are zones of tightly packed, short, hairy rootlets that occur widely in *Proteaceae* (where they are called proteoid roots) and in several other plant families [29]. Cluster roots provide a unique mechanism for acquiring phosphorus in extremely phosphorus-poor environments by concentrating the phosphorus-mobilizing mechanisms described above into a small volume of soil. Cluster root formation and attendant secretion of carboxylates, H+ ions, and acid phosphatase are promoted by phosphorus deficiency and in some species under other conditions such as Fe and N deficiency [25,29,30,153].

Only a few crop species form cluster roots, including white lupin (*Lupinus albus*, a nonmycorrhizal species) and some cucurbitaceae, for example, squash (*Cucurbita pepo*) [30,154]. Although white lupin has been studied extensively, the impact of cluster roots on other crops has received little attention. One report on squash implicates cluster root formation in Fe(III) reduction [154]. Cluster roots may be more important for other nutrients than for phosphorus acquisition for crops other than white lupin.

8.2.5 Microbial Symbioses

The majority of higher plant species have mycorrhizal symbioses with fungi that assist nutrient acquisition [9]. The ectomycorrhizas enhance phosphorus acquisition via mobilization of sparingly soluble phosphorus complexes, whereas both ectomycorrhizas and the arbuscular mycorrhizas common in many annuals and hardwood species enhance phosphorus acquisition because they increase the volume of soil explored beyond the depletion zone surrounding the root itself. In exchange for phosphorus supplied to the plant, the fungal symbiont obtains reduced carbon. The carbon cost of mycorrhizal symbioses is therefore one component of the cost of phosphorus acquisition in most species. In common bean, mycorrhizal colonization increased root phosphorus acquisition, but the resulting increase in shoot photosynthesis did not increase plant growth because of greater root respiration [22]. At high phosphorus supply, mycorrhizal colonization reduced the growth of citrus seedlings because of greater root carbon cost [117]. In general, the costs of the mycorrhizal symbiosis in various herbaceous and woody species ranges from 4 to 20% of daily net photosynthesis [22,155–159]. The greater metabolic burden of mycorrhizal roots may contribute to the nonbeneficial or even parasitic role that mycorrhizal fungi play in agroecosystems [160].

Mycorrhizal symbioses have attracted a great deal of attention by researchers in the past 30 years. The importance of mycorrhizal symbioses for phosphorus acquisition has led some mycorrhizal researchers to the belief that root traits are secondary or trivial in importance for phosphorus acquisition compared to fungal-assisted phosphorus acquisition. In this context it is useful to consider the strong correlations observed between phosphorus uptake and root traits such as root-hair length (Figures 8.9 and 8.10 [137], and references cited above) and root shallowness [17] even in the presence of mycorrhizas. This could signify that mycorrhizal foraging is incomplete and can be supplemented by direct root foraging, or that extraradical hyphae are restricted to the volume of soil near the root [161], so that root architectural patterns have a strong influence on foraging patterns by the fungal symbiont.

In our research with maize, soybean, and common bean, we typically observe similar genotypic rankings for plant growth in low phosphorus soil in the field where mycorrhizas are formed and in controlled environments without mycorrhizas [63,64,137]. This suggests that for these annual crops, mycorrhizal symbiosis changes the effective fertility status of the soil environment, but does not represent a selection criterion (either through natural selection or in plant breeding) among genotypes, possibly because it is ubiquitous. The primacy of root traits vs. mycorrhizal symbioses for phosphorus acquisition is also suggested by the fact that plant species adapted to extremely low phosphorus environments may be nonmycorrhizal [162].

8.2.6 Phenology

Some annual plants respond to phosphorus stress by delayed maturation [163–165]. This could be adaptive for phosphorus acquisition by permitting continued root growth, and by extending the period of time in which existing roots acquire phosphorus. Time is particularly important for phosphorus acquisition, because phosphorus diffusion through soil is slow, as is recharge of phosphorus-depleted soil [166]. We call this phenomenon "root foraging duration" by analogy with leaf area duration. In addition to possible benefits for phosphorus acquisition, an extended growing season would also increase the metabolic utility of acquired phosphorus, for example, by extending the time leaf phosphorus could be employed to generate photosynthates. In other words, the utility of phosphorus to the plant is dependent on the length of time the phosphorus is used by the plant, which in general would be greater with an extended growing season. Phenology is responsive to phosphorus availability in some plants and there is also a range of maturities available within crop species. If it is demonstrated that delayed maturation is a positive adaptation to low phosphorus availability, genotypic variation for this trait may have value in crop breeding programs, especially in tropical agroecosystems where temperature and moisture availability do not limit the effective growing season.

Another phenological response to low phosphorus availability in many annual species is accelerated leaf senescence [92,167,168]. Accelerated leaf senescence could be adaptive by permitting the resorption of phosphorus from senescing leaves to growing tissues including younger leaves, roots, and reproductive structures. It could also be viewed as an injury resulting from inadequate phosphorus supply to the older leaves, whose loss through accelerated senescence reduces photosynthate production. A direct test of this hypothesis is made possible by SAG-IPT transformants with suppressed leaf senescence [169]. Such transformants have reduced photosynthetic efficiency in young leaves under N stress [170]. Preliminary studies with phosphorus stress indicate that petunia (*Petunia hybrida* L.) SAG-IPT transformants are more resistant to phosphorus stress than wild-type plants [171]. If confirmed, this indicates that accelerated leaf senescence is detrimental rather than adaptive under conditions of phosphorus stress.

8.2.7 Trait Interactions

Several traits related to the efficiency of phosphorus acquisition and utilization may have functional interactions with each other or with other plant traits. These interactions could be positive or synergistic in improving phosphorus efficiency, or they may be antagonistic. An example of trait synergy in phosphorus acquisition is the interaction of four distinct root-hair traits: root-hair length, root-hair density, the distance from the root tip to the first appearance of root hairs, and the pattern of root-hair-bearing epidermal cells (trichoblasts) among nonhair-bearing cells (atrichoblasts). Low phosphorus availability causes coordinated increases in root-hair length and density in many species [10,12,172,173]. In *Arabidopsis*, low phosphorus availability also shortens the distance from the first root hair to the root tip, and changes the geometry of trichoblasts by increasing the number of trichoblast files, caused by cortical reorganization [12,110]. Geometric modeling showed that the combined effect of these four traits on phosphorus acquisition was 371% greater than their additive effects, demonstrating substantial morphological synergy [11]. Synergism among root-hair traits may account for their coordinated regulation.

Traits of individual root axes such as root hairs and root exudates may have synergism with root architectural traits, which locate root axes in soil domains with varying phosphorus availability. For example, longer root hairs would be expected to provide greater benefit to the plant if they were positioned in phosphorus-rich topsoil as opposed to phosphorus-poor subsoil. Phosphatases that mobilize phosphorus from soil organic matter would be more useful if exuded by shallow roots than by deep roots, because in most soils organic matter decreases with depth. In contrast, carboxylates that mobilize phosphorus from Fe and Al oxides may be more useful when released into deeper soil horizons where these forms of phosphorus predominate. Root architectural traits may themselves display interaction, by altering the extent of interroot competition, which is an important component of overall root foraging efficiency [66,74,75]. For example, root systems combining deep rooting (through, e.g., lateral branching from the taproot or deep basal roots) with shallow rooting (through adventitious roots or shallow basal roots) would be expected to be more complementary than root systems in which distinct root classes competed for the same soil niche [174]. This is especially relevant in the context of drought, because in many environments water is a deep soil resource whereas phosphorus is a shallow resource (see discussion below). We know very little about the interaction of traits related to phosphorus acquisition, despite the importance of trait interactions for whole-plant performance. This is pertinent to plant breeding, inasmuch as traits under distinct genetic control could be combined to maximize positive synergy.

8.2.8 Tradeoffs

Consideration of the utility of a trait for plants in low phosphorus environments must take into account potential tradeoffs of the trait for other plant processes. The most obvious tradeoff for many traits is simply the opportunity cost resulting from diversion of plant resources from other functions. For example, the production of adventitious roots reduces the development of basal roots, which in certain soils can be detrimental to overall plant phosphorus acquisition [174]. Because of the heterogeneity of soil resource distribution, architectural tradeoffs can also result if exploitation of one soil domain reduces exploitation of another soil domain with its attendant resources.

An important tradeoff or opportunity cost to topsoil foraging is increased sensitivity to drought stress, because water is a deep soil resource in many environments. A comparison of deep-rooted and shallow-rooted common bean genotypes showed that although shallower genotypes had superior growth under phosphorus stress, deep-rooted genotypes had superior growth under water stress (Figure 8.11) [63]. These results are consistent with economic optimization modeling of the relationship between root architecture and multiple resource acquisition, particularly water and phosphorus [175]. The general solution of the model states that a plant will locate its roots at a soil depth where the marginal benefit of water and phosphorus acquisition will exactly equal the marginal cost of interroot competition [175]. Indeed, common bean genotypes that are best adapted to low-phosphorus environments, where phosphorus is localized in the surface soil, tend to have a shallower basal root angle, whereas genotypes that are adapted to terminal drought environments have deeper root systems [63]. This example illustrates the importance of considering tradeoffs in assessing the adaptive importance of specific root traits, especially in crop breeding for distinct environments.

The large genotypic variation for root traits that appear to be positive adaptations for nutrient acquisition may be caused or maintained by tradeoffs incurred by certain phenotypes. For example, long, dense root hairs improve phosphorus acquisition at minimal metabolic cost (see references above), yet a large proportion of crop genotypes have few, sparse root hairs, and many genotypes display plasticity in root-hair traits, so that under high fertility root hairs are suppressed. Does this suggest that there are potential costs to root hairs, such as increased susceptibility

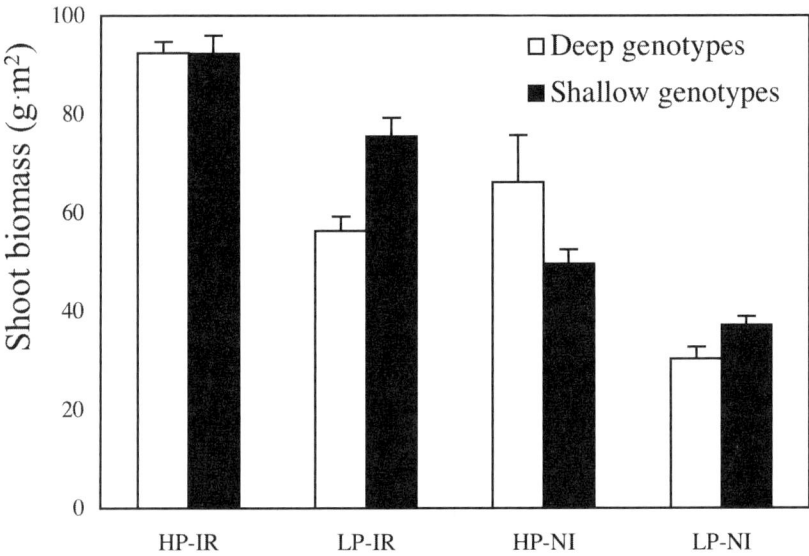

FIGURE 8.11 Shoot biomass at 44 days after planting for three shallow-rooted and three deep-rooted common bean (*Phaseolus vulgaris* L.) genotypes in the field. HP = high phosphorus availability, LP = low phosphorus availability, IR = irrigated, NI = nonirrigated. Each bar is the mean of four replicates; error bars = SEM.

to root pathogens? Similarly, cortical aerenchyma appears to reduce the metabolic costs of soil exploration (see references above), yet substantial intraspecific variation for constitutive aerenchyma formation exists, and aerenchyma formation is suppressed under high fertility. Does this suggest there are potential costs to aerenchyma formation, such as reduced radial transport of water and nutrients, or reduced mycorrhizal habitat? Such questions are largely unresolved.

8.2.9 Responses to Localized Availability of Phosphorus

In addition to the variability in phosphorus availability with depth discussed in Section 8.2.2.1, phosphorus availability may be heterogeneous in space and time as a result of organic matter decomposition, variation in soil composition, competition with the same or other root systems, water availability, and temperature [176]. Many plants have the ability to respond to these patches of higher phosphorus availability in ways that are expected to increase their ability to compete for phosphorus. The responses of root systems to heterogeneous nutrient distribution have been reviewed recently [177], so this topic is discussed here only in the context of developing more phosphorus-efficient crops.

When phosphorus-stressed plants encounter a patch of higher phosphorus availability, one advantageous response is to proliferate roots to enhance phosphorus acquisition from the patch. Root proliferation has been observed in nutrient patches and includes increased number, length, and branching of lateral roots. The extent to which this occurs varies among species, some showing very dramatic effects (e.g., barley [178]), and others showing little or no response [179,180]. To complicate matters further, plants may alter root development in nutrient patches when roots of another plant (even of the same species) are competing within the patch [181–183]. The available data justify the conclusion that root proliferation in nutrient patches is likely to be useful for plants competing for immobile nutrients in intercrop systems, as is the case for many crops grown in poor soils in the tropics and subtropics.

8.3 COMPETITION

The utility of traits for phosphorus efficiency will, in most cases, be manifest in competitive environments, in wild plants and subsistence agroecosystems usually in mixed stands with diverse taxa, and in commercial agriculture typically in high-density genetic monocultures. Traits influencing phosphorus efficiency will affect plant productivity, and thereby competitive performance, under phosphorus stress. An example of this is the positive effect of root hairs on plant performance in mixed stands of *Arabidopsis* at low phosphorus but not at high phosphorus [130]. Traits influencing phosphorus acquisition can also directly affect interplant competition by removing phosphorus from the soil that could be accessed by competitors. For example, common bean genotypes with shallow basal roots outcompete genotypes with deep basal roots in low phosphorus fields [75] because of enhanced topsoil exploitation and reduced competition among roots of the same plant [74].

At the population level, competition among root systems can be important in determining the utility of root traits for phosphorus efficiency. This appears to be the case for plasticity of basal root shallowness, for which genetic variation exists; that is, some genotypes respond to phosphorus stress by becoming more shallow, whereas others are unaffected or become deeper [63,64,175]. Plasticity of root shallowness would generally be considered to be a useful trait, because plasticity would permit a plant to modify its root architecture to adapt to the prevailing edaphic stress. However, if all plants in a population were equally plastic and therefore had the same root architecture, greater interplant competition would occur than if distinct root phenotypes existed in a population, thereby permitting complementary exploitation of distinct soil domains. Modeling showed that interplant competition could be important in determining an optimal balance of plastic and nonplastic root phenotypes under conditions of phosphorus stress and combined phosphorus and water stress (Figure 8.12) [184]. This suggests that genetic mixtures or multilines may have better performance in low phosphorus agroecosystems than genetic monocultures, especially in drought-prone environments.

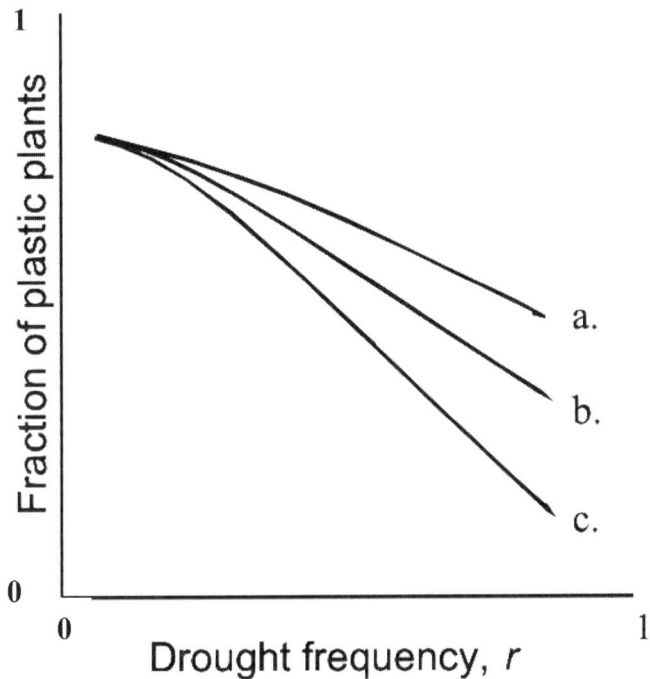

FIGURE 8.12 The theoretical relationship of the equilibrium steady-state fraction of plastic plants in a population and drought frequency, where spatial competition is considered, under (A) low, (B) intermediate, or (C) high drought intensity. Plastic plants in this model respond to low P availability by increasing topsoil foraging, which incurs costs in conditions of terminal drought, and when neighboring individuals have the same phenotype.

8.4 ECOSYSTEM ISSUES

A better understanding of plant adaptations to phosphorus stress is critically needed for two of the greatest challenges facing humanity in the 21st century: eliminating world hunger and understanding how natural and managed ecosystems will respond to global climate change.

The development of crops with superior growth in low phosphorus soil and with better responsiveness to applied phosphorus inputs would have tremendous value in many developing countries, where yields are limited by low soil fertility, and fertilizer use is minimal [7]. Because genotypic variation for phosphorus acquisition efficiency is much larger than variation for phosphorus use efficiency in crop plants, it is likely that phosphorus-efficient crops will have superior phosphorus acquisition compared with conventional genotypes [185]. Although such genotypes would extract more phosphorus from the soil than conventional genotypes, they may actually enhance soil fertility in the long term through beneficial effects on soil erosion and nutrient cycling, as well as benefits they accrue to farm income and thereby the use of fertility amendments [1].

Several genetic traits have been identified with potential utility in breeding phosphorus-efficient crops, as discussed above, including root exudates, root-hair traits, cortical aerenchyma, topsoil foraging through basal or adventitious rooting, and the use of multiline mixtures of root phenotypes. Deployment of these traits through plant breeding programs is resulting in progress in several crops including common bean [186] and soybean [68]. The success of this effort would constitute a second "Green Revolution," benefiting the resource-poor farmers who were largely left behind by the first Green Revolution, and who represent the single largest human labor occupation [7]. A better understanding of the biology of traits associated with phosphorus efficiency, especially how these traits combine and their tradeoffs in specific production environments, is needed to guide plant breeding efforts.

We will not be able to understand or manage ecosystem response to global change unless we learn more about how global change variables such as CO_2, temperature, and ozone interact with the edaphic stresses prevalent in most terrestrial ecosystems [8]. The vast majority of research on plant response to global change has focused on leaf responses and has not considered edaphic stresses other than water and possibly nitrogen with any rigor, despite the fact that plant responses to edaphic stresses are primary limitations to plant productivity in most forests and managed systems. Plants limited by low soil phosphorus availability may respond to elevated CO_2 by producing more exudates and by altered root growth and architecture, which may partially alleviate phosphorus stress, but interactions with other global change variables such as drought could be less beneficial, as discussed above. This topic merits research.

8.5 CONCLUSIONS

Low soil phosphorus availability is a primary constraint to plant growth on Earth. Accordingly, plants express a wide array of phenotypic traits that improve adaptation to low phosphorus availability, including increased biomass allocation to roots and to specific root classes within the root system, root architectural traits that enhance topsoil foraging, including basal root gravitropism, adventitious rooting, and lateral root branching, and reduced metabolic costs of soil exploration, via formation of cortical aerenchyma and altered respiratory pathways, the formation of finer roots and possibly root etiolation, root hairs, phosphorus-solublizing root exudates, mycorrhizal symbioses, phenological plasticity, and morphological plasticity. Ecological tradeoffs and interactions among these traits are poorly understood but are likely to be important in determining the functional utility of these traits, especially in competitive environments. A better understanding of these traits is needed to guide the development of more phosphorus-efficient crops for developing nations, and to understand how ecosystems will respond to global climate change.

REFERENCES

1. Lynch, J. and Deikman, J., Phosphorus in plant biology: Regulatory roles in molecular, cellular, organismic, and ecosystem processes. *Curr. Topics Plant Physiol.,* 19:401, 1998.
2. Walker, T., The significance of phosphorus in pedogenesis. In *Experimental Pedology*, Hallsworth, E. and Crawford, D., Eds., Butterworths, London, 1965.
3. Hartemink, A.E., *Soil Fertility Decline in the Tropics.* CABI, Wageningen, The Netherlands, 2003.
4. Oldeman, L., Hakkeling, R., and Sombroek, W., *World Map of the Status of Human-Induced Soil Degradation: An Explanatory Note*, 2d revised ed. ISRIC, Wageningen, The Netherlands, 1991.
5. Steen, I., Phosphorus availability in the 21st century. Management of a non-renewable resource. *Phosphor. Potass.,* 217:25–31, 1998.
6. Abelson, P.H., A potential phosphate crisis. *Science,* 283:2015, 1999.
7. Bank, W., *World Development Indicators.* World Bank, Washington, DC, 2004.
8. Lynch, J.P. and St Clair, S.B., Mineral stress: The missing link in understanding how global climate change will affect plants in real world soils. *Field Crops Res.,* 90:101, 2004.
9. Smith, S. and Read, D., *Mycorrhizal Symbiosis*, 2nd ed. Academic Press, San Diego, 1997.
10. Bates, T.R. and Lynch, J.P., Stimulation of root hair elongation in *Arabidopsis thaliana* by low phosphorus availability. *Plant Cell Environ.,* 19:529, 1996.
11. Ma, Z., Walk, T.C., Marcus, A., and Lynch, J.P., Morphological synergism in root hair length, density, initiation and geometry for phosphorus acquisition in *Arabidopsis thaliana*: A modeling approach. *Plant Soil,* 236:221, 2001.
12. Ma, Z., Bielenberg, D.G., Brown, K.M., and Lynch, J.P., Regulation of root hair density by phosphorus availability in *Arabidopsis thaliana. Plant Cell Environ.,* 24:459–467, 2001.
13. Jones, D.L., Organic acids in the rhizosphere — A critical review. *Plant Soil,* 205:25–44, 1998.
14. Ryan, P.R., Delhaize, E., and Jones, D.L., Function and mechanism of organic anion exudation from plant roots. *Ann. Rev. Plant Physiol. Plant Mol. Biol.,* 52:527–560, 2001.
15. Hinsinger, P., Bioavailability of soil inorganic P in the rhizosphere as affected by root-induced chemical changes: A review. *Plant Soil,* 237:173–195, 2001.
16. Hayes, J.E., Richardson, A.E., and Simpson, R.J., Phytase and acid phosphatase activities in extracts from roots of temperate pasture grass and legume seedlings. *Aust. J. Plant Physiol.,* 26:801–809, 1999.
17. Lynch, J.P. and Brown, K.M., Topsoil foraging — An architectural adaptation of plants to low phosphorus availability. *Plant Soil,* 237:225–237, 2001.
18. Niklas, K., *Plant Allometry: The Scaling of Form and Process.* University of Chicago Press, Chicago, 1994.
19. Gutschick, V., Nutrient-limited growth rates: Roles of nutrient-use efficiency and of adaptations to increase uptake rate. *J. Exp. Bot.,* 44:41–51, 1993.
20. Nielsen, K.L., Eshel, A., and Lynch, J.P., The effect of phosphorus availability on the carbon economy of contrasting common bean (*Phaseolus vulgaris* L.) genotypes. *J. Exp. Bot.,* 52:329–339, 2001.

21. Hansen, C.W., Lynch, J., and Ottosen, C.O., Response to phosphorus availability during vegetative and reproductive growth of chrysanthemum: I. Whole-plant carbon dioxide exchange. *J. Am. Soc. Hort. Sci.,* 123:215–222, 1998.
22. Nielsen, K.L., et al., Effects of phosphorus availability and vesicular-arbuscular mycorrhizas on the carbon budget of common bean (*Phaseolus vulgaris*). *New Phytol.,* 139:647–656, 1998.
23. Van der Werf, A., Welschen, R., and Lambers, H., Respiratory losses increase with decreasing inherent growth rate of a species and with decreasing nitrate supply: A search for explanations for these observations. In *Molecular, Biochemical, and Physiological Aspects of Plant Respiration*, Lambers, H. and Van der Plas, L., Eds., SPB Academic, The Hague, 1992.
24. Lynch, J. and Ho, M., Rhizoeconomics: Carbon costs of phosphorus acquisition. *Plant Soil,* 269:45–56, 2004.
25. Vance, C.P., Uhde-Stone, C., and Allan, D.L., Phosphorus acquisition and use: Critical adaptations by plants for securing a nonrenewable resource. *New Phytol.,* 157:423–447, 2003.
26. Ticconi, C.A. and Abel, S., Short on phosphate: Plant surveillance and countermeasures. *Trends Plant Sci.,* 9:548–555, 2004.
27. Abel, S., Ticconi, C.A., and Delatorre, C.A., Phosphate sensing in higher plants. *Physiol. Plant.,* 115:1–8, 2002.
28. Diem, H.G. et al., Cluster roots in Casuarinaceae: Role and relationship to soil nutrient factors. *Ann. Bot.,* 85:929–936, 2000.
29. Lamont, B.B., Structure, ecology and physiology of root clusters — A review. *Plant Soil,* 248:1–19, 2003.
30. Neumann, G. and Martinoia, E., Cluster roots — an underground adaptation for survival in extreme environments. *Trends Plant Sci.,* 7:162–167, 2002.
31. Smith, S.E., Smith, F.A., and Jakobsen, I., Mycorrhizal fungi can dominate phosphate supply to plants irrespective of growth responses. *Plant Physiol.,* 133:16–20, 2003.
32. Harrison, M.J., Molecular and cellular aspects of the arbuscular mycorrhizal symbiosis. *Ann. Rev. Plant Physiol. Plant Mol. Biol.,* 50:361–389, 1999.
33. Oldroyd, G.E.D., Harrison, M.J., and Udvardi, M., Peace talks and trade deals. Keys to long-term harmony in legume-microbe symbioses. *Plant Physiol.,* 137:1205, 2005.
34. Yan, X.L. et al., Induction of a major leaf acid phosphatase does not confer adaptation to low phosphorus availability in common bean. *Plant Physiol.,* 125:1901, 2001.
35. Lynch, J., The role of nutrient efficient crops in modern agriculture. *J. Crop Prod.* 1:241, 1998.
36. Lynch, J. and Brown, K., Root architecture and phosphorus acquisition efficiency in common bean. In *Phosphorus in Plant Biology: Regulatory Roles in Ecosystem, Organismic, Cellular, and Molecular Processes*, Lynch, J., Deikman, J., Eds. ASPP, Rockville, MD, 1998, pp. 148.
37. Whiteaker, G., Gerloff, G., Gabelman, W., and Lindgren, D., Intraspecific differences in growth of beans at stress levels of phosphorus. *J. Am. Soc. Hort. Sci.* 101:472, 1976.
38. Lynch, J., Lauchli, A., and Epstein, E., Vegetative growth of the common bean in response to phosphorus nutrition. *Crop Sci.,* 31:380, 1991.

39. Mollier, A. and Pellerin, S., Maize root system growth and development as influenced by phosphorus deficiency. *J. Exp. Bot.*, 50:487, 1999.
40. Radin, J. and Eidenbock, M., Hydraulic conductance as a factor limiting leaf expansion of phosphorus-deficient cotton plants. *Plant Physiol.*, 75:372, 1984.
41. Halsted, M. and Lynch, J., Phosphorus responses of C-3 and C-4 species. *J. Exp. Bot.*, 47:497, 1996.
42. Manske, G. et al., Traits associated with improved P-uptake efficiency in CIMMYT's semidwarf spring bread wheat grown on an acid Andisol in Mexico *Plant Soil*, 221:189, 2000.
43. Lynch, J.P., Root architecture and plant productivity. *Plant Physiol.* 109:7, 1995.
44. Borch, K. et al., Ethylene: A regulator of root architectural responses to soil phosphorus availability. *Plant Cell Environ.*, 22:425, 1999.
45. Robinson, D., Integrated root responses to variations in nutrient supply. In *Nutrient acquisition by plants. An Ecological Perspective*, BassiriRad, H., Ed., Springer-Verlag, Berlin, 2005, pp. 43–62.
46. Williamson, L.C. et al., Phosphate availability regulates root system architecture in Arabidopsis. *Plant Physiol.*, 126:875, 2001.
47. Lopez-Bucio, J. et al., Phosphate availability alters architecture and causes changes in hormone sensitivity in the *Arabidopsis* root system. *Plant Physiol.*, 129:244, 2002.
48. Al-Ghazi, Y. et al., Temporal responses of *Arabidopsis* root architecture to phosphate starvation: Evidence for the involvement of auxin signalling. *Plant Cell Environ.*, 26:1053, 2003.
49. Zhu, J.M. and Lynch, J.P., The contribution of lateral rooting to phosphorus acquisition efficiency in maize (*Zea mays*) seedlings. *Func. Plant Biol.*, 31:949, 2004.
50. Miller, C.R. et al., Genetic variation for adventitious rooting in response to low phosphorus availability: Potential utility for phosphorus acquisition from stratified soils. *Func. Plant Biol.*, 30:973, 2003.
51. Vartapetian, B.B. and Jackson, M.B., Plant adaptations to anaerobic stress. *Ann. Bot.*, 79:3, 1997.
52. Roman-Aviles, B., Snapp, S.S., and Kelly, J.D., Assessing root traits associated with root rot resistance in common bean. *Field Crops Res.*, 86:147, 2004.
53. Riedell, W.E. and Reese, R.N., Maize morphology and shoot CO_2 assimilation after root damage by western corn rootworm larvae. *Crop Sci.*, 39:1332, 1999.
54. Jeschke, W.D., Holobrada, M., and Hartung, W., Growth of *Zea mays* L. plants with their seminal roots only. Effects on plant development, xylem transport, mineral nutrition and the flow and distribution of abscisic acid (ABA) as a possible shoot-to-root signal. *J. Exp. Bot.*, 48:1229, 1997.
55. Pellerin, S., Mollier, A., and Plenet, D., Phosphorus deficiency affects the rate of emergence and number of maize adventitious nodal roots. *Agron. J.*, 92:690, 2000.
56. Barber, S., A diffusion and mass flow concept of nutrient availability. *Soil Sci.*, 93:39, 1962.
57. Anderson, G., Assessing organic phosphorus in soils. In *The Role of Phosphorus in Agriculture*, Khasawneh, F. E., Sample, E. C., and Kamprath, E. J., Eds. ASA, CSSA, SSSA, Madison, WI, 1980, pp. 411–431.
58. Pothuluri, J. et al., Phosphorus uptake from soil layers having different soil test phosphorus levels. *Agron. J.*, 78:991, 1986.
59. Chu, W. and Chang, S., Surface activity of inorganic soil phosphorus. *Soil Sci.* 101:459–464, 1966.

60. Barber, S., *Soil Nutrient Bioavailability: A Mechanistic Approach.* John Wiley, New York, 1984.
61. Joner, E.J. et al., P depletion and activity of phosphatases in the rhizosphere of mycorrhizal and nonmycorrhizal cucumber (*Cucumis sativus* L). *Soil Biol. Biochem.,* 27:1145, 1995.
62. Zhu, J., Kaeppler, S., and Lynch, J., Topsoil foraging and phosphorus acquisition efficiency in maize (*Zea mays* L.). *Func. Plant Biol.,* 32:749, 2005.
63. Ho, M. et al., Root architectural tradeoffs for water and phosphorus acquisition. *Func. Plant Biol.,* 32:737, 2005.
64. Bonser, A.M., Lynch, J., and Snapp, S., Effect of phosphorus deficiency on growth angle of basal roots in *Phaseolus vulgaris. New Phytol.,* 132:281, 1996.
65. Liao, H. et al., Effect of phosphorus availability on basal root shallowness in common bean. *Plant Soil,* 232:69–79, 2001.
66. Ge, Z., Rubio, G., and Lynch, J.P., The importance of root gravitropism for inter-root competition and phosphorus acquisition efficiency: results from a geometric simulation model. *Plant Soil,* 218:159, 2000.
67. Liao, H. et al., Genetic mapping of basal root gravitropism and phosphorus acquisition efficiency in common bean. *Func. Plant Biol.* 31:1, 2004.
68. Yan, X. and Lynch, J., Unpublished data, 2005.
69. Lambers, H., Atkin, O., and Millenaar, F.F., Respiratory patterns in roots in relation to their functioning. In *Plant Roots, the Hidden Half,* 3d edition, Waisel, Y., Eshel, A., and Kafkaki, K., Eds., Marcel Dekker, New York, 2002, pp. 521–552.
70. Bloom, A., Chapin, F., and Mooney, H., Resource limitation in plants — An economic analogy. *Ann. Rev. Ecol. System.,* 16:363, 1985.
71. Snapp, S., Koide, R., and Lynch, J., Exploitation of localized phosphorus-patches by common bean roots. *Plant Soil,* 177:211, 1995.
72. Koide, R.T., Goff, M.D., and Dickie, I.A., Component growth efficiencies of mycorrhizal and nonmycorrhizal plants. *New Phytol.,* 148:163, 2000.
73. Koide, R. and Elliott, G., Cost, benefit and efficiency of the vesicular-arbuscular mycorrhizal symbiosis. *Func. Ecol.,* 3:252, 1989.
74. Rubio, G. et al., Root gravitropism and below-ground competition among neighbouring plants: A modelling approach. *Ann. Bot.,* 88:929, 2001.
75. Rubio, G. et al., Topsoil foraging and its role in plant competitiveness for phosphorus in common bean. *Crop Sci.,* 43:598–607, 2003.
76. Bates, T.R. and Lynch, J.P., Plant growth and phosphorus accumulation of wild type and two root hair mutants of *Arabidopsis thaliana* (Brassicaceae). *Am. J. Bot.,* 87:958, 2000.
77. Bates, T.R. and Lynch, J.P., The efficiency of *Arabidopsis thaliana* (Brassicaceae) root hairs in phosphorus acquisition. *Am. J. Bot.,* 87:964, 2000.
78. Amthor, J.S., The McCree-de Wit-Penning de Vries-Thornley respiration paradigms: 30 years later. *Ann. Bot.,* 86:1, 2000.
79. Bouma, T.J., Broekhuysen, A.G.M., and Veen, B.W., Analysis of root respiration of Solanum tuberosum as related to growth, ion uptake and maintenance of biomass. *Plant Physiol. Biochem.,* 34:795, 1996.
80. Esau, K., *Anatomy of Seed Plants.* 2d ed., John Wiley and Sons, New York, 1977.
81. Jackson, M.B. and Armstrong, W., Formation of aerenchyma and the processes of plant ventilation in relation to soil flooding and submergence. *Plant Biol.,* 1:274, 1999.

82. Constable, J.V.H., Grace, J.B., and Longstreth, D.J., High carbon dioxide concentrations in aerenchyma of Typha latifolia. *Am. J. Bot.,* 79:415, 1992.
83. Constable, J. and Longstreth, D.J., Aerenchyma carbon dioxide can be assimilated in *Typha latifolia* L. leaves. *Plant Physiol.,* 106:1065, 1994.
84. Konings, H. and Verschuren, G., Formation of aerenchyma in roots of *Zea mays* in aerated solutions, and its relation to nutrient supply. *Physiol. Plant,* 49:265, 1980.
85. Drew, M., He, C., and Morgan, P., Decreased ethylene biosynthesis, and induction of aerenchyma, by nitrogen- or phosphate-starvation in adventitious roots of *Zea mays* L. *Plant Physiol.,* 91:266, 1989.
86. Fan, M.S. et al., Physiological roles for aerenchyma in phosphorus-stressed roots. *Func. Plant Biol.,* 30:493, 2003.
87. Eshel, A., Nielsen, K., and Lynch, J., Response of bean root systems to low level of P. In *Plant Roots — From Cells to Systems. 14th Long Ashton International Symposium.* IACR-Long Ashton Res. St. Bristol, England, 1995, p. 63.
88. Lu, Y. et al., Impact of phosphorus supply on root exudation, aerenchyma formation and methane emission of rice plants. *Biogeochemistry,* 47:203, 1999.
89. He, C.J., Morgan, P.W., and Drew, M.C., Enhanced sensitivity to ethylene in nitrogen-starved or phosphate-starved roots of *Zea mays* L. during aerenchyma formation. *Plant Physiol.,* 98:137, 1992.
90. Gillespie, I.M.M. and Deacon, J.W., Effects of mineral nutrients on senescence of the cortex of wheat roots and root pieces. *Soil Biol. Biochem.,* 20:525, 1988.
91. Lascaris, D. and Deacon, J.W., Relationship between root cortical senescence and growth of wheat as influenced by mineral nutrition, *Idriella bolleyi* (Sprague) von Arx and pruning of leaves. *New Phytol.,* 118:391, 1991.
92. Snapp, S.S. and Lynch, J.P., Phosphorus distribution and remobilization in bean plants as influenced by phosphorus nutrition. *Crop Sci.,* 36:929, 1996.
93. Kaeppler, S. et al., Variation among maize inbred lines and detection of quantitative trait loci for growth at low phosphorus and responsiveness to arbuscular mycorrhizal fungi. *Crop Sci.,* 40:358, 2000.
94. Kaeppler, S., Unpublished data, 2000.
95. Aguilar, E.A., Turner, D.W., and Sivasithamparam, K., Aerenchyma formation in roots of four banana (Musa spp.) cultivars. *Sci. Hort.,* 80:57, 1999.
96. Huang, B.R. et al., Root and shoot growth of wheat genotypes in response to hypoxia and subsequent resumption of aeration. *Crop Sci.,* 34:1538, 1994.
97. Garthwaite, A.J., von Bothmer, R., and Colmer, T.D., Diversity in root aeration traits associated with waterlogging tolerance in the genus Hordeum. *Func. Plant Biol.,* 30:875, 2003.
98. Lizaso, J.I., Melendez, L.M., and Ramirez, R., Early flooding of two cultivars of tropical maize. I. Shoot and root growth. *J. Plant Nutr.,* 24:979, 2001.
99. Ray, J.D. et al., Preliminary survey of root aerenchyma in Tripsacum, *Maydica* 43:49, 1998.
100. Visser, E.J.W. et al., Changes in growth, porosity, and radial oxygen loss from adventitious roots of selected mono- and dicotyledonous wetland species with contrasting types of aerenchyma. *Plant Cell Environ.,* 23:1237, 2000.
101. Ray, J.D., Kindiger, B., and Sinclair, T.R., Introgressing root aerenchyma into maize, *Maydica.* 44:113, 1999.
102. Setter, T.L. and Waters, I., Review of prospects for germplasm improvement for waterlogging tolerance in wheat, barley and oats. *Plant Soil,* 253:1, 2003.

103. Forde, B. and Lorenzo, H., The nutritional control of root development. *Plant Soil,* 232:51, 2001.
104. Eissenstat, D.M. et al., Building roots in a changing environment: Implications for root longevity. *New Phytol.,* 147:33, 2000.
105. Ryser, P. and Lambers, H., Root and leaf attributes accounting for the performance of fast-growing and slow-growing grasses at different nutrient supply. *Plant Soil,* 170:251, 1995.
106. Gahoonia, T.S. and Nielsen, N.E., Root traits as tools for creating phosphorus efficient crop varieties. *Plant Soil,* 260:47, 2004.
107. Drew, M.C. and Saker, L.R., Nutrient supply and the growth of the seminal root system in barley. *J. Exp. Bot.* 29:435, 1978.
108. Hackett, C., A method of applying nutrients locally to roots under controlled conditions, and some morphological effects of locally applied nitrate on the branching of wheat roots. *Aust. J. Biol. Sci.* 25:1169, 1972.
109. Xie, Y.J. and Yu, D., The significance of lateral roots in phosphorus (P) acquisition of water hyacinth (*Eichhornia crassipes*). *Aquat. Bot.,* 75 (4):311, 2003.
110. Ma, Z. et al., Regulation of root elongation under phosphorus stress involves changes in ethylene responsiveness. *Plant Physiol.,* 131:1381, 2003.
111. Bennie, A., Growth and mechanical impedance. In *Plant Roots: The Hidden Half,* Waisel, Y., Eshel, A., and Kafkafi, U., Eds., Marcel Dekker, New York, 1991, pp. 393–414.
112. Hoffmann, C. and Jungk, A., Growth and phosphorus supply of sugar-beet as affected by soil compaction and water tension. *Plant Soil,* 176:15, 1995.
113. Rychter, A.M. and Mikulska, M., The relationship between phosphate status and cyanide-resistant respiration in bean roots. *Physiol. Plant.,* 79:663, 1990.
114. Rychter, A.M. et al., The effect of phosphate deficiency on mitochondrial activity and adenylate levels in bean roots. *Physiol. Plant.,* 84:80, 1992.
115. Theodorou, M.E. and Plaxton, W.C., Metabolic adaptations of plant respiration to nutritional phosphate deprivation. *Plant Physiol.,* 101:339, 1993.
116. Malusa, E. et al., Free radical production in roots of *Phaseolus vulgaris* subjected to phosphate deficiency stress. *Plant Physiol. Biochem.,* 40:963, 2002.
117. Peng, S.B. et al., Growth depression in mycorrhizal citrus at high-phosphorus supply — Analysis of carbon costs. *Plant Physiol.,* 101:1063, 1993.
118. Fisher, M.C.T., Eissenstat, D.M., and Lynch, J.P., Lack of evidence for programmed root senescence in common bean (*Phaseolus vulgaris*) grown at different levels of phosphorus supply. *New Phytol.,* 153:63, 2002.
119. Eissenstat, D.M. and Yanai, R.D., The ecology of root lifespan. *Adv. Ecol. Res.,* 27:1, 1997.
120. Clarkson, D., Factors affecting mineral nutrient acquisition by plants. *Ann Rev Plant Physiol.,* 36:77, 1985.
121. Jungk, A., Root hairs and the acquisition of plant nutrients from soil. *J. Plant Nutr. Soil Sci.,* 164:121, 2001.
122. Peterson, R.L. and Farquhar, M.L., Root hairs: Specialized tubular cells extending root surfaces. *Bot. Rev.,* 62:1, 1996.
123. Bouldin, D., Mathematical description of diffusion process in the soil. *Soil Sci. Soc. Am. Proc.,* 25:476, 1961.
124. Lewis, D.G. and Quirk, J.P., Phosphate diffusion in soil and uptake by plants. *Plant Soil,* 26:445, 1967.

125. Bhat, K.K.S. and Nye, P.H., Diffusion of phosphate to plant roots in soil III. Depletion around onion roots without root hairs. *Plant Soil,* 41:383, 1974.
126. Itoh, S. and Barber, S., A numerical solution of whole plant nutrient uptake for soil root systems with root hairs. *Plant Soil,* 70:403, 1983.
127. Itoh, S. and Barber, S., Phosphorus uptake by six plant species as related to root hairs. *Agron. J.* 75:457, 1983.
128. Gahoonia, T.S. and Nielsen, N.E., Direct evidence on participation of root hairs in phosphorus (P-32) uptake from soil. *Plant Soil,* 198:147, 1998.
129. Gahoonia, T.S. and Nielsen, N.E., Phosphorus (P) uptake and growth of a root hairless barley mutant (bald root barley, brb) and wild type in low- and high-P soils. *Plant Cell Environ.,* 26:1759, 2003.
130. Bates, T.R. and Lynch, J.P., Root hairs confer a competitive advantage under low phosphorus availability. *Plant Soil,* 236:243, 2001.
131. Fohse, D., Claassen, N., and Jungk, A., Phosphorus efficiency of plants. 2. Significance of root radius, root hairs and cation-anion balance for phosphorus influx in 7 plant species. *Plant Soil,* 132:261, 1991.
132. Gahoonia, T.S., Nielsen, N.E., and Lyshede, O.B., Phosphorus (P) acquisition of cereal cultivars in the field at three levels of P fertilization. *Plant Soil,* 211:269, 1999.
133. Caradus, J., Effect of root hair length on white clover growth over a range of soil phosphorus levels. *N.Z. J. Agri. Res.,* 24:359, 1981.
134. Gahoonia, T.S., Care, D., and Nielsen, N.E., Root hairs and phosphorus acquisition of wheat and barley cultivars. *Plant Soil,* 191:181, 1997.
135. Gahoonia, T.S. and Nielsen, N.E., Variation in root hairs of barley cultivars doubled soil phosphorus uptake. *Euphytica,* 98:177, 1997.
136. Gahoonia, T.S., Nielsen, N.E., Joshi, P.A., and Jahoor, A., A root hairless barley mutant for elucidating genetic of root hairs and phosphorus uptake. *Plant Soil,* 235:211, 2001.
137. Miguel, M., Genotypic variation in root hairs and phosphorus efficiency in common bean *(Phaseolus vulgaris L.)*. MS, Pennsylvania State University, 2004.
138. Yan, X.L. et al., QTL mapping of root hair and acid exudation traits and their relationship to phosphorus uptake in common bean. *Plant Soil,* 265:17, 2004.
139. Wen, T.-J. and Schnable, P.S., Analysis of mutant of three genes that influence root hair development in *Zea mays* (Graminae) suggest that root hairs are dispensable. *Am. J. Bot.,* 81:833, 1994.
140. Zhu, J., Kaeppler, S., and Lynch, J., Mapping of QTL controlling root hair length in maize (*Zea mays* L.) under phosphorus deficiency. *Plant Soil,* 270:299, 2005.
141. Marschner, H., *Mineral Nutrition of Higher Plants*, 2d ed. Academic Press, San Diego, 1995.
142. Wouterlood, M., Lambers, H., and Veneklaas, E.J., Plant phosphorus status has a limited influence on the concentration of phosphorus-mobilising carboxylates in the rhizosphere of chickpea. *Func. Plant Biol.* 32:153, 2005.
143. Wouterlood, M. et al., Carboxylate concentrations in the rhizosphere of lateral roots of chickpea (*Cicer arietinum*) increase during plant development, but are not correlated with phosphorus status of soil or plants. *New Phytol.,* 162:745, 2004.
144. Lopez-Bucio, J. et al., Organic acid metabolism in plants: from adaptive physiology to transgenic varieties for cultivation in extreme soils. *Plant Sci.,* 160:1, 2000.
145. Kochian, L.V., Hoekenga, O.A., and Pineros, M.A., How do crop plants tolerate acid soils? Mechanisms of aluminum tolerance and phosphorous efficiency. *Ann. Rev. Plant Biol.,* 55:459, 2004.

146. Koyama, H. et al., Overexpression of mitochondrial citrate synthase in Arabidopsis thaliana improved growth on a phosphorus-limited soil. *Plant Cell Physiol.,* 41:1030, 2000.
147. Tesfaye, M. et al., Overexpression of malate dehydrogenase in transgenic alfalfa enhances organic acid synthesis and confers tolerance to aluminum. *Plant Physiol.,* 127:1836, 2001.
148. Tomscha, J.L. et al., Phosphatase under-producer mutants have altered phosphorus relations. *Plant Physiol.,* 135:334, 2004.
149. Li, S.M. et al., Acid phosphatase role in chickpea/maize intercropping. *Ann. Bot.,* 94:297, 2004.
150. Li, L. et al., Chickpea facilitates phosphorus uptake by intercropped wheat from an organic phosphorus source. *Plant Soil,* 248:297, 2003.
151. Yun, S.J. and Kaeppler, S.M., Induction of maize acid phosphatase activities under phosphorus starvation. *Plant Soil,* 237:109–115, 2001.
152. George, T.S. et al., Expression of a fungal phytase gene in Nicotiana tabacum improves phosphorus nutrition of plants grown in amended soils. *Plant Biotech. J.,* 3:129, 2005.
153. Skene, K.R., Cluster roots: model experimental tools for key biological problems. *J. Exp. Bot.,* 52:479ñ485, 2001.
154. Waters, B.M. and Blevins, D.G., Ethylene production, cluster root formation, and localization of iron(III) reducing capacity in Fe deficient squash roots. *Plant Soil,* 225:21–31, 2000.
155. Douds, D.D., Johnson, C.R., and Koch, K.E., Carbon cost of the fungal symbiont relative to net leaf-P accumulation in a split-root VA mycorrhizal symbiosis. *Plant Physiol.,* 86:491–496, 1988.
156. Eissenstat, D.M. et al., Carbon economy of sour orange in relation to mycorrhizal colonization and phosphorus status. *Ann. Bot.,* 71:1, 1993.
157. Harris, D. and Paul, E., Carbon requirements of vesicular-arbuscular mycorrhizae. In *Ecophysiology of VA Mycorrhizae*, Safir, G.R., Ed., CRC Press, Boca Raton, FL, 1987, pp. 93–105.
158. Jakobsen, I. and Rosendahl, L., Carbon flow into soil and external hyphae from roots of mycorrhizal cucumber plants. *New Phytol.,* 115:77, 1990.
159. Koch, K.E. and Johnson, C.R., Photosynthate partitioning in split root citrus seedlings with mycorrhizal and non-mycorrhizal root systems. *Plant Physiol.,* 75:26, 1984.
160. Ryan, M.H. and Graham, J.H., Is there a role for arbuscular mycorrhizal fungi in production agriculture? *Plant Soil,* 244:263, 2002.
161. Owusu-Bennoah, E., Zapata, F., and Fardeau, J.C., Comparison of greenhouse and P-32 isotopic laboratory methods for evaluating the agronomic effectiveness of natural and modified rock phosphates in some acid soils of Ghana. *Nutr. Cycling Agroecosys.,* 63:1, 2002.
162. Skene, K.R., Cluster roots: Some ecological considerations. *J. Ecol.,* 86:1060, 1998.
163. Rossiter, R., Phosphorus deficiency and flowering time in Subterranean Clover *Trifolium subterraneum*. *Ann. Bot.,* 42:325, 1978.
164. Chauhan, Y.S., Johansen, C., and Venkataratnam, N., Effects of phosphorus deficiency on phenology and yield components of short-duration pigeonpea. *Tropical Agric.,* 69:235, 1992.
165. Ma, Q.F., Longnecker, N., and Atkins, C., Varying phosphorus supply and development, growth and seed yield in narrow-leafed lupin. *Plant Soil,* 239:79, 2002.

166. Tinker, P. and Nye, P., *Solute Movement in the Rhizosphere.* Oxford University Press, New York, 2000.
167. Aerts, R., Nutrient resorption from senescing leaves of perennials: Are there general patterns? *J. Ecol.,* 84:597, 1996.
168. Van Heerwaarden, L.M., Toet, S., and Aerts, R., Current measures of nutrient resorption efficiency lead to a substantial underestimation of real resorption efficiency: Facts and solutions. *Oikos,* 101:664, 2003.
169. Gan, S.S. and Amasino, R.M., Inhibition of leaf senescence by autoregulated production of cytokinin. *Science,* 270:1986, 1995.
170. Jordi, W. et al., Increased cytokinin levels in transgenic PSAG12-IPT tobacco plants have large direct and indirect effects on leaf senescence, photosynthesis and N partitioning. *Plant Cell Environ.,* 23:279, 2000.
171. Jaramillo, R. and Lynch, J., Unpublished data, 2005.
172. Brewster, J., Bhat, K., and Nye, P., The possibility of predicting solute uptake and plant growth response from independently measured soil and plant characteristics. IV. The growth and uptake of rape in solutions of different phosphorus concentrations. *Plant Soil,* 44:279, 1976.
173. Foehse, D. and Jungk, A., Influence of phosphate and nitrate supply on root hair formation of rape, spinach and tomato plants. *Plant Soil,* 74:359, 1983.
174. Walk, T., Jaramillo, R., and Lynch, J., Architectural tradeoffs between adventitious and basal roots for phosphorus acquisition. *Plant Soil,* in press, 2006.
175. Ho, M.D., McCannon, B.C., and Lynch, J.P., Optimization modeling of plant root architecture for water and phosphorus acquisition. *J. Theor. Biol.,* 226:331, 2004.
176. Jackson, R.B. and Caldwell, M.M., The scale of nutrient heterogeneity around individual plants and its quantification with geostatistics. *Ecology* 74:612, 1993.
177. Hodge, A., The plastic plant: Root responses to heterogeneous supplies of nutrients. *New Phytol.,* 162:9, 2004.
178. Drew, M.C., Comparison of the effects of a localized supply of phosphate, nitrate, ammonium and potassium on the growth of the seminal root system, and the shoot, in barley. *New Phytol.,* 75:479, 1975.
179. Campbell, B.D., Grime, J.P., and Mackey, J.M.L., A trade-off between scale and precision in resource foraging. *Oecologia,* 87:532, 1991.
180. Farley, R.A. and Fitter, A.H., The responses of seven co-occurring woodland herbaceous perennials to localized nutrient-rich patches. *J. Eco.,* 87:849, 1999.
181. Maina, G.G., Brown, J.S., and Gersani, M., Intra-plant versus inter-plant root competition in beans: Avoidance, resource matching or tragedy of the commons. *Plant Ecol.,* 160:235, 2002.
182. Gersani, M. et al., Tragedy of the commons as a result of root competition. *J. Ecol.,* 89:660, 2001.
183. Robinson, D. et al., Plant root proliferation in nitrogen-rich patches confers competitive advantage. *Proc. Royal Soc. London Series B: Biol. Sci.,* 266:431, 1999.
184. Ho, M.D., Effects of root architecture, plasticity, and tradeoffs on water and phosphorus acquisition in heterogeneous environments. PhD Thesis, Pennsylvania State University, 2004.
185. Lynch, J.P. and Beebe, S.E., Adaptation of beans (*Phaseolus vulgaris* L.) to low phosphorus availability. *HortScience,* 30:1165, 1995.
186. CIAT, *Bean Project 1998 Annual Report.* CIAT, Cali, Colombia, 1999.

9 Physiological Effects of Heavy Metals on Plant Growth and Function

M.S. Liphadzi and M.B. Kirkham

CONTENTS

9.1 Introduction 243
9.2 Factors Affecting Heavy Metal Availability 245
9.3 Physiological Effects of Heavy Metals on Plants 247
    9.3.1 Copper 247
    9.3.2 Iron 249
    9.3.3 Manganese 249
    9.3.4 Molybdenum 251
    9.3.5 Zinc 252
    9.3.6 Cobalt 253
    9.3.7 Nickel 255
    9.3.8 Vanadium 256
    9.3.9 Cadmium 257
    9.3.10 Chromium 259
    9.3.11 Lead 260
    9.3.12 Mercury 262
    9.3.13 Other Heavy Metals 263
9.4 Summary 263
9.5 Future Research Perspectives 264
References 265

9.1 INTRODUCTION

There are 16 elements essential for plant growth and development. They can be remembered by the mnemonic CHOPK(I)NS CaFe Mg B Mn CuZn MoCl, pronounced "C. Hopkin's café managed by my cousin mocl" [Harriet B. Creighton, Department of Biological Sciences, Wellesley College, personal communication]. Iodine has not been shown to be essential for higher plant growth, but is added in parentheses to make the pronunciation clear. The typical concentration range varies with different elements in plants (Table 9.1). The first three elements in

TABLE 9.1
Typical Concentrations of Nutrient Elements in Plants

Nutrient Elements	Concentration (of Total Dry Matter)
N, K, Ca (%)	0.5–5
P, S, Mg (%)	0.5–5
Cl (%)	0.01–1
Fe, Mn (ppm)	25–300
Zn, B (ppm)	10–100
Cu (ppm)	4–15
Mo (ppm)	0.1–5

our mnemonic make up organic compounds, which are not discussed in this chapter. The last 13 elements are the mineral elements. Nitrogen, phosphorus, potassium, sulfur, calcium, and magnesium are considered to be macroelements. Chloride, iron, manganese, zinc, boron, copper, and molybdenum are microelements, also called trace elements [1]. Three other elements — cobalt, nickel, and vanadium — are not included among the 16 essential elements, but are discussed in this chapter because of their importance in some plants.

A heavy metal is a metal with a density greater than 5.0 g mL^{-1} [2]. Many references publish tables of concentrations of heavy metals in plants. Beeson [3] gives tables that show the maximum and minimum concentrations of cobalt, copper, iron, manganese, nickel, and zinc in dozens of plant species. The chapters in the book edited by Chapman [4] give tables of concentrations of chromium, cobalt, copper, iron, lead, manganese, molybdenum, nickel, vanadium, and zinc in many plants. These tables provide concentrations for the following levels: showing deficiency symptoms, low range, intermediate range, high range, and showing toxicity symptoms. Pictures of plants demonstrating toxicity symptoms, along with tables describing these symptoms, are given, too. Kirkham [5] presents information giving the concentrations of heavy metals in different plant parts of numerous crop species grown on heavy-metal contaminated soil (Table 9.2).

Although this chapter focuses on plants, in passing we consider some examples of heavy metal toxicity to animals and humans. The heavy metal requirements of animals are less well established than those of plants, mainly because of the difficulty of purifying diets [6, p. 114]. Bowen [6] gives concentrations of heavy metals in animal tissues including mammalian bones, organs, and blood, plasma, and red cells. There is much interest in the effect of heavy metals on diseases. For example, cadmium, lead, and copper are known to alter blood pressure or to induce cardio-cerebrovascular disease [7]. The reader is referred to Jerome O. Nriagu, who has written extensively concerning the effects of heavy metals on human health.

TABLE 9.2
Maximum Concentrations of Heavy Metals in Plants

Heavy Metals	Concentration (Parts per Million Dry Matter)
Cd	0.20
Co	0.30
Cr	0.50
Cu	15
Fe	300
Hg	0.01
Mn	100
Mo	1.0
Ni	1.0
Pb	5.0
V	1.0
Zn	150

This chapter is not intended to review mechanisms of uptake of heavy metals by plant roots. The physiology of micronutrient acquisition (iron, copper, zinc, manganese) and uptake through root membranes are discussed by Buchanan et al. [8]. Tinker and Nye [9] consider quantitative treatment of mass flow and diffusion near a single root. Bowen [6] defines the ranges that are often considered in elemental uptake: severe deficiency, mild deficiency, optimal range, luxury consumption, toxicity, and lethality. Instead, this chapter reviews factors affecting availability and physiological effects of various heavy metals. We look at studies that have considered excessive amounts of essential trace elements that are also heavy metals (in the following alphabetical order: copper, iron, manganese, molybdenum, and zinc), as well as cobalt, nickel, and vanadium, which are essential in some plants. We also consider the toxic, nonessential heavy metals (cadmium, chromium, lead, and mercury), because they are often of most concern in the environment. To put heavy metal toxicity in perspective, we include older literature containing basic information on heavy metals, which is often overlooked.

9.2 FACTORS AFFECTING HEAVY METAL AVAILABILITY

The availability to plants of heavy metals is influenced by many factors, including pH, organic matter, cation exchange capacity, the oxidation-reduction status of the soil, microorganisms, phosphorus, plant type, seasonal variation, and rhizosphere effects [1].

pH: The availability of most heavy metals decreases with increasing pH. The availability of molybdenum, however, increases with increasing pH.

Organic Matter: Applications of organic matter to the soil are used to increase or decrease heavy metal concentrations in plants. The presence of organic matter can promote the availability of heavy metals, presumably by supplying soluble complexing agents that interfere with their fixation. However, soils that are commonly deficient in certain metallic heavy metals, especially copper, are organic in nature. Organic matter chelates the metals and can make them less available to plants.

Cation Exchange Capacity: The cation exchange capacity of the soil is important in binding the cationic heavy metals. Therefore, soil horizons rich in clays or organic matter generally have higher contents of heavy metals than those containing silt or sandy horizons.

Oxidation Reduction: Many heavy metals (cobalt, copper, manganese, molybdenum, nickel, lead, vanadium, and zinc) are mobilized in poorly drained soils. The increased movement of heavy metals under conditions of even slightly impeded drainage is often so marked that changes in extractable heavy metal contents due to other causes are obscured. Drainage of wet soils can affect the availability of heavy metals, because more highly oxidized forms of these elements are formed. For example, toxicities of manganese that occur on poorly drained soils are remedied by drainage.

Microorganisms: Microorganisms can aggravate deficiency of heavy metals, because they either compete directly for essential heavy metals or decompose organically bound forms. They can also make heavy metals more available for plant uptake by releasing inorganic ions or soluble organic complexes during decomposition of organic materials.

Phosphorus: Phosphate is well known for decreasing injury caused by excessive levels of heavy metals. In agronomic practice, large applications of phosphorus fertilizer are used to reduce the availability of the metallic cation heavy metals (except cadmium, which is often present in phosphorus fertilizers).

Ion Competition: Competing elements are used to reduce the uptake and accumulation by plants of specific heavy metals. For example, plants generally take up more cadmium if the zinc content of the soil is low [5].

Plant Type: Plant species and varieties differ widely in tolerance to heavy metals. Vegetable crops sensitive to heavy metals are beets (*Beta vulgaris*), chard (*Beta vulgaris* var. *Cicla*), kale (*Brassica oleracea*), mustard (*Brassica* sp.), spinach (*Spinacia oleracea*), tomatoes (*Lycopersicon esculentum*), and turnip (*Brassica rapa*). Many farm crops such as corn (*Zea mays*), small grains, and soybeans (*Glycine max*) are moderately tolerant. Most grasses are tolerant. Highly tolerant ecotypes of grasses are found on ore outcrops or near mines that contain high concentrations of heavy metals. Plants also vary in amounts of heavy metals accumulated in roots, leaves, stems, flowers, and fruits.

Seasonal Variation: The availability of many elements in soils changes during the year. Manganese exhibits the most pronounced seasonal variation in availability, probably due to microbially induced oxidation and reduction. It is more available in the summer than other times of the year. Cobalt content of forage is high in spring. Copper and zinc are high in rainy, cool weather. Molybdenum is high in autumn.

Rhizosphere Effects: Plant roots exude a variety of compounds in quantities sufficient to change the availability of heavy metals in soils. Root exudates can alter the chemical environment of the root either directly through an interaction with heavy metals in the soil or indirectly through their influence on microorganisms.

Fourteen elements in the soil (aluminum, carbon, calcium, iron, hydrogen, potassium, magnesium, nitrogen, sodium, oxygen, phosphorus, sulfur, silicon, and titanium) constitute over 99% of the total elemental content. The remaining elements are the so-called trace elements. This term "trace elements" is generally used for those elements occurring in the soil in small amounts without regard to their requirement by organisms [1]. The term micronutrient is used for trace elements in the soil that are essential for healthy development of plants, animals, or microorganisms (boron, chlorine, cobalt, chromium, copper, fluorine, iodine, manganese, molybdenum, nickel, selenium, vanadium, and zinc and possibly barium, bromine, and strontium). Boron, chlorine, fluorine, iodine, selenium, barium, bromine, and strontium are not heavy metals.

9.3 PHYSIOLOGICAL EFFECTS OF HEAVY METALS ON PLANTS

9.3.1 COPPER

Copper is a micronutrient element involved in various enzymatic activities. It is active in enzymes such as polyphenol oxidase, monophenolase, ascorbic acid oxidase, laccase, and cytochrome oxidase. These are all called metalloenzymes.

Copper is also considered as a pollutant of the air and agricultural soils [10]. Intensive use of fungicides and herbicides, as well as sludge and manure, are the main causes of agricultural soil contamination by copper [11]. It has been used since the 1860s in Bordeaux mixture, a mixture of lime, water, and copper sulfate used as a spray on trees and plants to kill insects and fungi. The role of copper in agriculture has been the subject of much careful research since 1916 and 1917, when studies first showed that beneficial results could be obtained by application of copper sulfate to the soil and Bordeaux mixture as a foliage spray for the control of die-back, a widespread disorder of citrus in Florida [12]. Toxic concentrations of copper develop in agricultural soils as a result of accumulation over a period of many years of residual copper from Bordeaux fungicides or from copper sulfate fertilization. Toxic effects of high copper in citrus orchard soils are manifested in severe cases by a marked reduction in tree vigor and yield, severe chlorosis of foliage, and die-back of the twigs, associated with 150 to 200 mg kg^{-1} of total copper in the topsoil and a pH of 5.0 or less. Copper toxicity symptoms of several crops have been reported in old vegetable fields in Florida. The reports indicated a large amount (over 400 mg kg^{-1}) of total copper in the topsoil resulting from many years of frequent Bordeaux spraying of celery *Apium graveolens var. dulce* for control of fungal diseases [12].

In the early stages of copper toxicity, reduced plant growth is evident [12]. At the moderate to acute stages, excess copper in water-culture solutions, sand culture, or even in the field commonly induces iron-chlorosis symptoms in plants. Toxic amounts of copper in the soil or nutrient medium reduce growth and may depress the iron concentration in leaves, causing chlorosis [13]. Leaf chlorosis due to copper toxicity is strongly related to reduced volume and number of mesophyll cells [11] as well as displacement of iron from physiologically active centers [13].

In addition, copper excess in soil or nutrient medium is associated with stunting, reduced branching, thickening, and abnormally dark coloration of rootlets of many plants [12]. Grapevines (*Vitis* sp.) have been found most resistant to excess copper, whereas clover (*Trifolium* sp.), alfalfa (*Medicago sativa*), poppy (*Papaver* sp.), spinach, gladiolus (*Gladiolus* sp.), corn, bean (*Phaseolus* sp.), and squash (*Cucurbita* sp.) have been found to be sensitive to copper excess. Copper indicators (those plants universally or locally restricted to soils high in copper) are considerably more abundant than zinc indicators and are useful in locating copper ore. The copper indicators belong mainly to three plant groups: the Caryophyllaceae (pink family), the Labiatae (mint family), and the mosses. The blue-flowered basil, *Ocimum homblei*, will not grow in soil containing less than 100 mg kg^{-1} of copper [12].

Analysis of chlorotic foliage from affected citrus trees in Florida indicates that it has an abnormally low iron content, and that the sparse, dark, stubby fibrous roots in the topsoil have an exceedingly high copper content. In Florida citrus orchards, severe toxic effects occur when copper level in the soil reaches about 3 mg of copper per milliequivalent of exchange capacity in 100 grams of dry soil [12]. In the majority of cases, such copper toxicity symptoms of citrus are associated with an acid soil condition (pH 4.0 to 5.5) produced by the application of acid-forming fertilizers and large residues of sulfur used for pest control. If copper level in the soil is not too high, normal vigor of affected trees can be restored by the application of FeEDTA (iron ethylenediamine tetraacetate) and sufficient lime or sodium carbonate to raise the pH of the topsoil to near 7.0. The iron chelate quickly corrects chlorosis by supplying iron to the shoot through healthy roots in the subsoil that are unaffected by copper toxicity. Heavy liming reduces copper availability sufficiently to permit the gradual restoration of normal rooting in the topsoil [12]. Panou-Filotheou et al. [11] found that copper toxicity resulted in reduction of stem height and root volume in oregano (*Origanum vulgare*) plants. Any remarkable reduction of root volume due to copper toxicity also reduces water and nutrient uptake by the plant.

Copper toxicity may cause damage to the plasma membrane of both plants and animals [14,15]. Concentration of copper about 3–5 µmol L^{-1} increases nonspecific plasma membrane permeability, inhibits Cl$^-$ channels, and suppresses plasma membrane H⁺-ATPase [16]. Nonspecific conductance and H⁺-ATPase inhibition are destructive to a cell because they are accompanied by plasma membrane depolarization, disruption of ionic homeostasis, and subsequent perturbation of enzymatic reactions [14,15].

9.3.2 Iron

The essentiality of iron for plants was demonstrated in solution cultures more than a century and a half ago, after it was found in 1843 that foliar application of iron salts was beneficial to chlorotic grapevines [17]. Iron is still classified as a micronutrient element, although the requirements for manganese, zinc, copper, and molybdenum are smaller in most plants. Iron plays a role in chlorophyll formation because it has a central role in porphyrin systems. Iron and porphyrin combinations, which are prosthetic groups of enzymes, are known as hemes. Iron is a component of respiration's cytochrome oxidase and cytochromes, which are also in the chloroplasts and necessary for photosynthesis. Iron is in catalase and peroxidase, as well as the hemoglobin in the nodules of leguminous roots, and, therefore, it plays a role in nitrogen fixation. Iron is in nonporphyrin enzymes (ferrodoxins) and in phytoferritin, a storage compound.

Iron toxicity is not common under natural conditions [17]. Its toxicity has been observed in plants that have received soluble iron salts in excessive amounts, either as sprays or as soil amendments. The initial symptom of iron toxicity appears in the form of necrotic spots. Most iron that is added to soil, as a simple inorganic iron salt, is quickly rendered insoluble in alkaline soils and is no more available to plants than the native iron. The main problem with iron is iron deficiency and many methods have been developed to try to overcome it (e.g., use of chelates, injections of iron salts into stems of trees, spraying iron solutions directly on leaves).

Even though iron is classified as an element with low toxicity in plants [18], it is potentially noxious if taken up by plants in excess quantities. High levels of iron in plants promote the formation of reactive oxygen species, which damages vital cellular constituents, especially membranes that are known to be susceptible due to lipid peroxidation [19,20]. Above-optimal levels of iron may result in coalesced tissues, necrosis or bronzing, flaccidity, and blackening of the roots [21]. Iron has been given no regulation limit in sewage sludge (biosolids) by the U.S. Environmental Protection Agency [22], probably because of its low bioavailabilty, both to humans and to plants. However, iron can cause serious human disease. Most cases of acute iron poisoning occur in children. Poisoning can occur if children eat with rusty cooking utensils. Despite iron being considered as an element with low toxicity in many plants [18], iron toxicity is a common problem in rice plants.

9.3.3 Manganese

Manganese has been the subject of much careful study since it was shown in 1774 that pyrolusite, the most abundant manganese mineral, was the oxide of a new metal, altogether different from iron [23]. Before 1774, manganese compounds were mistaken for those of iron. Manganese is as widely distributed as iron in nature, but occurs in much smaller quantities in rocks, soils, plants, and animals. In 1788, the metal was isolated. At that time, it was proved that manganese was a constituent of rocks and soils and was assimilated by plants grown in the soil. Manganese was found in the ash of all plants examined, but no effort was made to ascertain if manganese performed a useful function in plants [23]. In 1865, it was proved that manganese could not replace iron in the growth of plants, but no one had showed that manganese was necessary for their growth. Studies in the late 1800s and early 1900s indicated that manganese was necessary for plant growth and might function as a component of the oxidase system of plants [23]. Since these early studies, many research reports pertaining to the biological importance of manganese have been published. Manganese is involved in the activation of respiratory enzymes, in some steps of nitrate reduction to ammonia, and in oxygen evolution in photosynthesis.

Manganese toxicity mostly occurs in waterlogged environments [18,24]. Manganese uptake is also affected by pH and becomes soluble along with aluminum at low pHs. The two elements increase each other's toxicity (communication from Brett H. Robinson). In the early stages of manganese excess, the visual symptoms on plants are not well defined. However, it has been observed by many research workers that too much available manganese in the soil or in sand and water cultures harms plant growth. Excessive manganese has been noted in soils of Kentucky, Connecticut, and Hawaii. Tobacco (*Nicotiana tabacum*), pineapple (*Ananas comosus*), and other crops grown on them have developed a severe chlorosis. Pineapple leaves show early stages of manganese toxicity by the uneven distribution of chlorophyll. In the moderate to acute stages, excess manganese produces iron deficiency in pineapple plants [23].

Excess of an essential heavy metal may induce deficiencies of others; for example, excess manganese may induce iron deficiency in addition to producing direct toxic effects similar to manganese deficiency. Because of the importance of manganese toxicity under soil acidity, the following plants can be regarded as especially sensitive to manganese excess: alfalfa, cabbage (*Brassica oleracea* var. *capitata*), cauliflower (*Brassica oleracea* var. *botrytis*), cereals, clover, pineapple, potato (*Solanum tuberosum*), runner bean (*Phaseolus* sp.), sugar beet (*Beta vulgaris*), and tomato. Indicator plants for manganese excess, such as barley (*Hordeum vulgare*), have been studied. Barley showed toxicity symptoms when the tops had 1000 mg kg^{-1} manganese in the dry matter. Soluble salts of manganese caused a chlorosis of leaves of barley in water culture. Both the chloride and the sulfate of manganese in high concentrations exerted a toxic effect on wheat (*Triticum aestivum*) [23].

A symptom of manganese toxicity is the occurrence of dark-brown spots on older leaves. These necrotic spots result from the local accumulation of oxidized manganese and phenolic compounds [25] and provide an index of the degree of manganese toxicity in plants [26,27]. Elevated concentrations of manganese in the growing medium can also interfere with the absorption, translocation, and utilization of other elements such as calcium, magnesium, iron, and phosphorus [24,27]. High concentrations of manganese in tissues can alter the activities of enzymes and hormones, which may render essential manganese-requiring processes nonfunctional or less active [25]. Effects of manganese toxicity on animals and humans are essentially not known.

9.3.4 Molybdenum

Except for chlorine, molybdenum is the most recently recognized micronutrient necessary for all plants. Two diseases of agricultural plants, "whiptail" of cauliflower and "yellow-spot" of citrus (*Citrus* sp.), have been described for many years. Careful work in 1939 and 1940 with controlled nutrient solutions demonstrated that molybdenum is essential for higher plants, and led to the association of "whiptail" and "yellow-spot" with molybdenum deficiency [28]. It then was demonstrated that molybdenum was important in pasture production in the process of nitrogen fixation by the *Rhizobium*-legume complex. Implication of molybdenum in the nitrogen metabolism of microorganisms was noted as early as 1930 [28].

The essential function of molybdenum in plants is related to nitrogen-fixation. Nitrate reductase is a molybdenum-containing flavoprotein. Other enzymes in higher plants that contain molybdenum as a cofactor are nitrogenase, xanthine oxidase/dehydrogenase, and, presumably, sulfite reductase [29]. Many nutritional disease symptoms of higher plants (including "whiptail" and "yellow-spot"), which have been shown to be amenable to treatment with molybdenum, result from an inability to utilize fixed nitrogen in the absence of adequate amounts of molybdenum [28].

Molybdenum sequestration in *Brassica* species involves the formation of a blue anthocyanin-molybdate complex. Metal complexation plays a role in determining anthocyanin color [30]. Even though tungsten is not considered an essential element for plants, microorgansisms, usually hyperthermoplic ones, contain enzymes with tungsten. The metal is associated with two pyranopterin–dithiolene co-factors, known as molybdopterin. Molybdopterin does not contain molybdenum, but is named for its common association with that metal. The hyperthermophilic microorganisms lack molybdenum enzymes, and tungsten apparently fills a role similar to that of molybdenum in other organisms. It is important in two-electron redox reactions [30]. Compared to molybdenum, tungsten is better suited to catalysis of kinetically rapid, low-potential redox reactions at high temperatures. The fact that tungsten and molybdenum can perform similar functions in different organisms reflects their chemical similarity. Because tungstate is chemically similar to molybdate, it is complexed by anthocyanins as well. Tungsten accumulation is correlated with anthocyanin content. In *Brassica* species, anthocyanins show a color change from pink to blue upon addition of tungsten [30].

The availability of most heavy metals decreases with increasing pH. The availability of molybdenum, however, increases with increasing pH. Indications of molybdenum excess in plants are rarely observed in the field. Plants appear to tolerate relatively high tissue concentrations of the element. Values as high as 372 mg kg^{-1} of molybdenum have been reported with no apparent foliar toxicity symptoms. Deficiency levels are usually under 0.1 mg kg^{-1}. Experimental demonstrations of foliar symptoms of molybdenum excess have been observed. For example, tomato plants under special conditions in solution culture may develop an intense golden-yellow color in their leaves at leaf molybdenum concentrations of 1000 to 2000 mg kg^{-1}. Seedlings of cauliflower turn an intense purple under similar conditions [28].

A unique feature of molybdenum nutrition is the wide variation between the critical deficiency and toxicity levels [29]. These levels may differ by a factor of up to 10^4 (e.g., 0.1 to1000 μg molybdenum per gram dry weight), as compared with a factor of 10 or less for boron or manganese. Under conditions of molybdenum toxicity, malformation of the leaves and the golden-yellow discoloration of the shoot tissue that occur are most likely due to the formation of molybdocatechol complexes in the vacuoles [29].

Molybdenum toxicity in plants is observed only under extreme experimental conditions. However, plants may accumulate large tissue concentrations of molybdenum and induce molybdenosis in ruminants consuming such material. In ruminants, especially, molybdenum excess is frequently serious. Molybdenum toxicity, variously referred to as molybdenosis, teart disease, and peat scours, has been reported from many parts of the world. In most instances, toxic amounts of molybdenum in forage consumed by ruminants have resulted from naturally occurring excess molybdenum in the soil or irrigation water. Applying molybdenum fertilizers or liming to release unavailable molybdenum must be done with the knowledge of possible development of molybdenosis in ruminants that may graze on such areas. Molybdenum toxicity in ruminants involves not only excess molybdenum but also low copper levels and high sulfate-sulfur concentrations in the forage [28]. Cattle and sheep afflicted with molybdenum-induced copper deficiency respond to copper supplements [31].

9.3.5 ZINC

The essentiality of zinc for plants was not fully accepted until the early 1930s, when scientists working with peaches and citrus were able to correct little-leaf of peaches and mottle-leaf of citrus with zinc compounds. In the mid to late 1930s, when experiments done in controlled cultures produced the same field symptoms correctible by zinc, it was conceded that zinc was an essential plant element. Zinc deficiency has been found in many soils throughout the world. But zinc toxicity has also been recognized since early studies [32]. Primary sources of zinc pollution are industrial wastes and sewage sludge [18]. Farm manures also have high concentrations of zinc [33].

Zinc deficiency is normally observed in the terminal parts of plants. They have short internodes, stunted growth, and small leaves with necrosis and chlorosis. Zinc is necessary for the synthesis of tryptophan, essential in the biosynthesis of indoleacetic acid, an auxin. Without zinc, auxin activity is markedly decreased. Zinc is involved in other enzymes, particularly dehydrogenases. It is important as a structural component of many proteins, for example, transcription factors, and as a critical co-factor in some superoxide dismutase isoforms.

Little information exists concerning the visual symptoms of zinc toxicity in the early stages. In the moderate to acute stages, it commonly produces iron chlorosis in plants. Indicator plants for zinc excess have been studied. A good indication of a zinc mineral outcrop is the presence of luxuriantly growing ragweed (*Ambrosia artemisiifolia*) when other vegetation is stunted. Chapman [32] reported that in a zinc slime pond, where the total zinc of the soil was 15,000 mg kg^{-1}, the zinc content ranged from 39 mg kg^{-1} in the fruit of false Solomon's seal (*Smilacina* sp.) to 5400 mg kg^{-1} in horsetail (*Equisetum*).

Zinc content varies with plant species, ranging from 10 mg kg^{-1} to 10,200 mg k g^{-1}. The higher concentrations indicate accumulator plants. Most plants have a zinc content under 20 ppm. However, zinc readily accumulates in the leaves of many plants. Deficiency levels are characterized by zinc levels of less than 20 to 25 ppm in the dry matter. Ample but not excessive levels commonly fall in the range of 25 to 150 ppm. Amounts greater than 400 ppm may indicate zinc excess [32].

When zinc supply is large, zinc toxicity can be induced readily in nontolerant plants. Root elongation is a sensitive parameter for zinc toxicity [29]. Chlorophyll content and root length are reduced with increased zinc concentrations in the growing media [34]. Khurana and Chatterjee [35] reported a reduction in biomass, seed number, seed weight, and soluble proteins in sunflower (*Helianthus annuus*) plants grown in zinc-laden soils.

Chlorosis induced by zinc toxicity may be related to an induced deficiency of magnesium or iron, because of the similar ion radius of Zn^{2+} to both Fe^{2+} and Mg^{2+} [29]. In bean (*Phaseolus vulgaris*) plants, zinc toxicity inhibits photosynthesis at various steps and through different mechanisms. Suppressed ribulose bisphosphate carboxylase activity is presumably caused by competition with magnesium, and inhibition of photosystem II activity could be due to the replacement of manganese in the thylakoid membranes. In the thylakoid membranes of control plants, about six atoms of both manganese and zinc are bound per 400 chlorophyll molecules. With zinc toxicity, this proportion shifts to two manganese and 30 zinc atoms [29].

9.3.6 COBALT

In 1954, Holm-Hansen et al. [36] reported that cobalt is essential for blue-green algae [37]. It is also essential for legumes when relying on nitrogen from nitrogen fixation and in nitrogen-fixing bacteria. This is because cobalt is in the co-enzyme form of vitamin B_{12}, which is not produced by higher plants. Vitamin B_{12} is present in nodules and *Rhizobia*.

Cobalt received little attention from the agricultural standpoint until 1935, when it was found to be of great importance in the diet of ruminants, and in 1948 it was shown to be a constituent of vitamin B_{12}, required by all animals. The elemental nature of cobalt was established in 1735, but, owing to the very small amounts involved and the lack of sensitive analytical procedures its presence in plants was not demonstrated until 1841, when its presence was reported in sweet pea (*Lathyrus odoratus*) plants [38]. However, even though many studies since then have found it in higher plants, there is no evidence that cobalt has any direct role in the metabolism of higher plants [29].

After cobalt was found essential for fresh water blue-green algae in 1954, it was found essential for marine blue-green algae in 1955. Work in the 1950s also showed that higher concentrations of cobalt occurred in the root nodules of legumes than in other parts of a plant. In the late 1950s and early 1960s, it was demonstrated that cobalt is required for the symbiotic fixation of nitrogen by soybean and alfalfa plants [38].

Cobalt exhibits toxicity to plants when the amounts available to the plant exceed certain low levels. Work in 1895 showed that 1 mg kg^{-1} in the culture solution was toxic to beans (*Phaseolus* sp.) and corn [38]. Later, it was shown that in solution cultures, small amounts of cobalt, sometimes as low as 0.1 mg kg^{-1}, produced adverse or toxic effects on many crop plants. The symptoms of cobalt excess include stunted growth, chlorosis, necrosis, and even death of the plant. The chlorosis is frequently described as resembling that of iron deficiency.

Cobalt behaves like other heavy metals. In a way similar to iron, manganese, zinc, and copper, it is bound by organic molecules as a chelate [31]. It can also displace other cations from binding sites and can thus affect their uptake and mode of action. Excess cobalt supply induces iron deficiency, and symptoms resemble manganese deficiency. Both these observations indicate that the toxicity of excess cobalt relates to the effect of cobalt in displacing other heavy metals from physiologically important centers. In extreme cases of cobalt toxicity, plants wither completely. The leaf symptoms of excess cobalt can be alleviated by the addition of 2 to 25 mg kg^{-1} of molybdenum to the culture solution or by painting leaves with solutions of iron salts [38].

Most plant species do not accumulate cobalt to any great extent. Values exceeding 1 mg kg^{-1} in the dry matter are rare and are, for the most part, reported from the Hawaiian Islands. There are, however, two plant species that are apparently remarkable accumulators of cobalt. A study showed that *Clethra barbienervis* contained more than 600 mg kg^{-1} cobalt in the ash of the leaves, whereas oak (*Quercus* sp.), chestnut (*Castanea* sp.), saxifrage (*Saxifraga* sp.), and dogwood (*Cornus* sp.) growing in the same area had only 2 to 5 mg kg^{-1} in the leaf ash. Cobalt content of the soil is not a sufficiently sensitive indicator of cobalt availability. Swamp black gum (*Nyssa sylvatica*) is an efficient accumulator and a good index of cobalt availability. Up to 845 mg kg^{-1} in the dry matter of the swamp black gum has been found, a value several hundred times as large as the cobalt content of broom sedge (*Andropogon virginicus*) growing alongside. A value less than 5 mg kg^{-1} in the leaves of the swamp black gum indicated an area deficient in cobalt for the needs of grazing cattle and sheep [38]. Other plant species in which cobalt accumulates serve to indicate the presence of cobaltiferous ores. One such species found in the Sharba region of Zaire is *Crotolaria cobalticola*; it occurs only in areas rich in cobalt. Values of 500 to 800 mg cobalt per kilogram dry matter have been observed in this species [31]. *Huamaniastrum katangense* is a hyperaccumulator that accumulates both copper and cobalt [39].

Because cobalt is apparently not required by higher plants, and because it is highly unlikely that an excess of this element exists in soils, cobalt would still be unimportant agriculturally, were it not for the fact that dietary deficiency of cobalt is known to be the cause of a progressive emaciation of ruminants, a disease variously known as pining, salt sickness, bush sickness, coast disease, and enzootic marasmus. The reason for the disease relates to lack of vitamin B_{12}, which, as noted, is an essential ingredient in the diet of all animals [38].

9.3.7 Nickel

The essentiality of nickel for plants has been debated, but is now accepted as an important element for plants growing using urea as the sole nitrogen source. Nickel is an integral part of the enzyme urease [31]. It is necessary for urease activity and prevents urea accumulation in legumes [40], and therefore, plants grown solely on urea nitrogen have a requirement for nickel. Many plant species, including nitrogen-fixing legumes, accumulate large amounts of ureides. For such plants to utilize nitrogen fully in these compounds in anaerobic reactions, urease is required and, therefore, so is nickel [41].

The elemental nature of nickel has been known since 1751, but its presence in plants was not proved for over a hundred years [42]. In 1855, nickel was found in oak wood. In 1905, the dimethyl glyoxime method was developed, which allowed determination of the minute amounts of this element normally present in most plants. In 1906, nickel was found in peat and brown coal ashes; in 1919, it was found in marine algae; and, in 1925, it was shown that traces of this element were present in many plants [42].

Contamination of the environment by nickel is mostly from traffic or refinery emissions and industrial or municipal wastes [18,43]. Also, nickel does occur in the soils of several widely scattered regions in amounts sufficient to have a deleterious effect on many crops. In 1893, the toxic properties of excessive amounts of nickel for plants were documented using culture solutions with corn and bean plants as test subjects. Since then, workers have demonstrated by means of culture solutions the high degree of nickel toxicity for many crops [42]. Barley was quickly killed with 8 mg kg^{-1} of nickel in the culture solution. About 0.5 mg kg^{-1} produced chlorosis in buckwheat (*Fagopyrum esculentum*) and tobacco. For bean, barley, and oats (*Avena sativa*), 2 mg kg^{-1} was found to be toxic. Nickel, when present in the chelated form, lost practically all its toxicity and was not appreciably taken up by mustard plants. Symptoms of nickel excess in flax (*Linum usitatissimum*) were alleviated by addition of molybdenum to the culture solution. In oats, if the iron supply in the culture solution was increased, toxicity symptoms of nickel were less severe. Also, with oats, nickel toxicity was aggravated by low calcium, magnesium, nitrogen, and potassium, and by high phosphorus in the culture medium [42].

In the early or incipient stages of nickel toxicity, there are no definitive symptoms, but only a dwarfing or repression of growth. Plants readily take up nickel. So leaf or soil analysis will furnish required information about excess.

In the moderate to acute stages, excess nickel produces a chlorosis that is usually described as resembling the symptoms of iron deficiency. In cereals, this shows as pale yellow stripes running the length of the leaves. Eventually the whole leaf may turn white and in extreme cases necrosis occurs at leaf margins. In dicotyledons, nickel toxicity appears as chlorotic markings between the leaf veins, and the symptoms are similar to those of manganese deficiency [31]. In citrus, symptoms of nickel toxicity resemble zinc deficiency. In cases where toxicity is severe, the chlorosis is followed by necrosis and even death of the plant [42].

Nickel toxicity inhibits cell division in the meristem of the roots and limits the root expansion zone [44]. In sensitive species, root growth is severely inhibited, especially if the calcium concentration is low [29]. Nickel interferes with the translocation of manganese, iron, copper, and zinc to the shoots [45]. This antagonistic effect causes symptoms typical of manganese and iron deficiency in leaves [45].

There are indicator plants for nickel excess. Because the symptoms of nickel toxicity simulate those of iron deficiency, the symptoms by themselves are of little usefulness for delineating areas of excess. Leaf analyses are the only sure means of detecting excessive nickel in soil. In 1948, scientists reported that 10% nickel oxide occurred in the ash of the leaves of *Alyssum bertholonii* growing on a serpentine soil in Tuscany, Italy [39,42]. No other plants growing in the same area contained so much nickel. Although they may not necessarily be tremendous accumulators, some plants, notably birch (*Betula* sp.) trees, conifers (cone-bearing trees, mostly evergreens), and grasses, are useful in botanical prospecting for nickel deposits. Nickel cannot replace the cobalt required in the diet of ruminants.

Plants like *Alyssum bertholonii* that accumulate large amounts of a heavy metal with no toxicity symptoms are called hyperaccumulators. There are 290 nickel hyperaccumulators [39, p. 4]. Also, 26 cobalt hyperaccumulators, 24 copper hyperaccumulators, 11 manganese hyperaccumulators, and 16 zinc hyperaccumulators are known [39, p. 4]. The one known cadmium hyperaccumulator is *Thlaspi caerulescens* in the Brassicaceae family [39, p. 4].

9.3.8 Vanadium

Vanadium is widely distributed in nature. Sea water contains vanadium in the concentration of a few parts per billion, and nearly all soils and plant materials contain small quantities. Soils usually contain 20 to 500 mg kg^{-1} of total vanadium. Vanadium contents of 62 plant materials ranged from 0.27 to 4.2 mg kg^{-1}, with an average of about 1 mg kg^{-1} [46]. There is no conclusive evidence that vanadium is an essential element for higher plants [29,31]. If vanadium is essential for higher plants, adequate levels in plant tissues are less than 2 mg kg^{-1} [29].

However, vanadium has been reported to be a requirement for the green algae, *Scenedesmus obliquus*. It has not been shown necessary for *Chlorella*, another green algae. Maximum growth rates for *Scenedesmus* were obtained when the vanadium concentration was 0.1 mg kg^{-1} in the nutrient solution. Vanadium has been shown to increase growth in other microorganisms, including fungi and bacteria [46]. The precise function of vanadium, however, is still unknown [31].

Some evidence suggests that vanadium may partially substitute for molybdenum in nitrogen fixation in microorganisms, but the evidence for the substitution in symbiotic nitrogen fixation is not conclusive.

Vanadium can be toxic to higher plants, but only under experimental conditions. In the field, vanadium deficiency or toxicity has not been reported [46]. In experiments, vanadium has been shown to be toxic to germinating seeds, but even more toxic at later stages of growth. When pea (*Pisum sativum*) and soybean were grown in solution culture, both plants showed a marked accumulation of vanadium in roots as compared with shoots. Soybean roots contained as high as 170 mg kg^{-1} vanadium when present in toxic quantities, whereas shoots had 2.3 mg kg^{-1}; pea plants had 510 mg kg^{-1} vanadium in roots, with only 17 mg kg^{-1} in shoots. In these experiments, addition of iron reduced the toxicity of vanadium. Plants were more sensitive to vanadium toxicity when iron availability was low. These data indicate that toxicity of vanadium might be indicated by 2 mg kg^{-1} vanadium in the shoots. Generally in a nutrient solution, vanadium at a concentration of 0.5 mg kg^{-1} or greater is toxic to plants [46].

Additions of vanadium to soils have produced toxicity for a variety of crops. For example, as little as 10 mg kg^{-1} of vanadium, added as calcium vanadate to a sandy soil, reduced growth of sour orange (*Citrus aurantinum*) seedlings; with 150 mg kg^{-1} vanadium, the plants died. In all cases, the leaves of seedlings had less than 1 mg kg^{-1} of vanadium [46].

9.3.9 CADMIUM

Cadmium is toxic and has not been shown to be essential for any aspect of metabolism of higher or lower plants or animals. A few reports have shown a slight stimulation of growth at low concentrations, but these are erratic and uncommon. Principal uses of cadmium are in electroplating, pigments, chemicals, and alloys. Household appliances, automobiles, trucks, agricultural implements, airplane parts, industrial machines, hand tools (wrenches, pliers, screwdrivers), and fasteners of all kinds (nuts, bolts, screws, rivets, and nails) are commonly cadmium-coated. Compounds of cadmium are used as heat and light stabilizers in the plastics industry. Sealed cadmium batteries are present in many convenience appliances such as toothbrushes, electric shavers, flashlights, and knives. Cadmium also is used for luminescent dials, in photography, lithography, process engraving, rubber curing, and as fungicides [2].

The fact that cadmium is toxic under any circumstance leads to its use to measure microbial respiration. Respiratory techniques are commonly used to study biological reactions in the soil. Production of oxygen by the bacteria is quantified using cadmium. In the technique, Cd^{2+} at concentrations from 0 to 100 µg mL^{-1} is added to the side arm of a flask attached to a respirometer. When cadmium is tipped into the flask with the bacterial suspension, the bacteria are killed in direct proportion to the amount of cadmium added (Paul M. White, Jr., Research Assistant, Department of Agronomy, Kansas State University, personal communication).

Cadmium is a toxic metal that can accumulate in the human body and has a half-life greater than 10 years. Elevated levels of cadmium in the body can cause kidney damage in humans [47]. Studies link renal dysfunction with a low level of cadmium content in the diet [48]. Other diseases associated with cadmium exposure are pulmonary emphysema and bone demineralization (osteoporosis) [49], because cadmium replaces calcium in bones. Similarly, calcium blocks both cadmium and lead transport into roots in plants [50].

Elevated cadmium concentration in soils is of concern because of its risk to human health and the productivity of plants and animals. Because of the ubiquity of cadmium, it occurs in sewage sludge. Sewage sludge contains an abundance of essential nutrients, and the U.S. Environmental Protection Agency advocates recycling of sludge back to land to conserve the nutrients [51]. However, even with careful source control (i.e., getting rid of industrial sources of cadmium), sludge applied to land contains cadmium because of the many domestic sources of the element. Since November 25, 1992, when the U.S. Environmental Protection Agency published in the Federal Register regulations for the application of sewage sludge to land (called the "40 CFR Part 503" regulations; CFR stands for Code of Federal Regulations), the amount of cadmium that can be added to soil, along with other heavy metals, is limited to safeguard health [52].

Even in agricultural soils without sludge, cadmium poses a threat. Cadmium naturally occurs with phosphorus, and, when phosphate fertilizers are used, cadmium accompanies phosphorus. Cadmium-polluted soils due to phosphate fertilization pose a nontariff trade barrier to animal exports in New Zealand. Many years of fertilization with cadmium-rich superphosphate on pastures in New Zealand have increased concentrations of cadmium in the soils, in some cases to levels above a self-imposed limit of 3 $\mu g\ g^{-1}$ in dry soils [53]. Sheep grazing on these soils then accumulate high levels of cadmium in their livers and kidneys, and the lamb for export is contaminated with the cadmium. Sunflowers in the United States have been banned from European markets because of high levels of cadmium in the seeds [54]. Sunflowers growing on high-clay soils and soils high in chlorine are the ones that have excessive levels of cadmium in the seed. Sunflowers need to be bred for low cadmium concentrations in their seeds if they are to be exported to Europe. Also, for protection of the food supply in the United States, the sunflower seeds need to be low in cadmium.

The accumulation of cadmium in water and soil has caused major environmental and human health problems [55]. Cadmium is usually less adsorbed by soil and organic matter than by several other heavy metals (e.g., lead, copper), which makes it more available to plants and more easily leached to ground water [18,56–58]. Gonzalez et al. [59] showed that the availability of cadmium in biosolids-amended soil is controlled by phosphatic clay instead of organic matter. Other studies indicate that cadmium is associated with iron-oxides or an iron-manganese oxide fraction in biosolids [60,61].

Plants show a disturbed water balance when grown on cadmium-laden soil [62]. The metal is readily taken up by roots and translocated to aerial organs, where it accumulates to high levels [63]. Cadmium affects stomatal function, water transport, and cell wall elasticity [64–66]. Poschenrieder et al. [62] reported an increase in the stomatal resistance of cadmium-treated plants, and similar results were reported by Kirkham [65] and Baryla et al. [63]. The increase in stomatal resistance strongly correlated with an increase of abscisic acid level in leaves [62].

Inhibition of photosynthesis is another toxic effect of cadmium, which is brought about by reduced stomatal conductance in response to metal toxicity and sensitivity of photosystem II to high cadmium concentration [63]. Cadmium may affect photosystem II on both the oxidizing (donor) and reducing (acceptor) side [67]. Rubisco activity in the Calvin cycle is inhibited by high cadmium [68]. The most apparent symptom of cadmium phytotoxicity is leaf chlorosis [63,65]. Replacement of iron by cadmium in the center of a precursor of the chlorophyll molecule was speculated as one of the causes of leaf chlorosis [69]. High cadmium concentration in the plant induces increased respiration and activities of the tricarboxylic acid cycle as well as other pathways of carbohydrate utilization [70].

Cadmium and zinc are chemically similar; thus cadmium is able to mimic the behavior of zinc in its uptake and metabolic functions. Unlike zinc, however, cadmium is toxic both to plants and animals. The basic cause of the toxicity probably lies in the much higher affinity of cadmium for thiol groupings (SH) in enzymes and other proteins. The presence of cadmium, therefore, disturbs enzyme activity. In plants, excess cadmium may also disturb iron metabolism and cause chlorosis [31]. Generally, the concentrations of copper, lead, nickel, and chromium are higher in roots than in shoots. Zinc and cadmium are easily translocated from roots to shoots [31].

The uptake mechanism for Cd^{2+} by plants has been studied. Presumably it is a facilitated diffusion across the plasmalemma. Data suggest that Cd^{2+} like Ca^{2+} and Zn^{2+} is translocated across the tonoplast by a proton antiport [31]. Accumulation of cadmium in the vacuole is related to the cadmium tolerance of plants. Uptake rates depend on the cadmium concentration in the growth medium. Tolerance also is related to phytochelatins, which are a major class of heavy metal chelating peptides that exist in plants [8]. They are low molecular weight, enzymatically synthesized cysteine-rich peptides known to bind cadmium and are important for cadmium detoxification [8,48].

Cadmium can be transported readily from the soil via the plant root to the upper plant parts. The particular hazard with cadmium is that the plant does not necessarily act as an indicator of levels toxic to humans and animals, because plants tolerate higher levels of cadmium than do animals. The same is true for mercury. Plants can appear healthy but may contain a high concentration of cadmium and mercury, which are completely unacceptable in an animal and human diet [31].

9.3.10 Chromium

Chromium is widely distributed in soils, waters, and biological materials. In soils, chromium has been reported usually to be in the range of 5 to 1000 mg kg^{-1} [71].

Although there is no conclusive evidence that chromium is essential for the growth of plants, some investigators have reported growth stimulation from the application of small amounts of chromium in salts. The apparently beneficial effects of added chromium salts may have resulted from response to the elements added with the chromium, or to such indirect effects as pathogen control. Until more evidence has accumulated, chromium should be considered a nonessential element in plants [31]. However, in humans and animals, it is essential and participates in glucose metabolism. In plant-production media, the use of chromium salts should be avoided, because of their toxic effects.

A large number of investigators have found that chromium, despite being relatively insoluble, is toxic to plants. Early studies showed that dichromate ($Cr_2O_7^{2-}$) was more toxic than chromate (CrO_4^{2-}). In an experiment with corn seedlings in solution culture, a stimulation of growth was observed when there was 0.5 mg kg^{-1} of chromium as chromic sulfate; toxicity occurred at 5 mg kg^{-1}, and little growth occurred at 50 mg kg^{-1}. When sugar beets were grown in sand culture with 8 and 16 mg kg^{-1} chromium added as the chromic ion and chromate ion, iron chlorosis was produced. But when the leaves were sprayed with ferrous sulfate, the chlorosis disappeared. The chromate ion was more toxic than the chromic ion. In an experiment with oats, chromium at 5 and 10 mg kg^{-1} in nutrient solution produced iron chlorosis. Tea (*Thea sinensis*) plants affected by "witches' broom" disease had more chromium than normal leaves, and chromium increased as the severity of the disease increased. High chromium content of citrus was associated with "yellow branch" disease in the Transvaal citrus-growing area of South Africa [71].

Symptoms of chromium excess depend upon the plant. In corn, plants are severely stunted, and the leaves have a tendency to roll around the shoot. The leaves are narrow and purple-green, with an intense purple color on the lower blades. Oats affected by chromium toxicity are stunted, with narrow, brownish-red leaves containing small necrotic areas with poorly developed roots. In tobacco, there are no specific leaf symptoms, but no inflorescence develops and stem development is retarded. Acute infertility in some serpentine soils may result from chromium toxicity [31]. Their infertility results from several factors, including a high Mg:Ca ratio, lack of nutrients, and high nickel content [71].

Chromium toxicity is only part of the problem. The metal tends to be accumulated in roots, where the main toxic effect is exerted. In experiments with orange (*Citrus senensis*) seedlings and corn affected by chromium toxicity, no more chromium in shoots of plants affected by chromium toxicity occurred than in healthy plants. The chromium content of roots of tobacco showing toxicity symptoms was more than 20 times higher than that in the leaves [71].

9.3.11 Lead

The Earth's lithosphere contains approximately 16 mg kg^{-1} lead. The total lead content of agricultural soils may vary between 2 and 200 mg kg^{-1}. In general, the soluble or available lead concentrations in soil vary between 0.05 and 5 mg kg^{-1} lead. Most of the lead in soils is sparingly soluble and largely unavailable to plants [72].

Soil pollution by lead has occurred through activities such as spraying lead arsenate as an insecticide [72], mining, smelting, land application of biosolids, and the past use of antiknock gasoline additives such as tetramethyl and tetraethyl lead [73]. Lead in gasoline is no longer allowed in the United States. Paint and toothpaste tubes also used to have lead, but lead is banned from them, too. Lead shot is a major source of soil contamination. People are usually exposed to lead through drinking water, breathing lead-laden dust, and consuming food that accumulates high concentrations of lead, because it has been grown on the soil contaminated by lead [74]. Lead impairs the nervous system and has effects on the fetus, infants, and young children that result in a low intelligence quotient [75]. Lead, along with nickel, if present in jewelry, can cause allergic reactions. Lead is classified as a possible human carcinogen because it can cause cancer. Low levels of exposure to lead may cause ailments such as heart disease, abnormalities in children, testicular atrophy, anemia, and nephritis [75].

Lead has not been shown to be essential to plant growth, although a few workers have reported apparent benefits from additions of lead in fertilizers. In an experiment where lead carbonate was applied to a Washington soil, roots of barley contained up to 1,475 mg kg^{-1} lead oxide. Most of the lead added to the soil was fixed within a few days and the remainder was usually fixed within the first growing season [72]. In reports of lead toxicity to plants, the results come from solution culture experiments. Cation fixation studies indicate that lead is more tenaciously fixed by humus soils than mercury, barium, magnesium, manganese, zinc, cobalt, nickel, or copper [72].

Lead analyses have been obtained on vegetable crops grown on nonorchard soils and on old orchard soils containing accumulated lead arsenate from 20 to 25 years of commercial orchard spraying. The data show that the accumulated lead arsenate did result in increased lead uptake by plants, but in most cases the increase in lead content of aboveground parts were only a few parts per million. In young citrus leaves, the normal lead concentration has been reported to be 0.5 to 4 mg kg^{-1}. In old citrus leaves, the concentration is around 10 mg kg^{-1} lead [72].

Vegetation near the side of roads may have levels of 50 mg lead per kg dry matter but at a distance of only 150 m away from the highway, the level is normally about 2 to 3 mg kg^{-1} dry matter. Contamination occurs only on the outer part of plant seeds, leaves, and stems, and a high proportion can be removed by washing. Levels of lead in grain, tubers, and roots are little affected and do not deviate much from normal levels for such tissues of about 0.5 mg lead per kg dry matter [31]. This is true if the lead is provided in inorganic form such as Pb^{2+}. Organic lead, however, such as lead tetraethyl, lead triethyl, and lead diethyl, as well as other alkyl derivates of lead, are extremely mobile in the soil and are taken up by plants much more rapidly than Pb^{2+}. Such organic lead forms may be released in gas combustion fumes if combustion is incomplete. Lead trialkyl is a powerful mutagenic agent and is known to derange the spindle of the fiber mechanism of cell division in both plant and animal cells. In corn plants, lead is first concentrated in dictyosome vesicles, which fuse together to encase lead deposits. These are then removed from the cytoplasm to outside the plasmalemma to fuse with the cell wall where much lead can be accumulated [31].

Lead is toxic because it mimics aspects of the metabolic behavior of calcium, and inhibits many enzyme systems. Lead toxicity can cause plasma membrane alteration in plants, as Pb^{2+} is physiologically similar to Ca^{2+} [13]. Elevated lead interferes with chlorophyll formation and the normal metabolism of iron [76]. High concentration of lead has been linked to poor seed germination, high stomatal resistance, inhibited carbon dioxide uptake, and low photosynthetic rate [77].

9.3.12 MERCURY

Mercury occurs in both organic and inorganic forms. All forms of mercury, except calomel, are toxic and the organic form of mercury is most poisonous. The vapor from volatilized mercury is also toxic to animals and humans [18]. Coal combustion, metal refineries, and waste incineration are the main anthropogenic sources of mercury [78]. Crematoria are also a major source of mercury volatilized from tooth fillings (communication from Brett H. Robinson, June 8, 2005). Elevated concentrations of mercury in soil are strongly correlated with soil organic matter content [18]. The toxicity of mercury to humans is now taken seriously in the developed countries. For instance, states in the United States have issued mercury-related advisories on fish consumption and ban mercury fever thermometers.

Methylated forms of mercury in the environment accumulate at the apices of food webs, which pose a health risk to children and pregnant women, especially those who eat fish. Organic mercury enters the food chain mainly by its ingestion by fish and other aquatic organisms, which, when consumed by humans, is easily absorbed by the gastric and intestinal organs and then transported in the blood to the brain, liver, kidney, and fetus [74]. Toxicity effects of mercury are confined primarily to the human central nervous system. Its effects are characterized by numbness and unsteadiness in the legs and hands, awkward movements, tiredness, ringing in the ears, narrowing of the field of vision, loss of hearing, sense of smell and taste, slurred speech, forgetfulness, and kidney damage [22,74]. Minamata is the Japanese name for the disease caused by eating mercury-contaminated fish or shellfish [74].

Mercury in surface soils is generally retained in the solid phase or volatilizes [4]. Although mercury toxicity in plants is essentially not known, plant-based systems are being used for the remediation of mercury-polluted soils. Two methods are being employed. The first involves the use of transgenic plants encoding the bacterial mercury ion reductase (merA) gene. These plants have been shown to grow in and volatilize mercury from soils [79]. Microorganisms can be manipulated genetically to remove not only mercury but also other toxic elements from the environment [80]. The second approach uses sulfur-containing solutions, such as ammonium thiosulfate, to induce mercury accumulation into aboveground tissues of high-biomass plant species [81]. In the latter system, mercury accumulates in the plant causing the plant to die. However, the mercury-laden plant, including roots, can be removed from the soil, thereby allowing mercury removal from the polluted soil. Root mercury accumulation and root area and length are related [82,83]. Plants with large root systems, therefore, are desirable for removal of mercury in contaminated soils.

9.3.13 Other Heavy Metals

Thallium is widespread and is taken up by some plants [84]. It is extremely toxic and has been used as rat poison and for the control of ants. It also has uses in the electronics industry for semiconductors, switches, and fuses. Rare earths, the name of the 14 metallic elements having atomic numbers 58 to 71, are being introduced to soil from electronic waste or so-called "e-waste" such as old computer components. Tyler [85] gives a review of the rare earth elements in soil and plant systems. More research is needed concerning these new sources of heavy metals and their potential toxicity.

9.4 SUMMARY

We have considered toxicity to plants of 12 heavy metals: copper, iron, manganese, molybdenum, zinc, cobalt, nickel, vanadium, cadmium, chromium, lead, and mercury. The first five in the list are essential for plants in trace amounts; the next three are essential in only certain plants; the last four in the list are not essential.

Toxic amounts of copper in soil or nutrient medium usually reduce growth and depress iron concentration in leaves causing chlorosis due to displacement of iron from physiologically active centers. Iron toxicity is not much in evidence under natural conditions. Most iron added to soil as an inorganic salt is quickly rendered insoluble in many soils and is no more available to plants than the native iron. Manganese toxicity mostly occurs in waterlogged environments and in acid soils. Excess manganese induces iron deficiency. The availability of most heavy metals decreases with increasing pH; the availability of molybdenum, however, increases with increasing pH. Indications of molybdenum excess in plants are rarely observed in the field. But large tissue concentrations of molybdenum induce molybdenosis in ruminants.

Zinc readily accumulates in the leaves of many plants, and zinc toxicity can be induced in nontolerant plants. It produces iron chlorosis. Cobalt is essential for blue-green algae and legumes when relying on nitrogen from nitrogen fixation and nitrogen-fixing bacteria. This is because cobalt is in the co-enzyme form of vitamin B_{12}, which is not produced by higher plants. Vitamin B_{12} is present in nodules and *Rhizobia*. Most plant species do not accumulate cobalt to any great extent and values exceeding 1 mg kg^{-1} dry matter are rare. Even though higher plants do not require cobalt, animals require it. Nickel is necessary for urease activity and prevents urea accumulation in legumes. Toxic properties of excessive amounts of nickel for plants have been documented using culture solutions. Nickel is readily taken up by plants and excess nickel produces chlorosis. Vanadium has not been shown to be essential for higher plants, but some green algae require it. Vanadium can be toxic to higher plants, but only under experimental conditions.

Cadmium is usually less adsorbed by soil and organic matter than by several other heavy metals (e.g., lead, copper), which makes it more available to plants and more easily leached to ground water. Cadmium can be transported readily from the soil to the root and to the upper plant parts. The particular hazard with cadmium is that the plant does not necessarily act as an indicator of levels toxic to humans and animals, because plants tolerate higher levels of cadmium than do animals. The same is true for mercury. Plants can appear healthy but may contain high concentrations of cadmium and mercury that are toxic to animals.

Chromium is toxic to plants. The dichromate ($Cr_2O_7^{2-}$) has been found to be more toxic than the chromate (CrO_4^{2-}) ion. It tends to accumulate in roots and the main toxic effect is exerted in roots. Most of the lead added to soil is fixed within a few days. Lead contaminates the outer plant parts along highways due to atmospheric deposition. Organic lead is mobile in the soil and is taken up by plants much more rapidly than inorganic lead. Mercury in surface soils is generally retained in the solid phase or volatilizes. Mercury toxicity in plants is essentially not known. However, mercury is highly toxic to animals and humans. Mercury can be removed from soil by adding sulfur-containing solutions, such as ammonium thiosulfate, to induce mercury accumulation into above-ground tissues of plants.

9.5 FUTURE RESEARCH PERSPECTIVES

Many questions remain regarding effects of heavy metals on plant growth. Are cobalt, nickel, and vanadium essential in plants other than the ones so far studied? Do heavy metals that are considered toxic at high concentration promote growth at lower concentrations? For example, erratic reports throughout the decades have indicated that cadmium promotes growth at low concentrations. If so, should it be considered an essential element at low concentrations? As analytical techniques become more sensitive, small concentrations of heavy metals can be determined under carefully controlled conditions. Why does a plant take up more of one toxic metal than another? For example, why is a species a hyperaccumulator of nickel but not of cadmium? How important is adsorption of metals onto the surface of a root? Once taken up, how is a metal partitioned between cell walls and the transpiration stream? How does the transpiration rate affect the uptake of metals? Are heavy metals taken up at night when stomata are closed? Can food plants be identified that grow in lunar soils, which contain deleterious concentrations of heavy metals? Such plants might be grown on space stations on the moon. What is the danger of the new sources of heavy metals, such as those from the electronic wastes? Can we identify plants that can take them up? Can we identify more hyperaccumulator plants, for example, those that might clean up metal-contaminated land around abandoned mines? Can we find hyperaccumulators that have high biomass and, therefore, will take up large amounts of heavy metals? Can we do this through traditional breeding methods? If so, then genetic modification, with its environmental concerns, might not be needed for hyperaccumulation.

REFERENCES

1. Kirkham, M.B., Trace elements. In *Encyclopedia of Earth Sciences Series*, Vol. 12, *The Encyclopedia of Soil Science, Part 1. Physics, Chemistry, Biology, Fertility, and Technology*, Fairbridge, R.W. and Finkl, C.W., Eds., Dowden, Hutchinson and Ross, Stroudsburg, PA, 1979, p. 571.
2. Page A.L., *Fate and Effects of Trace Elements in Sewage Sludge When Applied to Agricultural Lands. A Literature Review Study*, EPA-670/2-74-005, Office of Research and Development, National Environmental Research Center, U.S. Environmental Protection Agency, Cincinnati, OH, 1974, p. 98.
3. Beeson, K.C., *The Mineral Composition of Crops with Particular Reference to the Soils in Which They Were Grown*, Miscellaneous Publication No. 369, U.S. Department of Agriculture, Washington, DC, 1941, p. 164.
4. Chapman, H.D., Ed. *Diagnostic Criteria for Plants and Soils*. 2d printing, Quality Printing, Abilene, TX, 1973, p.793.
5. Kirkham, M.B., Trace elements in sludge on land: Effects on plants, soils, and ground water. In *Land as a Waste Management Alternative*, Loehr, R.C., Ed., Ann Arbor Science, Ann Arbor, MI, 1977, 209.
6. Bowen, H.J.M., *Trace Elements in Biochemistry*, Academic, London, 1966, p. 241.
7. Nomiyama, K., Nomiyama, H., Liu, S., Ishimaru, Y., and Hirai, M., Trace elements in cardio-cerebrovascular diseases. In *The Third International Conference on Nutrition in Cardio-Cerebrovascular Diseases*, Lee, K.T., Oike, Y., and Kanazawa, T., Eds., New York Academy of Sciences, New York, 1993, p. 308.
8. Buchanan, B.B., Gruissem, W., and Jones, R.L., Eds., *Biochemistry and Molecular Biology of Plants*. American Society of Plant Physiology, Rockville, MD, 2000, p. 1367.
9. Tinker, P.B. and Nye, P.H., *Solute Movement in the Rhizosphere*. Oxford University Press, New York, 2000, p. 444.
10. Alloway, B.J., Cadmium. In *Heavy Metals in Soils*, Alloway, B.J., Ed., Blackie, Glasgow, 1990, p. 100.
11. Panou-Filotheou, H., Bosabalids, A.M., and Karataglis, S., Effects of copper toxicity on leaves of oregano (*Origanum vulgare* susp *Hirtum*). Ann. Bot., 88, 207, 2001.
12. Reuther, W. and Labanauskas, C.K., Copper. In *Diagnostic Criteria for Plants and Soils*, Chapman, H.B., Ed., Quality Printing, Abilene, TX, 1973, p. 157.
13. Srivastava, P.C. and Gupta, U.C., *Trace Elements in Crop Production*. Science Publishers, Lebanon, NH, 1996.
14. Hall, J.L., Cellular mechanisms for heavy metal detoxification and tolerance. *J. Exp. Bot.*, 53:1, 2002.
15. Demidchik, V.V., Sokolik, A.L., and Yurin, V.M., Characteristics of non-specific permeability and H^+-ATPase inhibition induced in the plasma membrane of *Nitella flexilis* by excessive Cu^{2+}. *Planta*, 212:583, 2001.

16. Demidchik, V.V., Sokolik, A., and Yurin, V., The effect of Cu^{2+} on ion transport systems of the plant cell plasmalemma. *Plant Physiol.*, 114:1313, 1997.
17. Wallihan, E.F., Iron. In *Diagnostic Criteria for Plants and Soils*, Chapman, H.D., Ed., Quality Printing, Abilene, TX, 1973, p. 203.
18. McBride, M.B., *Environmental Chemistry of Soils*, 1st ed. Oxford University Press, New York, 1994, p. 406.
19. Schmidt, W., Review: Mechanisms and regulation of reduction-based iron uptake in plants. *New Phytol.*, 141:1, 1999.
20. Schützendübel, A. and Polle, A., Plant responses to abiotic stresses: Heavy metal-induced oxidative stress and protection by mycorrhization. *J. Exp. Bot.*, 53:1351, 2002.
21. Laan, P., Smolders, A., and Blom, C.W.P.M., The relative importance of anaerobiosis and high iron levels in the flood tolerance of *Rumex* species. *Plant Soil*, 136,:53, 1991.
22. United States Environmental Protection Agency, *Technical Support Document for Land Application of Sewage Sludge*, Vol. I (Revised). Office of Water, U.S. Environmental Protection Agency, Washington, DC, 1992, p. 821. (Available from the National Technical Information Service, Springfield, Virginia 22161. Order document no. PB93-110575.)
23. Labanauskas, C.K., Manganese. In *Diagnostic Criteria for Plants and Soils*, Chapman, H.D., Ed., Quality Printing, Abilene, TX, 1973, p. 264.
24. Hopkins, W.G., *Introduction to Plant Physiology*, 1st ed., Wiley, New York, 1995, p. 79.
25. Horst, W.J., The physiology of Mn toxicity. In *Manganese in Soils and Plants*, Graham, R.D., Hannam, R.J., and Uren, N.C., Eds., Kluwer Academic, Dordrecht, The Netherlands, 1988, p. 175.
26. Horst, W.J. and Fecht, M., Physiology of manganese toxicity and tolerance in *Vigna unguiculata* Walp. *Soil Sci. Plant Nutrition*, 162:263, 1999.
27. Wang, Y.X., Wu, P., Wu, Y.R., and Yan, X.L., Molecular marker analysis of manganese toxicity tolerance in rice under greenhouse conditions. *Plant Soil*, 238:227, 2002.
28. Johnson, C.M.. Molybdenum. In *Diagnostic Criteria for Plants and Soils*, Chapman, H.D., Ed., Quality Printing, Abilene, TX, 1973, p. 286.
29. Marschner, H., *Mineral Nutrition of Higher Plants*, 2nd ed., Academic, An Elsevier Science Imprint, San Diego, 2002, p. 889.
30. Hale, K.L., Tufan, H.A., Pickering, I.J., George, G.N., Terry, N., Pilon, M., and Pilon-Smits, E.A.H., Anthocyanins facilitate tungsten accumulation in *Brassica*. *Physiol. Plant*, 116:351, 2002.
31. Mengel, K. and Kirkby, E.A., *Principles of Plant Nutrition*, 5th ed., Kluwer Academic, Dordrecht, The Netherlands, 2001, p. 849.
32. Chapman, H.D., Zinc. In *Diagnostic Criteria for Plants and Soils*, Chapman, H.D., Ed., Quality Printing, Abilene, TX, 1973, p. 484.
33. Mikkelsen, R.L., Nutrient management for organic farming: A case study. *J. Nat. Res. Life Sci. Edu.*, 29:88, 2000.
34. Bekiaroglou, P. and Karataglis, S., The effect of lead and zinc on *Mentha spicata*. *J. Agron. Crop Sci.,* 18:201, 2002.
35. Khurana, N. and Chatterjee, C., Influence of variable zinc on yield, oil content, and physiology of sunflower. *Commun. Soil Sci. Plant Anal.*, 32:3023, 2001.
36. Holm-Hansen, O., Gerloff, G.C., and Skoog, F., Cobalt as an essential element for blue-green algae. *Physiol. Plant*, 7:665, 1954.

37. Gerloff, G.C., Comparative mineral nutrition of plants. *Ann. Rev. Plant. Physiol.*, 14:107, 1963.
38. Vanselow, A.P., Cobalt. In *Diagnostic Criteria for Plants and Soils*, Chapman, H.D., Ed., Quality Printing, Abilene, TX, 1973, p. 142.
39. Brooks, R.R., General introduction. In *Plants that Hyperaccumulate Heavy Metals*, Brooks, R.R., Ed., CAB International, Wallingford, Oxon, UK, 1998, p. 1.
40. Eskew, D.L., Welch, R.M., and Norvell, W.A., Nickel in higher plants. Further evidence for an essential role. *Plant. Physiol.*, 76:691, 1984.
41. Welch, R.M., The biological significance of nickel. *J. Plant Nutrition*, 3:345, 1984.
42. Vanselow, A.P., Nickel. In *Diagnostic Criteria for Plants and Soils*, Chapman, H.D., Ed., Quality Printing, Abilene, TX, 1973, p. 302.
43. Barbafiere, M., The importance of nickel phytoavailable chemical species characterization in soil for phytoremediation applicability. *Int. J. Phytoremed.*, 2:105, 2000.
44. Robertson, A.I., The poisoning of roots of Zea mays by nickel ions, and the protection afforded by Mg and Ca. *New Phytol.*, 100:173, 1985.
45. Anderson, A.J., Meyer, D.R., and Mayer, F.K., Heavy metal toxicities; levels of Ni, Co, and Cr in the soil and plants associated with visual symptoms and variation in growth of an oat crop. *Aust. J. Agric. Res.*, 24:557, 1973.
46. Pratt, P.F., Vanadium. In *Diagnostic Criteria for Plants and Soils*, Chapman, H.D., Ed., Quality Printing, Abilene, TX, 1973, p. 480.
47. Salt, D.E., Pickering, I.J., Prince, R.C., Gleba, D., Dushenkov, S., Smith, R.D., and Raskin, I., Metal accumulation by aquacultured seedlings of Indian mustard. *Environ. Sci. Technol.*, 31:1636, 1997.
48. Salt, D.E., Prince, R.C., Pickering, I.J., and Raskin, I., Mechanisms of cadmium mobility and accumulation in Indian mustard. *Plant Physiol.*, 109:1427, 1995.
49. Bhattacharyya, M.H., Whelton, B.D., Peterson, D.P., Carnes, B.A., Moretti, E.S., Toomey, J.M., and Williams, L.L., Skeletal changes in multiparous mice fed a nutrient-sufficient diet containing cadmium. *Toxicology*, 50:193, 1998.
50. Kim, Y.-Y., Yang, Y.-Y., and Lee, Y., Pb and Cd uptake in rice roots. *Physiol. Plant*, 116:368, 2002.
51. Bastian, R.K., Biosolids management in the United States, *Water Environ. Technol.*, 9(5):45, 1997.
52. United States Environmental Protection Agency, *Land Application of Sewage Sludge, A Guide for Land Appliers on the Requirements of the Federal Standards for the Use or Disposal of Sewage Sludge*, 40 CFR Part 503, EPA/831-B-93-002b, Office of Enforcement and Compliance Assurance, U.S. Environmental Protection Agency, Washington, DC, 1994.
53. Robinson, B.H., Mills, T.M., Petit, D., Fung, L.E., Green, S.R., and Clothier, B.E., Natural and induced cadmium-accumulation in poplar and willow: Implications for phytoremediation. *Plant Soil*, 227:301, 2000.
54. United States Department of Agriculture, Getting cadmium out of sunflower seeds. *Agric. Res.*, 43(1):21, 1995.
55. Salt, D.E., Blaylock, M., Kumar, N.P.B.A., Viatcheslav, D., and Ensley, B.D., Phytoremediation: A novel strategy for the removal of toxic metals from the environment using plants. *Biotechnology*, 13:468, 1995.
56. Basta, N.T. and Sloan, J.J., Bioavailability of heavy metals in strongly acidic soils treated with exceptional quality biosolids. *J. Environ. Quality*, 28:633, 1999.

57. McLaughlin, M.J., Hamon, R.E., McLaren, R.G., Speir, T.W., and Rogers, S.L., Review: A bioavailability-based rationale for controlling metal and metalloid contamination of agricultural land in Australia and New Zealand. *Aust. J. Soil Res.* 38:1037, 2000.
58. Perronnet, K., Schwartz, C., Gérard, E., and Morel, J.L., Availability of cadmium and zinc accumulated in the leaves of *Thlaspi caerulescens* incorporated into soil. *Plant Soil*, 227:257, 2000.
59. Gonzalez, R.X., Sartain, J.B., and Miller, W.L., Cadmium availability and extractability from sewage sludge as affected by waste phosphatic clay. *J. Environ. Quality*, 21:272, 1992.
60. Dudka, S. and Chlopecka, A., Effects of solid-phase speciation on metal mobility and phytoavailability in sludge-amended soil. *Water Air Soil Pollution*, 51:153, 1990.
61. Bell, P.F., James, B.R., and Chaney, R.L., Heavy metal extractability in long-term sewage sludge and metal salt-amended soils. *J. Environ. Quality*, 20:481, 1991.
62. Poschenrieder, C., Gunsé, B., and Barceló, J., Influence of cadmium on water relations, stomatal resistance, and abscisic acid content in expanding bean leaves. *Plant Physiol.*, 90:1365, 1989.
63. Baryla, A., Carrier, P., Franck, F., Coulomb, C., Sahut, C., and Havaux, M., Leaf chlorosis in oilseed rape plants (*Brassica napus*) grown on cadmium-polluted soils: Causes and consequences for photosynthesis and growth. *Planta*, 212:696, 2001.
64. Bazzaz, F.A., Rolfe, G.L., and Carlson, R.W., Effects of cadmium on photosynthesis and transpiration of excised leaves of corn and sunflower. *Physiol. Plant.*, 32:373, 1974.
65. Kirkham, M.B., Water relations of cadmium-treated plants. *J. Environ. Quality*, 7:334, 1978.
66. Baszynski, T., Wajda, L., Krol, M., Wolinska, D., Krupa, Z., Tukendorf, A., Photosynthetic activities of cadmium-treated tomato plants. *Physiol. Plant.*, 48:365, 1980.
67. Haag-Kerwer, A., Schafer, H.J., Heiss, S., Walter, C., and Rausch, T., Cadmium exposure in *Brassica juncea* causes a decline in transpiration rate and leaf expansion without effect on photosynthesis. *J. Exp. Bot.*, 50:1827, 1999.
68. Rivera-Becerril, F., Calantzis, C., Turnau, K., Caussanel, J.-P., Belimov, A.A., Gianinazzi, S., Strasser, R.J., and Gianinazzi-Pearson, V., Cadmium accumulation and buffering of cadmium-induced stress by arbuscular mycorrhiza in three *Pisum sativum* L. genotypes. *J. Exp. Bot.*, 53:1177, 2002.
69. Kupper, H., Kupper, F., and Spiller, M., *In situ* detection of heavy metals substituted chlorophylls in water plants. *Photosyn. Res.*, 58:123, 1998.
70. Arisi, A.-C.M., Mocquot, B., Lagriffoul, A., Mench, M., Foyer, C.H., and Jouanin, L., Responses to cadmium in leaves of transformed poplars overexpressing -glutamylcysteine synthetase. *Physiol. Plant.*, 109:143, 2000.
71. Pratt, P.F., Chromium. In *Diagnostic Criteria for Plants and Soils*, Chapman, H.D., Ed., Quality Printing, Abilene, TX, 1973, p. 136.
72. Brewer, R.F., Lead. In *Diagnostic Criteria for Plants and Soils*, Chapman, H.D., Ed., Quality Printing, Abilene, TX, 1973, p. 213.
73. Badawy, S.H., Helal, M.I.D., Chaudri, A.M., Lawlor, K., and McGrath, S.P., Soil solid-phase controls lead activity in soil solution. *J. Environ. Quality*, 31:162, 2002.
74. Ogola, J.S., Mitullah, W.V., and Omjula, M.A., Impact of gold mining on the environment and human health: A case study in the Migori gold belt, Kenya. *Environ. Geochem. Health*, 24:141, 2002.

75. United Nations, *Global Opportunities for Reducing Use of Leaded Gasoline*. Interorganization for the Sound Management of Chemicals/United Nations Environmental Program/CHEMICALS/98/9, United Nations, Geneva, Switzerland, 1998, p. 59.
76. Kacabova, P. and Natr, L., Effect of Pb on growth characteristics and chlorophyll content in barley seedlings. *Photosynthetica*, 20:411, 1986.
77. Poskuta, J.W., Parys, E., and Romanovska, E., Effects of lead on the gaseous exchange and photosynthetic carbon metabolism of pea seedlings. *Acta Bot. Pol.*, 57:149, 1987.
78. Liphadzi, M.S. and Kirkham, M.B., Phytoremediation of soil contaminated with heavy metals: A technology for rehabilitation of the environment. *South Afr. J. Bot.*, 71:24, 2005.
79. Meagher, R.B., Rugh, C.L., Kandasamy, M.K., Gragson, G., and Wang, N.J., Engineered phytoremediation of mercury pollution in soil and water using bacterial genes. In *Phytoremediation of Contaminated Soil and Water*, Terry, N. and Bañuelos, G., Eds., Lewis, Boca Raton, FL, 2000, p. 201.
80. Wood, J.M. and Wang, H.-K., Microbial resistance to heavy metals. *Environ Sci. Technol.*, 17:582A, 1983.
81. Moreno, F.N., Anderson, C.W.N., Stewart, R.B., and Robinson, B.H., Phytoremediation of mercury-contaminated mine tailing by induced plant-mercury accumulation. *Environ. Pract.*, 6:57, 2004.
82. Cocking, D., Rohrer, M., Thomas, R., Walker, T.J., and Ward, D., Effects of root morphology and Hg concentration in the soil on uptake by terrestrial vascular plants. *Water Air Soil Pollution*, 80:1113, 1995.
83. Heeraman, D.A., Claassen, V.P., and Zasoski, R.J., Interaction of lime, organic matter and fertilizer on growth and uptake of arsenic and mercury by Zorro fescue (*Vulpia myuros* L.). *Plant Soil*, 234:215, 2001.
84. Leblanc, M., Petit, D., Deram, A., Robinson, B.H., and Brooks, R.R., The phytomining and environmental significance of hyperaccumulation of thallium by Iberis intermedia from Southern France. *Environ. Geol.*, 94:109, 1999.
85. Tyler, G., Rare earth elements in soil and plant systems — A review. *Plant Soil*, 267:191, 2004.

10 Growth and Function of Roots under Abiotic Stress in Soils

Hong Wang and Akira Yamauchi

CONTENTS

10.1 Introduction 271
10.2 Morphological Adaptation of Roots to Abiotic Stresses
      and Its Regulation 272
      10.2.1 Drought 272
      10.2.2 Soil Compaction 274
      10.2.3 Waterlogging 275
      10.2.4 Salinity 280
10.3 Physiological and Biochemical Adaptation of Roots to Abiotic
      Stresses and Its Regulation 282
      10.3.1 Stress Signaling 282
              10.3.1.1 Drought 282
              10.3.1.2 Soil Compaction 284
              10.3.1.3 Waterlogging 285
              10.3.1.4 Salinity 287
      10.3.2 Osmotic Adjustment 288
      10.3.3 Metabolism of Reactive Oxygen Species (ROS) 290
      10.3.4 Uptake of Nutrient Ions 294
              10.3.4.1 Drought 294
              10.3.4.2 Soil Compaction 295
              10.3.4.3 Waterlogging 296
              10.3.4.4 Salinity 297
10.4 Summary 298
References 299

10.1 INTRODUCTION

Soil is an extremely complex growth medium with enormous biodiversity and wide variation in mineral and organic matter compositions. Soil provides plants with water, nutrients, and a secure anchorage, but may also restrict root growth in unfavorable physical or chemical conditions, as roots are the underground structure of plants directly in contact with soils. Plant roots exhibit various adaptive responses to soil stresses. They are the first to encounter stress factors in the soil and thus are potentially the first sites of damage or the first lines of defense in the plants exposed to soil stress. Abiotic stresses of the soil include the availability of water and oxygen, high mechanical resistance, heavy metal toxicities, nutrient deficiencies, and salinity toxicity around roots. A better understanding of the mechanisms of root morphological, physiological, and molecular responses would expand our understanding of whole plant responses to abiotic stresses. In this chapter, we review recent research on morphological, physiological, and cellular responses of roots to major soil abiotic stresses, including water (drought, submergence), compaction, and salinity stresses.

10.2 MORPHOLOGICAL ADAPTATION OF ROOTS TO ABIOTIC STRESSES AND ITS REGULATION

Root growth is a complicated process, including cell division in the apical meristem just behind the root tip and cell expansion in the zone behind the root apex which is driven by cell turgor pressure produced by water influx into root cells.

10.2.1 Drought

Root growth is most commonly reduced by soil drying, but is usually less inhibited than shoot growth, and may even be promoted, resulting in an increase of root-to-shoot ratio under limited water content of soils or at low soil water potentials [1–4]. Sharp and Davies [5] attributed such effects to relatively more dry matter allocation into the root fraction, whereas shoot growth is limited by water deficit in the absence of any effect on carbon gain. The discrepancy in the literature on the effects of water stress on root growth is attributable to the differences in genetics (species or cultivars), stress intensities, as well as types of roots examined among the studies. In general, both root and shoot growth are reduced at water potentials lower than –1.5 MPa [6,7].

Moreover, there are spatial and temporal variations in growth rates within different parts of the root [8–10]. The elongation rate was unaffected by a low water potential in the apical 3 mm of primary root of maize (*Zea mays*) seedling, but it was progressively inhibited at more basal locations compared with those of the well-watered roots, and increased to a peak at 4.5 mm and then gradually declined beyond 11 mm [10]. Cell expansion in longitudinal and radial directions of the root can be regulated independently [10]. In fact, water-stressed roots became substantially thinner than well-watered roots, suggesting that root thinning is an adaptive response to water stress [7,8,10]. The thinning of roots is suggested to be caused by the decrease in rates of tangential and radial expansion in both the stele and cortex under water stress, but only in the apical 5 mm of a root; the lateral expansion rates at the basal part were similar in well-watered and water-stressed roots [10].

Other important responses of root systems to soil drying include enhanced geotropism, increased branching, and deep rooting, which facilitates plants to exploit larger soil volume [11,12]. Serraj et al. [13] suggested that deep and prolific root systems have been associated with enhanced avoidance of drought stress in chickpea (*Cicer arietinum*). In fact, many field studies with various crops have shown that dense root systems that may extract more water in upper soil layers, along with longer root systems that may extract soil moisture from deeper soil layers, are both important for plants to adapt to drought stress [14,15]. In rice (*Oryza sativa*) traits such as deep and fine roots have been associated with increased water extraction during progressive water stress [16–19]. A high ratio of deep root-to-shoot weight was also found to be associated with higher plant water potentials and have a positive effect on yield under water stress [20].

The extensiveness of a root system can be quantified by root-length density (RLD) defined as root length per unit soil volume (cm root cm^{-3} soil) [21]. Water uptake rate of root systems and plant drought resistance may be highly correlated with RLD. Plants with a greater RLD in the deep soil layer are better able to maintain water status and stomatal conductance when the soil becomes dry than those with a lower RLD [22,23]. Siopongco et al. [24] proposed that the critical RLD for water uptake differs with rice plants that are adapted to different ecosystems. Specific root length (SRL, root length per unit root mass) of cauliflower (*Brassica oleracea*) was lower under drought stress conditions, which was associated with a higher dry matter accumulation in the fine root fraction; in addition, the vertical increment of rooting depth per day almost doubled under drought stress conditions [25].

Yamauchi et al. [26] suggested that the plastic response in root branching triggered by water stress is an important adaptive mechanism. Ogawa et al. [27] showed that sugar accumulation in the seminal root axis of maize plants in response to osmotic stress might be one of the possible physiological bases for plastic lateral root development. In addition, under field conditions, soil moisture usually fluctuates rather than simply progressively drying. For example, under rainfed lowland conditions that are characterized by soil moisture fluctuation between aerobic and anaerobic conditions, Bañoc et al. [4] and Kamoshita et al. [18] found the cultivars that are well adapted to rainfed lowland conditions exhibit plastic development in seminal roots and their lateral branches in response to moisture fluctuations. Similarly, Pardales et al. [28] and Pardales and Yamauchi [29] also suggested that the developmental responses of adventitious roots and their laterals to fluctuating soil moisture were the determining traits for the crop adaptation to such conditions in sweet potato (*Ipomoea batatas*) and cassava (*Manihot esculenta* Crantz).

The initiation and growth of root hairs are stimulated when soil dries [30–32]. This may be an adaptive response to soil water shortage [23]. Increase in root hairs in dry soil can increase total root surface area, which in turn, enhances water uptake from soil. In addition, root hairs can be sites for extensive mucilage production [33]. Mucilage can enhance the ability of root hairs to attach to soil particles to form soil sheath, thereby preventing air gaps from developing between soil and root surface when soil dries. There may also be a reduction in water efflux from plants into drying soils, which may ultimately delay root desiccation [34–36].

10.2.2 Soil Compaction

Soil compaction is a major soil stress that restricts root growth [37–39]. This stress is often associated with soil drought and anaerobiosis. Limited root elongation in compacted soils may be due to a decrease in both the rate at which new cells are added onto a file (i.e., cell flux) and final cell length [38]. Soil compaction seems to mostly affect cells within the root elongation zone. The reduction in cell elongation may result from reduced cell turgor either from high cell wall rigidity or from a direct effect of mechanical forces on cell wall rheology [40]. Roots that encounter mechanical impedance tend to increase the accumulation of solutes in cells, which can cause increases in cell turgor pressure [41,42]. Turgor pressure is considered to play an important role in the penetration of the roots into the soil [43].

Roots may become thicker in response to soil compaction [38,44,45]. Materechera et al. [37] reported that seedling root diameter of 22 plant species increased by up to twofold under mechanical impedance. Bengough et al. [43] showed that root diameter increased at only one location behind the root apex. They found that root diameter increased rapidly within 2 mm of the apex and remained relatively constant up to 10 mm from the apex in unimpeded roots of pea (*Pisum sativum*); however, in mechanically impeded roots, diameter continued to increase for up to 10 mm behind the apex. This radial thickening in mechanically stressed roots is due mainly to the increase in the cell diameter of the cortex, but extra periclinal divisions in the cortex may also lead to extra cell files. The cells in the outer layers of root cortex increase in diameter much more than cells in inner layers, probably due to the confining pressure of the cells surrounding the inner layers [46]. The increase in root diameter in response to soil compaction is caused by cortical cells enlarging radially rather than axially, with a corresponding change in the longitudinal orientation of the cellulose microfibrils in the cell walls [47].

Differences in both root elongation and root diameter were also observed among many plant species. Generally, roots of dicotyledons (with large diameters) penetrate compacted medium more easily than those of graminaceous monocotyledons (with smaller diameters). There was a significant positive correlation between root diameter and elongation in different plant species exposed to compacted soils [37,44]. Lupin (*Lupinus angustifolius*), medic (*Medicago scutelata*), and faba bean (*Vicia faba*) were ranked higher in relative root elongation and root thickening compared to wheat (*Triticum aestivum*), rhodesgrass (*Chloris gayana*), and barley (*Hordeum vulgare*). Seed weight did not seem to influence either root thickening or elongation [37].

The root system growing in hard soils appears stunted, but often has increased lateral branching, compared with those grown in loose soils [38]. Lateral roots are finer than the parent root axis and are therefore capable of penetrating smaller pores.

This response presumably optimizes the exploratory capabilities of the root system as a whole in compacted soils [39]. Primary root axes may be impeded by pore size that permits the smaller-diameter laterals to penetrate relatively easily [48]. Thickening of root tips is less likely to assist penetration of lateral roots in hard soils [49]. More lateral roots per unit length of parent root axis were found in pea and wheat seedlings grown in compacted sandy loam soils [50]. A similar tendency was found in the roots of lupins growing in compacted sandy clay loam soil, despite the small change in total number of lateral roots per unit length of the main axis [51]. However, the total number of lateral roots decreased to half in barley grown in compacted glass beads [48]. In young maize plants growing in a rhizotron experiment, increasing soil bulk density reduced seminal root growth only in the lowest soil layer and the number of first-order lateral roots originating from these seminal roots, whereas it did not influence the length of individual first-order lateral roots [52].

Bengough [38] and Iijima et al. [44] suggested that there were compensatory adjustments in growth between the main root axis and its lateral roots in response to soil compaction. For example, in barley, the main axial root length was reduced by a uniform increase in soil strength to a greater extent than lateral root length. The reduction in main root length under high soil strength was associated with a 29% increase in mean length per lateral root [53]. The ratio of first-order laterals producing higher-order laterals on their axes in rice was greater in compacted soils, whereas that of maize was not significantly increased, but the main root axes of maize penetrated deeper than those of rice during four weeks growth [44].

Root hairs and root caps also are useful for roots to penetrate into compacted soils by providing anchorage and mucilage [54–56]. The decapped roots of maize were significantly more sensitive than intact roots to the effect of mechanical impedance [56]. It has been suggested that the root cap reduces soil mechanical impedance by means of its secretion of slimy mucilage and by the sloughing of border cells [45,57,58].

Inukai et al. [59] found a recessive rice mutant, *rrl3*, which was highly sensitive to the mechanical stimulus and developed short roots in response to the mechanical stimulus. No significant difference was observed between the seminal roots of *rrl3* mutant and wild-type plants in the mean axial and radial length of mature cortical cells. However, the meristematic zone of the root was smaller and the cortical cell flux in the growing zone of the root was significantly lower in the mutant than in the wild-type plant. In addition, the *rrl3* mutant and the wild-type plant did not differ in their sensitivity to ethylene, IAA, or ABA. These results suggested that the *RRL3* gene specifically regulates the cell production process in root meristematic zone in response to mechanical stimuli and does not regulate the sensitivities to ethylene, IAA, and ABA [59].

10.2.3 Waterlogging

Oxygen diffusion rate is 10,000 times slower in water than in air [62], which is one of the main causes of injury to plant roots in waterlogged soils [60,61]. Oxygen depletion is further exacerbated by competition with aerobic microorganisms around the roots. Oxygen deficiency or hypoxia due to soil flooding limits root growth in both root length and mass [63]. Restricted root growth in flooded soil may also be associated with the production of ethylene in root apices [64,65] and the accumulation of CO_2 that is produced by roots and soil microorganisms. Flooding impedes the diffusive escape or oxidative breakdown of ethylene or CO_2 [66]. Oxygen depletion also decreases the redox potential in saturated soils and favors the activities of facultative anaerobes that chemically reduce nitrate to nitrite, then nitrous oxide, and finally to nitrogen gas by denitrification process, diminishing nitrate availability to roots [67]. Highly soluble and toxic Mn^{2+} and Fe^{2+} ions [68] and poisonous gas H_2S [69] are also produced in flooded soil, which interfere with enzyme activities and damage membranes in roots. Flooding may also increase the incidence of soil-borne fungal and bacterial diseases [70,71].

A short exposure to anoxic (no oxygen) conditions, such as a few minutes or hours, may lead to severe growth inhibition and cell death [61,63], which are generally attributed to either (1) the lack of ATP supply to roots or (2) self-poisoning of roots by products from anaerobic metabolism. First, ATP supply in anoxic conditions may not meet the demand of roots. ATP synthesis is achieved by mitochondria-based aerobic respiration via the oxidation of carbon sources using oxygen as an electron acceptor in an aerobic environment. In anaerobic roots, complete substrate oxidation is restricted by the lack of an electron acceptor. Production of ethanol and lactic acid through the fermentation pathways is one mechanism that organisms use to provide glycolytic substrate oxidation and ATP synthesis. This pathway yields only two ATPs from each glucose molecule, with only about 6% of the ATP being generated via the Krebs' cycle and the mitochondrial electron transport chain [63]. Second, anaerobic roots may also die from self-poisoning byproducts of anaerobic metabolism [63]. Low cytoplasmic pH is suggested to be the most toxic factor, which acidifies the cytoplasm and vacuole [72].

However, the sources of these excess protons and the mechanisms that regulate their concentration remain unclear [63]. Acetaldehyde [73] and nitric oxide [74] are also possible potential toxic factors. Subbaiah and Sachs [75] suggested that death of the root tip may be a less cell-autonomous process but more of a necrotic process. They suggest that acceleration of the process as well as making it more cell-autonomous, that is, pushing the process more toward a genetically programmed cell death, would provide a definite advantage during postanoxic recovery of maize seedlings. The loss of the root tip allows the remainder of the root (with their dormant lateral root primordia) to survive longer.

In response to the low oxygen concentration in the root zone, root apices acquire O_2 from other sources; otherwise, they will eventually die. Many plant species form aerenchyma tissues in roots, rhizome stems, and submerged leaves, through which oxygen moves from aerial shoots to organs submerged in water with less resistance [62,76]. Aerenchyma in roots provides an internal diffusion pathway from the aerial shoot to supply O_2 to the root apex for aerobic respiration, which allows roots to continue to grow [60,62,76]. Aerenchyma formation is widespread among wetland and some nonwetland species [77]. In roots

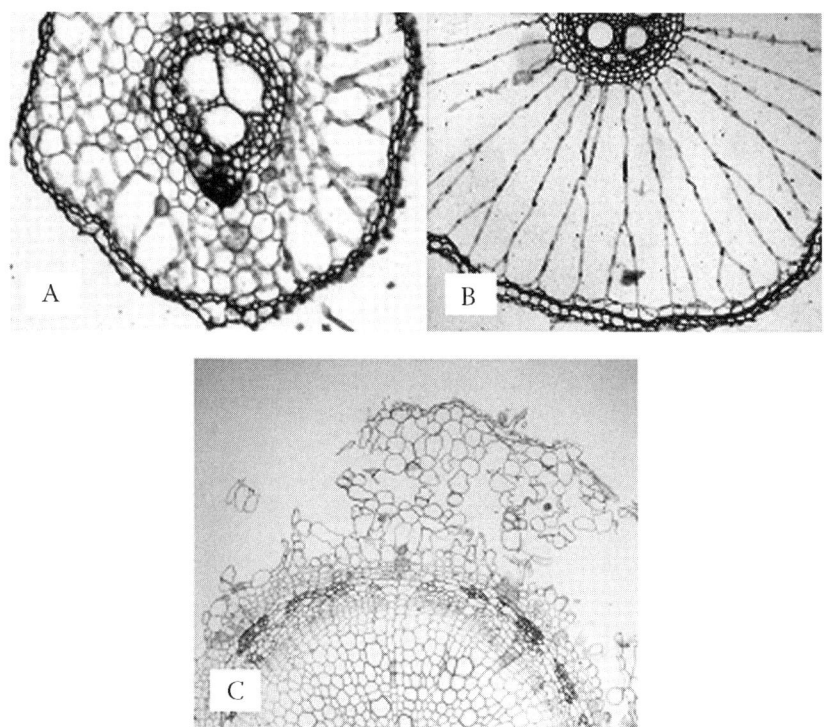

FIGURE 10.1 Light microscope photographs showing cortical aerenchyma of rice roots under drought (A) and flooded (B) conditions (photograph by Sularta and Yamauchi) and secondary aerenchyma in the taproot of *Sesbania cannabina* plant after 24 hours of flooding (C). (Photograph by Daimon.)

of nonwetland species such as wheat [78], barley [79], and maize [80], aerenchyma tissue is believed to be induced when plants are exposed to low O_2 concentrations. In wetland species, aerenchyma formation is constitutive; for example, in rice, cortical aerenchyma is formed regardless of soil moisture conditions, although flooding tends to promote the formation as shown in Figures 10.1A,B. Armstrong [81] and Colmer et al. [82] found that aerenchyma volume increased when rice roots were subjected to low O_2 concentrations. Kono et al. [83,84] proposed that when cortical aerenchyma is formed, some chemicals may be released from the cortex of the parent root, which may stimulate lateral root formation.

There are two types of aerenchyma in roots according to the position and the type of roots [85]. The most common type is cortical aerenchyma that is formed lysogenically or schizogenically and typically exhibited in many plant species such as maize [80], rice [86], barley [87], wheat [88], and some *Rumex* species [89]. The cortical aerenchyma provides a pathway with low resistance for the

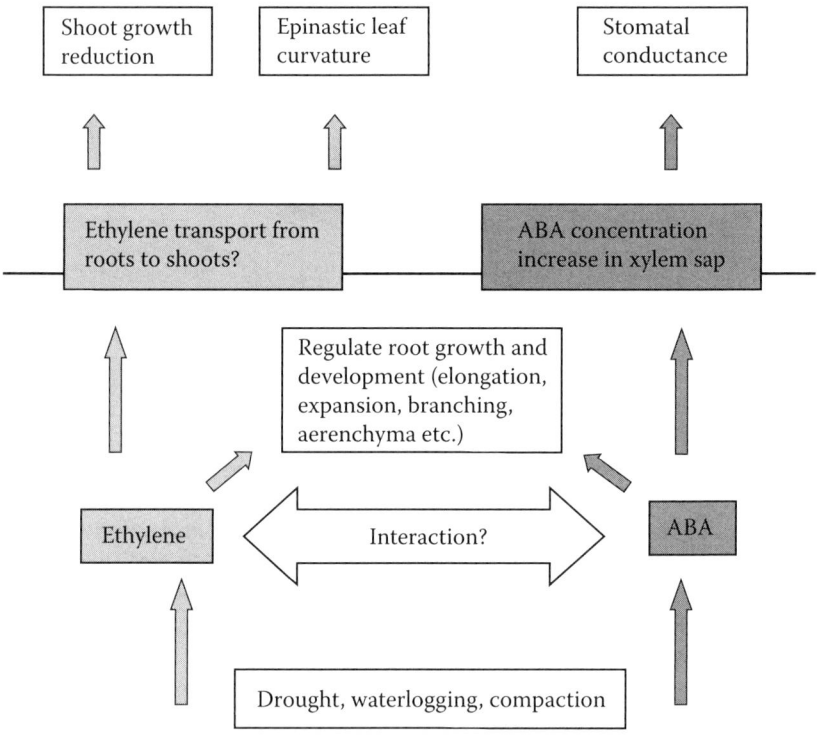

FIGURE 10.2 Schematic diagram showing the root-borne signals mediating root growth and shoot responses under soil abiotic stresses.

transport of oxygen. Ventilating pressure, which is the resistance to air flow through aerenchyma extending from shoots to roots in the plant body, is extremely low in plants with a high volume of aerenchyma [90]. The rate of radial oxygen loss (releasing oxygen) from the roots of intact rice plants was higher in very porous roots than in low porosity roots [81]. Lysigenous air-space formation arises from programmed cell death of targeted cells in a process mediated by ethylene accumulation under low (but not zero) O_2 concentration [63,91–93]. Evolution or application of ethylene leads to aerenchyma formation in maize [94] and wheat [88,95]. Jackson and Ricard [63] and Evans [93] showed several possible stages of lysigenous aerenchyma formation in roots of maize induced by partial oxygen shortage external to the root. The first stage is an increase of ethylene synthesis in roots when an oxygen shortage is perceived. Ethylene then signals a transduction cascade that probably involves Ca^{2+}, protein kinases, or others, and at last induces a form of programmed cell death in target cells of the cortex (see Chapter 7 for detailed information).

The other type of aerenchyma is a spongy tissue filled with gas spaces and is formed in the taproot, adventitious roots, and root nodules, which was often found in *Glycine soja* [96] and *Glycine max* [85], *Sesbania spp.* [97,98], *Lotus uliginosus* [99], and *Viminaria juncea* [100] when grown under flooded conditions. This type of aerenchyma is called secondary aerenchyma (Figure 10.1C) because it differentiated from secondary meristem (phellogen) [91]. Secondary aerenchyma has large intercellular spaces and consists of living cells whose walls do not become suberized, whereas phellem has no intercellular spaces and consists of dead cells whose walls become suberized [101,102]. Shiba and Daimon [98] observed the development of roots of *Sesbania cannabina* seedlings during the periods of up to 48 h of flooding. The outermost cells of the phellogen of the taproot of *S. cannabina* expanded and elongated during the first 12 h of flooding. After 18 h, the outermost of these regions was composed of cells that had expanded radially to form a spongy zone inside the endodermis. After 36 h, these elongated cells were radially connected to each other and formed the secondary aerenchyma surrounding the stele of taproot (Figure 10.1C).

Nitrogen and phosphorous starvation that commonly occur in flooded soils can also induce aerenchyma formation, such as found in maize and common bean genotypes [103–105]. Exposure of only a portion of adventitious root tips to deoxygenated agar nutrient solution while the rest of the root system was in aerated nutrient solution triggered the development of aerenchyma along the entire main root axes in wheat [106]. This study suggested that factors other than oxygen supply might promote aerenchyma formation.

The effectiveness of aerenchyma for oxygen transport may be affected by oxygen loss by outward radial diffusion in a stagnant deoxygenated solution [82]. Significant genotypic variations for high root porosity and low radial oxygen loss (ROL) were observed in rice when grown in stagnant deoxygenated solution [62]. A tight barrier to ROL in the basal regions of adventitious roots was observed, which under high root porosity provided a low resistance pathway for O_2 movement to the root tip and effective longitudinal O_2 diffusion toward the apex [62]. Some wetland species contained large volumes of aerenchyma exceeding 55%, and a barrier to ROL also occurred at the basal zones [107]. Low ROL in roots under flooded conditions may also prevent an influx of soil-derived gases such as CO_2, methane, and ethylene, potentially toxic substances (i.e., reduced metal ions), and even nutrient and water uptake [107].

Significant barriers to ROL induced by stagnant solution have been found in many wetland species, but not all wetland species developed ROL barriers [108]. Most dryland species, such as wheat [109], barley, *Secale cereale* [110], sorghum (*Sorghum bicolor*), and *Avena sativa* [111] may not develop ROL barriers, with the exception of *Critesion marinum*) [110]. When ROL occurs in the basal zones, it would substantially decrease longitudinal diffusion of oxygen to the root apex, therefore limiting the potential penetration depth of roots in an anaerobic soil [107]. However, for flood-tolerant species as demonstrated by *Typha domingensis*, ROL occurred at such a minimal amount that it did not contribute to internal deficiencies of oxygen associated with higher rate of alcohol fermentation [112]. Although oxygen leakage compromises the internal aeration of roots [82] and respiration efficiency, it has been considered as an adaptation mechanism for plants to reduce root susceptibility to toxic reduced compounds that accumulate in waterlogged soils.

Higher root porosity seems to correlate well with deep rooting ability [77]. Wetland species with high root porosity tend to have higher root penetration ability than nonwetland species with low root porosity in flooded soils. The production of aerenchymatous lateral roots in surface soils is also a strategy for long-term flood tolerance [77]. Rice is a typical example of a wetland species with a root system consisting of coarse, aerenchymatous, roots with gas-impermeable walls, which conduct O_2 to short, fine, and gas-permeable lateral roots. These lateral roots provide a greater absorbing surface per unit of aerated mass than thick main roots [113]. Greater root porosity and a barrier to radial oxygen loss presumably contributed to greater adventitious root length of *Critesion marinum* [110] and wheat [114] when grown in stagnant solution. In addition, aerenchyma formation was closely related with the production of new roots, and the presence of aerenchyma in the developing roots determined flooding tolerance in *Rumex* species [89].

Some flood-tolerant plants develop adventitious roots in the original root systems or submerged portions of stems when exposed to flooding [115,116]. Flood-induced adventitious roots may help plants to survive flooded conditions by increasing water and nutrient uptake from the surface soil [117], oxidation of the rhizosphere, and conversion of soil-borne toxins to less toxic compounds through increasing oxygen uptake [118], and supply of root-synthesized gibberellins and cytokinins [119]. Formation of adventitious roots has been attributed to increased ethylene production in flooded conditions [120]. Using transgenic plants (*etr1-1* gene from *Arabidopsis thaliana*; encoding for a defective ethylene receptor; *Tetr*) of *Nicotiana tabacum*, McDonald and Visser [121] suggested that interaction between ethylene and auxin may play an important role in waterlogging-induced adventitious root formation, most likely at the level of polar auxin transport. However, production of ethylene may not always correlate well with the formation of adventitious roots in some flooded plants [120].

10.2.4 SALINITY

Salt accumulation in soils can cause damages in root growth due to osmotic or water-deficit effect of salinity [122] or ion-excess effect of salinity [122,123]. Roots are usually less sensitive to salt stress than shoots, and thus root-to-shoot ratio often increases in plants grown in saline soils. This response may diminish the potential demand for nutrient and water supply to the shoot, and increase the ability of roots to take up more nutrient and water from soils [122,124].

Plants develop some root characteristics that facilitate tolerance or avoidance to salt stress [21,26,125–127]. For example, large root systems may retain salts in roots, which may reduce salt damage in shoots. Roots of some species control the loading of salts in the xylem, reducing salt uptake from the external solution [128,129]. Root architecture may also be altered during plant adaptation to salinity. The main root axis and lateral root development are affected differently by salinity [130]. NaCl at high concentrations may severely inhibit the growth extension of the main root in many plants [131], but hardly affect [132,133] or may even stimulate lateral root formation and growth [130,134]. However, Rubinigg et al. [130] showed that in hydroponically growing seedlings of halophyte *Plantago maritima*, primary root length increased by NaCl at concentrations ranging from 50 to 200 mM. Total lateral root length increased at 50 mM, was unaffected at 100 mM, and was considerably reduced at 200 mM NaCl. Moreover, NaCl severely reduced root branching pattern of the first-, second-, and third-order laterals [130], which might be the result of the inhibition of the initiation of lateral root primordia [135].

Root growth is usually restricted to the root axis near the tip. The apical meristem of a root is less affected by salt stress than the root elongation zone. Salinity may inhibit root elongation by a combined effect on the size of the growth zone as well as the magnitude of the localized tissue elongation [131,136,137]. Bernstein and Kafkafi [122] proposed four factors for the stress-induced inhibition of root elongation: rate of cell division, rate of cell expansion, duration of cell growth, and orientation of cell expansion. In principle, any of these processes could underlie root growth response to salinity. As cells are first produced prior to their expansion, it is plausible that the rate of cell production is a determining factor for root growth rate.

Salt stress inhibited the increase in cell number in cortical tissue in seminal roots of rice plants [138], cell production in the primary root of *Arabidopsis* [139], and apparent root cell production rate in cotton (*Gossypium hirsutum*) [140] and maize [141]. DNA synthesis in the S phase of the cell cycle was inhibited, and nuclei in the meristem were deformed and degraded in root cells cultured under salt stress [142,143]. A protein AtHAL3a is suggested to be involved in cell cycle regulation in growth rate. Transgenic *Arabidopsis* plants constitutively overexpressing *AtHAL3a* gene showed salt tolerance [144]. Trigoneline, a salt-stress osmoregulator in legumes, was found to function as a cell cycle regulator [145].

The inhibition of cell expansion in salt-stressed roots has also been demonstrated through different types of studies: (1) reduction of relative elemental growth rate in the root elongation zone; (2) rapid reduction of organ growth rates under salinity [146]; (3) rapid increase of root growth after removal of salt stress [147]; (4) changes in root growth rate following salinization and removal of salt stress; and (5) reduction in root cell size [140] and shortening of roots under salt stress [141,148].

Salt alters the mechanical properties of cell walls and affects cell expansion [122,149]. However, some researchers [138,139] consider that NaCl may stimulate the transition phase from cell division to cell elongation, resulting in the shortening of the meristematic zone and stopping of cell division at a smaller size. Moreover, early expansion of cells in roots could be an adaptive response to NaCl as it involves vacuolation [150,151]. Compartmentalization of sodium and chloride ions in the vacuole within the cytoplasm provides space for accumulation of excess ions and helps reduce their toxic effect. In excised pea roots, vacuolated cells were located much closer to the root apex under salt stress [152].

Salt might affect the orientation of root cell growth (growth anisotropy). Salinity caused roots to become thicker and shorter in barley [153] and cotton

[131], and retarded cell elongation and altered cell shape but not cell volume in cotton roots [140]. Alteration of cell growth anisotropy implies that NaCl affects the cytoskeleton [122].

In addition to effects on root morphology, salinity may also affect root anatomical features. Salt stress stimulated early development of cotton root endodermis and induced development of exodermis [133]. In halophyte *Suaeda maritime*, the Casparian strip was developed closer to the root apex [154]. Shannon et al. [155] showed that salinity promotes suberization of hypodermis and endodermis. Under saline conditions, rice roots exhibited cell-wall thickening in the epidermis, an increase in endoplasmic reticulum, myelin figures, less compact golgi bodies, and inhibited production of golgi vesicles [151].

Koyro et al. [137] found that the salt-induced changes in physiological parameters were correlated with the structural and ultrastructural changes in root cells of the xerophytes, but not in salt-sensitive sorghum plants. Salt treatment also increased the membrane surface in root cells and the quantity of vesicles in the epidermis and in the middle cortex cells. Additionally, some of the epidermal cells in salt-treated plants revealed a characteristic buildup of transfer cells, suggesting an increase in membrane surfaces to increase the uptake and storage of substances. The number of mitochondria increased in the epidermal and cortical cells under salt stress, indicating an additional supply of energy for osmotic adaptation and for selective uptake and transport of salt.

10.3 PHYSIOLOGICAL AND BIOCHEMICAL ADAPTATION OF ROOTS TO ABIOTIC STRESSES AND ITS REGULATION

10.3.1 STRESS SIGNALING

In the plants subjected to stressful conditions, a series of signaling molecules are activated and flow among signaling cascades. In addition, stress perception and signal transduction are complicated processes. The magnitude, duration, and frequency of the signaling cascades in cellular responses may vary with different stresses [156].

10.3.1.1 Drought

Abscisic acid (ABA) has been demonstrated to act as a chemical signal mediating plant responses to drought [157,158] (also see Chapter 5). As soil dries, ABA is believed to be produced in the roots and transported through the transpiration stream to the shoot where stomatal conductance and leaf growth can be regulated, independently of shoot water status [158,159].

Davies and Bacon [12] suggested that the increase in xylem sap pH may also be a root-borne signal but its effect may be exerted through modified ABA distribution. Bacon et al. [160] reported that the pH of xylem sap obtained from stems cut closer to the roots increased as soil dried. Increases in xylem sap pH from the root can be detected within hours, whereas significant rises in root-borne ABA may occur only after several days in contact with drying soil.

Why does xylem sap pH increase as soil dries? One explanation is that ATPase activity, which is responsible for maintaining the acidic pH of the apoplast in roots, decreases with soil drying. Another possible reason is the change in nitrate reductase activity, which is responsible for converting nitrate to nitrite in the root. Changes in nitrate reductase activity may change the ionic sap composition. Such changes may also influence the pattern of ionic exchange with the walls of root xylem and hence modify xylem sap pH and may affect exchange of other molecules (e.g., ABA) with the xylem parenchyma [161].

Other plant hormones, such as cytokinins and ethylene may also be involved in root-shoot signaling, either actingalone or in combination with others [162]. One example of the combined action of hormones in root-shoot communication is that increased concentration of cytokinins in the xylem sap promoted stomatal opening directly and decreased stomatal sensitivity to ABA (see review by Wilkinson and Davies [159]).

Ethylene production is suggested to play an important role in plant adaptation to drought stress [12,163]. Yang and Hoffman [164] reported that drought stress enhanced ethylene evolution in plants. Soil drying can increase the level of 1-aminocyclopropane-1-carboxylic acid (ACC), a precursor of ethylene, in roots and the xylem of shoots [165]. ACC may be transported from roots to shoots through transpiration streams [166]. Feeding ACC via the xylem to detached barley shoots can inhibit leaf growth [163], which may result in the reduction in transpirational demand. However, the possibility that a process in leaf tissues increases ethylene evolution in the absence of ACC supplied from the root cannot be excluded. In sunflower (*Helianthus annuus*) plants that maintained leaf turgor in drying soil (leaf turgor declined at -0.1 MPa d^{-1}), a twofold upregulation of ACC oxidase gene expression was observed in the leaves [167], which might be expected to enhance ethylene evolution. However, it is not yet clear whether export of the ethylene precursor, ACC, from roots acts as a long-distance signal of drying soil.

It seems likely that soil-derived chemicals play a role in signaling. In fact, Hartung et al. [168] detected abscisic acid in aqueous extracts of different soils that were used to grow various crop, pasture, and forest species. Under natural conditions, ABA and its conjugates were present in the soil solution of moist soils at concentrations up to 30 nM [168,169]. ABA in soils may be released from plant roots or produced by microorganisms such as soil fungi. Hartung et al. [168] assumed that the concentrations of ABA dissolved in the soil solution ranged from 0.6–2.8 nM. This range is suggested to be required in order to prevent ABA release from the root hairs to soils according to the computer simulations by Slovik et al. [170].

Maintenance of root growth during water deficits is under genetic control [171,172], and is regulated by factors such as ABA [7]. Saab et al. [173] reported that ABA accumulation within the root elongation zone of primary maize roots was required for the maintenance of apical root growth. The effect of ABA accumulation on root growth [7] has been studied using: (1) fluridone that blocks carotenoid synthesis, and thereby inhibits ABA synthesis at an early step of the pathway [174]; (2) the *vp5* mutant, defective in the step that is blocked by fluridone; and (3) the *vp14* mutant, which has a defect in the synthesis of xanthoxin [175]. Xanthoxin synthesis is considered to be a key regulatory step in water-stress-induced ABA production [176].

The studies that used these methods have showed that ABA deficiency was associated with the reduction in root cell expansion. Root elongation in those plants treated with fluridone or in the *vp5* mutant or *vp14* mutants was inhibited more severely than that in wild-type or untreated seedlings at low water potential, which was associated with reduced ABA accumulation. Root elongation rates can be fully recovered by exogenous ABA application to the growth zone, further supporting the conclusion that the accumulation of ABA is necessary for root growth under water stress [7]. Studies reported interactions between ethylene and ABA in their effects on root growth. A low concentration of ethylene stimulates root elongation and thinning of the root [177]. Ethylene evolution is excessive in ABA-biosynthesis mutants of maize plants [178]. ABA suppresses the synthesis of ethylene especially at low water potentials.

10.3.1.2 Soil Compaction

Root-borne signals may affect the rate of development of the apical meristem, cell division, and cell expansion in shoots. Root signals are either electrical or hormonal. Ethylene, ABA, auxin, cytokinins, or in combination may be involved in regulating plant growth in compact soil [40].

Clark et al. [39] suggested that ethylene plays an important role in mediating root morphological responses to mechanical impedance, but a role of ABA has yet to be demonstrated. However, both ABA and ethylene play important roles in mediating shoot responses to compaction, although some of the observed responses might not be due to direct effects of mechanical impedance.

Mechanical impedance stimulates ethylene production [179,180]. A decrease in elongation rate and an increase in root diameter were found in those unimpeded roots when they were treated with exogenous ethylene [181,182]. Sarquis et al. [180] found that pretreatment with inhibitors of ethylene biosynthesis before application of mechanical impedance decreased the morphological responses to the impedance. The combined application of aminoethoxyvinyl glycine (AVG, which inhibits ACC formation) and silver thiosulphate (Ag^+ ions inhibit the formation of a functional ethylene-receptor complex) was able to maintain root elongation rate of the impeded roots at 90% of the unimpeded value. The increase in ethylene production in response to impedance also preceded measurable root morphological changes by at least 1 h. Ethylene production in impeded roots was due partly to increased ACC synthase activity, leading to increased ACC concentrations in the tissue [183]. However, there was also an increase in the concentration of malonyl-ACC and that of the ethylene-forming-enzyme (EFE) complex, suggesting that regulation of ethylene production under mechanical stress involves the coordinated action of ACC synthase, the EFE complex, and ACC *N*-malonyltransferase [94]. Ethylene may affect plant growth in compacted soils by inhibiting polar transport of auxin in shoots and roots. Several studies suggested that the inhibition of root growth by ethylene was largely mediated by internal auxin accumulation [184–186].

Lachno et al. [187] and Moss et al. [179] did not find an increase in ABA content of mechanically impeded maize roots. However, Tardieu et al. [188] reported that soil compaction increased ABA concentration in xylem sap, but concluded that this was a response to decreased water availability rather than a direct response to mechanical impedance. In young maize seedlings, soil compaction caused transient increases in ABA concentration in xylem sap, but this response appeared to be related to changes in water relations [189]. In young barley seedlings, ABA concentration in xylem sap appeared to control stomatal conductance in response to compaction [190], which seems to act as a signal at moderate levels of compaction [191]. Mulholland et al. [190] observed that soil compaction caused a transient increase in ABA concentration in xylem sap, but no changes in leaf water potential. Stomatal conductance decreased when roots encountered compacted soil but this response was not seen in ABA-deficient mutants [192].

Ethylene and ABA are interactive in regulating root growth in compacted soils. Ethylene production in response to compaction was associated with an inhibition of shoot growth, but ABA limited ethylene production, and therefore leaf growth was maintained at moderate levels of compaction in tomato (*Lycopersicon esculentum* Mill.) plants [193]. It remains to be elucidated to what extent these signaling responses reflect the change in mechanical impedance or change in aeration of the soil [194]. In response to water stress, ABA functions to maintain root elongation by restricting ethylene production [195]. It is possible that a similar antagonistic relationship between ABA and ethylene may control root growth of mechanically impeded roots, but this is yet to be determined.

10.3.1.3 Waterlogging

Waterlogging initially affects the underground part of the plant, because the roots initially sense the stress. How this stress is perceived is still unclear. Soil waterlogging must trigger a cellular signal transduction pathway leading to physiological and morphological changes, and the physical stress (waterlogging) must be converted into a biochemical response [156]. Plant growth under flooding is characterized by changes in the level of several plant growth regulators and other signaling molecules [156], although it is still unclear which signals and which sensory mechanisms are responsible for triggering physiological responses. It is now established that flooding responses are characterized by enhanced ethylene production, accompanied by a signaling cascade that includes a network of hormones and other common secondary signaling molecules [63,156].

Signal transport from the root to shoot can be seen in dryland species such as peas, tomato, and castor bean (*Ricinus communis*) during the first hours of soil flooding [63,65]. Epinastic leaf curvature (cell expansion on the adaxial surface of petioles resulting in downward leaf growth) and stomatal closure are two well-known responses to signals from flooded roots. Both processes begin within a few hours after the start of flooding [63].

Severely hypoxic or anaerobic roots of tomato generate a message that stimulates epinastic leaf curvature, which is suggested to be associated with increased ethylene biosynthesis in plants under waterlogging stress [196]. Biosynthesis of ethylene was found to increase rapidly (within 4 h) during hypoxia in several species [80,197]. It is believed that ethylene production rate in leaves of flooded plants is closely correlated with the induction of ACC synthase and the accumulation of ACC in the roots of plants exposed to flooding [198–200]. ACC can be rapidly synthesized in the roots, and transported to the shoot within a short time (within 6–12 h) [201]. ACC produced by enhanced ACC synthase activity is converted to ethylene in the leaves, and causes epinasty [196,200].

Ethylene derived from the ACC may be supplemented by ethylene taken up from the soil [202], which is presumably generated from microorganisms. Ethylene may act as a signal response to soil waterlogging or flooding, and also may be involved in the stimulation of adventitious root initiation [197] and aerenchyma development [198,203]. A role of ethylene in plant responses is crucial, but how and where the ethylene is positioned in the signaling cascade is still unknown. Furthermore, the question of how oxygen-deprived roots promote ACC oxidase activity in leaves is still unresolved [156].

In addition to the involvement of ethylene in root-to-shoot signaling events during flooding, there is increasing interest in the role of ABA, gibberellic acid, auxin, and cytokinins [65,156]. Cytokinin and gibberellin activity in xylem sap was depressed by flooding [204–206]. The epinasty in response to flooding can also be inhibited by foliar supply of cytokinins and gibberellins [207]. Commonly observed flooding-induced morphological changes, such as lateral root formation and adventitious root development at the base of the shoot, suggest the involvement of auxin in flooding response [208,209]. In addition, a transient increase in transcript levels of two auxin-responsive genes (*IAA2, IAA3*) was found in hypoxia-treated *Arabidopsis* roots [210].

Stomatal closure started within 2 h after the start of soil flooding and was more effective than epinasty in offsetting foliar dehydration [65]. ABA is involved in stomatal closure. ABA-deficient mutants of pea have impaired stomatal responses to flooding [211,212]. Root-borne ABA may not be involved in this response because ABA delivery to the xylem sap is strongly decreased by flooding [213]. However, the loss of root hydraulic conductance resulting from oxygen deficiency under waterlogged soil probably induces ABA production in leaves by the dehydration mechanism operating in plants subjected to soil drying [176]. Evidence of this has been found in *Ricinus communis* in which pneumatic pressure applied to flooded roots to negate the negative hydraulic message delayed stomatal closure [214]. However, in flooded tomato plants, root pressure did not prevent stomata closure [215], indicating that some nonhydraulic signal (increased pH?) may initiate stomatal closure. Sharp et al. [178] suggested the interaction between ABA and ethylene in regulating plant growth under abiotic stress. A large decrease in ABA delivery was found soon after roots were flooded [216], which could affect leaf responses to increased ethylene accumulation.

Ca^{2+} ion-dependent signal transduction pathways play an important role in signaling cascades and interactive networks involved in the response to flooding [156] (also see Chapter 7). A transient increase in cytosolic Ca^{2+} concentration was observed within a few minutes of flooding in seedlings and cultured cells of maize [217,218] and *Arabidopsis* [219]. Cytosolic response to Ca^{2+} occurred much earlier than any detectable anoxia-induced changes in gene expression in maize suspension-cultured cells. Ruthenium red blocked the Ca^{2+} elevation in cytoplasm and also the induction of anaerobic gene expression such as *adh1* (encoding alcohol dehydrogenase) and *sh1* (encoding sucrose synthase) mRNA whereas Ca^{2+} chelators failed to influence anoxic gene expression, suggesting that cytosolic Ca^{2+} is a physiological transducer of anoxia signals in plants [217,220].

Flooding stress initiates a signal transduction pathway in which cytosolic Ca^{2+} stimulates Ca^{2+}-calmodulin-dependent glutamate decarboxylase (GAD) and gamma-aminobutyric acid (GABA) synthesis [156,220]. GABA is synthesized by irreversible decarboxylation of glutamic acid in a reaction catalyzed by GAD. In fact, GABA increases severalfold during anoxia and is believed to regulate acidosis of the cytoplasm because GAD activity consumes H^+ ion [156,220,221].

Efflux transporters (transporters that remove Ca^{2+} from the cytoplasm) as well as influx transporters (transporters that allow Ca^{2+} into the cytoplasm) play important roles in Ca^{2+} signaling. Calmodulins are involved in regulating Ca^{2+} homeostasis by activating plasma and endo-membrane Ca^{2+}-ATPase [222]. Studies with a cDNA clone, *CAP1*, isolated from anoxic maize roots, which encode a Ca^{2+}-ATPase, showed that its transcripts were induced during the first 4–6 h of anoxia [222], indicating the regulation of Ca^{2+} homeostasis may be involved in the O_2-deprived maize root cells [222]. A role of Ca^{2+} signaling has been demonstrated during aerenchyma development in maize roots [92].

10.3.1.4 Salinity

Salt stress affects cellular ion homeostasis as well as osmotic homeostasis. Excess ions (Na^+ and Cl^-) and osmotic stress-induced turgor change may act as inputs for salt-stress signaling [223,224]. Excess Na^+ entering cells through nonspecific ion channels under high Na^+ concentrations may result in membrane depolarization in roots. A change in membrane polarization could signal salt stress, as it is known to activate Ca^{2+} channels [225].

Cell turgor loss due to osmotic stress may alter the cell volume and the retraction of the plasma membrane from the cell wall, which cause conformational changes of the membrane-bound receptor kinases, ion transporters/channels, transmembrane proteins, and integrin-like proteins, and hence these proteins may act as sensors of osmotic stress [223,224].

Because the cytoskeleton connects different organelles of the cell with the plasma membrane, it can sense cell-volume change under osmotic stress and transduce it to internal Ca^{2+} channels or other signaling components. Salinity induces the biosynthesis and accumulation of ABA [226] and also induces accumulation of reactive oxygen species (ROS) [227–229]. Current evidence suggests that the primary salt stress signals (ionic and osmotic stress) are transduced through Ca^{2+} as well as receptor kinase pathways, whereas the secondary salt-stress signals such as ABA and H_2O_2 also regulate plant salt tolerance [223,224]. For example, ABA and salt stress trigger an increase in the cytosolic Ca^{2+} level in plant cells [230,231]. As a second messenger, Ca^{2+} activates signaling pathways and therefore influences multiple aspects of cellular functions [232,233].

There are three major families of Ca^{2+}-binding proteins that serve as transducers or sensors of the Ca^{2+} signal in plants, including calmodulin [234,235], Ca^{2+}-dependent protein kinases [236,237], calcineurin B-like proteins [235], and salt overly sensitive 3 (SOS3) [238]. These proteins may regulate ion transportation and homeostasis, and gene expression in response to salt stress and enhance plant tolerance to salt stress. Further information in detail can be seen in other reviews [223,224].

10.3.2 Osmotic Adjustment

Expansion of root cells by water absorption may be controlled by at least two factors: osmotic potential of cell sap and mechanical properties of cell walls [136,239]. Osmotic adjustment, defined as the active accumulation of osmotic solutes, during the development of water stress in plants allows lowering osmotic potential of the cell to maintain cell turgor in plant tissues at low water potential [240]. The accumulation of osmotic solutes in cells can be achieved by either an increase in the net rate of accumulation of these substances (including the effects on solute synthesis, transport, and distribution) or decrease in the rate of tissue expansion, and therefore, in the rate of osmoticum dilution. The former is more likely to represent an adaptive response that could contribute to growth maintenance [7].

Roots of many plant species have a substantial capacity for active osmotic adjustment when exposed to osmotic stress caused by salt or water stress. Osmotic adjustment is believed to be a major mechanism of root survival of cellular dehydration. Some researchers suggest that osmotic adjustment in root tips is important to sustain root growth and root elongation under soil-drying and salt-stress conditions [7,241–245].

Sharp and co-workers [7] found that the tips of nodal roots [5] and primary roots [246] exhibit a strong osmotic adjustment capacity under soil drought stress in maize. Cell turgor maintained by osmotic adjustment can sustain root growth under salt or drought stress. Moreover, this capacity of roots is considered to be greater than the comparable capacity of leaves [5,247].

Hsiao and Xu [248] and Ogawa and Yamauchi [244] suggested that the growth zone of roots adjusted the osmotic potential rapidly to cope with reductions in water potential, whereas the leaf adjusted relatively slowly or not at all, which could be attributed to the lower sensitivity of root growth to water deficits, compared to the leaf. Matyssek et al. [249] suggested that high osmotic adjustment in root tips could divert water from other plant organs into the root tips, which results in sustained root growth in dry soils. The cost to synthesize organic solutes for osmotic adjustment is relatively small; for example, the number of moles of ATP needed to use one mole of NaCl as an osmoticum is approximately 4 in root cells, and 7 in leaf cells [250].

The osmolytes generally found in higher plants include inorganic compounds such as low molecular weight sugars, organic acids, polyols, nitrogen containing compounds (i.e., amino acids, amides, ectoine, proteins, and quaternary ammonium compounds), and some inorganic ions such as K^+, Ca^{2+}, Na^+, and Cl^- through ion selectivity and compartmentalization of salt ions. Their relative contribution to osmotic adjustment varies with plant species/cultivar and tissues within the same plant [7,242,243,244]. In maize primary roots, hexoses were a primary contributor to osmotic adjustment in the basal region of the growth zone, but the accumulation of hexose can be accounted for by the reduced rates of volume expansion in this region; in fact, hexose deposition rates in the region beyond 3 mm from the apex were decreased by water stress [246]. In contrast, in the apical few millimeters, proline concentration increased dramatically in water-stressed roots, and contributed up to 50% of the osmotic adjustment. This response involved a severalfold increase in the net rate of proline deposition [241] and an increase in the rate of proline transport to the root tip [251]. The accumulation of ABA was required for the increase in proline deposition at a low water potential [252], suggesting that ABA plays a role in regulating proline transport to the root tip. Other solutes (sucrose, other amino acids, potassium) may make a relatively minor contribution to osmotic adjustment in this root tip region [241].

Under saline stress, cell pressure potential or turgor would be decreased by osmotic withdrawal of water from enlarging cells. Therefore, cells must then develop a sufficiently low osmotic potential to reverse the flow of water by osmotic adjustment in order to maintain their enlargement [140]. It is assumed that organic solutes are compartmented mainly in the cytoplasm, whereas inorganic ions (principally Na^+, K^+, and Cl^-) are sequestered in the vacuoles or allocated to the vacuole and cytoplasm [241,253,254]. Organic solutes in the cytosol contribute to an intracellular osmotic balance when inorganic ion concentrations are high in the vacuole, and they may also protect cytosolic enzymes when the ion concentrations increase [123].

Organic solutes may play an important role in root osmotic adjustment, as K^+ is often lower in roots than in shoots [255]. However, proline and glycine betaine concentrations on a fresh weight basis were five times lower in roots than in shoots of barley cultured in a solution with over 100 ~ 200 mM NaCl [256]. Organic solutes are more likely to accumulate in roots, and organic solutes alone are involved in osmotic adjustment when roots receive an osmotic stress by exposure to mannitol [257]. In contrast, Rodriguez et al. [258] reported different results, in which inorganic ions contributed relatively more to osmotic potential than did organic solutes.

Rapid uptake of Cl⁻ upon exposure to NaCl assisted in adjustment of osmotic potential during the initial hours of salt shock.

The extent of osmotic adjustment in maize primary root tip, although substantial, was insufficient to maintain turgor in roots growing under severe water deficits [7]. Cell-wall extension properties may also be involved in maintenance of root elongation rate under water stress [7,12] or salt stress [122]. Experimental results have demonstrated that the original turgor is not necessary to fully recover root elongation from the stressed root, but it is necessary for the turgor to exceed the wall "yield threshold", which depends on cell-wall properties that may change under stress [259]. Pritchard et al. [260] suggested that a drop in the pressure potential or turgor may account for the inhibition of root extension in wheat roots with osmotic stress induced by mannitol, although cell walls in the proximal region of the growing zone became more rigid, that is, yield threshold increased or wall extensibility decreased [261].

Spollen and Sharp [262] found that turgor was reduced by over 50% throughout the growth zone at a water potential of −1.6 MPa, but the elongation rate within the root apical few millimeters was unchanged, which may be associated with the enhancement of longitudinal cell wall extensibility in this region under water stress. Wu et al. [263] showed a large increase in acid-induced extensibility in the apical 5 mm and a decrease in the 5–10 mm region under a water-stressed condition compared with well-watered roots.

The turgor and extension rate of maize roots immediately decreased upon exposure to mannitol or KCl. However, the extension was resumed with parallel recovery of turgor in the next 30 min [264]. However, Neumann et al. [265] suggested that the change in turgor was not associated with lowered root extension in maize seedlings but rather a greater hardening of cell walls under salinity stress because turgor remained, compared with unsalinized conditions. In this experiment, seedlings had been exposed to 100 mM NaCl for 24 h before the observation began, so that the early responses to salt shock were not measured. Osmotic adjustment facilitates water extraction from drying or saline soils. Data obtained with wheat [266,267] lines differing in the capacity of osmotic adjustment clearly demonstrated this point.

10.3.3 Metabolism of Reactive Oxygen Species (ROS)

Plants, like other aerobic organisms, require oxygen for efficient production of energy. During the reduction of O_2 to H_2O, especially under stressful conditions, ROS, such as superoxide radical ($O_2^{\bullet-}$), hydrogen peroxide (H_2O_2), and hydroxyl radical ($OH^{\bullet}$) are formed. Plants possess complex systems, including enzymatic scavengers and nonenzymatic antioxidants, to protect cells from oxidative damage by scavenging ROS (for reviews see Foyer and Noctor [268], Mittler [269], and Mittler et al. [270]). Oxidative damage may occur in leaves and roots of a plant. Antioxidant mechanisms have been examined extensively in leaves [268–270], but limited information is available for roots.

Superoxide dismutase (SOD) catalyzes the reaction of the dismutation of superoxide into H_2O_2 and O_2 [271]. SODs are classified into three types according to the metal ion it contains: Mn-, Fe-, and Cu/Zn-SOD. The enzymes are localized in plants in mitochondria, chloroplasts, and cytosol and chloroplasts, respectively. For example, mitochondria of pea root cells contain MnSOD [272]. H_2O_2 is scavenged by peroxidases, such as ascorbate peroxidase (APX), and catalase (CAT). CAT removes H_2O_2 formed in photorespiration or in fl-oxidation of fatty acids in the glyoxysomes in leaf peroxisomes [273]. However, CAT activity was also found in roots [274,275]. APX, which is the most important plant peroxidase in the first step of the ascorbate-glutathione cycle, uses two molecules of ascorbate (AsA) as a reductant to reduce H_2O_2 to water, with the concomitant generation of two molecules of monodehydroascorbate (MDHA) [276]. MDHA is a short life-time radical and can be reduced to AsA by MDHA reductases using NADPH as electron donors. Dehydroascorbate (DHA) is then reduced to AsA by the action of DHA reductase by using reduced glutathione (GSH), which is generated from oxidized glutathione (GSSG) by glutathione reductase (GR) at the expense of NADPH [269,270,276]. Besides APX, guaiacol-dependent peroxidase (POD) is also involved in the scavenging of soluble hydroperoxides in plants [277], and its presence has been reported in maize root mitochondria by Sukalovic and Vuletic [278]. ROS is also scavenged nonenzymatically by some low molecular weight hydrophilic antioxidants such as AsA and GSH, and these antioxidants were detected in bean and tomato roots [275,279].

Accelerated production of ROS has been observed in plants exposed to abiotic stresses [280–282]. The mitochondrial electron transport chain is considered to be the most probable candidate for intracellular ROS formation in nonphotosynthetic tissues [283,284]. The production of ROS in mitochondria has been shown to be attributed to electron leakage at the ubiquinone site–the ubiquinone:cytochrome*b* region [285] and at the matrix side of complex I (NADH dehydrogenase [283]. Besides mitochondria [286], plastids [287] and peroxisomes [286] are sites of ROS generation in roots under salt stress. Healthy plants possess a number of enzymatic and nonenzymatic detoxification mechanisms to efficiently scavenge ROS and their reaction products [269,270]. Under environmental stress, however, these detoxification systems can be dysfunctional, resulting in an excessive production of ROS [269]. Thus, the increase of ROS levels in root cells is due not only to extracellular production of ROS, but also to downregulation of ROS scavenging mechanisms.

Salt stress manifests the toxic effects of oxidative stress [229,275,288]. Khan and Panda [289] suggested that a salt-sensitive rice cultivar exposed to NaCl stress was damaged by oxidative stress, which was related to an increase in peroxide content and lipid peroxidation accompanied with the decrease in catalase, guaiacol peroxidase, and superoxide dismutase activities in root tissues. Panda and Upadhyay [290] also found that NaCl caused an increase in lipid peroxidation and loss of membrane integrity in the root apices of *Lemna minor*, but both nonenzymatic antioxidants, ascorbate and glutathione, also increased with the increase of NaCl concentration in the roots. However, an increase in SOD, POD, and GR activities were observed in the roots under NaCl stress. Bandeoglu et al. [291] showed that salt stress caused a significant increase in SOD activity in roots mainly due to the increase in Cu/ZnSOD isoforms to protect from NaCl stress, but there was no significant change in the activity of APX, CAT, and GR in roots of 14-day-old lentil (*Lens culinaris* M.) seedlings under salt stress.

Lee et al. [292] showed that NaCl treatment resulted in an accumulation of H_2O_2 in the leaves but not in the roots of rice plants. However, Tsai et al. [293] showed that H_2O_2 level increased in NaCl-treated roots of rice seedlings, but NaCl had no effect on SOD and CAT activities, which may be mediated by NADPH oxidase in rice roots. In addition, NaCl-induced cell wall-bound NADH peroxidase and diamine oxidase activities, which are devoted to H_2O_2 generation, have also been detected in the roots of rice seedlings [294]. Tsai et al. [293] found an interesting result that Cl⁻, rather than Na⁺, may be responsible for the NaCl-increased GSH and GSSG levels in roots of rice seedlings.

It is well accepted that oxygen deficiency or low oxygen concentration causes the formation of ROS, which may lead to damage of plant tissues subjected to waterlogging or flooding. Without oxygen, mitochondrial respiration and the tricarboxylic acid cycle are blocked, which results in decreased formation of ATP and accumulation of NAD(P)H [295]. Eventually, all energy-consuming processes such as synthesis of macromolecules are inhibited and growth is prevented. Electron-saturated redox chains prevail under oxygen deficiency, favoring the formation of ROS [296]. Re-oxygenation damage occurs after re-aeration in the plants deprived of oxygen due to flooding. This is considered to be mainly brought about by a burst formation of ROS [297–301]. Aeration after anaerobiosis seems to be favorable for the production of ROS [302,303], and even re-aeration after a short-term hypoxic treatment (for 2, 12, or 24 h) caused oxidative stress in lupine roots [304]. The reason for the increase in ROS generation may be the high level of reductants accumulated during the period of restricted oxygen availability.

ROS formation was increased in plant tissues after a period of anoxic or hypoxic treatment (postanoxic or posthypoxic injury) [303], which may induce both protective responses and cellular damage manifested by the symptoms of oxidative stress observed in roots [305,306]. The concentrations of glutathione and ascorbic acid were increased and the antioxidant enzymatic defense systems, such as SOD and APX, were activated during posthypoxia [305,307]. Postanoxic or posthypoxic injury may be caused by toxic acetaldehyde deriving from catalase-dependent oxidation of ethanol, which accumulates under oxygen-deprived conditions [297,308].

Garnczarska et al. [306] suggested that re-aeration following hypoxia imposes an oxidative stress in lupine roots, which was indicated by an increase in the level of ROS and the induction of SOD activity immediately after re-oxygenation of hypoxically pretreated roots. However, the capacity of lupine roots to keep the balance between the formation and detoxification of ROS seemed to be sufficient to counteract harmful effects induced by re-aeration because the hydrogen peroxide content and MDA concentration decreased even after longer hypoxic pretreatment. Biemelt et al. [305] reported an antioxidative stress-defense system in wheat roots, which responded differently to hypoxia or anoxia and subsequent re-aeration. Although hypoxia in the root environment followed by re-aeration caused only little change in the activities of enzymes in the ascorbate-glutathione cycle, activities of APX and GR progressively decreased with duration of anoxic stress.

The generation of ROS may be of general importance for plant signaling and development in response to abiotic stress [270,280,282,309]. ROS generation in cellular compartments such as the mitochondria and chloroplasts results in changes of the nuclear transcriptome, indicating that information must be transmitted from these organelles to the nucleus, but the identity of the transmitting signal remains unknown [282]. Apel and Hirt [282] suggested that there are three principal modes of action that indicate how ROS affects gene expression: (1) extracellular and intracellular ROS could be sensed by ROS sensors such as membrane-localized histidine kinases. A series of signaling cascades would be induced by the activation of ROS sensors ultimately affecting gene expression; (2) the components of signaling pathways could be directly oxidized by ROS. Intracellular ROS can influence mitogen-activated protein kinase (MAPK) signaling pathway through inhibition of MAPK phosphatases (PPases) or downstream transcription factors; and (3) ROS might change gene expression by targeting and modifying the activity of transcription factors. MAP kinases regulate gene expression by altering transcription factor activity through phosphorylation of serine and threonine residues, and ROS generation is regulated by oxidation of cysteine residues. ROS sensors could be activated to induce signaling cascades that ultimately impinge on the gene expression. Alternatively, components of signaling pathways could be directly oxidized by ROS. Finally, ROS might change gene expression by targeting and modifying the activity of transcription factors [282].

Liszkay et al. [310] hypothesized that the wall-loosening reaction controlling root elongation is affected by the production of reactive oxygen intermediates, which is initiated by a NAD(P)H oxidase-catalyzed formation of O_2^- at the plasma membrane, culminating in the generation of polysaccharide-cleaving hydroxyl radicals (^-OH) by cell wall peroxidase. They tested this hypothesis using primary roots of maize seedlings, and obtained the following results: (1) production of O_2^-, H_2O_2, and ^-OH could be monitored in the growing zone; (2) auxin-induced inhibition of growth was accompanied by a reduction of O_2^- production; (3) generation of ^-OH in the cell walls with the Fenton reaction caused wall loosening (cell wall creep), specifically in the growing zone. Moreover, cell wall loosening could be induced by OH, produced by endogenous cell wall peroxidase in the presence of NADH and H_2O_2; and (4) inhibition of endogenous ^-OH formation by O_2^- and ^-OH scavengers and inhibitors of NAD(P)H oxidase or peroxidase activity suppressed root elongation growth.

The *root hair defective2 (rhd2)* mutant of *Arabidopsis*, defect in the third NADPH oxidase catalytic subunit, contained lower levels of ROS in growing root hairs than did wild-type plants. NADPH oxidases controlled root development by producing ROS that controls cell elongation and root-hair development through the activation of Ca^{2+} channels. This defect leads to a distortion of Ca^{2+} uptake and a disruption of root cell expansion [311,312].

AtrbohE and AtrbohF NADPH oxidases are required for the production of ROS during ABA-induced inhibition of root cell expansion in *Arabidopsis* [313]. Abscisic acid accumulation and oxidative stress are two common responses of plants to drought stresses. However, little is known about their relationships. Sharp et al. [7] hypothesized that ABA accumulation may promote the antioxidant system in the growth zone of water-stressed roots, and thereby preventing high levels of ROS, excess ethylene production, and growth inhibition in roots under water stress. This hypothesis is based on three lines of evidence. First, more ROS is produced in stressed tissues. Under well-watered conditions, ROS levels were low in roots of both the wild-type and *vp14* mutant seedlings in which ABA levels were deficient in water-stressed but not well-watered roots. Under water-stressed conditions, ROS levels were slightly higher in the growth zone of wild-type roots and increased dramatically in *vp14*. The increase in ROS levels in *vp14* was prevented when ABA was restored to the wild-type level by exogenous application. The effect of ABA deficiency on ROS level was striking in the region 1–3 mm from the root apex where cell elongation is normally maintained under a water deficit, but is inhibited by ABA deficiency [9,314,315]. Second, ABA treatment increases the expression of genes for antioxidant enzymes, for example, catalase in maize leaves [316]. Finally, excess ROS levels can cause increased ethylene synthesis [317].

However, Zhao et al. [318] revealed that ROS and nitric oxide play important roles in drought-induced abscisic acid synthesis in plants, because the activities of superoxide synthases and nitric oxide synthase in drought-stressed root tips of wheat increased earlier than abscisic acid accumulation. Moreover, the induction of abscisic acid by drought was strongly blocked by pretreating the root tips with reactive oxygen species eliminators and enhanced by reactive oxygen species generators. Kwak et al. [313] suggested that ROS are rate-limiting second messengers in ABA signaling.

10.3.4 Uptake of Nutrient Ions

10.3.4.1 Drought

Drought influences plant nutrient uptake and utilization through its effects on root morphology, architecture, and anatomy, as discussed earlier. In addition, many other processes besides changes in roots may be affected by drought stress, which influences nutrient uptake. Among these processes, nutrient movement from the soil to the root surface is limited by soil drying, thereby leading to low nutrient availability. Stimulation of mucilage secretion from the root cap in dry soil [320] might contribute to maintaining nutrient ion uptake by facilitating ion transport from embedded soil particles to the root surface [321,322]. This process may be further promoted by increasing hydraulic transfer of water from the subsoil and subsequent release into the dry topsoil layer [323,324].

Soil drought may decrease the demand for nutrients by plants due to their low growth and inhibit translocation of nutrients from the root to shoot. Arbuscular mycorrhizal fungi infection is also affected under low soil moisture conditions and the content of nutrients such as phosphorus and micronutrients in the host plant decreases [325,326], whereas the arbuscular mycorrhizal symbiosis may alleviate plant responses to moisture deficit by several mechanisms including improvement of nutrient supply to the host plant by hyphae in the soil [327–329].

10.3.4.2 Soil Compaction

Uptake of nutrients such as phosphorus [330] is impaired by soil compaction. The mobility of nutrients in soils decreases with compaction [331]. Several factors or some integrated mechanisms should be taken into account to better understand the effects of soil compaction on nutrient availability and uptake by roots.

Ion movements from soil to plant root surface occur through mass flow or diffusion [331,332]. The transport of nutrients to roots in the soil is affected because compaction normally increases mass flow transport [333] and the diffusion coefficient at given gravimetric water content [334–336]. However, Nye and Tinker [337] reviewed the influences of changes of soil bulk density on water diffusion in soil cores and found that both increases and decreases in diffusion rates have been recorded in response to increasing soil bulk density, depending upon soil type, water content, and diffusing ions. Arvidsson [338] suggested that the uptake of nutrients transported by diffusion was more affected by compaction than for nutrients transported by mass flow. Soil compaction can increase root-to-soil contact, which may facilitate nutrient uptake [339]. However, root-to-soil contact is affected by the changes of root growth response to soil compaction.

Aeration affects availability of nutrients involved in redox reactions, such as nitrogen, iron, manganese, and sulphur, and the growth and function of roots [331,340]. The mobility of these nutrients increases because they can be reduced chemically to more soluble forms under low oxygen conditions, and sometimes, Fe^{2+}, Mn^{2+}, and H_2S concentrations may dramatically increase and reach to detrimental levels [331]. Hoffmann and Jungk [336] found that shoot P concentration of sugar beet (*Beta vulgaris*) remained almost unchanged with the increase of soil bulk density, which was attributed to the fact that the plants in compacted soil partly substituted for the reduction of root size by increasing the P uptake efficiency per unit of root.

Mucilage secretion from the root cap has been shown to be stimulated by increased mechanical impedance [58,320]. As discussed earlier, mucilage facilitates soil and root contact in compacted soils. This mechanism may also increase ion transport from smaller soil pores to the root surface [341].

Yano et al. [342] found that an increase in soil compaction reduced mycorrhizal formation with *Gigaspora margarita* and P uptake in *Cajanus cajan* in a greenhouse study. Nadian et al. [343] obtained similar results in *Trifolium subterraneum*, but soil compaction had no significant effect on the proportion of roots containing arbuscules and vesicles. Li et al. [344,345] also found that increasing soil compaction significantly reduced mycorrhizal formation with G. *mosseae* and P uptake in *Trifolium pratense*.

In irrigated lowland rice, Kundu et al. [346] found that deep tillage increased the mineral nitrogen availability. Ishaq et al. [347] observed a 12–35% and a 23% decrease in N uptake by soil compaction in wheat and sorghum, respectively. Lower N uptake by roots may contribute to reduced soil water availability and water infiltration, impaired root growth, decreased root-soil contact, a smaller proportion of macropores in compacted soil, and also possible increased soil N loss. Several studies showed that the rate of denitrification or N_2O production was increased by compaction [348–350]. Moreover, nitrogen runoff from the soil may be reduced by soil compaction due to lower water infiltration.

10.3.4.3 Waterlogging

Soil flooding affects the chemical transformation of some nutrients due to a strongly reduced environment where molecular oxygen is depleted. For example, facultative anaerobes chemically reduce nitrate to nitrite, nitrous oxide, and nitrogen gas (denitrification) rendering nitrate unavailable to roots under waterlogging stress. Fe^{3+}, Mn^{4+}, and SO_4^{2-} are chemically reduced to form highly soluble Mn^{2+}, Fe^{2+}, and gaseous H_2S, which are toxic to roots by interfering with enzyme activities and damaging membranes [63,68,69,351].

The nutrient demand of waterlogging-tolerant plants is supposed to be high under soil anoxia inasmuch as plants are able to maintain or even increase their biomass production under these conditions [352]. The reduction in root-to-shoot ratio caused by waterlogging could be detrimental for the nutrient foraging capacity of plants in the absence of a compensation mechanism (e.g., increase in specific root length). When oxygen loss occurs in roots, it aerates the rhizosphere of plants growing in waterlogged soils resulting in significant changes in soil chemistry within the rhizosphere [62].

Although rice can tolerate NH_4^+ acquisition as sole N source [353], its lateral roots that are sufficiently used to nitrify NH_4^+ to NO_3 have been found to leak some amount of O_2 [354], thereby allowing plants to absorb half of its N requirements as NO_3 whereas the rest came from NH_4^+ nutrition [355], as well as to oxidize toxins such as Fe^{2+} [113,356].

Oxygen released from roots also mobilized phosphorus [356] from the surrounding reduced soil, thereby increasing P uptake [357]. Rubio et al. [358] showed that waterlogging increased P uptake per unit root as a consequence of changes in root morphology of waterlogged plants, which was more favorable to nutrient uptake (finer roots), having a higher physiological capacity to absorb P (higher Vmax; lower Km), increased soil P availability, and an effective diffusion coefficient. This compensation through an increase in nutrient uptake per unit root could allow plants to keep their nutrient concentration at the same level as nonwaterlogged plants [357].

Zinc can also be mobilized by oxidation from its insoluble form [359] and is accessible to plants in acid soluble form by re-adsorption on $Fe(OH)_3$ and organic matter. All of these were achieved by inducing acidification in the rhizosphere caused by H^+ generation in Fe^{2+} oxidation near the leaking roots [360] and cation–anion intake balance. Presence of two strains of ammonium-oxidizing bacteria (AOB) around the surrounding roots of rice was also related to leakage of oxygen from roots to soil [361], although composition and activities of AOB depend on the amount of leaked oxygen.

10.3.4.4 Salinity

The toxic effects of ions and the imbalanced uptake of essential nutrients are the main deleterious effects of salt stress on plant growth [122,362]. At the cellular level, toxic Na^+ and Cl^- increase, and the levels of beneficial K^+ and Ca^{2+} decrease under salt stress [122,123,362–367]. Mechanisms that control ion balance facilitate plant tolerance to salinity.

Membrane proteins in root cells play a significant role in controlling and mediating selective uptake, distribution, and exclusion of salt ions in cells, which is generally associated with salt tolerance of plants [362,367–370]. Primary H^+-ATPase drives active transport of Na^+, Cl^-, and other ions through generating the membrane potential and proton gradient as the driving force [371–373]. ATPase activity is suggested to have a distinct advantage with respect to salt tolerance through depolarization of the plasma membrane potential, which, in turn, decreases Na^+ absorption under salinity [374]. Under salt stress, an increase of the tonoplast ATPase activity was observed, whereas the plasma membrane-bound-ATPase activity decreased in the medial and basal regions of the root of *Lycopersicon esculentum* [375]. Yang et al. [376] suggested that thiol oxidation and a decrease in the enzyme content was responsible for the inhibition of the H^+-ATPase activity in root plasma membrane of one wheat cultivar (Y424) in response to salt stress and the increased plasma membrane H^+-ATPase activity in the roots of another wheat cultivar (L-Ch20) might be due to upregulation of the expression of the plasma membrane H^+-ATPase. The sodium efflux from the cell and its concentration in the cytoplasm may be reduced via Na^+/H^+ antiporters coupled with outward movement of H^+ from the cell [122]. Na^+/H^+ antiports in the plasma membrane are responsible for removing excess Na^+ from the cell and Na^+/H^+ antiports in the tonoplast function possibly for extruding Na^+ into the vacuole [362,366,367].

A large body of molecular information has been obtained in *Arabidopsis* [224,366]. The *SOS1* (salt overly sensitive 1) gene encodes a plasma membrane Na^+-H^+ antiporter [377]. The *sos1* mutation accumulated high levels of Na^+ under salt stress due to impaired Na^+ efflux [378,379]. The transcript level of *SOS1* in seedlings was very low or undetectable under nonstress growth conditions, which appeared primarily in root epidermal cells and cells surrounding the vascular tissues in response to Na^+ treatment [377,380]. The activity of the SOS1 antiporter is regulated by the SOS2 Ser/Thr protein kinase [379,381], which is in turn regulated by the SOS3 Ca^{2+}-binding protein [238,382]. Another member of the family of Na^+-H^+ antiporters to which SOS1 belongs is AtNHX1, which uses the proton gradient existing across the tonoplast to move Na^+ into the vacuole [383–386]. Overexpression of *AtNHX1* increased salt tolerance in several plant species [383,387,388].

The uptake of Na^+ by roots from the saline soils is a passive process, and ion channels provide significant conduits for entry of Na^+ into roots [389]. The AtHKT1, a Na^+-H^+ symporter, was capable of transporting Na^+ across the plasma membrane of *Arabidopsis* root cells [390–392]. *AtHKT1* was expressed in the root stele and leaf vasculature of *Arabidopsis*, and also had a high-affinity K^+ uptake [393]. OsHKT transporters were involved in Na^+ movements in rice, and OsHKT1 specifically mediated Na^+ uptake in rice roots when the plants were K^+-deficient [394].

Cl^- is thought to traverse the root by a symplastic pathway. Passive process and active systems may be responsible for Cl^- uptake by roots [395]. In saline environments, sodium uptake into the cell may alter the electrochemical potential to allow passive Cl^- entry into cells through the Cl^- efflux channel [395]. Electrophysiological studies have demonstrated the presence of an electrogenic $Cl^-/2H^+$ symporter in the plasma membrane of root-hair cells and Cl^- channels mediating either the influx or efflux of Cl^- across the plasma membrane. Similarly, both biochemical and electrophysiological evidence exhibited that Cl^- channels mediate Cl^- fluxes in either direction across the tonoplast and that a Cl^-/NH^+ antiport mediates Cl^- influx to the vacuole [395].

In addition to the ion transporters or channels mediating Na^+ antagonism with K^+ uptake in root cells, other factors may affect the uptake of K^+ and Ca^{2+} by roots. In saline soils, excess exchangeable sodium decreased soil exchangeable K and Ca levels [396]. The Ca^{2+} ion reduction induced by salts [397] in the nutrient solution had profound effects on root growth [242]. In the presence of 10 mM Ca^{2+}, root elongation of maize was unaffected by salinity under 80 mM NaCl. In the absence of Ca^{2+}, salinity suppressed root growth, and subsequent supply of Ca^{2+} did not readily reverse the suppression [398]. A saline condition as high as 150 mM NaCl suppressed the elongation of sorghum roots by only 20% in the presence of Ca^{2+}, but by 80% in the absence of Ca^{2+} [399].

10.4 SUMMARY

Plants are often exposed to various soil abiotic stresses, such as drought, waterlogging, soil compaction, and salinity. Roots are the first to encounter these constraints as they are directly in contact with soils. Roots exhibit some adaptive responses to adverse soil conditions, including growth and development features, such as changes in root morphology, architecture, and anatomy, and various physiological characteristics. Osmotic adjustment is one of the important physiological responses for roots to adapt to osmotic stress resulting from drought or salinity. The consequences of soil abiotic stresses on roots are reduced nutrient and water uptake, and altered hormonal synthesis.

Some hormones such as ABA and ethylene generated in roots act as signals in response to stresses, which not only regulate root growth and development, but also cause shoot responses such as shoot growth reduction, epinastic leaf curvature, and stomatal closure. The generic signaling transduction pathways in root responses to drought, waterlogging, salinity, and compaction have been proposed, which may start with signal perception by sensors, followed by the production of second messengers (e.g., inositol phosphates and ROS). Second messengers can regulate intracellular Ca ion levels, resulting in a protein phosphorylation cascade, and finally target proteins directly involved in cellular protection or transcription factors controlling specific sets of stress-regulated genes, or increased synthesis of hormones such as ABA and ethylene. Most current research in root tolerance to soil abiotic stress focuses on root-level or cellular responses. Further works are required to examine molecular bases of morphological and physiological responses of roots to soil physical and chemical constraints.

REFERENCES

1. Wilson, J.B., A review of evidence on the control of shoot, root ratio, in relation to models. *Ann. Bot.*, 61:433, 1988.
2. Sharp, R.E. and Davies, W.J., Regulation of growth and development of plants growing with a restricted supply of water. In *Plants Under Stress*, Jones, H.G., Flowers, T.L., and Jones, M.B., Eds., Cambridge University Press, Cambridge, 1989, pp. 71–93.
3. Setter, T.L., Transport/harvest index, photosynthate partitioning in stressed plants. In *Stress Response in Plants, Adaption and Acclimation Mechanisms*, Cumming, J.R., Ed., Wiley-Liss, Inc., New York, 1990, pp. 17–36.
4. Bañoc, D.M et al., Genotypic variations in response of lateral root development to fluctuating soil moisture in rice. *Plant Prod. Sci.*, 3:335, 2000.
5. Sharp, R.E. and Davies, W.J., Solute regulation and growth by roots and shoots of water-stressed maize plants. *Planta*, 147:43, 1979.
6. Spollen, W.G. et al., Regulation of cell expansion in roots and shoots at low water potentials. In *Water Deficits. Plant Responses from Cell to Community*, Smith, J.A.C. and Griffiths, H., Eds., Bios Scientific, Oxford, 1993, 37–52.
7. Sharp, R.E. et al., Root growth maintenance during water deficits, physiology to functional genomics. *J. Exp. Bot.*, 55:2343, 2004.
8. Sharp, R.E., Silk, W.K., and Hsiao, T.C, Growth of the maize primary root at low water potentials. I. Spatial distribution of expansive growth. *Plant Physiol.*, 87:50, 1988.
9. Saab, I.M., Sharp, R.E., and Pritchard, J., Effect of inhibition of abscisic acid accumulation on the spatial distribution of elongation in the primary root and mesocotyl of maize at low water potentials. *Plant Physiol.*, 99:26, 1992.
10. Liang, B.M., Sharp, R.E., and Baskin, T.I., Regulation of growth anisotropy in well-watered and water-stressed maize roots. I. Spatial distribution of longitudinal, radial and tangential expansion rates. *Plant Physiol.*, 115:101, 1997.

11. Sharp, R.E. and Davies, W.J., Root growth and water uptake by maize plants in drying soil. *J. Exp. Bot.*, 36:1441, 1985.
12. Davies, W.J. and Bacon, M.A., Adaptation of roots to drought. In *Root Ecology*, de Kroon, H. and Visser, E.J.W., Eds., Springer Verlag, Berlin, Heidelberg, 2003, pp. 173–192.
13. Serraj, R. et al., Variation in root traits of chickpea (*Cicer arietinum* L.) grown under terminal drought. *Field Crops Res.*, 88:115, 2004.
14. Ludlow, M.M. and Muchow, R.C., A critical evaluation of traits for improving crop yields in water-limited environments. *Adv. Agron.*, 43:107, 1990.
15. Turner, N.C., Wright, G.C., and Siddique, K.H.M., Adaptation of grain legumes (pulses) to water limited environments. *Adv. Agron.*, 7:193, 2001.
16. Fukai, S. and Cooper, M., Development of drought-resistant cultivars using physio-morphological traits in rice. *Field Crops Res.*, 40:67, 1995.
17. Azhiri-Sigari, T. et al., Genotypic variation in response of rainfed lowland rice to drought would help to elucidate how rooting depth and deep root and rewatering. II. Root growth. *Plant Prod. Sci.*, 3:180, 2000.
18. Kamoshita A., Wade. L.J., and Yamauchi, A., Genotypic variation in response of rainfed lowland rice to drought and rewatering. III. Water extraction during drought period. *Plant Prod. Sci.* 3:189, 2000.
19. Kamoshita, A. et al., Mapping QTLs for root morphology of a rice population adapted to rainfed lowland conditions. *Theor. Appl. Genet.*, 104:880, 2002.
20. Mambani, B. and Lal, R., Response of upland rice varieties to drought stress. I. Relation between root system development and leaf water potential. *Plant Soil*, 73:59, 1983.
21. Taylor, H.M., Modifying root systems of cotton and soybean to increase water absorption. In *Adaptation of Plants to Water and High Temperature Stress*, Turner, N.C. and Kramer, P.J., Eds., John Wiley & Sons, New York, 1980, pp. 75–84.
22. Huang, B., Duncan, R.R., and Carrow, R.N., Drought-resistance mechanisms of seven warm-season turfgrasses under surface soil drying, I. Shoot response. *Crop Sci.*, 37:1858, 1997.
23. Huang, B.R., Role of root morphological and physiological characteristics in drought resistance of plants. In *Plant–Environment Interactions*, 2d ed. Wilkinson, R.E., Ed., Marcel Dekker, New York, 2000, pp. 39–64.
24. Siopongco, J.D.L.C. et al., Root growth and water extraction response of double-haploid rice lines to drought and rewatering during the vegetative stage. *Plant Prod. Sci.*, 8:497–508, 2005.
25. Kage, H., Kochler, M., and Stutzel, H., Root growth and dry matter partitioning of cauliflower under drought stress conditions, measurement and simulation. *Eur. J. Agron.*, 20:379, 2004.
26. Yamauchi, A., Pardales, J.R., Jr., and Kono, Y., Root system structure and its relation to stress tolerance. In *Dynamics of Roots and Nitrogen in Cropping Systems of the Semi-arid Tropics*, Ito, O. et al., Eds., Japan International Research Center for Agricultural Sciences, Tsukuba, 1996, pp. 211–233.
27. Ogawa, A., Kawashima, C., and Yamauchi, A., Sugar accumulation along the seminal root axis, as affected by osmotic stress in maize, a possible physiological basis for plastic lateral root development. *Plant Prod. Sci.*, 8:173, 2005.
28. Pardales, .J.R., Jr., et al., The effect of soil moisture fluctuation on root development during the establishment phase of sweetpotato. *Plant Prod. Sci.*, 3:134, 2000.

29. Pardales, J.R., Jr., and Yamauchi, A., Regulation of root development in sweet-potato and cassava by soil moisture during their establishment period. *Plant Soil*, 255:201, 2003.
30. Mckay, A.D. and Barber, S.A., Effect of cyclic wetting and drying of soil on root hair growth of maize roots. *Plant Soil*, 104:291, 1987.
31. Huang, B.R. and Fry, J.D., Root anatomical, physiological, and morphological responses to drought stress for fall fescue cultivars. *Crop Sci.*, 38:1017, 1998.
32. Vasellati, V. et al., Effects of flooding and drought on the anatomy of *Paspalum dilatatum*. *Ann. Bot.*, 88:355, 2001.
33. Dawes, C.J. and Bowler, E., Light and electron microscope studies of the cell wall structure of the root hairs of *Raphanus sativus*. *Am. J. Bot.*, 46:561, 1959.
34. Greaves, M.P. and Darbyshire, J.F., The ultrastructure of the mucilaginous layer on plant roots. *Soil Biol. Biochem.*, 4:443, 1972.
35. Sprent, J.L., Adherence of sand particulars to soybean roots under water stress. *New Phytol.*, 74:461, 1975.
36. Huang, B.R. and Nobel, P.S., Hydraulic conductivity and anatomy along lateral roots of cacti, changes with soil water status. *New Phytol.*, 123:499, 1993.
37. Materechera, S.A., Dexter, A.R., and Alston, A.M., Penetration of very strong soils by seedling roots of different plant species. *Plant Soil*, 135:31, 1991.
38. Bengough, A.G., Root growth and fuction in relation to soil structure, composition, and strength. In *Root Ecology*, de Kroon, H. and Visser, E.J.W, Eds., Springer Verlag, Berlin, Heidelberg, 2003, pp. 151–171.
39. Clark, L.J., Whalley, W.R., and Barraclough, P.B., How do roots penetrate strong soil? *Plant Soil*, 255, 93, 2003.
40. Masle, J., High soil strength, mechanical forces at play on root morphogenesis and in root, shoot signaling. In *Plant Roots, the Hidden Half* ,3d ed. Waisel, Y., Eshel, A., and Kafkafi, U., Eds., Marcel Dekker, New York, 2002, pp. 807–819.
41. Clark, L.J. et al., Complete mechanical impedance increases the turgor of cells in the apex of pea roots. *Plant Cell Environ.*, 19:1099, 1996.
42. Clark, L.J., Whalley, W.R., and Barraclough, P.B., Partial mechanical impedance can increase the turgor of seedling pea roots. *J. Exp. Bot.*, 52:167, 2001.
43. Bengough, A.G., Croser, C., and Pritchard, J., A biophysical analysis of root growth under mechanical stress. *Plant Soil*, 189:155, 1997.
44. Iijima, M. et al., Effects of soil compaction on the development of rice and maize root systems. *Environ. Exp. Bot.*, 31:333, 1991.
45. Iijima, M. and Kono, Y., Development of golgi apparatus in the root cap cells of maize (*Zea mays* L.) as affected by compacted soil. *Ann. Bot.*, 70:207, 1992.
46. Wilson, A.J., Robards, A.W., and Goss, M.J., Effects of mechanical impedance on root-growth in barley, *Hordeum-vulgare* L. 2. Effects on cell development in seminal roots. *J. Exp. Bot.*, 28:1216, 1977.
47. Veen, B.W., The influence of mechanical impedance on the growth of maize roots. *Plant Soil*, 66:101, 1982.
48. Goss, M.J., Effects of mechanical impedance on root-growth in barley (*Hordeum-vulgare*-L). 1. Effects on elongation and branching of seminal root axes. *J. Exp. Bot.*, 28:96, 1977.
49. Misra, R.K., Maximum axial growth pressures of the lateral roots of pea and eucalypt. *Plant Soil*, 188:161, 1997.
50. Barley, K.P., Farrell, D.A., and Greacen, E.L., The influence of soil strength on the penetration of a loam by plant roots. *Aust. J. Soil Res.*, 3:69, 1965.

51. Atwell, B. J., Physiological responses of lupin roots to soil compaction. *Plant Soil*, 111:277, 1988.
52. Kuchenbuch, R.O. and Ingram, K.T., Effects of soil bulk density on seminal and lateral roots of young maize plants (*Zea mays* L.). *J. Plant Nutr. Soil Sci.*, 167:229, 2004.
53. Bingham, I.J. and Bengough, A. G., Morphological plasticity of wheat and barley roots in response to spatial variation in soil strength. *Plant Soil*, 250:273, 2003.
54. Ennos, A.R., The mechanics of anchorage in seedlings of sunflower, Helianthusannuus L. *New Phytol.*, 113:185, 1989.
55. Iijima, M., Barlow, P.W., and Bengough, A. G., Root cap structure and cell production rates of maize (*Zea mays* L.) roots in compacted sand. *New Phytol.*, 160:127, 2003.
56. Iijima, M. et al., Root cap removal increases root penetration resistance in maize (*Zea mays* L.). *J. Exp. Bot.*, 54:2105–2109, 2003.
57. Iijima, M., Griffith, B., and Bengough, A.G., Sloughing of cap cells and carbon exudation from maize seedling roots in compacted sand. *New Phytol.*, 145:477, 2000.
58. Iijima, M., Higuchi, T., and Barlow, P.M., Contribution of root cap mucilage and presence of an intact root cap in maize (*Zea mays* L.) to the reduction of the soil mechanical impedance. *Ann. Bot.*, 94:473, 2004.
59. Inukai, Y. et al., Mechanical stimulus-sensitive mutation, *rrl3*, affects the cell production process in the root meristematic zone in rice. *Plant Prod. Sci.*, 6:265, 2003.
60. Vartapetian, B.B. and Jackson, M.B., Plant adaptations to anaerobic stress. *Ann. Bot.*, 79:3, 1997.
61. Jackson, M.B., The impact of flooding stress on plants and crops. http://www.plantstress.com/Articles/index.asp.
62. Colmer, T.D., Long-distance transport of gases in plants, *a perspective on internal aeration and radial oxygen loss from roots. Plant Cell Environ.*, 26:17, 2003.
63. Jackson, M.B. and Ricard, B., Physiology, biochemistry and molecular biology of plant root systems subjected to flooding of the soil. In *Root Ecology*, de Kroon, H. and Visser, E.J.W., Eds., Springer Verlag, Berlin, Heidelberg, 2003, pp. 193–213.
64. Brailsford, R.W. et al., Enhanced ethylene production by primary roots of *Zea mays* L. in response to sub-ambient partial pressures of oxygen. *Plant Cell Environ.*, 16:1071, 1993.
65. Jackson, M.B., Long-distance signalling from roots to shoots assessed: the flooding story. *J. Exp. Bot.*, 53:175, 2002.
66. Arshad, M. and Frankenberger, W.T.J., Production and stability of ethylene in soil. *Biol. Fert. Soils.*, 10:29, 1990.
67. Setter, T.L. and Belford, R., Waterlogging — How it reduces plant growth and how plants can overcome its effects. *J. Agric. West Aust.*, 31:51, 1990.
68. Laanbroek, H.J., Bacterial cycling of minerals that affect plant growth in waterlogged soils, a review. *Aquatic Bot.*, 38:109, 1990.
69. Ponnamperuma, F.N., The chemistry of submerged soils. *Adv. Agron.*, 24:29, 1972.
70. Walker, G.E., Chemical, physical and biological control of carrot seedling diseases. *Plant Soil*, 136:31, 1991.
71. Yanar, Y., Lipps, P.E., and Deep, I.W., Effect of soil saturation, duration and water content on root rot of maize caused by Pythium arrhenomanes. *Plant Dis.*, 81:475, 1997.

72. Gerendás, J. and Ratcliffe, R.G., Root pH control. In *Plant Roots, the Hidden Half*, 3d ed., Waisel, Y., Eshel, A., and Kafkafi, U., Eds., Marcel Dekker, New York, 2002, pp. 553–570.
73. Boamfa, E.I. et al., Dynamic aspects of alcoholic fermentation of rice seedlings in response to anaerobiosis and to complete submergence, relationship to submergence tolerance. *Ann. Bot.*, 91:279, 2003.
74. Dordas, C., Rivoal, J., and Hill, R.D., Plant haemoglobins, nitric oxide and hypoxic stress. *Ann. Bot.* 91:173, 2003.
75. Subbaiah, C.C. and Sachs, M.M., Molecular and cellular adaptations of maize to flooding stress. *Ann. Bot.*, 91:119, 2003.
76. Armstrong, W., Aeration in higher plants. *Adv. Bot. Res.*, 7:225, 1979.
77. Justin, S.H.F.W. and Armstrong, W., The anatomical characteristics of roots and plant response to soil flooding. *New Phytol.*, 106:465, 1987.
78. Trought, M.C.T. and Drew, M.C., The development of waterlogging damage in young wheat plants in anaerobic cultures. *J. Exp. Bot.*, 31:1573, 1980.
79. Benjamin, L.R. and Greenway, H., Effect of a range of O_2 concentrations on porosity of barley roots and on their sugar and protein concentrations. *Ann. Bot.*, 43:383, 1979.
80. Drew, M.C., Jackson, M.B., and Gaffard, S., Ethylene-promoted adventitious rooting and development of cortical air space (aerenchyma) in roots may be adaptive responses to flooding in *Zea mays* L. *Planta*, 147:83, 1979.
81. Armstrong, W., Radial oxygen losses from intact rice roots as affected by distance from the apex, respiration and waterlogging. *Physiol. Plant.*, 25:192, 1971.
82. Colmer, T.D. et al., The barrier to radial oxygen loss from roots of rice (*Oryza sativa* L.) is induced by growth in stagnant solution. *J. Exp. Bot.*, 49:1431, 1998.
83. Kono, Y., Igeta, M., and Yamada, N., Studies on the developmental physiology of the lateral roots in the rice seminal roots. *Proc. Crop Sci. Soc. Japan*, 41:192, 1972.
84. Kono, Y. and Yamada, N., Studies on the developmental physiology of the relationship between the cortical disintegration and lateral root growth in rice seminal roots. *Proc. Crop Sci. Soc. Japan*, 41:256, 1972.
85. Shimamura, S. et al., Formation and function of secondary aerenchyma in hypocotyl, roots and nodules of soybean (*Glycine max*) under flooded conditions. *Plant Soil*, 251:351, 2003.
86. Kawai, M. et al., Cellular dissection of the degradation pattern of cortical cell death during aerenchyma formation of rice roots. *Planta*, 204:277, 1998.
87. Arikado, H. and Adachi, Y., Anatomical and ecological responses of barley and some forage crops to the flooding treatment. *Bull. Fac. Agr. Mie Univ.*, 11:1, 1955.
88. Huang, B.R. et al, Root and shoot growth of wheat genotypes in response to hypoxia and subsequent resumption of aeration. *Crop Sci.*, 34:1538, 1994.
89. Laan, P. et al., Root morphology and aerenchyma formation as indicators of the flood-tolerance of Rumex species. *J. Ecol.*, 77:693, 1989.
90. Arikado, H., Supplementary studies on the development of the ventilating system in various plants growing on lowland and on upland. *Bull. Fac. Agr. Mie Univ.*, 20:1, 1959.
91. Jackson, M.B. and Armstrong, W., Formation of aerenchyma and the processes of plant ventilation in relation to soil flooding and submergence. *Plant Biol.*, 1:274, 1999.
92. Drew, M.C., He, C.J., and Morgan, P.W., Programmed cell death and aerenchyma formation in roots, *Trends. Plant Sci.* 5:123, 2000.
93. Evans, D.E., Aerenchyma formation. *New Phytol.*, 161:35, 2003.

94. He, C.J. et al., Ethylene biosynthesis during aerenchyma formation in roots of maize subjected to mechanical impedance and hypoxia. *Plant Physiol.*, 112:1679, 1996.
95. Erdmann, B., Hoffman, P., and Wiedenroth, E.M., Changes in the root system of wheat seedlings following root anaerobiosis I. Anatomy and respiration in *Triticum sativum* L. *Ann. Bot.*, 58:597, 1986.
96. Arikado, H., Different responses of soybean plants to an excess of water with special reference to anatomical observations. *Proc. Crop Sci. Soc. Japan*, 23:28, 1954.
97. Saraswati, R., Matoh, T., and Sekiya, J., Nitrogen fixation of Sesbania rostrata, contribution of stem nodules to nitrogen acquisition. *Soil Sci. Plant Nutr.*, 38:775, 1992.
98. Shiba, H. and Daimon, H., Histological observation of secondary aerenchyma formed immediately after flooding in *Sesbania cannabina* and *S. rostrata*. *Plant Soil*, 255:209, 2003.
99. James, E.K. and Sprent, J.I., Development of N_2-fixing nodules on the wetland legume Lotus uliginosus exposed to conditions of flooding. *New Phytol.*, 142:219, 1999.
100. Walker, B.A., Pate, J.S., and Kuo, J., Nitrogen fixation by nodulated roots of *Viminaria juncea* (Schrad. and Wendl.) Hoffmans (Fabaceae) when submerged in water. *Aust. J. Plant Physiol.*, 10:409, 1983.
101. Arber, A., *Water Plants, a Study of Aquatic Angiosperms*, Cambridge University Press, Cambridge, 1920, p. 436.
102. Metcalfe, C.R., The "aerenchyma" of *Sesbania* and *Neptunia* In *Bulletin of Miscellaneous Information, His Majesty's Stationery Office*, London, 1931, pp. 151–154.
103. Konings, H. and Verschuren, G., Formation of aerenchyma in roots of Zea mays in aerated solutions, and its relation to nutrient supply. *Physiol. Plant.*, 49:265, 1980.
104. He, C.J., Morgan, P.W., and Drew, M.C., Enhanced sensitivity to ethylene in nitrogen- or phosphate-starved roots of *Zea mays* L. during aerenchyma formation. *Plant Physiol.*, 98:137, 1992.
105. Fan, M.S. et al., Physiological roles for aerenchyma in phosphorus-stressed roots. *Funct. Plant Biol.*, 30:493, 2003.
106. Malik, A.I. et al., Aerenchyma formation and radial O_2 loss along adventitious roots of wheat with only the apical root portion exposed to O_2 deficiency. *Plant Cell Environ.*, 26:1713, 2003.
107. Colmer, T.D., Aerenchyma and an inducible barrier to radial oxygen loss facilitate root acration in upland, paddy and deepwater rice (*Oryza sativa* L.). *Ann. Bot.*, 91:301, 2003.
108. Visser, E.J.W. et al., Changes in growth, porosity, and radial oxygen loss from adventitious roots of selected mono-and dicotyledonous wetland species with contrasting types of aerenchyma. *Plant Cell Environ.*, 23:1237, 2000.
109. Thomson, C.J. et al., Tolerance of wheat (*Triticum aestivum* cvs. Gamenya and Kite) and Triticale (Triticosecale cv. Muir) to waterlogging. *New Phytol.*, 120:335, 1992.
110. McDonald, M.P., Galwey, N.W., and Colmer, T.D., Waterlogging tolerance in the tribe Triticeae, the adventitious roots of *Critesion marinum* have a high porosity and a barrier to radial oxygen loss. *Plant Cell Environ.*, 24:585, 2001.
111. McDonald, M.P., Galwey, N.W., and Colmer, T.D., Similarity and diversity in adventitious root anatomy as related to root aeration among range of wet- and dry-land grass species. *Plant Cell Environ.*, 25:441, 2002.

112. Chabbi, A., McKee, K.L., and Mendelssohn, I.A., Fate of oxygen losses from *Typha domingensis* (Typhaceae) and *Cladium jamaicensi* (Cyperaceae) and consequences for root metabolism. *Am. J. Bot.*, 87:1081, 2000.
113. Kirk, G.J.D., Rice root properties for internal aeration and efficient nutrient acquisition in submerged soil. *New Phytol.*, 159:185, 2003.
114. Watkin, E.L.J., Thomson, C.J., and Greenway, H., Root development and aerenchyma formation in two wheat cultivars and one Triticale cultivar grown in stagnant agar and aerated nutrient solution. *Ann. Bot.*, 81:349, 1998.
115. Kozlowski, T.T. and Pallardy, S.G., Acclimation and adaptive responses of woody plants to environmental stresses. *Bot. Rev.*, 68:270, 2002.
116. Sena Gomes, A.R. and Kozlowski, T.T., Effects of flooding in growth of *Eucalyptus camaldulensis* and E. *globules seedlings*. *Oecologia*, 46:139, 1980.
117. Tsukahara, H. and Kozlowski, T.T., Importance of adventitious roots to growth of flooded Platanus occidentalis seedlings. *Plant Soil*, 88:123, 1985.
118. Hook, D.D. and Brown, C.L., Root adaptations and relative flood tolerance of five hardwood species. *Forest. Sci.*, 19:225, 1973.
119. Reid, D.M. and Bradford, K.J., Effects of flooding on hormone relations. In *Flooding and Plant Growth*, Kozlowski, T.T., Ed., Academic, Orlando, FL, 1984, pp. 195–219.
120. Tang, Z.C. and Kozlowski, T.T., Ethylene production and morphological adaptations of woody plants to flooding. *Can. J. Bot.*, 62:1659, 1984.
121. McDonald, M.P. and Visser, E.J.W., A study of the interaction between auxin and ethylene in wild type and transgenic ethylene-insensitive tobacco during adventitious root formation induced by stagnant root zone conditions. *Plant Biol.* 5:550, 2003.
122. Bernstein, N. and Kafkafi, U., Root growth under salinity stress. In *Plant Roots, the Hidden Half*, 3d ed., Waisel, Y., Eshel, A., and Kafkafi, U., Eds., Marcel Dekker, New York, 2002, pp. 787–805.
123. Greenway, H. and Munns, R., Mechanisms of salt tolerance in non-halophytes. *Annu. Rev. Plant Physiol.*, 31:149, 1980.
124. Cheeseman, J.M., Mechanisms of salinity tolerance in plants. *Annu. Rev. Plant Physiol.*, 87:547, 1988.
125. Kummerow, J., Adaptation of roots in water-stressed native vegetation. In *Adaptation of Plants to Water and High Temperature Stress*, Turner, N.C. and Kramer, P.J., Eds., John Wiley & Sons, New York, 1980, pp. 57–73.
126. Passioura, J.B., Water collection by roots. In *The Physiology and Biochemistry of Drought Resistance in Plants*, Paleg, L.G. and Aaspinall, D., Eds., Academic, Sydney, 1981, pp. 39–53.
127. Passioura, J.B., Roots and drought resistance. *Agr. Water Manage.*, 1:265, 1983.
128. Gorham, J., Genetics and physiology of enhanced K^+/N^+ discrimination. In *Developments in Plant and Soil Sciences*, Vol. 50, Kluwer Academic, The Netherlands, 1993, pp. 151–158.
129. Gorham, J. et al., Genetic analysis and physiology of a trait for enhanced K^+/Na^+ discrimination in wheat. *New Phytol.*, 137:109, 1997.
130. Rubinigg, M. et al., NaCl salinity affects lateral root development in *Plantago maritime*. *Funct. Plant Biol.*, 31:775, 2004.
131. Zhong, H. and Läuchli, A. Spatial and temporal aspects of growth in the primary root of cotton seedlings, effect of NaCl and $CaCl_2$. *J. Exp. Bot.*, 44:763, 1993.
132. Waisel, Y. and Breckle, S.W., Differences in responses of various radish roots to salinity. *Plant Soil*, 104:191, 1987.

133. Reinhardt, D.H. and Rost, T.L., Salinity accelerated endodermal development and induces an exodermis in cotton seedling roots. *Environ. Exp. Bot.*, 35:563, 1995.
134. Kramer, D., Transfer cells in the epidermis of roots. In *Plant Membrane Transport, Current Conceptual Issues*, Spanswick, R.M., Lucas, W.J., and Dainty, J., Eds., Elsevier, Amsterdam, 1980, pp. 393–394.
135. Burssens, S. et al., Expression of cell cycle regulatory genes and morphological alterations in response to salt stress in *Arabidopsis thaliana*. *Planta*, 211:632, 2000.
136. Pritchard, J., The control of cell expansion in roots. Tansley review No. 68. *New Phytol.*, 127:3, 1994.
137. Koyro, H.W., Ultrastructural and physiological changes in root cells of sorghum plants (*Sorghum bicolor* x *S. sudanensis* cv. Sweet Sioux) induced by NaCl. *J. Exp. Bot.*, 48:693, 1997.
138. Samarajeewa, P.K. et al., Cortical cell death, cell proliferation, macromolecular movements and rTip1 expression pattern in roots of rice (*Oryza sativa* L.) under NaCl stress. *Planta*, 207:354, 1999.
139. West, G., Inze, D., and Beemster, G.T.S., Cell cycle modulation in the response of the primary root of Arabidopsis to salt stress. *Plant Physiol.*, 135:1050, 2004.
140. Kurth, E. et al., Effects of NaCl and $CaCl_2$ on cell enlargement and cell production in cotton roots. *Plant Physiol.*, 82:1102, 1986.
141. Zidan, I., Azaizeh, H., and Neumann, P.M., Does salinity reduce growth of maize root epidermal cells by inhibiting their capacity for cell wall acidification? *Plant Physiol.*, 93:7, 1990.
142. Katsuhara, M. and Kawasaki, T., Salt stress induced nuclear and DNA degradation in meristematic cells of barley roots. *Plant Cell Physiol.*, 37:169, 1996.
143. Richardson, K.V.A., Wetten, A.C., and Caligari, P.D.S., Cell and nuclear degradation in root meristems following exposure of potatoes (*Solanum tuberosum* L.) to salinity. *Potato Res.*, 44:389, 2001.
144. Espinosa-Ruiz, A. et al., *Arabidopsis thaliana* AtHAL3, a flavoprotein related to salt and osmotic tolerance and plant growth. *Plant J.*, 20:529, 1999.
145. Tramontano, W.A. and Jouve, D., Trigonelline accumulation in salt-stressed legumes and the role of other osmoregulators as cell cycle control agents. *Phytochemistry*, 44:1037, 1997.
146. Cramer, G.R. and Bowman, D.C., Kinetics of maize leaf elongation. 1. Increased yield threshold limits short-term, steady-state elongation rates after exposure to salinity. *J. Exp. Bot.*, 42:1417, 1991.
147. Rawson, H.M. and Munns, R., Leaf expansion in sunflower as influenced by salinity and short-term changes in carbon fixation. *Plant Cell Environ.*, 7:207, 1984.
148. Azaizeh, H., Gunse, B., and Steudle, E., Effects of NaCl and $CaCl_2$ on water transport across root-cells of maize (*Zea mays* L) seedlings. *Plant Physiol.*, 99:886, 1992.
149. Volkmar, K.M., Hu, Y., and Steppuhn, H., Physiological responses of plants to salinity, a review. *Can. J. Plant Sci.*, 78:19, 1998.
150. Huang, C.X. and Van Steveninck, R.F.M., Salinity induced structural changes in meristematic cells of barley roots. *New Phytol.*, 115:17, 1990.
151. Rahman, M.S. et al., Effects of salinity stress on the seminal root tip ultrastructures of rice seedlings (*Oryza sativa* L.). *Plant Prod. Sci.*, 4:103, 2001.
152. Solomon, M. et al., Changes induced by salinity to the anatomy and morphology of excised pea roots in culture. *Ann. Bot.*, 68:47, 1986.
153. Huang, J. and Redmann, R.E., Responses of growth, morphology, and anatomy to salinity and calcium supply in cultivated and wild barley. *Can. J. Bot.*, 73:1859, 1995.

154. Hajibagheri, M.A., Yeo, A.R., and Flowers, T.J., Salt tolerance in *Suaeda maritima* (L.) Dum. Fine-structure and ion concentrations in the apical region of roots. *New Phytol.*, 99:331, 1985.
155. Shannon, M.C., Grieve, C.M., and Francois, L.E., Whole plant response to salinity. In *Plant–Environment Interaction*, Wilkinson, R.E., Eds., Marcel Dekker, New York, 1994, pp. 199–244.
156. Dat, J.F. et al., Sensing and signalling during plant flooding. *Plant Physiol. Biochem.*, 42:273, 2004.
157. Zhang, J. and Davies, W.J., Increased synthesis of ABA in partially dehydrated root-tips and ABA transport from roots to leaves. *J. Exp. Bot.*, 38:2015, 1987.
158. Davies, W.J. and Zhang, J.H., Root signals and the regulation of growth and development of plants in drying soil. *Annu. Rev. Plant Physiol. Plant Mol. Biol.*, 42:55, 1991.
159. Wilkinson, S. and Davies, W.J., ABA-based chemical signalling, the co-ordination of responses to stress in plants. *Plant Cell Environ.*, 25:195, 2002.
160. Bacon, M.A., Wilkinson, S., and Davies, W.J., pH-regulated leaf cell expansion in droughted plants is abscisic acid dependent. *Plant Physiol.*, 118:1507, 1998.
161. Hartung, W., Sauter, A., and Hose, E., Abscisic acid in the xylem, where does it come from, where does it go to? *J. Exp. Bot.*, 53:27, 2002.
162. Chaves, M.M. and Oliveira, M.M., Mechanisms underlying plant resilience to water deficits, prospects for water-saving agriculture. *J. Exp. Bot.*, 55:2365, 2004.
163. Sobeih, W.Y. et al., Long-distance signals regulating stomatal conductance and leaf growth in tomato (*Lycopersicon esculentum*) plants subjected to partial root-zone drying. *J. Exp. Bot.*, 55:2353, 2004.
164. Yang, S.F. and Hoffman, N.E., Ethylene biosynthesis and its regulation in higher-plants. *Annu. Rev. Plant Physiol. Plant Mol. Biol.*, 35:155, 1984.
165. Tudela, D. and Primo-Millo, E., 1-aminocyclopropane-1-carboxylic acid transported from roots to shoots promotes leaf abscission in cleopatra mandarine (*Citrus reshni* Hort ex Tan) seedlings rehydrated after water-stress. *Plant Physiol.*, 100:131, 1992.
166. Else, M.A. and Jackson, M.B. Transport of 1-aminocyclopropane-1-carboxylic acid (ACC) in the transpiration stream of tomato (*Lycopersicon esculentum*) in relation to foliar ethylene production and petiole epinasty. *Aust. J. Plant Physiol.*, 25:453, 1998.
167. Ouvrard, O. et al., Identification and expression of water stress- and abscisic acid-regulated genes in a drought-tolerant sunflower genotype. *Plant Mol. Biol.*, 31:819, 1996.
168. Hartung, W. et al., Abscisic acid in soils, What is its function and which factors and mechanisms influence its concentration? *Plant Soil*, 184:105, 1996.
169. Sauter, A. and Hartung, W., Radial transport of abscisic acid conjugates in maize roots, its implication for long distance stress signals. *J. Exp. Bot.*, 51:929, 2000.
170. Slovik, S., Daeter, W., and Hartung, W., Compartmental redistribution and long-distance transport of abscisic-acid (ABA) in plants as influenced by environmental changes in the rhizosphere — A biomathematical model. *J. Exp. Bot.*, 46:881, 1995.
171. O'Toole, J. C. and Bland, W. L., Genotypic variation in crop plant root systems. *Adv. Agron.*, 41:91, 1987.
172. Sponchiado, B.N. et al., Root growth of four common bean cultivars in relation to drought tolerance in environments with contrasting soil types. *Exp. Agri.*, 25:249, 1989.

173. Saab, I.N. et al., Increased endogenous abscisic acid maintains primary root growth and inhibits shoot growth of maize seedlings at low water potentials. *Plant Physiol.*, 93:1329, 1990.
174. Taylor, I.B, Burbidge, A., and Thompson, A.J., Control of abscisic acid synthesis. *J. Exp. Bot.*, 51:1563, 2000.
175. Tan, B.C. et al., Genetic control of abscisic acid biosynthesis in maize. *Proc. Natl. Acad. Sci. USA*, 94:12235, 1997.
176. Qin, X. and Zeevaart, J.A.D., The 9-cis-epoxycarotenoid cleavage reaction is the key regulatory step in abscisic acid biosynthesis in water-stressed bean. *Proc. Natl. Acad. Sci. USA*, 96:15354, 1999.
177. Spollen, W.G. et al., Abscisic acid accumulation maintains maize primary root elongation at low water potentials by restricting ethylene production. *Plant Physiol.*, 122:967, 2000.
178. Sharp, R.E. et al., Endogenous ABA maintains shoot growth in tomato independently of effects on plant water balance, evidence for an interaction with ethylene. *J. Exp. Bot.*, 51:1575, 2000.
179. Moss, G.I., Hall, K.C., and Jackson, M.B., Ethylene and the responses of roots of maize (*Zea mays* L.) to physical impedance. *New Phytol.*, 109:303, 1988.
180. Sarquis, J.I., Jordan, W., and Morgan, P.W., Ethylene evolution from maize (*Zea mays* L.) seedling roots and shoots in response to mechanical impedance. *Plant Physiol.*, 96:1171, 1991.
181. Baluök, F. et al., Cellular dimorphism in the maize root cortex-involvement of microtubules, ethylene and gibberellin in the differentiation of cellular behavior in postmitotic growth zones. *Bot. Acta.*, 106:394, 1993a.
182. Osborne, D.J., Control of cell shape and cell size by the dual recognition of auxin and ethylene. In *Perspectives in Experimental Biology*, Vol. 2., Sunderland, N., Ed., Pergamon, Oxford, 1976, pp. 89–102.
183. Sarquis, J.I., Morgan, P.W., and Jordan, W., Metabolism of 1-aminocyclopropane-1-carboxylic acid in etiolated maize seedlings grown under mechanical impedance. *Plant Physiol.*, 98:1342, 1992.
184. Timpte, C. et al., The AXR1 and AUX1 genes of Arabidopsis function in separate auxin-response pathways. *Plant J.*, 8:561, 1995.
185. Luschnig, C. et al., EIR1, a root-specific protein involved in auxin transport, is required for gravitropism in *Arabidopsis thaliana*. *Genes Dev.*, 12:2175, 1998.
186. Harper, R.M. et al., The NPH4 locus encodes the auxin response factor ARF7, a conditional regulator of differential growth in aerial *Arabidopsis* tissue. *Plant Cell*, 12:757, 2000.
187. Lachno, D.R., Harrisonmurray, R.S., and Audus, L.J., The effects of mechanical impedance to growth on the levels of ABA and IAA in root-tips of *Zea mays* L. *J. Exp. Bot.* 33:943, 1982.
188. Tardieu, F. et al., Maize stomatal conductance in the field — Its relationship with soil and plant water potentials, mechanical constraints and ABA concentration in the xylem sap. *Plant Cell Environ.*, 14:121, 1991.
189. Hartung, W., Zhang, J.H., and Davies, W.J., Does abscisic acid play a stress physiological-role in maize plants growing in heavily compacted soil. *J. Exp. Bot.*, 45:221, 1994.
190. Mulholland, B.J. et al., Effect of soil compaction on barley (*Hordeum vulgare* L.) growth. 1. Possible role for ABA as a root-sourced chemical signal. *J. Exp. Bot.*, 47:539, 1996.

191. Mulholland, B.J. et al., Effect of soil compaction on barley (*Hordeum vulgare* L) growth. 2. Are increased xylem sap ABA concentrations involved in maintaining leaf expansion in compacted soils? *J. Exp. Bot.*, 47:551, 1996.
192. Hussain, A. et al., Novel approaches for examining the effects of differential soil compaction on xylem sap abscisic acid concentration, stomatal conductance and growth in barley (*Hordeum vulgare* L.). *Plant Cell Environ.*, 22:1377, 1999.
193. Hussain, A. et al., Does an antagonistic relationship between ABA and ethylene mediate shoot growth when tomato (*Lycopersicon esculentum* Mill.) plants encounter compacted soil? *Plant Cell Environ.*, 23:1217, 2000.
194. Roberts, J.A. et al., Use of mutants to study long-distance signalling in response to compacted soil. *J. Exp. Bot.*, 53:45, 2002.
195. Sharp, R.E., Interaction with ethylene, changing views on the role of abscisic acid in root and shoot growth responses to water stress. *Plant Cell Environ.*, 25:211, 2002.
196. Jackson, M.B., Hormones from roots as signals for the shoots of stressed plants. *Trends Plant Sci.*, 2:22, 1997.
197. Lorbiecke, R. and Sauter, M., Adventitious root growth and cell-cycle induction in deepwater rice. *Plant Physiol.*, 119:21, 1999.
198. Kende, H., Ethylene biosynthesis. *Annu. Rev. Plant Physiol. Plant Mol. Biol.*, 44:283, 1993.
199. Olson, D.C., Oetiker, J.H., and Yang, S.F., Analysis of LE-ACS3, a 1-aminocyclopropane-1-carboxylic acid synthase gene expressed during flooding in the roots of tomato plants. *J. Biol. Chem.*, 270:14056, 1995.
200. Peng, H.P. et al., Signaling events in the hypoxic induction of alcohol dehydrogenase gene in Arabidopsis. *Plant Physiol.*, 126:742, 2001.
201. Shiu, O.Y. et al., The promoter of LE-ACS7, an early flooding-induced 1-aminocyclopropane-1-carboxylate synthase gene of the tomato, is tagged by a Sol3 transposon. *Proc. Natl. Acad. Sci. USA*, 95:10334, 1998.
202. Jackson, M.B. and Campbell, D.J., Movement of ethylene from roots to shoots, a factor in the responses of tomato plants to waterlogged soil conditions. *New Phytol.*, 7:397, 1975.
203. Voesenek, L.A.C.J. et al., Submergence-induced ethylene synthesis, entrapment, and growth in two plant species with contrasting flooding resistances. *Plant Physiol.*, 103, 783, 1993.
204. Burrows, W.J. and Carr, D.J., Effects of flooding root system of sunflower plants on cytokinin content in xylem sap. *Physiol. Plant.*, 22, 1105, 1969.
205. Reid, D.M. and Crozier, D.M. Effects of waterlogging on the gibberellin content and growth of tomato plants. *J. Exp. Bot.*, 22, 39, 1971.
206. Neuman, D.S., Rood, S.B., and Smit, B.A., Does cytokinin transport from root-to-shoot in the xylem sap regulate leaf responses to root hypoxia? *J. Exp. Bot.*, 41:1325, 1990.
207. Jackson, M.B. and Campbell, D.J., Effects of benzyladenine and gibberellic-acid on the responses of tomato plants to anaerobic root environments and to ethylene. *New Phytol.*, 82:331, 1979.
208. Visser, E.J.W. et al., An ethylene-mediated increase in sensitivity to auxin induces adventitious root formation in flooded *Rumex palustris* Sm. *Plant Physiol.*, 112:1687, 1996.
209. Chhun, T. et al., Interaction between two auxin-resistant mutants and their effects on lateral root formation in rice (*Oryza sativa* L.). *J. Exp. Bot.*, 54:2701, 2003.

210. Klok, E.J. et al., Expression profile analysis of the low-oxygen response in Arabidopsis root cultures. *Plant Cell*, 14:2481, 2002.
211. Jackson, M.B. and Hall, K.C., Early stomatal closure in waterlogged pea plants is mediated by abscisic acid in the absence of foliar water deficits. *Plant Cell Environ.*, 10:121, 1987.
212. Jackson, M.B., Regulation of water relationships in flooded plants by ABA from leaves, roots and xylem sap. In *Abscisic Acid. Physiology and Biochemistry*, Davies, W.J. and Jones, H.G., Eds., Bios Scientific, Oxford, 1991, pp. 217–226.
213. Else, M.A. et al., Concentrations of abscisic acid and other solutes in xylem sap from root systems of tomato and castor-oil plants are distorted by wounding and variable sap flow rates. *J. Exp. Bot.*, 45:317, 1994.
214. Else, M.A. et al., Decreased root hydraulic conductivity reduces leaf water potential, initiates stomatal closure and slows leaf expansion in flooded plants of castor oil (*Ricinus communis*) despite diminished delivery of ABA from roots to shoots in xylem sap. *Physiol. Plant.*, 111:46, 2001.
215. Else, M.A. et al., A negative hydraulic message from oxygen-deficient roots of tomato plants? Influence of soil flooding on leaf water potential, leaf expansion and the synchrony of stomatal conductance and root hydraulic conductivity. *Plant Physiol.*, 109:1017, 1995.
216. Else, M.A. et al., Export of abscisic acid, 1-aminocyclopropane-1-carboxylic acid, phosphate, and nitrate from roots to shoots of flooded tomato plants-accounting for effects of xylem sap flow rate on concentration and delivery. *Plant Physiol.*, 107:377, 1995.
217. Subbaiah, C.C., Bush, D.S., and Sachs, M.M., Elevation of cytosolic calcium precedes anoxic gene expression in maize suspension cultured cells. *Plant Cell*, 6:1747, 1994.
218. Subbaiah, C.C., Zhang, J., and Sachs, M.M., Involvement of intracellular calcium in anaerobic gene expression and survival of maize seedlings. *Plant Physiol.*, 105:369, 1994.
219. Sedbrook, J.C. et al., Transgenic AEQUORIN reveals organ specific cytosolic Ca^{2+} responses to anoxia in Arabidopsis thaliana seedlings. *Plant Physiol.*, 111:243, 1996.
220. Subbaiah, C.C. and Sachs, M.M., Calcium-mediated responses of maize to oxygen deprivation, *Russ. J. Plant Physiol.*, 50, 2003.
221. Shelp, B.J., Bown A.W., and McLean, M.D., Metabolism and functions of gamma-aminobutyric acid. *Trends Plant Sci.*, 4:446, 1999.
222. Subbaiah, C.C. and Sachs, M.M., Maize cap1 encodes a novel SERCA-type calcium-ATPase with a calmodulin-binding domain. *J. Biol. Chem.*, 275:21678, 2000.
223. Zhu, J.K., Salt and drought stress signal transduction in plants. *Annu. Rev. Plant Biol.*, 53:247, 2002.
224. Chinnusamy, V. and Zhu, J.K., Plant salt tolerance. In *Plant Responses to Abiotic Stress. Topics in Current Genetics*, Vol. 4, Hirt, H. and Shinozaki, K., Eds., Springer-Verlag, Berlin, Heidelberg, 2003, pp. 241–170.
225. Sanders, D., Brownlee, C., and Harper, J.F., Communicating with calcium. *Plant Cell.*, 11:691, 1999.
226. Jia, W.S. et al., Salt-stress-induced ABA accumulation is more sensitively triggered in roots than in shoots. *J. Exp. Bot.*, 53:2201, 2002.
227. Smirnoff, N., The role of active oxygen in the response of plants to water deficit and desiccation. *New Phytol.*, 125:27, 1993.

228. Gomez, J.M. et al., Differential response of antioxidative enzymes of chloroplast and mitochondria to long term NaCl stress of pea plants. *Free Radic. Res.*, 31:S11, 1999.
229. Hernández, J.A. et al., Antioxidant systems and O_2^-/H_2O_2 production in the apoplast of pea leaves. Its relation with salt-induced necrotic lesions in minor veins. *Plant Physiol.*, 127:817, 2001.
230. Cramer, G.R. and Jones, R.L., Osmotic stress and abscisic acid reduce cytosolic calcium activities in roots of *Arabidopsis thaliana*. *Plant Cell Environ.*, 19:1291, 1996.
231. Knight, H. and Knight, M.R., Abiotic stress signalling pathways, specificity and cross-talk. *Trends Plant Sci.*, 6:262, 2001.
232. Knight, H., Trewavas, A.J., and Knight, M.R., Calcium signalling in *Arabidopsis thaliana* responding to drought and salinity. *Plant J.*, 12:1067, 1997.
233. Trewavas, A.J. and Malho, R., Ca^{2+} signalling in plant cells, the big network! *Curr. Opin. Plant Biol.*, 1:428, 1998.
234. Zielinski, R.E., Calmodulin and calmodulin-binding proteins in plants. *Annu. Rev. Plant Physiol. Plant Mol. Biol.*, 49:697, 1998.
235. Luan, S. et al., Calmodulins and calcineurin B-like proteins: Calcium sensors for specific signal response coupling in plants. *Plant Cell*, 14:S389, 2002.
236. Harmon, A.C., Gribskov, M., and Harper, J.F., CDPKs-a kinase for every Ca^{2+} signal? *Trends Plant Sci.*, 5:154, 2000.
237. Romeis, T. et al., Calcium-dependent protein kinases play an essential role in a plant defence response. *EMBO J.*, 20:5556, 2001.
238. Liu, J.P. and Zhu, J.K., A calcium sensor homolog required for plant salt tolerance. *Science*, 280:1943, 1998.
239. Kramer, P.J. and Boyer, J.S., *Water Relations of Plants and Soils*, Academic Press, New York, 1995.
240. Blum, A., *Plant Breeding for Stress Environments*, CRC Press, Boca Raton, FL, 1988.
241. Voetberg, G.S. and Sharp, R.E., Growth of the maize primary root at low water potentials. III. Role of increased proline deposition in osmotic adjustment. *Plant Physiol.*, 96:1125, 1991.
242. Munns, R., Comparative physiology of salt and water stress. *Plant Cell Environ.*, 25:239, 2002.
243. Serraj, R. and Sinclair, T.R., Osmolyte accumulation, can it really help increase crop yield under drought conditions? *Plant Cell Environ.*, 25:333, 2002.
244. Ogawa, A. and Yamauchi, A., Root osmotic adjustment under osmotic stress conditions in maize seedlings. 1. Transient response of growth and water relations in roots to osmotic stress. *Plant Prod. Sci.*, 9:27–38, 2006.
245. Ogawa, A. and Yamauchi, A., Root osmotic adjustment under osmotic stress conditions in maize seedlings. 2. Comparison of time-course accumulation of several solutes for osmotic adjustment in root. *Plant Prod. Sci.*, 9:39–46, 2006.
246. Sharp, R.E., Hsiao, T.C., and Silk, W.K., Growth of the maize primary root at low water potentials. II. The role of growth and deposition of hexose and potassium in osmotic adjustment. *Plant Physiol.*, 93:1337, 1990.
247. Westgate, M.E. and Boyer, J.S., Osmotic adjustment and the inhibition of leaf, root, stem, and silk growth at low water potentials in maize. *Planta*, 164:540, 1985.
248. Hsiao, T.C. and Xu, L.K., Sensitivity of growth of roots versus leaves to water stress, biophysical analysis and relation to water transport. *J. Exp. Bot.*, 51:1595, 2000.

249. Matyssek, R., Tang, A.C., and Boyer, J.S., Plants can grow on internal water. *Plant Cell Environ.*, 14:925, 1991.
250. Raven, J.A., Regulation of pH and generation of osmolarity in vascular plants: A cost-benefit analysis in relation to efficiency of use of energy, nitrogen and water. *New Phytol.*, 25:25, 1985.
251. Verslues, P.E. and Sharp, R.E., Proline accumulation in maize (*Zea mays* L.) primary roots at low water potentials. II. Metabolic source of increased proline deposition in the elongation zone. *Plant Physiol.*, 119:1349, 1999.
252. Ober, E.S. and Sharp, R.E., Proline accumulation in maize (*Zea mays* L.) primary roots at low water potentials. I. Requirement for increased levels of ABA. *Plant Physiol.*, 105:981, 1994.
253. Jeschke, W.D., Aslam, Z., and Greenway, H., Effects of NaCl on ion relations and carbohydrate status of roots and on osmotic regulation of roots and shoots of Atriplex amnicola. *Plant Cell Environ.*, 9:559, 1986.
254. Hajibagheri, M.A., Harvey, D.M.R., and Flowers, T.J., Quantitative ion distribution within root cells of salt-sensitive and salt tolerant maize varieties. *New Phytol.*, 105:367, 1987.
255. Gorham, J. et al., Salt tolerance in the triticeae-K/Na discrimination in barley. *J. Exp. Bot.*, 41:1095, 1990.
256. Wyn Jones, R.G. and Storey, R., Salt stress and comparative physiology in the Gramineae. IV. Comparison of salt stress in Spartina by Townsendii and three barley cultivars. *Aust. J. Plant Physl.*, 25:839, 1978.
257. Pritchard, J. and Tomos, A.D., Correlating biophysical and biochemical control of root expansion. In *Water Deficits, Plant Responses from Cell to Community*, Smith, J.A.C. and Griffiths, H., Eds., Bios Scientific, Oxford, 1993, pp. 53–72.
258. Rodrìgue, H.G. et al., Growth, water relations, and accumulation of organic and inorganic solutes in roots of maize seedlings during salt stress. *Plant Physiol.*, 113:881, 1997.
259. Lockhart, J.A., An analysis of irreversible plant cell elongation. *J. Theor. Biol.*, 8:264, 1965.
260. Pritchard, J., Wyn Jones, R.G., and Tomos, A.D., Measurements of yield threshold and cell wall extensibility of intact wheat roots under different ionic, osmotic and temperature treatments. *J. Exp. Bot.*, 277:669, 1990.
261. Pritchard, J., Wyn Jones, R.G., and Tomos, A.D., Turgor, growth, and rheological gradients of wheat roots following osmotic stress. *J. Exp. Bot.*, 42:1043, 1991.
262. Spollen, W.G. and Sharp, R.E., Spatial distribution of turgor and root growth at low water potentials. *Plant Physiol.*, 96:438, 1991.
263. Wu, Y.J. et al., Growth maintenance of the maize primary root at low water potentials involves increases in cell wall extension properties, expansin activity and wall susceptibility to expansins. *Plant Physiol.*, 111:765, 1996.
264. Frensch, J. and Hsiao, T.C., Transient responses of cell turgor and growth of maize roots as affected by changes in water potential. *Plant Physiol.*, 104:247, 1994.
265. Neumann, P.M., Azaizeh, H., and Leon, D., Hardening of root cell walls, a growth inhibitory response to salinity stress. *Plant Cell Environ.*, 17:303, 1994.
266. Morgan, J.M. and Condon, A.G., Water-use, grain-yield, and osmoregulation in wheat. *Aust. J. Plant Physiol.*, 13:523, 1986.

267. Morgan, J.M., Growth and yield of wheat lines with differing osmoregulative capacity at high soil-water deficit in seasons of varying evaporative demand. *Field Crops Res.*, 40:143, 1995.
268. Foyer, C.H. and Noctor, G., Oxygen processing in photosynthesis, regulation and signaling. *New Phytol.*, 146:359, 2000.
269. Mittler, R., Oxidative stress, antioxidants and stress tolerance. *Trends Plant Sci.*, 7:405, 2002.
270. Mittler, R. et al., Reactive oxygen gene network of plants. *Trends Plant Sci.*, 9:490, 2004.
271. Scandalios, J.G., Molecular genetics of superoxide dismutase in plants. In *Oxidative Stress and the Molecular Biology of Antioxidant Defenses*, Scandalios, J.G., Ed., Cold Spring Harbor Laboratory Press, New York, 1997, pp. 527–568.
272. Malecka, A., Jarmuszkiewicz, W., and Tomaszewska, B., Antioxidative defense to lead stress in subcellular compartments of pea root cells. *Acta Biochim. Pol.*, 48:687, 2001.
273. Dat, J. et al., Dual action of the active oxygen species during plant stress responses. *Cell. Mol. Life Sci.*, 57:779, 2000.
274. Corpas, F.J. et al., Purification of catalase from pea leaf peroxisomes, identification of five different isoforms. *Free Radical Res.*, 31:S235, 1999.
275. Shalata, A. et al., Response of the cultivated tomato and its wild salt-tolerant relative Lycopersicon pennellii to salt-dependent oxidative stress, the root antioxidative system. *Physiol. Plant.*, 112:487, 2001.
276. Noctor, G. and Foyer, C.H., Ascorbate and glutathione, keeping active oxygen under control. *Annu. Rev. Plant Physiol. Plant Mol. Biol.*, 49:249, 1998.
277. Otter, T. and Polle, A., The influence of apoplastic ascorbate on the activities of cell wall-associated peroxidase and NADH oxidase in needles of Norway spruce (*Picea abies* L). *Plant Cell Physiol.*, 35:1231, 1994.
278. Sukalovic, V.H.T. and Vuletic, M., The characterization of peroxidases in mitochondria of maize roots. *Plant Sci.*,164:999, 2003.
279. Cuypers, A., Vangronsveld, J., and Clijsters, H., Biphasic effect of copper on the ascorbate-glutathione pathway in primary leaves of Phaseolus vulgaris seedlings during the early stages of metal assimilation. *Physiol. Plant*, 110:512, 2000.
280. Vranová, E., Inzé, D., and Van Breusegem, F., Signal transduction during oxidative stress. *J Exp. Bot.*, 53:1227, 2002.
281. Pastori, G.M. and Foyer, C.H., Common components, networks, and pathways of cross-tolerance to stress. The central role of "redox" and abscisic acid-mediated controls. *Plant Physiol.*, 129:G7460, 2002.
282. Apel, K. and Hirt, H., Reactive oxygen species, metabolism, oxidative stress, and signal transduction. *Annu. Rev. Plant Biol.*, 55:373, 2004.
283. Møller, I.M., Plant mitochondria and oxidative stress, electron transport, NADPH turnover, and metabolism of reactive oxygen species. *Annu. Rev. Plant Physiol. Plant Mol. Biol.*, 52:561, 2001.
284. Fleury, C., Mignotte, B., and Vayssiere, J.L., Mitochondrial reactive oxygen species in cell death signaling. *Biochimie*, 84:131, 2002.
285. Gille, L. and Nohl, H., The ubiquinol/bc$_1$ redox couple regulates mitochondrial oxygen radical formation. *Arch. Biochem. Biophys.*, 388:34, 2001.
286. Mittova, V. et al., Salinity up-regulates the antioxidative system in root mitochondria and peroxisomes of the wild salt-tolerant tomato species *Lycopersicon pennellii*. *J. Exp. Bot.*, 55:1105, 2004.

287. Mittova, V. et al., Response of the cultivated tomato and its wild salt-tolerant relative *Lycopersicon pennellii* to salt-dependent oxidative stress, increased activities of antioxidant enzymes in root plastids. *Free Radical. Res.*, 36:195, 2002.
288. Gueta-Dahan, Y. et al., Salt and oxidative stress, similar and specific responses and their relation to salt tolerance in citrus. *Planta*, 203:460, 1997.
289. Khan, M.H. and Panda, S.K., Induction of oxidative stress in roots of *Oryza sativa* L. in response to salt stress. *Biol. Plant*, 45:625, 2002.
290. Panda, S.K. and Upadhyay, R.K., Salt stress injury induces oxidative alterations and antioxidative defence in the roots of *Lemna minor. Biol. Plant*, 48:249, 2004.
291. Bandeoglu, E. et al., Antioxidant responses of shoots and roots of lentil to NaCl-salinity stress. *Plant Growth Regul.*, 42:69, 2004.
292. Lee, D.H., Kim, Y.S., and Lee, C.B., The inductive responses of the antioxidant enzymes by salt stress in the rice (*Oryza sativa* L.). *J. Plant Physiol.*, 158:737, 2001.
293. Tsai, Y.C. et al., Relative importance of Na^+ and Cl^- in NaCl-induced antioxidant systems in roots of rice seedlings. *Physiol. Plant.*, 122:86, 2004.
294. Lin, C.C. and Kao, C.H., Cell wall peroxidase activity, hydrogen peroxide level and NaCl-inhibited root growth of rice seedlings. *Plant Soil*, 230:135, 2001.
295. Drew, M.C., Oxygen deficiency and root metabolism, Injury and acclimation under hypoxia and anoxia. *Annu. Rev. Plant Physiol.*, 48:223, 1997.
296. VanToai, T.T. and Bolles, C.S., Postanoxic injury in soybean (*Glycine max*) seedlings. *Plant Physiol.*, 97:588, 1991.
297. Monk, L.S., Brände, R., and Crawford, R.M.M., Catalase activity and post-anoxic injury in monocotyledonous species. *J. Exp. Bot.*, 38:233, 1987.
298. Monk, L.S., Fagerstedt, K.V., and Crawford, R.M.M., Superoxide dismutase as an anaerobic polypeptide. A key factor in recovery from oxygen deprivation in Iris pseudacorus? *Plant Physiol.*, 85:1016, 1987.
299. Crawford, R.M.M., Walton, J.C., and Wollenweber-Ratzer, B., Similarities between post-ischaemic injury to animal tissues and post-anoxic injuries in plants. *Proc. Roy. Soc. Edinb.*, Section B., 102:325, 1994.
300. Chirkova, T.V., Novitskaya, L.O., and Blokhina, O.B., Lipid peroxidation and antioxidant systems under anoxia in plants differing in their tolerance to oxygen deficiency. *Russ. J. Plant Physiol.*, 45:55, 1998.
301. Blokhina, O.B., Fagerstedt, K.V., and Chirkova, T.V., Relationships between lipid peroxidation and anoxia tolerance in a range of species during post-anoxic reaeration. *Physiol. Plant*, 105:625, 1999.
302. Amor, Y., Chevion, M., and Levine, A., Anoxia pretreatment protects soybean cells against H_2O_2-induced cell death possible involvement of peroxidases and of alternative oxidase. *FEBS Lett.*, 477:175, 2000.
303. Blokhina, O., Virolainen, E., and Fagerstedt, K.V., Antioxidants, oxidative damage and oxygen deprivation stress. *Ann. Bot.*, 91:179, 2003.
304. Garnczarska, M. and Bednarski, W., Effect of a short-term hypoxic treatment followed by re-aeration on free radicals level and antioxidative enzymes in lupine roots. *Plant Physiol. Biochem.*, 42:233, 2004.
305. Biemelt, S., Keetman, U., and Albrecht, A., Re-aeration following hypoxia or anoxia leads to activation of the anioxidative defense system in root of wheat seedlings. *Plant Physiol.*, 116:651, 1998.

306. Garnczarska, M., Bednarski, W., and Morkunas, I., Re-aeration-induced oxidative stress and antioxidative defenses in hypoxically pretreated lupine roots. *J. Plant Physiol.*, 161:415, 2004.
307. Albrecht, G. and Wiedenroth, E.M., Protection against activated oxygen following re-aeration of hypoxically pretreated wheat roots. The response of glutathione system. *J. Exp. Bot.*, 273:449, 1994.
308. Zuckermann, H. et al., Dynamics of acetaldehyde production during anoxia and post-anoxia in red bell pepper studied by photoacoustic techniques. *Plant Physiol.*, 113:925, 1997.
309. Laloi, C., Apel, K., and Danon, A., Reactive oxygen signalling, the latest news. *Curr. Opin. Plant Biol.*, 7:323, 2004.
310. Liszkay, A., van der Zalm, E., and Schopfer, P., Production of reactive oxygen intermediates (O_2^-, H_2O_2, and $\cdot OH$) by maize roots and their role in wall loosening and elongation growth. *Plant Physiol.*, 136:3114, 2004.
311. Demidchik, V. et al., Free oxygen radicals regulate plasma membrane Ca^{2+}- and K^+-permeable channels in plant root cells. *J. Cell Sci.*, 116:81, 2003.
312. Foreman, J. et al., Reactive oxygen species produced by NADPH oxidase regulate plant cell growth. *Nature*, 422:442, 2003.
313. Kwak, J.M. et al., NADPH oxidase AtrbohD and AtrbohF genes function in ROS-dependent ABA signaling in *Arabidopsis*. *EMBO J.*, 22:2623, 2003.
314. Sharp, R.E. et al., Confirmation that abscisic acid accumulation is required for maize primary root elongation at low water potentials. *J. Exp. Bot.* 45:1743, 1994.
315. Ober, E.S. and Sharp, R.E., Electrophysiological responses of maize roots to low water potentials, relationship to growth and ABA accumulation. *J. Exp. Bot.*, 54:813, 2003.
316. Guan, L.M., Zhao, J., and Scandalios, J. G., *Cis*-elements and *trans*-factors that regulate expression of the maize Cat1 antioxidant gene in response to ABA and osmotic stress, H_2O_2 is the likely intermediary signaling molecule for the response. *Plant J.*, 22:87, 2000.
317. Overmyer, K. et al., Ozone-sensitive Arabidopsis rcd1 mutant reveals opposite roles for ethylene and jasmonate signaling pathways in regulating superoxide-dependent cell death. *Plant Cell.*, 12:1849, 2000.
318. Zhao, Z.G., Chen, G.C., and Zhang, C.L., Interaction between reactive oxygen species and nitric oxide in drought-induced abscisic acid synthesis in root tips of wheat seedlings. *Aust. J. Plant Physiol.*, 28:1055, 2001.
319. Marschner, H., *In Mineral Nutrition of Higher Plants*, 2d ed., Academic, Boston, 1995.
320. Czarnes, S. et al., Root-and microbial-derived mucilages affect soil structure and water transport. *Eur. J. Soil Sci.*, 51:435, 2000.
321. Nambiar, E.K.S., Uptake of Zn^{65} from dry soil by plants. *Plant Soil*, 44:267, 1976.
322. Nambiar, E.K.S., The uptake of zinc-65 by oats in relation to water content and root growth. *Aust. J. Soil Res.*, 14:67, 1976.
323. Vetterlein, D. and Marschner, H., Use of a microtensionmeter technique to study hydraulic lift in a sandy soil planted with pearl millet (*Pennisetum americanum* [L.] Leeke). *Plant Soil*, 149:275, 1993.
324. Sekiya, N. and Yano, K., Do pigeon pea and sesbania supply groundwater to intercropped maize through hydraulic lift? Hydrogen stable isotope investigation of xylem waters. *Field Crops Res.*, 86:167, 2004.

325. Al-Karaki, G.N. and Al-Raddad, A., Effects of arbuscular mycorrhizal fungi and drought stress on growth and nutrient uptake of two wheat genotypes differing in drought resistance. *Mycorrhiza*, 7:83, 1997.
326. Bryla, D.R. and Duniway, J.M., Growth, phosphorus uptake and water relations of safflower and wheat infected with an arbuscular mycorrhizal fungus. *New Phytol.*, 136:581, 1997.
327. Johnson, C.R. and Hummel, R.L., Influence of mycorrhizae and drought stress on growth of Poncirus x Citrus seedlings. *Hort. Sci.*, 20:754, 1985.
328. Fitter, A.H., Water relations of red clover *Trifolium partense* L. as affected by VA mycorrhizal colonization of phosphorus supply before and during drought. *J. Exp. Bot.*, 39:595, 1988.
329. Augé, R.M., Water relations, drought and vesicular-arbuscular mycorrhizal symbiosis. *Mycorrhiza*, 11:3, 2001.
330. Dolan, M.S. et al., Corn phosphorus and potassium uptake in response to soil compaction. *Agron. J.*, 84:639, 1992.
331. Jungk, A.O., Dynamics of nutrient movement at the soil-root interface. In *Plant Roots, the Hidden Half*, 3d ed., Waisel, Y., Eshel, A., and Kafkafi, U., Eds., Marcel Dekker, New York, 2002, pp. 587–616.
332. Barber, S.A., *Soil Nutrient Bioavailability*, 2d ed., Wiley, New York, 1995.
333. Kemper, W.D., Stewart, B.A., and Porter, L.K., Effects of compaction on soil nutrient status. In *Compaction in Agricultural Soils*, Barnes, K.K. et al., Eds., Am. Soc. Agr. Eng., St. Joseph, MI, 1971, pp. 178–189.
334. So, H.B. and Nye, P.H., The effect of bulk density, water content and soil type on the diffusion of chloride in the soil. *J. Soil Sci.*, 40:743, 1989.
335. Bhadoria, P.B.S. et al., Impedance factor for chloride diffusion in soil as affected by water content and bulk density. *Z. Pflanzernaehr. Bodenkd.*, 154:69, 1991.
336. Hoffmann, C. and Jungk, A., Growth and phosphorus supply of sugar-beet as affected by soil compaction and water tension. *Plant Soil*, 176:15, 1995.
337. Nye, P.H. and Tinker, P.B., *Solute Movement in the Soil-root System*, Blackwell, Oxford, 1977.
338. Arvidsson, J., Nutrient uptake and growth of barley as affected by soil compaction. *Plant Soil*, 208:9, 1999.
339. Veen, B.W. et al., Root-soil contact of maize, as measured by a thin-section technique. III. Effects on shoot growth, nitrate and water uptake. *Plant Soil*, 139:131, 1992.
340. Lipiec, J. and Stepniewski, W., Effects of soil compaction and tillage systems on uptake and losses of nutrients. *Soil Till. Res.*, 35:37, 1995.
341. Read, D.B. et al., Plant roots release phospholipids surfactants that modify the physical and chemical properties of soil. *New Phytol.*, 157:315, 2003.
342. Yano, K. et al., Arbuscular mycorrhizal formation in undisturbed soil counteracts compacted soil stress for pigion pea. *Appl. Soil. Ecol.*, 10:95, 1998.
343. Nadian, H. et al., Effects of soil compaction on plant growth, phosphorus uptake and morphological characteristics of vesicular-arbuscular mycorrhizal colonization of *Trifolium subterraneum*. *New Phytol.*, 135:303, 1997.
344. Li, X.L., George, E., and Marschner, H., Extension of the phosphorus depletion zone in VA-mycorrhizal white clover in a calcareous soil. *Plant Soil*, 136:41, 1997.
345. Li, X.L. et al., Phosphorus acquisition from compacted soil by hyphae of a mycorrhizal fungus associated with red clover (*Trifolium pratense*). *Can. J. Bot.*, 75:723, 1997.

346. Kundu, D.K., Ladha, J.K., and Lapitan de Guzman, E., Tillage depth influence on soil nitrogen distribution and availability in a rice lowland. *Soil Sci. Soc. Am. J.*, 60:1153, 1996.
347. Ishaq, M. et al., Subsoil compaction effects on crops in Punjab, Pakistan. II. Root growth and nutrient uptake of wheat and sorghum. *Soil Till. Res.*, 60:153, 2001.
348. Linn, D.M. and Doran, J.W., Effect of water-filled pore space on carbon dioxide and nitrous oxide production in tilled and non-tilled soils. *Soil Sci. Soc. Am. J.*, 48:1267, 1984.
349. Bakken, L.R., BØrresen, T, and NjØs, A, Effect of soil compaction by tractor traffic on soil structure, denitrification, and yield of wheat (*Triticum aestivum* L.). *J. Soil Sci.*, 38:541, 1987.
350. Torbert, H.A., and Wood, C.W., Effects of soil compaction and water-filled porespace on soil microbial activity and N losses. *Commun. Soil Sci. Plant Anal.*, 23:1321, 1992.
351. Narteh, L.T. and Sahrawat, K.L., Influence of flooding on electrochemical and chemical properties of West African soil. *Geoderma*, 87:179, 1999.
352. Rubio, G., Casasola, G., and Lavado, R.S., Adaptations and biomass production of two grasses in response to waterlogging and soil nutrient enrichment. *Oecologia*, 102:102, 1995.
353. Kronzucker, H.J. et al., Effects of hypoxia on $^{13}NH_4^+$ fluxes in rice roots-kinetics and compartmental analysis. *Plant Physiol.*, 116:581, 1998.
354. Kronzucker, H.J. et al., Comparative kinetic analysis of ammonium and nitrate acquisition by tropical lowland rice, implications for rice cultivation and yield potential. *New Phytol.*, 145:471, 2000.
355. Kronzucker, H.J. et al., Nitrate ammonium synergism in rice, a subcellular flux analysis. *Plant Physiol.*, 119:1041, 1999.
356. Saleque, M.A. and Kirk, G.J.D., Root induced solubilization of phosphate in the rhizosphere of lowland rice. *New Phytol.*, 129:325, 1995.
357. Rubio, G. and Lavado, R.S., Acquisition and allocation of resources in two waterlogging-tolerant grasses. *New Phytol.*, 143;539, 1999.
358. Rubio, G. et al., Mechanisms for the increase in phosphorus uptake of waterlogged plants, soil phosphorus availability, root morphology and uptake kinetics. *Oecologia*, 112:150, 1997.
359. Kirk, G.J.D. and Bajita, J.B., Root-induced iron oxidation, pH changes and zinc solubilization in the rhizosphere of lowland rice. *New Phytol.*, 131:129, 1995.
360. Begg, C.B.M. et al., Root induced iron oxidation and pH changes in the lowland rice rhizosphere. *New Phytol.*, 128:269, 1994.
361. Briones, A.M. et al., Influence of different cultivars on population of ammonium-oxidizing bacteria in the root environment of rice. *Appl. Environ. Microb.*, 68:3067, 2002.
362. Mansour, M.M.F., Salama, K.H.A., and Al-Mutawa, M.M., Transport proteins and salt tolerance in plants. *Plant Sci.*, 164:891, 2003.
363. Flowers, T.J., Troke, P.F., and Yeo, A.R., Mechanism of salt tolerance in halophytes. *Annu. Rev. Plant Physiol.*, 28:89, 1977.
364. Cramer, G.R. et al., Influx of Na^+, K^+, and Ca^{2+} into roots of salt-stressed cotton seedlings — effects of supplemental Ca^{2+}. *Plant Physiol.*, 83:510, 1987.
365. Munns, R., Physiological processes limiting plant growth in saline soils, some dogmas and hypotheses. *Plant Cell Environ.*, 16:15, 1993.

366. Zhu, J.K., Regulation of ion homeostasis under salt stress. *Curr. Opin. Plant Biol.*, 6:441, 2003.
367. Ashraf, M. and Harris, P.J.C., Potential biochemical indicators of salinity tolerance in plants. *Plant Sci.*, 166:3, 2004.
368. DuPont, F.M., Salt induced changes in ion transport, regulation of primary pumps and secondary transporters. In *Transport and Receptor Proteins of Plant Membranes, Molecular Structure and Function*, Cooke, D.T. and Clarkson, D.T., Eds., Plenum, New York, 1992, pp. 91–100.
369. Jacoby, B., Mechanisms involved in salt tolerance of plants. In *Handbook of Plant and Crop Stress*, 2d ed., Pessarakli, M., Ed., Marcel Dekker, New York, 1999, pp. 97–123.
370. Grattan, S.R. and Grieve, C.M., Mineral nutrient acquisition and response by plants grown in saline environments. In *Handbook of Plant and Crop Stress*, 2d ed., Pessarakli, M., Ed., Marcel Dekker, New York, 1999, pp. 203–229.
371. Ratner, A. and Jacoby, B., Effect of K^+, its counter anion, and pH on sodium efflux from barley root tips. *J. Exp. Bot.*, 27:843, 1976.
372. Niu, X., Bressan, R.A., and Hasegawa, P.M., Halophytes up-regulate plasma membrane H^+-ATPase gene more rapidly than glycophytes in response to salt stress. *Plant Physiol.*, 102, Suppl.:133, 1993.
373. Shi, H. et al., Overexpression of a plasma membrane Na^+/H^+ antiporter gene improves salt tolerance in Arabidopsis thaliana. *Nat. Biotechnol.*, 21:81, 2003.
374. Suhayda, C.G. et al., Electrostatic changes in Lycopersicon esculentum root plasma membrane resulting from salt stress. *Plant Physiol.*, 93:471, 1990.
375. Sanchezaguayo, I., Gonzalezutor, A.L., and Medina, A., Cytochemical-localization of ATPase activity in salt-treated and salt-free grown *Lycopersicon-esculentum* roots. *Plant Physio*l., 96:153, 1991.
376. Yang, Y.L. et al., NaCl induced changes of the H^+-ATPase in root plasma membrane of two wheat cultivars. *Plant Sci.*, 166:913, 2004.
377. Shi, H.Z. et al., The *Arabidopsis thaliana* salt tolerance gene SOS1 encodes a putative Na^+/H^+ antiporter. *Proc. Natl. Acad. Sci. USA*, 97;6896, 2000.
378. Wu, S.J., Ding, L., and Zhu, J.K., SOS1, a genetic locus essential for salt tolerance and potassium acquisition. *Plant Cell*, 8:617, 1996.
379. Qiu, Q.S. et al., Regulation of SOS1, a plasma membrane Na^+/H^+ exchanger in *Arabidopsis thaliana*, by SOS2 and SOS3. *Proc. Natl. Acad. Sci. USA*, 99;8436, 2002.
380. Shi, H.Z. et al., The putative plasma membrane Na^+/H^+ antiporter SOS1 controls long-distance Na^+ transport in plants. *Plant Cell*, 14:465, 2002.
381. Quintero, F.J. et al., Reconstitution in yeast of the *Arabidopsis* SOS signaling pathway for Na^+ homeostasis. *Proc. Natl. Acad. Sci. USA*, 99:9061, 2002.
382. Liu, J.P. et al., The *Arabidopsis thaliana* SOS2 gene encodes a protein kinase that is required for salt tolerance. *Proc. Natl. Acad. Sci. USA*, 97:3730, 2000.
383. Apse, M.P. et al., Overexpression of a vacuolar Na^+/H^+ antiport confers salt tolerance in *Arabidopsis*. *Science*, 285:1256, 1999.
384. Apse, M.P., Sottosanto, J.B., and Blumwald, E., Vacuolar cation/H^+ exchange, ion homeostasis, and leaf development are altered in a T-DNA insertional mutant of AtNHX1, the *Arabidopsis* vacuolar Na^+/H^+ antiporter. *Plant J.*, 36:229, 2003.
385. Gaxiola, R.A. et al., The *Arabidopsis thaliana* proton transporters, AtNhx1 and Avp1, can function in cation detoxification in yeast. *Proc. Natl. Acad. Sci. USA*, 96:1480, 1999.

386. Darley, C.P. et al., *Arabidopsis thaliana* and *Saccharomyces* cerevisiae NHX1 genes encode amiloride sensitive electroneutral Na^+/H^+ exchangers. *Biochem. J.*, 351:241, 2000.
387. Zhang, H.X. and Blumwald, E., Transgenic salt-tolerant tomato plants accumulate salt in foliage but not in fruit. *Nature Biotechnol.*, 19:765, 2001.
388. Zhang, H.X. et al., Engineering salt-tolerant *Brassica* plants, Characterization of yield and seed oil quality in transgenic plants with increased vacuolar sodium accumulation. *Proc. Natl. Acad. Sci. USA*, 98:12832, 2001.
389. Demidchik, V., Davenport, R.J., and Tester, M., Nonselective cation channels in plants. *Annu. Rev. Plant Biol.*, 53:67, 2002.
390. Schachtman, D.P. and Schroeder, J.I., Structure and transport mechanism of a high-affinity potassium uptake transporter from higher plants. *Nature*, 370:655, 1994.
391. Rubio, F., Gassmann, W., and Schroeder, J.I., Sodium-driven potassium uptake by the plant potassium transporter HKT1 and mutations conferring salt tolerance. *Science*, 270:1660, 1995.
392. Uozumi, N. et al., The *Arabidopsis* HKT1 gene homolog mediates inward Na^+ currents in Xenopus laevis oocytes and Na^+ uptake in *Saccharomyces cerevisiae*. *Plant Physiol.*, 122:1249, 2000.
393. Maser, P. et al., Altered shoot/root Na^+ distribution and bifurcating salt sensitivity in Arabidopsis by genetic disruption of the Na^+ transporter AtHKTI1. *FEBS. Lett.*, 531:157, 2002.
394. Garciadeblas, B. et al., Sodium transport and HKT transporters, the rice model. *Plant J.*, 34:788, 2003.
395. White, P.J. and Broadley, M.R., Chloride in soils and its uptake and movement within the plant, a review. *Ann. Bot.*, 88:967, 2001.
396. Qadir, M. and Schubert, S., Degradation processes and nutrient constraints in sodic soils. *Land Degrad. Dev.*, 13:275, 2002.
397. Cramer, G.R. and Läuchli, A., Ion activities in solution in relation to Na^+-Ca^{2+} interactions at the plasmalemma. *J. Exp. Bot.*, 25:321, 1986.
398. Cramer, G.R., Epstein, E., and Läuchli, A., Kinetics of root elongation of maize in response to short-term exposure to NaCl and elevated calcium concentration. *J. Exp. Bot.*, 25:1513, 1988.
399. Colmer, T.D. et al., Interactive effects of Ca^{2+} and NaCl salinity on the ionic relations and proline accumulation in the primary root tip of *Sorghum bicolor*. *Physiol. Plant*, 25:421, 1996.

11 Physiological and Biochemical Indicators for Stress Tolerance

Shimon Rachmilevitch, Michelle DaCosta, and Bingru Huang

CONTENTS

- 11.1 Introduction ... 322
- 11.2 Photosynthesis ... 322
  - 11.2.1 Chlorophyll Fluorescence ... 323
    - 11.2.1.1 Principles and Methodology of Chlorophyll Fluorescence ... 323
    - 11.2.1.2 Chlorophyll Fluorescence and Heat Stress ... 324
    - 11.2.1.3 Chlorophyll Fluorescence and Drought Stress ... 325
    - 11.2.1.4 Chlorophyll Fluorescence and Cold Stress ... 325
  - 11.2.2 Gas Exchange Parameter ... 326
    - 11.2.2.1 Principles and Methodology ... 326
    - 11.2.2.2 Photosynthetic Gas Exchange and Stress Tolerance ... 327
- 11.3 Water Relations ... 329
  - 11.3.1 Relative Water Content ... 329
    - 11.3.1.1 RWC Determination ... 329
    - 11.3.1.2 Relative Water Content and Stress Tolerance ... 330
  - 11.3.2 Osmotic Adjustment ... 331
    - 11.3.2.1 Determination of Osmotic Adjustment ... 332
    - 11.3.2.2 Regression Method ... 332
    - 11.3.2.3 Full Turgor Adjustment Method ... 333
    - 11.3.2.4 Rehydration Method ... 333
  - 11.3.3 Osmotic Adjustment and Stress Tolerance ... 333
- 11.4 Membrane Stability ... 335
  - 11.4.1 Electrolyte Leakage Analysis ... 335
  - 11.4.2 Membrane Stability and Stress Tolerance ... 335
- 11.5 Lipid Peroxidation ... 337
  - 11.5.1 Antioxidant Enzyme Defense System ... 338

	11.5.2	Determination of Lipid Peroxidation..338
	11.5.3	Lipid Peroxidation and Stress Tolerance..................................340
11.6	Stress-Induced Proteins 340
	11.6.1	Heat Shock Proteins...341
	11.6.2	Late Embryogenesis Abundant Proteins...................................342
11.7	Summary...343
References 345

11.1 INTRODUCTION

Plants exposed to environmental stresses often exhibit a variety of symptoms or indications. Stress indicators are signs of disturbance, either as visible, growth or morphological modification or invisible, physiological or biochemical changes that relate to repair and resistance mechanisms. Evaluating physiological and cellular parameters for assessing the severity of stress injury or level of stress tolerance is not always trivial; visual performance ratings are most often subjective. Your neighbor's lawn always looks greener, although objectively it is not always true. The scientific reasoning behind this claim is that it all depends literally on the angle you are looking at, whether you are looking directly above your lawn, seeing all the open spaces, or looking at your neighbor's lawn at a certain angle, seeing only the upper green parts.

Plant species and cultivars vary in their responses to environmental stresses. The nature and intensity of the response of individual plants to a particular stress factor may also vary considerably, depending on plant or tissue age, annual season, and even diurnal activity [1]. Although there is often a good correlation between the intensity of a specific stress factor and the elicited response, the degree of impairment suffered by the plant is not necessarily always proportional to the harmful degree of the stressor. Therefore, it is necessary to consider various factors for proper assessment of stress tolerance of plants to different environmental stresses. Using quantitative measurements to evaluate plant stress and the level of stress tolerance is of significance in order to assess objectively the severity of stress damage on plants and to be able to screen stress tolerant species and cultivars.

The current chapter reviews the methodology and principles of several major physiological and biochemical indicators commonly used to evaluate stress tolerance or injury: photosynthesis, water relations, cell membrane stability (CMS), lipid peroxidation, and stress-induced proteins. Multiple morphological, physiological, and biochemical changes may occur simultaneously in response to an environmental stress, and sometimes are difficult to distinguish from one another. However, for an orderly discussion, each indicator is presented independently in this chapter.

11.2 PHOTOSYNTHESIS

Photosynthesis is the basic process affecting plant growth, productivity, and survival under stressful environments. It is the reduction of CO_2 by plants that makes them autotrophic and virtually enables the complexity of life on earth.

Physiological and Biochemical Indicators for Stress Tolerance

In this section, we review the use of chlorophyll fluorescence and gas exchange rates as indicators of plant tolerance to environmental stresses. The former indicates photochemical efficiency of the light reaction and the latter quantifies the photosynthetic rate.

11.2.1 Chlorophyll Fluorescence

11.2.1.1 Principles and Methodology of Chlorophyll Fluorescence

Chlorophyll fluorescence serves as an intrinsic indicator of photochemical processes in photosynthetic organisms. The relationship between chlorophyll fluorescence and photosynthesis has been a subject of a great number of studies since 1931 when Kautsky and Hirsch [2] reported that fluorescence intensity in green leaves displayed characteristic changes upon illumination (Kautsky effect). Since then, for many years fluorescence has been used as a tool for biophysical studies concerning the primary processes of photosynthesis. However, practical use of fluorescence in plant physiology and ecophysiology was limited by the availability of suitable equipment for measuring fluorescence in sunlight and by the complexity of the fluorescence information obtained in vivo. In the last 20 years advanced techniques have become available that permit relatively simple monitoring of fluorescence in sunlight and the understanding of the complex information obtained from fluorescence [3]. In recent years, the technique of chlorophyll fluorescence measurements has become ubiquitous in whole-plant physiology studies, a trend that has been fueled by the introduction of a number of user-friendly and portable chlorophyll fluorometers [4]. In spite of the simplicity of the measurements, the theory behind and the interpretation of the data remains complex (for reviews see [4–6]).

The basic principle behind fluorescence measurements relies on the fact that the light energy that is absorbed by chlorophyll molecules in photosynthetic organisms can be used in any one of three different methods of energy dissipation: it can be used to drive photosynthesis (photochemical quenching); excess energy can be dissipated as heat (nonphotochemical quenching); or it can be re-emitted as light, chlorophyll fluorescence. Each quantum of light absorbed by a chlorophyll molecule raises an electron to an excited state. Upon de-excitation, a small proportion (1–4%) of the excitation energy is dissipated as red fluorescence. The use of fluorescence as an indicator arises from the fact that fluorescence emission is complementary to the other two dissipation mechanisms, the photochemical quenching (photosynthesis) and the heat dissipation. Generally, fluorescence yield is highest when photochemistry and heat dissipation are lowest; therefore, changes in fluorescence reflect photochemical efficiency and heat dissipation.

The energy emitted from a leaf as fluorescence is lower than the light energy absorbed, and consequently, the fluorescence emission peak has a longer wavelength than that of the light absorbed. This phenomenon enables quantification of fluorescence yield by exposing a leaf to light of specific wavelengths and by measuring the re-emitted light at longer wavelengths. A major modification to the basic fluorometers that enabled their use in sunlight was the use of modulated light [7]. These fluorometers are called PAM (pulse modulated fluorometers). In PAM, the fluorometer uses a flashing (modulated) light source at high frequency to measure fluorescence and the detector is tuned to measure only fluorescence excited by the measuring light. In consequence, the relative yield of fluorescence can also be measured in the presence of background light such as sunlight.

One of the most common parameters used for measuring chlorophyll fluorescence is the Fv/Fm ratio. When a leaf is kept in the dark all the residual energy is processed and the amount of fluorescence is small; this is considered the minimal fluorescence signal (Fo). After flashing the leaf with bright high-intensity light, the fluorescence signal will increase to a maximum (Fm), as it cannot photochemically use all this energy. The difference between the maximum and minimum fluorescence is called the variable fluorescence (Fv). The ratio of Fv/Fm indicates the proportion of the maximum possible fluorescence that was used for photosynthesis, and is an estimate of the PSII maximum efficiency. This efficiency in healthy plants is usually about 80% or 0.8. Fv/Fm is considered to be inversely proportional to stress; therefore, values of Fv/Fm that are lower than ~0.8 are considered as injurious levels [8].

11.2.1.2 Chlorophyll Fluorescence and Heat Stress

Fluorescence methods have been used to evaluate heat stress effects on photosynthetic apparatus in many studies [8]. Fluorescence changes may reflect structural changes within the thylakoid membranes due to heat stress. Given that photosynthesis is considered to be the physiological process most sensitive to high-temperature damage and that PSII appears to be a very heat-sensitive component of the photosynthetic apparatus, using chlorophyll fluorescence to measure thermotolerance of PSII in plants may provide early stress indication.

Haldimann and Feller [9] examined the impact of heat stress on the functioning of the photosynthetic apparatus in pea (*Pisum sativum* L.), and found chlorophyll a fluorescence started to decrease when leaf temperature increased above 35°C before the decline of net photosynthetic rate. Thermotolerance of PSII in leaves of salt-adapted *Artemisia anethifolia* Weber ex Stechm. plants was evaluated after exposure to heat stress (30–45°C) for 30 min [10]. High thermotolerance of salt-adapted plants was associated with an improvement in thermotolerance of the PSII reaction centers, oxygen-evolving complexes, and the light-harvesting complex as measured by chlorophyll fluorescence. Salvucci and Crafts-Brandner [11] examined chlorophyll fluorescence and Rubisco activase activity in species from contrasting environments. Nonphotochemical quenching of chlorophyll fluorescence and the maximum yield of PSII (Fv/Fm) were more sensitive to temperature in Antarctic hairgrass (*Deschampsia antarctica* E. Desv.) and two other species endemic to cold regions (i.e., *Lysipomia pumila* [(Wedd.) E.Wimm.]) L. and spinach (*Spinacia oleracea*)) compared with

Physiological and Biochemical Indicators for Stress Tolerance

creosote bush [*Larrea tridentata* (Sessé and Moc. ex DC.) Coville] and three species adapted to relatively high temperatures (i.e., jojoba [*Simmondsia chinensis* (Link) C. K. Schneid.]), tobacco (*Nicotiana tabacum* L.), and cotton (*Gossypium hirsutum* L.).

11.2.1.3 Chlorophyll Fluorescence and Drought Stress

Water stress has long been shown to cause substantial changes to chlorophyll fluorescence [5]. Water stress experiments performed by Flexas et al. [12] with grapevines (*Vitis vinifera* L.) and other C-3 plants indicated that the ratio of normalized steady-state chlorophyll fluorescence (Fs) to dark-adapted intrinsic fluorescence (Fo) inversely correlated with nonphotochemical quenching (NPQ). Also, at high irradiance, the ratio Fs/Fo was positively correlated with CO_2 assimilation in air, electron transport rate calculated from fluorescence, and stomatal conductance.

Souza et al. [13] studied the responses of photosynthetic gas exchange and chlorophyll fluorescence in cowpea (*Vigna unguiculata* Walp.) during water stress and recovery. During the initial phase of stress, photochemical activity was not affected, as revealed by lack of alterations in fluorescence parameters associated with PSII activity. Development of nonradiative energy dissipation mechanisms was evident during prolonged stress by increases in nonphotochemical quenching and decreases in efficiency of excitation capture by open centers. At an advanced phase of stress, a downregulation of PSII activity was observed along with some impairment of photochemical activity, as revealed by decreases in Fv/Fm. Complete recovery of all gas exchange and fluorescence parameters occurred three days after rewatering. However, on the first day after water stress relief, assimilation rates only partially recovered in spite of the availability of internal CO_2, suggesting that nonstomatal limitation of photosynthesis, such as downregulation of PSII activity observed during stress, may limit photosynthetic recovery.

11.2.1.4 Chlorophyll Fluorescence and Cold Stress

Chlorophyll fluorescence has been used to measure the damage of chilling stress to photosynthesis in various studies. In a study by Baker et al. [14], young maize (*Zea mays*) plants were chilled for 6 h at 5°C in high light intensity, and on return to 20°C, the leaves showed a 45% decrease in the apparent quantum yield of photosynthetic oxygen evolution. The effects of this chilling treatment on chlorophyll fluorescence indicated a 25% decrease in the primary photochemical quantum yield of PSII. The fluorescence emission spectra of these leaves demonstrated a marked modification in the distribution of excitation energy within the photochemical apparatus of the thylakoid membranes. The chill-induced reduction in the quantum efficiency of PSII and the modification of excitation energy distribution between PSI and PSII in favor of PSI would produce a decrease in the noncyclic electron transport and account for the 45% decrease in the apparent quantum yield of photosynthetic oxygen evolution in the chill-treated plants. Such a chill-induced modification of membrane photofunction in vivo when plants are returned to ambient temperature would clearly be a major factor in reducing carbon assimilation.

11.2.2 Gas Exchange Parameter

11.2.2.1 Principles and Methodology

Gas exchange involves entry and loss of gases such as CO_2, H_2O, and O_2 in plant tissues. During photosynthetic gas exchange, CO_2 is taken up or consumed by leaves and O_2 is evolved or produced. During the respiratory gas exchange, O_2 is consumed whereas CO_2 is evolved. When more CO_2 is consumed in photosynthesis than is simultaneously evolved in respiratory processes, the net uptake of CO_2 is considered to be net photosynthesis, whereas the total CO_2 assimilated in the chloroplasts is the gross amount of photosynthesis. Net photosynthetic rate is often used as a measure of plant activity.

The twice Nobel Laureate Otto Warburg pioneered measurements of photosynthetic rates using manometric techniques [15]. Warburg's manometers measure pressure changes in a closed vessel that occur when a photosynthesizing or respiring organism exchanges CO_2 and O_2 with the atmosphere. Most manometric methods involve the use of aqueous suspensions of cells, due to difficulties in measuring tissues that are not aqueous and, in addition, manometric methods are not useful for measurements of rapid changes [16]. Technological advances in microelectronics have led to the development of sophisticated systems for measuring photosynthesis and respiration. Currently numerous companies offer competing systems for measuring CO_2 and O_2 gas exchange [16].

Use of polagraphic O_2 electrodes for physiological studies dates back to Clarck [17]. Typically an O_2 electrode consists of a platinum cathode and a silver anode linked by an electrolyte. The electrodes and electrolyte are separated from the sample by an oxygen-permeable membrane (usually Teflon). The latter gas exchange methods are suitable mainly for time-integrated rather than continuous measurements and require the samples to be placed in closed chambers. Closed gas exchange systems are subject to errors due to the effects of changes in the partial pressure of gases on the plants physiological and biochemical processes.

An alternative method, and in many ways advantageous, for measuring gas exchange is the use of open flow systems. In open flow systems, gas is placed in a chamber through which gas is flowed at a measured flow rate and then vented to the atmosphere, maintaining constant partial pressure of gases. Other advantages of open flow systems include: continuous monitoring, measurements of intact tissues and of whole plants, easy control of environmental conditions, and noninvasive analysis. The most common analyzers of CO_2 gas exchange are infrared gas analyzers (IRGA). IRGA are based on the fact that heteroatomic molecules such as CO_2 and H_2O absorb infrared radiation. The main absorbance of CO_2 is at 4.25 μM [18]. Absorption at any wavelength follows the Beer–Lambert law, so that any gas containing CO_2 will be detected by the absorbance of CO_2.

A large number of commercial, user-friendly analyzers are available and all have the ability to resolve CO_2 concentrations of less than 1 μmol mol^{-1} (ppm), which is necessary for investigations of photosynthesis and respiration [16]. CO_2 gas exchange measurements of leaves and whole plants are fairly easy due to the relatively low background of CO_2 in the atmosphere (~370 μmol mol^{-1}). However, O_2 gas exchange measurements of intact leaves and whole plant are much more complicated due to the relatively high background of O_2 in the atmosphere (~210,000 μmol mol^{-1}), and therefore there are significantly fewer studies looking at O_2 gas exchange in whole and intact plants as compared to CO_2 gas exchange.

CO_2 gas exchange in photosynthesis is involved in two main interactive reactions: diffusion of CO_2 to the sites of photosynthesis through stomata and metabolic fixation of CO_2 in the Calvin cycle utilizing light energy. CO_2 assimilation during photosynthesis depends on the partial pressure of CO_2 in the intercellular air space (Ci). Ci can be calculated from measured rates of stomatal conductance and net photosynthesis [19]. One of the most commonly used parameters, in addition to net photosynthetic rate, for characterizing photosynthesis on the basis of gas exchange is a CO_2 assimilation (A) to Ci response curve (A/Ci curve). The response of A follows saturation kinetics with respect to Ci. The maximal photosynthetic rate (Vmax) at saturating Ci and light intensity is limited by RuBP, the CO_2 acceptor. The initial slope of the curve represents the carboxylation efficiency [20]. The CO_2 compensation point is the point where net photosynthesis equals zero and is determined both by enzyme kinetics and by mitochondrial respiration in the light. These three parameters (Vmax, carboxylation efficiency, and the compensation point) may be utilized as photosynthetic indicators, and may vary markedly as the result of environmental stress.

Another important tool in addition to A/Ci curves is the light-response curve. Similar to the A/Ci curve, at low-light intensities when photosynthetic CO_2 uptake and respiratory CO_2 evolution are in equilibrium, the irradiance level is called the light compensation point. Plants with high respiration rates require higher light intensities for compensation as compared to plants that show lower respiration rates. The linear slope after the compensation point equals the quantum yield of the leaf; thus a steep slope of a light curve indicates a higher quantum yield. At the saturation level of the slope, photosynthesis is not limited by photochemical but rather by enzymatic processes and by CO_2. Monitoring CO_2 assimilation and net O_2 evolution simultaneously enables the measurements of assimilatory quotient (AQ), the ratio of net CO_2 assimilation to net O_2 evolution. The AQ depends upon NO_3^- assimilation because transfer of electrons to NO_3^- and then to NO_2^- increases O_2 evolution from the light-dependent reactions of photosynthesis, whereas CO_2 assimilation remains unchanged or decreases [21–23].

11.2.2.2 Photosynthetic Gas Exchange and Stress Tolerance

Photosynthesis is generally positively correlated to stress tolerance and, therefore, photosynthetic gas exchange parameters are widely used as stress indicators [24]. A/Ci curves can be a useful tool for studying plant responses to temperature changes.

The CO_2 compensation point increases with increasing temperatures, mainly due to the increase in respiration. Extreme temperatures markedly stimulate the rate of photosynthesis and the Vmax is also usually stimulated [25].

Many studies correlate net photosynthetic rate to stress tolerance. A study on the interactive effects of drought, heat, and elevated CO_2 on photosynthesis illustrated the importance of CO_2 gas exchange in stress tolerance [26]. One-year-old creosote bush seedlings [*Larrea tridentata* (Sessé and Moc. ex DC.) Coville] were exposed to high temperatures (53°C) under three atmospheric CO_2 concentrations (360, 550, and 700 μmol mol^{-1}) and two water regimes (well watered and drought subjected) [26]. This study found that increasing CO_2 concentrations to 700 μmol mol^{-1} improved net photosynthetic rate under heat stress. Net photosynthetic rate was also higher under 550 and 700 μmol mol^{-1} CO_2 than in ambient CO_2-grown plants exposed to drought stress. Exposure of plants to 700 μmol mol^{-1} CO_2 also resulted in a more rapid recovery of net photosynthetic rate after relief from heat stress. These findings suggested that the improved photosynthetic activity under elevated CO_2 conditions could enhance the potential for creosote bush to survive heat and drought conditions.

The sensitivity of stomata to water stress has been well documented [27]. Therefore, it has been long accepted that a major part of the reduction in net photosynthetic rate during water stress is due to stomatal closure, which prevents the inward passage of CO_2. In addition to the induction of stomatal closure, drought can have a major effect on biochemical processes involved in CO_2 fixation [28].

A comparison of the effects of a rapid and a slowly imposed water deficit on photosynthesis in *Setaria sphacelata* (Schumach.), a C_4 grass, demonstrated the relationship of photosynthesis and drought stress [29]. Gas exchange was measured in rapidly and slowly dehydrated mature leaves either under atmospheric CO_2 partial pressure with an infrared gas analyzer or under saturating CO_2 partial pressure with a leaf disc oxygen electrode. The decrease of net photosynthetic rate and leaf conductance was more pronounced under rapid stress than in slow stress. However, photosynthesis was mainly limited by stomata in both types of stress, albeit the contribution of nonstomatal limitations increased at severe water deficits in slow stress experiments. The substomatal CO_2 partial pressure significantly increased in both types of stress, suggesting an increased resistance due to an internal barrier to CO_2 diffusion.

Salinity stress tolerance has also been associated with photosynthetic gas exchange in various studies. Gucci et al. [30] studied the effect of NaCl stress and relief on gas exchange properties of two olive (*Olea europaea* L.) cultivars differing in tolerance to salinity. Plants were exposed to NaCl concentrations between 0 and 200 mM for 34–35 days followed by 30–34 days of relief from stress. Salinity stress resulted in a reduction in net CO_2 assimilation and stomatal conductance in both cultivars, but the effect was more pronounced in salt-tolerant "Frantoio" than in the salt-sensitive "Leccino" cultivar. However, recovery in gas exchange parameters during relief from stress was faster for the salt-tolerant than for the salt-sensitive cultivar. The results of this study indicated that gas exchange rates could be used to evaluate plant survival of salinity stress, but should be used with caution in reference to specific conditions.

11.3 WATER RELATIONS

The maintenance of favorable plant water status is an essential strategy for plant tolerance to stresses that result in cellular water deficit and turgor loss. In this section, we review the use of relative water content and osmotic adjustment for assessing changes in plant water relations in response to stresses and for the evaluation of stress tolerance. The advantages and disadvantages of these methods as water-stress indicators is also discussed.

11.3.1 Relative Water Content

Water potential and relative water content (RWC) are two fundamental concepts that characterize water relations of plants and are widely used as indicators for plant water status. There has been much debate over which measurement more accurately describes plant water status [31–33]. Water potential refers to the thermodynamic properties of water and describes the energy status and water fluxes through the plant, whereas RWC describes the actual content of water in the plant based on maximal water content it can hold at full turgidity. Water potential may be a good measurement of water movement or water uptake through the plant. Different experiments have demonstrated a closer correlation of RWC than water potential with various physiological and biochemical activities (i.e., photosynthesis, protein synthesis, leaf senescence, and other physiological parameters) under environmental stresses [33]. Furthermore, osmotic adjustment is a mechanism that is conserved among many plant species and involved in water retention and cell turgor maintenance under dehydration stresses. Osmotic adjustment in leaf tissues may lower the water potential at which a particular RWC level is reached [33]. Therefore, it has been argued that RWC, through its inclusion of osmotic adjustment and its relationship to other physiological activities, may be a better indicator of water status and physiological activity, particularly under stresses that cause cellular water deficit [33–35].

11.3.1.1 RWC Determination

The protocol for RWC measurement was originally developed by Barrs and Weatherly [36] (previously referred to as the relative turgidity technique). The following section describes briefly the method of RWC analysis, which is adapted from Kirkham [37].

Fully expanded leaves are usually used for RWC analysis. Multiple subsamples or leaves may be taken from a single treatment or plant. The size of leaf samples depends on the size and shape of individual leaves or plant species: leaf discs (approximately 1.5 cm diameter) or strips (5–10 cm^2) from large, broad-leaved species; or whole leaves, such as from grasses or conifers. Whole leaves or leaf discs or segments should be of similar size, and sampling should be done quickly in order to minimize any tissue deterioration. Fresh weight of a leaf sample is determined as soon as leaves are detached from the plant. Samples are then immediately placed in de-ionized water in capped containers to maintain approximately 100% relative humidity in the container in order for leaves to reach full hydration. It typically takes approximately 4–6 hours for leaves to be fully rehydrated [38]; however, longer times for full hydration are not uncommon [39]. The time for leaves to be fully hydrated depends on leaf age, leaf surface characteristics, and development stage as discussed above. Samples are typically allowed to hydrate to full turgidity under diffuse light and room temperature conditions or refrigeration. Once samples are fully rehydrated, samples are removed from water and lightly blotted dry to remove all surface moisture. Samples are weighed for turgid weight. Dry weight is then determined for samples dried in an oven at 75–85 C for approximately 48–72 hours. Percent RWC is calculated as:

$$\text{RWC \%} = \frac{100 \times \text{Fresh weight} - \text{dry weight}}{\text{Turgid weight} - \text{dry weight}}$$

The RWC method has several advantages, including simplicity of protocol, small amount of required leaf tissues, no specialized costly equipment other than an analytical balance and oven, and many samples can be processed within a short period of time. According to Barrs [38], some potential problems exist with determination of RWC.

Occurrence of growth in the sample during the rehydration period
Infiltration of water into intercellular spaces
Respiratory-induced changes in dry mass
Inadequate drying of sample prior to measuring turgid weight

These problems can be avoided by reducing the amount of re-hydration time to minimize additional growth and intercellular-water infiltration, placing samples at low temperatures to reduce respiration rates, and utilizing whole leaves rather than leaf discs or small segments.

11.3.1.2 Relative Water Content and Stress Tolerance

RWC has been identified as a reliable trait for screening drought tolerance in winter wheat (*Triticum aestivum* L.) [40,41], durum wheat (*Triticum turgidum* subsp. *durum* (Desf.) Husn.) [42], and oat (*Avena sativa* L.) [43]. Several studies also reported positive correlations between grain yield in cereals and RWC under drought stress [41,42,44]. RWC was found to be a better tool for evaluating genotypic differences in drought tolerance in soybean (*Glycine max* (L.)) compared to water potential [45]. Schonfeld et al. [40] found greater genetic variation in RWC related to drought tolerance in wheat, with little variation in water potential components.

Enhanced heat tolerance and turfgrass quality in Kentucky bluegrass (*Poa pratensis* L.) [46] and creeping bentgrass (*Agrostis stolonifera* L.) [47] were related to maintenance of leaf water status under heat stress. Under salt treatment, salt-sensitive lines of sugar beet (*Beta vulgaris* L.) had lower levels of RWC than tolerant lines, resulting in greater growth inhibition in sensitive lines. Quartacci et al. [48] evaluated two wheat cultivars different in drought tolerance and found that drought-tolerant cultivars exhibited higher RWC, whereas leaf water potential values decreased similarly between drought-tolerant and sensitive cultivars when subjected to drought stress. In transgenic soybean subjected to drought stress, significantly higher RWC was related to increased drought tolerance [49].

Plant maturity may affect how successful the RWC technique is at screening for differences in stress tolerance. A good correlation was obtained with barley (*Hordeum vulgare* L.) seedlings for RWC and field drought responses [50], but screening for RWC was better associated with tolerance in more mature plants of wheat [51]. Tahara et al. [41] investigated the association between grain yield and winter wheat under drought stress and found that the optimum time for RWC measurement occurred after heading (anthesis and midgrain fill) rather than prior to heading, where no differences in RWC were observed between drought-divergent populations. Chan and Fowler [52] utilized the RWC method in crambe (*Crambe abyssinica* Hochst.), an oil seed crop, to determine rehydration kinetics of leaf discs differing in leaf age and stage of plant development at different levels of water stress. They found that both leaf age and maturity level influenced the rate of rehydration of leaf discs, where rate of water uptake increased with leaf age and in plants in vegetative versus reproductive state. This highlights the importance of utilizing leaves of similar age as well as similar developmental stages when RWC values are used to compare water status of different plant populations.

11.3.2 Osmotic Adjustment

Cellular dehydration results from various major environmental stresses, including drought, salinity, and extremes in temperature. Osmotic adjustment (OA) is a mechanism of acclimation to cellular dehydration in which compatible solutes are actively accumulated within cells. This causes a reduction in osmotic potential, and leads to water movement into the cell or prevention of water efflux out of cells [53–55]. In turn, plants maintain higher turgor potential and water retention. In plants that exhibit OA, loss in turgor can occur at lower water potentials than in nonosmotically adjusted leaf tissues [56]. It is important to note that this active accumulation of solutes triggered by decreases in cellular water content differs from accumulation of solutes simply due to loss of water and concentrating effects [57].

Osmotically active solutes involved in OA include amino acids (e.g., proline), ammonium compounds (e.g., glycine betaine), sugars (e.g., fructans, sucrose), polyols (e.g., mannitol), inorganic ions (e.g., potassium), and organic acids (e.g., malate) [56,58]. In addition to effects on turgor potential and water retention, the accumulation of these solutes has been associated with the maintenance of membrane and protein structures, as well as protection against oxidative damage [59–62]. These compounds may function by replacing water in some biochemical reactions, or by associating with macromolecules such as lipids and proteins in order to prevent membrane disintegration, dissociation of protein complexes, or inactivation of enzymes [61], thus leading to higher structural stabilization under water deficit.

OA occurs in many different organisms, including higher plants, fungi, and bacteria, suggesting a relatively conserved mechanism for the active accumulation of solutes under dehydration stresses [56,63]. The mechanism for OA has been extensively investigated in leaves, but may also occur in roots [64–66], which contributes to maintenance of root turgor and growth during drought [67–69]. OA is commonly categorized as a dehydration tolerance mechanism, because plants may continue with metabolic and physiological functions even at low water potentials; however it may also be considered a dehydration avoidance mechanism, whereby greater soil water extraction is promoted by increasing the water potential gradients between roots and soil through root OA [70].

11.3.2.1 Determination of Osmotic Adjustment

Several methods have been used to determine osmotic adjustment. Babu et al. [71] conducted a comprehensive investigation on the comparison of different methods for evaluation of OA in rice (*Oryza sativa* L.) genotypes differing in capacity for OA. They found that absolute OA values and correlations of OA across genotypes varied among the different methods. This reflects the difficulty in direct comparison of when different studies were used.

11.3.2.2 Regression Method

OA can be calculated based on the regressions of relative water content (RWC) on osmotic potential, which are obtained at several intervals throughout the duration of stress [72,73]. One regression line is based on the measurement of both RWC and osmotic potential, and the other based on RWC and a predicted osmotic potential based on the concentration effect. Predicted osmotic potential is based on the inverse relationship between osmotic potential and volume of cell solution, or RWC, according to Morgan [73] and Wright et al. [74]. OA is then calculated as the difference between measured and estimated osmotic potential at a RWC at wilting point, typically reported in some species to be approximately 60–70% RWC. Therefore, measurement requires calculation of RWC and osmotic potential at many intervals throughout the duration of stress. This extensive sampling may be a disadvantage of this method, which can be time consuming especially for screening a large number of plant populations. However, it is generally considered the most inclusive of methods for estimating OA due to thoroughness of data acquired throughout the drying cycle [71]. Furthermore, this method separates active solute accumulation from the simple concentration effect due to water loss from tissues.

11.3.2.3 Full Turgor Adjustment Method

OA is estimated from the differences in osmotic potential at full turgor (100% RWC) between stressed and nonstressed plants [75–77]. Osmotic potential of nonstressed plants consists of a point measurement following the last irrigation and prior to stress imposition (day 0), and measurement for osmotic potential in stressed plants is taken at various times in the drying cycle until reaching the general wilting point, such as 60% of RWC. In this method, RWC is used to convert a given osmotic potential at different treatment periods to osmotic potential under 100% RWC using a correction factor for tissue apoplastic water [78]. Although this method may be more rapid than the regression method, if using leaf sap to measure osmotic potential utilizing an osmometer, there might be difficulty in expressing enough sap in dry leaf tissues. This, however, is not a problem with the rehydration method (described below) inasmuch as sap extraction is performed with rehydrated tissue.

11.3.2.4 Rehydration Method

This method is similar to the full turgor adjustment method, except that osmotic potential of fully rehydrated tissues from stressed plants is measured instead of being estimated using RWC and osmotic potential measured at different points of the stress period [79–82]. Osmotic potential of leaves at full turgor is determined after leaves (stressed or nonstressed) are soaked in water for full rehydration. Sap of leaves can be extracted readily and measured for osmotic potential using an osmometer. OA is calculated as osmotic potential of stressed leaves at full turgor minus osmotic potential of nonstressed leaves at full turgor. In this method during the rehydration period, there is a potential risk of dilution of solutes or reduced osmolality due to respiration rates. However, Steponkus et al. [83] found that OA was stable for at least 24 hours during the rehydration process. As recommended by Babu et al. [71], a shorter rehydration period may eliminate some of the problems associated with a long rehydration period. Compared to the regression method, this method is faster and less labor intensive. OA values from this method are often comparable to those from the full turgor adjustment method, with the advantages that the measurement of RWC is not required and sap is easily extracted for osmotic potential determination. This method is a good choice for quantifying OA in a large amount of samples.

11.3.3 OSMOTIC ADJUSTMENT AND STRESS TOLERANCE

The accumulation of compatible solutes in plant shoots during OA has been positively correlated with stress tolerance in many species, including cereal crops [84–90], perennial grasses [46,91], vegetables [92], and woody species [70,93,94]. Nguyen-Queyrens and Bouchet-Lannat [70] found that high OA capacity in maritime pine (*Pinus pinaster* Aiton) contributed to the maintenance of positive turgor at low water potentials, which was related to the maintenance of growth of this species under drought conditions. Ghoulam et al. [95] investigated effects of salt stress on sugar beet and established that inorganic ions contributed to osmotic adjustment and increased salinity tolerance. OA also contributed to heat tolerance in Kentucky bluegrass *(Poa pratensis* L.*)*, whereby maintenance of OA was associated with higher leaf RWC and enhanced turfgrass quality [46].

OA also contributes to recovery from stress [58]. Elmi and West [96] suggested that OA-enhanced turgor maintenance increased the viability of the meristematic and leaf elongation regions in tall fescue (*Festuca arundinacea* Shreb.), which was a critical factor for survival and recovery after drought. White et al. [97] and Volaire [98] also found that differential osmotic adjustment of turf and forage grasses, particularly within leaf bases, resulted in contrasted tiller survival and recovery after drought stress.

There is genetic diversity among species and within species in the capacity for OA [71,73,91,92,99]. The degree of OA may also be influenced by other factors, including tissue type and age [100–103], the rate of dehydration and stress development [104], and uniformity of evaporative demand [105]. Basnayake et al. [106] utilized different pot sizes to manipulate the rate of water deficit in sorghum (*Sorghum bicolor* (L.)moench*)* genotypes, where small pots resulted in more rapid rates of water depletion than larger pots. Investigators found that the expression of maximum OA was best under conditions of slower rates of dehydration, suggesting that gradual and uniform stress development is crucial for full expression of OA, particularly in experiments where screening of genotypes for stress tolerance is required. Under conditions of rapid dehydration, two forage grasses had also decreased in magnitude of osmotic adjustment compared to the same plants under a slower development of drought stress [107]. Slower rates of water stress allow for more gradual acclimation of physiological parameters such as photosynthesis and osmotic potential, whereas faster rates of dehydration mostly cause rapid cellular injury and thus result in the inability of plants to adjust osmotically before they are killed [108].

OA may not always be correlated with increased stress tolerance, depending on plant species and severity of drought stress [109]. In a comparison between cocksfoot (*Dactylis glomerata* L.) and perennial ryegrass (*Lolium perenne* L.), cocksfoot had the least osmotic adjustment even though it was more drought tolerant compared to ryegrass plants [107], suggesting mechanisms other than OA may contribute to the difference in drought tolerance between these two species. Similar findings have been reported in sorghum [110] and rice [111]. Among three cultivars of vegetable amaranth (*Amaranthus* spp.), the degree of OA was negatively correlated to the relative yield reduction; however a fourth cultivar also exhibiting low yield reductions under soil drying had a very low degree of OA [92]. Under these circumstances, other morphological or physiological responses may play a more important role compared to osmotic adjustment to allow the plant to withstand prolonged drought stress.

11.4 MEMBRANE STABILITY

A major impact of plant environmental stress is cellular membrane modification, which may result in impaired function or total dysfunction. The exact structural and functional modification caused by stress is not fully understood. However, the cellular membrane dysfunction due to stress is expressed as increased permeability and leakage of ions, which can be readily measured by the efflux of electrolytes. Hence, the estimation of membrane stability under stress by measuring cellular electrolyte leakage from affected plant tissues into an aqueous medium has been widely used as a screening tool for stress acclimation and tolerance. Stuart [112] recommended expressing electrolyte leakage as an index percentage of total electrolytes in the tested tissues. Based on Stuart's method Flint et al. [113] developed an index of measuring electrolyte leakage for freezing injury. This method has since been used for evaluation of various stresses such as chilling injuries [114–116], high temperature [117,118], dehydration [119], salinity [120], and metal toxicity such as copper [121].

11.4.1 Electrolyte Leakage Analysis

The procedure of electrolyte leakage analysis is quite simple. Fresh tissues (leaves or roots) are soaked in distilled, de-ionized water and solutes are allowed to efflux from the tissue into the solution until leakage ceases, at which time the electrical conductivity of the solution is measured (initial leakage, C_i). Following this measurement, samples are usually killed by autoclaving or rapid freezing in liquid nitrogen and are allowed time for complete leakage of solutes from the killed tissues to the solution. The conductivity of the solution is then measured and is considered as the maximum electrolyte leakage (C_m). Membrane stability or index of electrolyte leakage can be expressed as $C_i/C_m \times 100$. This technique is simple; however, it has also been criticized [122]. The two major problems with the method are that it does not measure membrane stability directly, and it does not take into account the effects of differences in anatomy between different species, which in turn can affect the conductance. However, the use of electrolyte leakage as an indicator for assessing the effect of environmental stresses on membranes remains a strong and an important tool, especially when comparing samples for the same species.

11.4.2 Membrane Stability and Stress Tolerance

Although numerous processes change during stress, various studies suggest that membranes are among the first affected by environmental stresses, and membrane changes may constitute the initial response of a plant to stressful conditions [123,124]. Maintenance of membrane integrity in terrestrial organisms represents a major mechanism of desiccation tolerance [60]. The role of membranes in the survival of bacteria under severe stresses such as extreme temperatures, salinity, and drought has been extensively studied [125,126].

Heat stress typically is characterized by direct damage to the integrity of cells, leading to various physiological changes [1]. Cell membranes, which are structurally made up of larger polyunsaturated fatty acids, are highly susceptible to high temperature and consequently changes in membrane fluidity, permeability, and cellular metabolic functions. Cell membranes become more fluid as temperatures rise, increasing the chance for membrane leakage [127]. Electrolyte leakage was used to measure ecotypic differences in heat tolerance of aspen (*Populus tremuloides* Michx.) leaves. Leaves were obtained from trees growing at different elevations from 1960 to 2454 m. Heat tolerance was greatest in trees growing in the lower sites and trees propagated from these sites that were grown in even lower elevations showed some increase in heat tolerance as measured by electrolyte leakage. Shanahan et al. [118] found electrolyte leakage to be a useful screening procedure for selecting heat-tolerant genotypes of spring wheat (*Triticum aestivum*). Jiang and Huang [128] examined the effect of heat stress alone and in combination with drought on two cool season grasses, tall fescue and perennial ryegrass. Increase in electrolyte leakage was much more severe and occurred earlier for ryegrass than fescue subjected either to heat alone or combined with drought. Wallner et al. [129] used electrolyte leakage to measure the effect of duration and level of imposed heat stress on turfgrass leaf segments. Electrolyte leakage revealed that differences between species in heat tolerance were most apparent when injury was monitored over time at 50°C.

These studies suggest that electrolyte leakage could be used for selecting stress-tolerant grass species and cultivars. Quantitative differences in heat tolerance, evaluated using electrolyte leakage, were consistent with qualitative descriptions of drought resistance in some species [129], but not in other plant species. Although wheat cultivars exhibited higher drought tolerance than sorghum, maize, or millet (*Pennisetum glaucum*), electrolyte leakage of wheat cultivars was significantly higher than sorghum, maize, and millet under high temperatures [130].

Tripathy et al. [131] used cell membrane stability as a major indicator for drought stress when searching for quantitative trait loci (QTL) associated with drought tolerance in rice. There was no significant difference in RWC between different lines. A significant difference in cell membrane stability was observed between the different lines subjected to drought stress. The continuous distribution of membrane stability and its broad-sense heritability (34%) indicated that this parameter should be polygenic in nature. Saneoka et al. [132] assessed the effects of nitrogen nutrition on membrane stability under water-stress conditions. Plants showed greater adaptation to water stress at higher nitrogen (N) levels, as evidenced by increased membrane stability. The results suggested that higher levels of N nutrition might have contributed to drought tolerance by preventing cell membrane damage. Electrolyte leakage has been found to be correlated to other stress indicators such as osmotic potential and accumulation of water-soluble organic solutes during drought stress [133].

Similarly to effects of heat and drought stress, one of the major events in chilling injuries is alteration of the cellular membrane. Plants injured by chilling show inhibition of photosynthesis and carbohydrate translocation, protein synthesis, and increased degradation of existing proteins. These responses largely depend on a common mechanism involving partial or complete loss of membrane function during cold stress. Increased solute leakage was found in leaves of chill-sensitive *Passiflora maliformis* floated on water at 0°C but not in chill-resistant *Passiflora caerulea*. [134]. Under cold stress the cellular membrane alters from a relatively fluid state to a more rigid state [135].

This change decreases membrane permeability and changes the activity of membrane-associated enzymes [136–138]. At the cellular level the accumulation of cryprotectants in the cytoplasm is important in the development of frost hardiness, which involves modifications in cell membranes that affect membrane stability [139]. Plants that are capable of acquiring frost hardiness also exhibit an increase in the content of unsaturated membrane lipids, whereas in plants with lower capacity for hardening, these changes are less pronounced [139]. Increased leakage of solutes was first observed as one aspect of chilling injury in sweet potato (*Ipomoea batatas* (L.) Lam) [140].

Since then, chilling has been shown to increase permeability of the plasmalemma membrane in various species, as measured by increased solute or electrolyte leakage [141–143]. King and Ludford [114] used electrolyte leakage to study chilling sensitivity in tomato (*Lycopersicon esculentum* Mill.) and correlated it to injury susceptibility in mature green fruit from two chilling-sensitive and two chilling-tolerant cultivars. Mature green chilling-sensitive tomatoes showed higher electrolyte leakage as compared to the tolerant fruit. These results show indications of chilling injury before the development of visual symptoms, at least with some cultivars. Similar results have been shown in cucumbers (*Cucumis sativus* L.) [144]; however, not all chilling-sensitive fruits appear to show such differences, as electrolyte leakage was reported to be unaffected in both peach (*Prunus persica* (L.) Batsch) [145] and bell-pepper (*Capsicum annuum* L.) [145]. Yelenosky [116] used electrolyte leakage to test the cold hardening ability of orange (*Citrus sinensis* (L.) Osbeck) trees on cold-sensitive rootstocks in severe freezing conditions in order to obtain basic information on the freeze survival of specific cultivars and their tolerance to low temperatures. Electrolyte leakage from unhardened frozen leaves increased over 20-fold as compared to frozen leaves on hardened trees and nonfrozen leaves from unhardened trees. These studies confirmed the relationship between electrolyte leakage and chilling-stress injury or tolerance.

11.5 LIPID PEROXIDATION

Oxidative injury in plants can be induced by many different biotic and abiotic stresses, including disease [146], drought [147–150], salinity [151,152], temperature extremes [153–157], nutrient deficiencies [158], high irradiance levels [159], and herbicides [160,161]. These environmental factors are common in that they can all lead to formation of active oxygen and oxidative damage within cells. Excessive energy from reduced electron transport during stress can lead to an accumulation of excitation energy that can be dissipated by reducing molecular oxygen, whereby electrons leaked from chloroplasts and mitochondria interact with oxygen and generate active oxygen species (AOS) [162]. AOS include singlet oxygen ($^1O_2$), superoxide ($O_2{\cdot^-}$), hydrogen peroxide (H_2O_2), and hydroxyl radical (OH·). These species interact with cellular molecules and cause severe damage to lipids, nucleic acids, and proteins. Halliwell [163] produced the first extensive review on active oxygen and its consequences, and several other overviews have subsequently been presented [149,159,164–167].

11.5.1 Antioxidant Enzyme Defense System

One assessment of the antioxidant system is through measurement of antioxidant enzymatic activities. Increases in enzyme activity may be induced by the presence of active oxygen species, thus providing indirect evidence for the extent of generation of active oxygen species or the greater requirement for this type of defense response [149]. Some common antioxidant enzymes include catalase (CAT), superoxide dismutase (SOD), peroxidase (POD), glutathione reductase, and ascorbate peroxidase (APX). Tolerance to the aforementioned stresses is associated with maintenance or an increase in antioxidant enzyme activities [168–173].

However, not all antioxidant enzymes change in their activities with the same pattern in response to stresses, or they may have differential effects on stress tolerance [154,173–175]. Sairam et al. [173] indicated that there were differences in the increase in activities of various antioxidants among tolerant genotypes of wheat, such that one tolerant genotype had very high levels of ascorbic acid and APX, whereas another tolerant genotype exhibited higher SOD and CAT and intermediate ascorbic acid activities. They concluded that not all of the antioxidant enzymes may increase uniformly within tolerant selections, and that different antioxidant enzymes may be more significant for imparting tolerance in some genotypes compared to others. Queiroz et al. [176] studied cold stress in *Coffea arabica* L. roots and found that APX and CAT did not change considerably, but POD increased to a greater extent under chilling stress compared to control plants. In some cases, antioxidant enzyme activity is not related to stress tolerance. Cavalcanti et al. [177] concluded that SOD, POD, and CAT did not appear to contribute to survival under high salinity levels in salt-stressed cowpea leaves. Although antioxidant enzymes generally do play a role in the antioxidant capabilities of plants, there is obviously some variability in enzyme activities among species and genotypes. Therefore, this section focuses on lipid peroxidation, an indicator of stress injury that has a relatively consistent response among plant species.

11.5.2 Determination of Lipid Peroxidation

The extent of lipid peroxidation can be calculated by measurement of total 2-thiobarbituric acid (TBA) reactive substances (TBARS) and expressed as equivalents of malondialdehyde (MDA) content, which is a secondary breakdown product of lipid peroxidation [159]. The more accurate estimation of lipid peroxidation is through direct quantification of the primary hydroperoxide products, however, this approach proves to be more difficult due to the labile nature of these primary products as well as a more involved protocol [178]. The TBARS assay has been utilized extensively in both plant and animal systems as a consistent means for estimation of lipid peroxidation.

Many of the studies evaluating lipid peroxidation follow the original protocol of Heath and Packer [179]. Briefly, leaf or root samples (typically 0.2–0.5 g fresh weight) are homogenized in a solution of approximately 0.1% trichloroacetic acid (TCA) at 4°C, and then centrifuged at approximately 15,000 g for 10 minutes. For every 1 ml of supernatant, 4 ml of a thiobarbituric acid (TBA) solution (0.5% TBA in 20% TCA) is added for the reaction. Samples are incubated at 95°C for approximately 30 minutes, with the reaction then quickly terminated in an ice bath. The samples are centrifuged again at 10,000 g for 10 minutes. Finally, the specific absorbance of the product is recorded at 532 nm, with the value for nonspecific, background absorbance at 600 nm subtracted from 532 nm. MDA content is calculated using its extinction coefficient of 155 mM^{-1} cm^{-1}.

Different modifications of this protocol exist, particularly in regard to tissue sample amounts and the reaction ratio of supernatant:TBA solution. For extraction of compounds in addition to MDA with limited plant tissue, different extraction methods are employed, such as those discussed in Ali et al. [180] and Wang and Huang [181]. These factors may be tested on an individual species basis to determine the most consistent levels of MDA.

MDA can also be directly quantified using high-performance liquid chromatography or gas chromatography [182]; however, the spectrophotometric method for testing the TBA–MDA complex may be less costly. Additional benefits for the TBARS assay include its convenient and straightforward protocol, reliability for estimation of small changes in MDA, and ease of processing large numbers of samples [178].

In some cases, however, there may be potential for interference of non-MDA substances in plant tissues resulting in an overestimation of MDA content and lipid peroxidation [182,183], such as in tissues containing high amounts of carbohydrates and pigments, or where additional compounds also contribute to absorbance at 532 nm [178,184,185]. Therefore, even though many current investigations utilize the original method, recent modifications in the Heath and Packer protocol have been made in order to improve the estimation of plant tissue lipid peroxidation. For example, Du and Bramlage [184] adapted the basic TBARS measurement to subtract the sugar absorbance maximum at 440 nm from that at 532 nm in plant tissues high in sugar content. In order to correct for compounds that also absorb at 532 nm, Hodges et al. [178] subtracted the absorbance of solutions extracted without TBA from the same plant extract reacted with TBA. The investigators found that TBA–MDA levels could be significantly over-estimated in plants such as red cabbage (*Brassica oleracea* L. var. *capitata* L.), eggplant (*Solanum melongena* L.), radish (*Raphanus sativus* L.), and highbush blueberry (*Vaccinium corymbosum* L.) by interfering compounds like anthocyanins if correction for these compounds were not made. Therefore, the modified protocol provided greater accuracy for quantification of MDA with little resulting loss in ease or rapidity of the original Heath and Packer procedure. Taulavuori et al. [183] tested the MDA correction method by Hodges et al. [178] in bilberry (*v. myrtillus* L.) under field and growth chamber conditions and also found that the original, uncorrected method overestimated lipid peroxidation levels during cold acclimation. Additional studies have also corrected for interference by sucrose [186] and polyols [187].

11.5.3 LIPID PEROXIDATION AND STRESS TOLERANCE

Tolerance to conditions resulting in oxidative stress has been associated with low levels of lipid peroxidation. Lipid peroxidation is caused by the oxidation of phospholipids and other unsaturated lipids when production of active oxygen species overwhelms the scavenging ability of the antioxidant defense system. Peroxidation leads to breakdown of these lipids and membrane function by causing loss of fluidity, lipid crosslinking, and inactivation of membrane enzymes. Lipid peroxidation is characterized by increasing production of MDA. Therefore, the level of lipid peroxidation is often expressed as MDA content, and is a commonly utilized measurement for assessing oxidative damage in both leaves and roots [156,176], and its maintenance of low levels has been associated with increased stress tolerance in many species [148,188]. Luna et al. [189] assessed MDA and chlorophyll retention as selection tools for salt tolerance in Rhodes grass (*Chloris gayana* Kunth). MDA values were consistently and significantly lower in tolerant clones compared to salt-susceptible plants whether using whole plants or detached leaf segments, whereas no consistent response was observed with chlorophyll content. They concluded that MDA was negatively correlated with survival capability, and could be used as an appropriate selection tool for salt tolerance that did not require whole plants or much leaf tissue to assess salt tolerance.

Decreased levels of lipid peroxidation or MDA content have been associated with increased salt tolerance in cotton [151], tomato [190], and wheat [188,191]. In a comparison between cultivated beet and a salt-tolerant wild beet (*Beta vulgaris* L. subsp. *maritima* (L.) Arcang.), the wild beet exhibited inherently lower lipid peroxidation under control conditions, as well as under increasing salt conditions compared to cultivated beet [192]. Increased lipid peroxidation was measured in drought-sensitive genotypes of *Coffea canephora* Pierre ex A. Froehner [193], wheat [148,194], and pea [195]. Temperature-tolerant genotypes of wheat also had lower lipid peroxidation levels under high temperatures than did temperature-sensitive genotypes [173].

11.6 STRESS-INDUCED PROTEINS

To cope with environmental stress, plants activate a large set of genes leading to the accumulation of specific proteins, which are generally considered stress-induced proteins. Some stress-induced proteins, such as heat shock proteins (HSP) and late embryogenesis abundant proteins (LEA) are required for stress tolerance, and their accumulation has a role in protecting plant tissues from possible damages caused by environmental stresses [196]. The alteration of the protein level is a reflection of both transcriptional and translational regulation. The current section discusses the effect of environmental stress conditions on HSPs and LEAs, which function in protecting cells from stress injury and are the two major types of stress-inducible proteins that accumulate upon water, salinity, and extreme temperature stress [54,197].

11.6.1 Heat Shock Proteins

Over two decades ago it was shown that when seedlings are shifted to temperatures 5°C or more above their optimal growth temperature, synthesis of most proteins and mRNAs is repressed, and the transcription and translation of a small set of proteins, which are named heat shock proteins (HSPs), are initiated (for reviews see [198–200]). Production of high levels of HSPs can also be triggered by exposure to other environmental stress conditions, such as drought, cold, salinity, and hypoxia (oxygen deprivation). Consequently, HSPs are also referred to as stress proteins and their upregulation is sometimes described more generally as part of the stress response [201]. New HSP transcripts (mRNA) can be detected three to five minutes after heat stress [196]. Heat shock proteins also occur under nonstressful conditions, simply "monitoring" the cell's proteins. Some examples of their role as "monitors" are that they carry old proteins to the cell's "recycling bin" and they help newly synthesized proteins fold properly. These activities are part of a cell's own repair system, called the "cellular stress response." One of the important roles of HSPs involves stabilization of proteins in a particular state of folding. Through this mechanism, HSPs such as HSP90 and HSP70 facilitate a wide diversity of important processes including folding and transport of proteins across membranes and therefore these HSPs are also called "molecular chaperones" [200,202].

The molecular masses of HSPs range from 15 to 104 kDA. Among five conserved families of HSPs—HSP100, HSP90, HSP70, HSP60, and small HSP (sHSP)—the small HSPs are found to be most prevalent in plants. sHSPs vary in size, with a molecular weight between 15 to 42 kDa [203]. sHSPs are produced ubiquitously in procaryotic and eucaryotic cells upon environmental stresses such as temperature, light, salinity, and drought [197]. Under unstressed growth conditions, most sHSPs cannot usually be detected in most plant tissues, however, upon stress there is an accumulation of sHSPs. The induction of sHSPs depends on the severity of the stress and its duration [204].

However, in the resurrection plant *Craterostigma plantagineum* Hochst., sHSPs have been shown to be expressed consitiutively in vegetative tissues [205]. Resurrection plants are a group of angiosperms that can tolerate extreme dehydration. Resurrection plants are able to dehydrate, remain quiescent during long periods of drought, and then resurrect upon rehydration. Constitutive expression of HSPs in this species demonstrates the importance of sHSsP in long-term adaptation to environmental stresses.

High levels of sHSPs, as a result of overproducing HSF's, increased the basal level of thermotolerance in *Arabidopsis* [206]. The correlation between sHSP synthesis and stress response led to the hypothesis that sHSPs protect cells from detrimental effects of stress, however, the mechanisms in which they are involved in cell protection are still not fully understood [203]. Hamilton and Heckathorn [207] suggested that sHSPs might act as antioxidants in protecting complex-I electron transport in the mitochondria during salinity stress. sHSPs as well as other HSPs are regarded as stress proteins with a potential to protect cells from stress damage. An increasing number of studies show the existence of cross-tolerance in plants, where an exposure of tissue to moderate stress induces tolerance to other stresses. Although cross-tolerance has been demonstrated in several plant species, a common mechanism has not yet been found; however, HSPs have been demonstrated to play an important role in cross protection [201].

Several recent publications have highlighted the importance of the molecular chaperone HSP90 complex in plant development and responsiveness to external stimuli. In particular, HSP90 is crucial for R-protein-mediated defense against pathogens. Other facets of the HSP90 function in plants include its involvement in phenotypic plasticity, developmental stability, and buffering of genetic variation [208]. Senthil-Kumar et al. [209] screened sunflower (*Helianthus annuus*) hybrid parents for high temperature tolerance. Seedlings of parental lines showed considerable genetic variability for thermotolerance. Thus, the existing variability formed the basis for identifying thermotolerant lines. The identified parental inbred lines were selected and established in the field and crossed to get F-1 hybrid seeds. The hybrid developed from selected variants of parental lines was compared with the original for thermotolerance. The selected hybrid was more tolerant compared with the original hybrid. The selected hybrid showed enhanced expression of HSP90 and HSP104 and also accumulated higher levels of the heat shock transcription factor.

In another study, Srikanthbabu et al. [210] exposed pea seedlings to a moderately high temperature prior to exposure to stressful high temperature. Plants that were acclimated to high temperature exhibited better growth compared to seedlings that were directly exposed to high temperature. The acclimated seedlings accumulated higher levels of hsp18.1 and hsp70 transcripts as well as HSP104 and HSP90 proteins. Those studies all point out the importance of using HSPs as stress indicators. Detailed discussion in the involvement of HSPs in heat tolerance is presented in a previous chapter in this book.

11.6.2 LATE EMBRYOGENESIS ABUNDANT PROTEINS

Late embryogenesis abundant (LEA) genes encode a diverse group of stress protection proteins during embryo maturation in all angiosperms [211]. LEA proteins were first identified and characterized in cotton and represent one of the dominant proteins and mRNA species in mature embryos [211,212]. Accumulation of LEA proteins correlates with increased levels of ABA and with acquisition of desiccation tolerance [213]. Similarly to HSPs, LEA proteins are usually not expressed in nonstressed tissues, but can be induced by osmotic stress or an exogenous application of ABA [54]. Goyal et al. [214] proposed that LEA proteins might act as a novel form of molecular chaperone, a "molecular shield," to help prevent the formation of damaging protein aggregates during water stress. Figueras et al. [215] analyzed the effect of Rab17 expression, which is a LEA protein from maize, under a constitutive promoter in vegetative tissues of transgenic *Arabidopsis* plants. These transgenic plants have higher sugar and proline contents, and also higher water loss rate under water stress. In addition, these plants were more tolerant than nontransformed controls to high salinity and recovered faster from mannitol treatment. The results pointed to a protective effect of Rab17 protein in vegetative tissues under osmotic stress conditions. In another study, Bernacchia and Furini [216] observed an induction in the expression of a large number of LEA transcripts in resurrection plants (*Selaginella lepidophylla* (Hook. and Grev.) Spring). Gene products with a putative protective function such as LEA proteins have been identified; they were expressed at high levels in the cytoplasm or in chloroplasts upon dehydration and/or ABA treatment of vegetative tissue.

Jayaprakash et al. [217] reported the extent of genetic variability in the level of expression of LEA2 and LEA3 under stress in finger millet (*Eleusine coracana*) and rice seedlings. In both species, the expression of LEA genes was seen in the mesophytic tissue in response to salinity, partial dehydration, and abscisic acid. Tolerant genotypes exhibited higher expression of rab16A and M3 that code for LEA2 proteins than susceptible genotypes. A positive correlation was found between LEA2 and LEA3 protein levels and the growth of seedlings during stress and recovery in both rice and finger millet, suggesting that the expression of LEA proteins could be a strong indication of plant tolerance to dehydration-related stresses.

11.7 SUMMARY

Plants are subjected to a variety of environmental stresses that restrict their growth and survival, and the capacity of plants to survive under environmental stress can be termed as stress tolerance. Environmental stresses frequently interact with one another in both natural and agricultural ecosystems. The identification of traits associated with stress tolerance is complicated by the interactions of these different stresses and other factors, including plant genetics, phenological stage, stress duration and intensity, and seasonal and daily variation in climatic conditions. In the current chapter, we covered major physiological and biochemical traits, including carbon metabolism, water relations, membrane stability, and protein induction associated with stress tolerance. Different approaches or tools have been developed to evaluate stress tolerance.

Carbon metabolism is the major factor affecting plant growth productivity and survival. Chlorophyll fluorescence and gas exchange measurements are the most common tools for assessing changes in photosynthesis and carbon metabolism in response to environmental stress in plants. Chlorophyll fluorescence serves as an intrinsic indicator of photochemical efficiency in photosynthetic organisms. Fluorescence methods have been used as screening tools in evaluating various stresses. Environmental stresses cause change in the photosynthetic apparatus as well as changes in chloroplast ultrastructure and in the photochemical functioning of plants, which can easily be assessed by fluorometers. Photosynthetic gas exchange is affected by many stresses including drought, flooding, salinity, chilling, high temperature, and inadequate nutrition. Gas exchange in plants implies the exchange of CO_2 and O_2 between the interior of the plant and its surroundings, and it is an essential tool for measuring both photosynthesis and respiration rates. Measuring gas exchange can be done either in closed gas exchange systems or in open flow systems. Open systems are usually more advantageous for measuring intact and whole plants.

The maintenance of favorable plant water status is an essential strategy for plant tolerance to stresses that result in cellular water deficit and loss of turgor pressure. Relative water content (RWC) and osmotic adjustment (OA) are both major tools for assessing changes in plant water relations for studying plant responses to stress and subsequent relation to stress tolerance. RWC has been used as a trait for screening different environmental stresses, where plants that maintain higher RWC tend to be more tolerant to stress compared to plants that cannot efficiently control leaf water status. OA is a mechanism of acclimation to cellular dehydration in which compatible solutes are actively accumulated within cells, resulting in higher cellwater retention and maintenance of membrane and protein structures and protection against oxidative damage under stress.

Membranes provide key metabolic interfaces between the plant and its environment by participating both actively and passively in discrimination and transport processes at the level of tissues and organelles. A major impact of plant environmental stress is cellular membrane modification, which results in its impaired function or total dysfunction. Different environmental stresses such as extreme temperatures, drought, and salinity have major effects on membranes. Cellular electrolyte leakage from affected leaf and root tissues into an aqueous medium is used for estimation of membrane dysfunction under stress. Electrolyte leakage is widely used to study and screen for stress tolerance in plants.

Membrane dysfunction is also caused by lipid peroxidation. Environmental stresses can cause oxidative injury due to the generation of active oxygen species. These species interact with cellular molecules and cause severe damage to lipids, nucleic acids, and proteins. Lipid peroxidation, expressed as malondialdehyde (MDA) content, is a commonly utilized measurement for evidence of oxidative damage in both leaves and roots, and its maintenance of low levels has been associated with increased stress tolerance in many species.

Stress-inducible proteins such as heat shock proteins (HSP) and late embryogenesis abundant (LEA) proteins play important roles in protecting plant tissues from stress damages. Their expression is generally positively related to stress tolerance, and it is also an indication of stress response in some cases. Selecting for upregulation of HSP and LEA expression has been proved useful for improving plant tolerance to heat and dehydration tolerance.

REFERENCES

1. Larcher, W., *Physiological Plant Ecology*. Springer-Verlag, Heidelberg, 2003.
2. Kautsky, H. and Hirsch, A., Neue Versuche zur Kohlenstoffassimilation. *Naturwissenschaften*, 19:964, 1931.
3. Schreiber, U., Schliwa, U., and Bilger, W., Continuous recording of photochemical and non-photochemical fluorescence quenching with a new type of modulation fluorometer. *Photosyn. Res.*, 10:51, 1986.
4. Maxwell, K. and Johnson, G.N., Chlorophyll fluorescence: A practical guide, *J. Exp. Bot.*, 51:659, 2000.
5. Govindjee, 63 Years since Kautsky — Chlorophyll-a fluorescence. *Aust. J. Plant Physiol.*, 22:131, 1995.
6. Lichtenthaler, H.K., *Applications of Chlorophyll Fluorescence in Photosynthesis Research, Stress Physiology, Hydrobiology, and Remote Sensing*. Kluwer Academic, Norwell, MA, 1988.
7. Quick, W.P. and Horton, P., Studies on the induction of chlorophyll fluorescence quenching by redox state and transthykaloid pH gradient. *Proc. R. Soc. London Ser. B*, 217:405, 1984.
8. Schreiber, U. and Bilger, W., Rapid assessment of stress effects on plant leaves by chlorophyll fluorescence measurements. In *Plant Response to Stress, NATO ASI Series*, Springer-Verlag, Berlin, 1987.
9. Haldimann, P. and Feller, U., Inhibition of photosynthesis by high temperature in oak (*Quercus pubescens* L.) leaves grown under natural conditions closely correlates with a reversible heat-dependent reduction of the activation state of ribulose-1,5-bisphosphate carboxylase/oxygenase. *Plant Cell Environ.*, 27:1169, 2004.
10. Wen, X.G. et al., Enhanced thermotolerance of photosystem II in salt-adapted plants of the halophyte *Artemisia anethifolia*. *Planta*, 220:486, 2005.
11. Salvucci, M.E. and Crafts-Brandner, S.J., Relationship between the heat tolerance of photosynthesis and the thermal stability of rubisco activase in plants from contrasting thermal environments. *Plant Physiol.*, 134:1460, 2004.
12. Flexas, J. et al., Steady-state chlorophyll fluorescence (Fs) measurements as a tool to follow variations of net CO_2 assimilation and stomatal conductance during water-stress in C_3 plants. *Physiol. Plant.*, 114:231, 2002.
13. Souza, R.P. et al., Photosynthetic gas exchange, chlorophyll fluorescence and some associated metabolic changes in cowpea (*Vigna unguiculata*) during water stress and recovery. *Environ. Exp. Bot.*, 51:45, 2004.
14. Baker, N.R., East, T.M., and Long, S.P., Chilling damage to photosynthesis in young Zea mays. II. Photochemical function of thylakoids in vivo. *J. Exp. Bot.*, 34:189, 1983.
15. Warburg, O., Uber die Geschwindigkeit der photochemischen Kohlensaurezersetzung in lebenden Zellen. *Biochem. Z.*, 100:230, 1919.
16. Hunt, S., Measurements of photosynthesis and respiration in plants. *Physiol. Plant.*, 117:314, 2003.
17. Clarck, L.C., Monitor and control of blood and tissue oxygen tension. *Trans. Am. Soc. for Artificial Organs*. 2:42, 1956.
18. Hill, D.W. and Powell, T., *Non-dispersive Infrared Gas Analysis in Science*. Adam Hilger, London, 1968.
19. von Caemmerer, S. and Farquhar, G.D., Some relationships between the biochemistry of photosynthesis and the gas exchange of leaves. *Planta*, 153:376, 1981.

20. Farquhar, G.D. and von Caemmerer, S., Modeling of photosynthetic response to environmental conditions. In *Encyclopedia of Plant Physiology NS*. Nobel, P.S., Osmond, C.B., and Ziegler, H., Eds., Springer-Verlag, Berlin, 1982.
21. Myers, J., The pattern of photosynthesis in *Chlorella*. In *Photosynthesis in Plants*, Franck, J. and Loomis, W.E., Eds., Iowa State College Press, Ames, IA, 1949.
22. Bloom, A.J. et al., Nitrogen assimilation and growth of wheat under elevated carbon dioxide. *Proc. Natl. Acad. Sci. USA*, 99:1730, 2002.
23. Rachmilevitch, S., Cousins, A.B., and Bloom, A.J., Nitrate assimilation in plant shoots depends on photorespiration. *Proc. Natl. Acad. Sci. USA*, 101:11506, 2004.
24. Long, S.P. and Woodward, F.I., *Plants and Temperature*. Company of Biologists, London, 1988.
25. Sage, R.F. and Reid, C.D., Photosynthetic response mechanisms to environmental changes. In *Plant–Environment Interactions*. Wilkinson, R. E., Ed., Marcel Dekker, New York, 1994.
26. Hamerlynck, E.P. et al., Effects of extreme high temperature, drought and elevated CO_2 on photosynthesis of the Mojave Desert evergreen shrub, *Larrea tridentata*. *Plant Ecol.*, 148:183, 2000.
27. Hsiao, T.C., Plant responses to water stress. *Annu. Rev. Plant Physiol.*, 24:519, 1973.
28. Farquhar, G.D. et al., Photosynthesis and gas exchange. In *Plants Under Stress*, Jones Hamlyn, G., Flowers, T.J., and Jones, M.B., Eds., Cambridge University Press, 1989.
29. Da Silva, J.M. and Arrabaca, M.C., Photosynthetic enzymes of the C-4 grass *Setaria sphacelata* under water stress: A comparison between rapidly and slowly imposed water deficit. *Photosynthetica*, 42:43, 2004.
30. Gucci, R., Lombardini, L., and Tattini, M., Analysis of leaf water relations in leaves of two olive (*Olea europaea*) cultivars differing in tolerance to salinity. *Tree Physiol.*, 17:13, 1997.
31. Kramer, P.J., Changing concepts regarding plant water relations. *Plant Cell Environ.*, 11:565, 1988.
32. Passioura, J.B., Response to Dr P.J. Kramer's article, 'Changing concepts regarding plant water relations,' Volume 11, Number 7, pp. 565–568. *Plant Cell Environ.*, 11:569, 1988.
33. Sinclair, T.R. and Ludlow, M.M., Who taught plants thermodynamics? The unfulfilled potential of plant water potential. *Aust. J. Plant Physiol.*, 12:213, 1985.
34. Kaiser, W.M., Correlation between changes in photosynthetic activity and changes in total protoplast volume in leaf tissue from hygro-, meso-, and zerophytes under osmotic stress. *Planta*, 154:538, 1982.
35. Winter, S.R., Musick, J.T., and Porter, K.B., Evaluation of screening techniques for breeding drought-resistant winter wheat. *Crop Sci.*, 28:512, 1988.
36. Barrs, H.D. and Weatherley, P.E., A re-examination of the relative turgidity technique for estimating water deficit in leaves. *Aust. J. Biol. Sci.*, 15:413, 1962.
37. Kirkham, M.B., *Principles of Soil and Plant Water Relations*. Elsevier Academic, New York, 2005.
38. Barrs, H.D., Determinations of water deficits in plant tissues. In *Water Deficits and Plant Growth*. Kozlowski, T.T., Ed., Academic, New York, 1968.
39. Diaz-Perez, J.C., Shackel, K.A., and Sutter, E.G., Relative water content and water potential of tissue-cultured apple shoots under water deficits. *J. Exp. Bot.*, 46:111, 1995.

40. Schonfeld, M.A. et al., Water relations in winter wheat as drought resistance indicators. *Crop Sci.*, 28:526, 1988.
41. Tahara, M. et al., Relationship between relative water content during reproductive development and winter wheat grain yield. *Euphytica*, 49:225, 1990.
42. Merah, O., Potential importance of water status traits for durum wheat improvement under Mediterranean conditions. *J. Agric. Sci.*, 137:139, 2001.
43. Peltonen-Sainio, P. and Makela, P., Comparison of physiological methods to assess drought tolerance in oats. *Acta Agric. Scand.*, 45:32, 1995.
44. Tadasse, N. et al., Genetic variation of oat (*Avena sativa* L.) characters related to water-stress tolerance. *Agron. Abstr.*, Crop Science Society of America, 97, 1988.
45. Carter, Jr., T.E. and Patterson, R.P., Use of relative water content as a selection tool for drought tolerance in soybean. *Agron. Abstr.*, Crop Science Society of America, Madison, WI, 1985.
46. Jiang, Y. and Huang, B., Osmotic adjustment and root growth associated with drought preconditioning-enhanced heat tolerance in Kentucky bluegrass. *Crop Sci.*, 41:1168, 2001.
47. Lehman, V.G. and Engelke, M.C., Heritability of creeping bentgrass shoot water content under soil dehydration and elevated temperature. *Crop Sci.*, 33:1061, 1993.
48. Quartacci, M.F. et al., Lipid composition and protein dynamics in thylakoids of two wheat cultivars differently sensitive to drought. *Plant Physiol.*, 108:191, 1995.
49. de Ronde, J.A. et al., Comparative study between transgenic and non-transgenic soybean lines proved transgenic lines to be more drought tolerant. *Euphytica*, 138:123, 2004.
50. Matin, M.A., Brown, J.H., and Ferguson, H., Leaf water potential, relative water content, and diffusive resistance as screening techniques for drought resistance in barley. *Agron. J.*, 81:100, 1989.
51. Singh, J. and Patel, A.L., Water status, gas exchange, proline accumulation and yield of wheat in response to water deficit. *Ann. Biol.*, 12:77, 1996.
52. Chan, J.L. and Fowler, J.L., Validation of relative water content for studying plant water relations in crambe. *Industrial Crops Prod.*, 1:21, 1992.
53. Hare, P.D., Cress, W.A., and Van Staden, J., Dissecting the roles of osmolyte accumulation during stress. *Plant Cell Environ.*, 21:535, 1998.
54. Ingram, J. and Bartels, D., The molecular basis of dehydration tolerance in plants. *Annu. Rev. Plant Physiol. Plant Mol. Biol.*, 47:377, 1996.
55. Bohnert, H.J. and Jensen, R.G., Strategies for engineering water-stress tolerance in plants. *Trends Biotech.*, 14:89, 1996.
56. Zhang, J., Nguyen, H.T., and Blum, A., Genetic analysis of osmotic adjustment in crop plants. *J. Exp. Bot.*, 50:291, 1999.
57. Blum, A., *Plant Breeding for Stress Environments*. CRC Press, Boca Raton, FL, 1988.
58. Chaves, M.M., Maroco, J.P., and Pereira, J.S., Understanding plant responses to drought-from genes to the whole plant. *Funct. Plant Biol.*, 30:239, 2003.
59. Arakawa, T., Protein-solvent interactions in pharmaceutical formulations. *Pharm. Res.*, 8:285, 1991.
60. Crowe, J.H., Hoekstra, F.A., and Crowe, L.M., Anhydrobiosis. *Annu. Rev. Physiol.*, 54:579, 1992.

61. Hoekstra, F.A., Golovina, E.A., and Buitink, J., Mechanisms of plant desiccation tolerance. *Trends Plant Sci.*, 6:431, 2001.
62. Rhodes, D. and Hanson, A.D., Quaternary ammonium and tertiary sulfonium compounds in higher plants. *Annu. Rev. Plant Physiol. Plant Mol. Biol.*, 44:357, 1993.
63. Zhu, J., Hasegawa, P.M., and Bressan, R.A., Molecular aspects of osmotic stress in plants. *Crit. Rev. Plant Sci.*, 16:253, 1997.
64. Liang, J. et al., Can differences in root responses to soil drying and compaction explain differences in performances of trees growing on landfill sites? *Tree Physiol.*, 19:619, 1999.
65. Zou, C., Sands, R., and Sun, O., Physiological responses of radiata pine roots to soil strength and soil water deficit. *Tree Physiol.*, 20:1205, 2000.
66. Nguyen, A. and Lamant, A., Variation in growth and osmotic regulation of roots of water-stressed maritime pine (*Pinus pinaster* Ait.) provenances. *Tree Physiol.*, 5:123, 1989.
67. Westgate, M.E. and Boyer, J.S., Osmotic adjustment and the inhibition of leaf, root, stem and silk growth at low water potentials in maize. *Planta*, 164:540, 1985.
68. Sharp, R.E. and Davies, W.J., Solute regulation and growth by roots and shoots of water-stressed maize plants. *Planta*, 14: 43, 1979.
69. Sharp, R.E., Hsiao, T.C., and Silk, W.K., Growth of the maize primary root at low water potentials II. Role of growth and deposition of hexose and potassium in osmotic adjustment. *Plant Physiol.*, 93:1337, 1990.
70. Nguyen-Queyrens, A. and Bouchet-Lannat, F., Osmotic adjustment in three-year-old seedlings of five provenances of maritime pine (*Pinus pinaster*) in response to drought. *Tree Physiol.*, 23:397, 2003.
71. Babu, R.C. et al., Comparison of measurement methods of osmotic adjustment in rice cultivars. *Crop Sci.*, 39:150, 1999.
72. Morgan, J.M., Growth and yield of wheat lines with differing osmoregulative capacity at high soil water deficit in seasons of varying evaporative demand. *Field Crops Res.*, 40:143, 1995.
73. Morgan, J.M., Osmotic components and properties associated with genotypic differences in osmoregulation in wheat. *Aust. J. Plant Physiol.*, 6:67, 1992.
74. Wright, G.C., Smith, R.C.G., and Morgan, J.M., Differences between two grain sorghum cultivars in adaptation to drought stress. III. Physiological responses. *Aust. J. Agric. Res.*, 34:637, 1983.
75. Flower, D.J. and Ludlow, M.M., Contribution of osmotic adjustment to the dehydration tolerance of water-stressed pigeonpea (*Cajanus cajan* (L.) Millsp.) leaves. *Plant Cell Environ.*, 9:33, 1986.
76. Ludlow, M.M. et al., Adaptation of species of *Centrosema* to water stress. *Aust. J. Plant Physiol.*, 10:119, 1983.
77. Ludlow, M.M., Santamaria, F.J., and Fukai, S., Contribution of osmotic adjustment to grain yield of *Sorghum bicolor* (L.) Moench under water-limited conditions. *Aust. J. Agric. Res.*, 41:67, 1990.
78. Wilson, J.R. et al., Comparison between pressure-volume and dew point hygrometry technique to determining the water relation characteristics of grass and legume leaves. *Oecologia*, 41:77, 1979.
79. Jones, M.M. and Turner, N.C., Osmotic adjustment in leaves of sorghum in response to water deficits. *Plant Physiol.*, 61:122, 1978.
80. Wilson, J.R. et al., Adaptation to water stress of the leaf water relations of four tropical forage species. *Aust. J. Plant Physiol.*, 7:207, 1980.

81. Turner, N.C., Crop water deficits: A decade of progress. *Adv. Agron.*, 39:1, 1986.
82. Blum, A. and Sullivan, C.Y., The comparative drought resistance of landraces of sorghum and millet from dry and humid regions. *Ann. Bot.*, 57:835, 1986.
83. Steponkus, P.L., Role of the plasma membrane in freezing injury and cold acclimation. *Annu. Rev. Plant Physiol.*, 35:543, 1984.
84. Lauchli, A. et al., Solute regulation by calcium in salt-stressed plants. In *Biochemical and Cellular Mechanisms of Stress Tolerance in Plants*. Cherry, J.H., Ed., 1994.
85. Kusaka, M., Lalusin, A.G., and Fujimura, T., The maintenance of growth and turgor in pearl millet (*Pennisetum glaucum* [L.] Leeke) cultivars with different root structures and osmo-regulation under drought stress. *Plant Sci.*, 168:1, 2005.
86. Karakas, B. et al., Salinity and drought tolerance of mannitol-accumulating transgenic tobacco. *Plant Cell Environ.*, 20:609, 1997.
87. Dorffling, K. et al., Heritable improvement of frost tolerance in winter wheat by in vitro-selection of hydroxyproline-resistant proline overproducing mutants. *Euphytica*, 93:1, 1997.
88. Romero, I., Maldonado, A.M., and Eraso, P., Glucose-independent inhibition of yeast plasma-membrane H+-ATPase by calmodulin antagonists. *Biochem. J.*, 322:823, 1997.
89. Saneoka, H. et al., Salt tolerance of glycinebetaine-deficient and -containing maize lines. *Plant Physiol.*, 107:631, 1995.
90. Sheveleva, E. et al., Increased salt and drought tolerance by D-ononitol production in transgenic *Nicotiana tabacum* L. *Plant Physiol.*, 115:1211, 1997.
91. Qian, Y. and Fry, J.D., Water relations and drought tolerance of four turfgrasses. *J. Amer. Soc. Hort. Sci.*, 122:129, 1997.
92. Liu, F. and Stutzel, H., Leaf water relations of vegetable amaranth (*Amaranthus* spp.) in response to soil drying. *Europ. J. Agron.*, 16:137, 2002.
93. Badalotti, A., Anfodillo, T., and Grace, J., Evidence of osmoregulation in *Larix decidua* at Alpine treeline and comparative responses to water availability of two co-occuring evergreen species. *Ann. Forest Sci.*, 57:623, 2000.
94. Grieu, P., Guehl, J.M., and Aussenac, G., The effects of soil and atmospheric drought on photosynthesis and stomatal control of gas exchange in three coniferous species. *Physiol. Plant.*, 73:97, 1988.
95. Ghoulam, C., Foursky, A., and Fares, K., Effects of salt stress on growth, inorganic ions and proline accumulation in relation to osmotic adjustment in five sugar beet cultivars. *Environ. Exp. Bot.*, 47:39, 2002.
96. Elmi, A.A. and West, C.P., Endophyte infection effects on stomatal conductance, osmotic adjustment and drought recovery of tall fescue. *New Phytol.*, 131:61, 1995.
97. White, R.H. et al., Competitive turgor maintenance in tall fescue. *Crop Sci.*, 32:251, 1992.
98. Volaire, F., Growth, carbohydrate reserves and drought survival strategies of contrasting *Dactylis glomerata* populations in a Mediterranean environment. *J. Appl. Ecol.*, 32:56, 1995.
99. Premachandra, G.S. et al., Osmotic adjustment and stomatal response to water deficits in maize. *J. Exp. Bot.*, 4:1451, 1992.
100. Bajji, M., Lutts, S., and Kinet, J.M., Water deficit effects on solute contribution to osmotic adjustment as a function of leaf ageing in three durum wheat (*Triticum durum* Desf.) cultivars performing differently in arid conditions. *Plant Sci.*, 160:669, 2001.

101. Chaves, M.M., Effects of water deficits on carbon assimilation. *J. Exp. Bot.*, 42:1, 1991.
102. Kameli, A. and Losel, D.M., Contribution of carbohydrates and other solutes to osmotic adjustment in wheat leaves under water stress. *J. Plant Physiol.*, 145:363, 1995.
103. Veneklaas, E. and Van den Boogaard, R., Leaf-age structure effects on plant water use and photosynthesis of two wheat cultivars. *New Phytol.*, 128:331, 1994.
104. Thomas, H., Accumulation and consumption of solutes in swards of *Lolium perenne* during drought and after rewatering. *New Phytol.*, 118:35, 1991.
105. Glinka, Z. and Ludlow, M.M., Comparative osmotic adjustment to water deficit in Texas 671 and E57. In *Proceedings of the 2nd Australian Sorghum Conference — Australian Institute of Agricultural Science*, Foale, M. A., Henzell, R. G., and Vance, P. N., Eds., Melbourne, 1992.
106. Basnayake, J. et al., Influence of rate of development of water deficit on the expression of maximum osmotic adjustment and desiccation tolerance in three grain sorghum lines. *Field Crops Res.*, 49:65, 1996.
107. Thomas, H., Effect of rate of dehydration on leaf water status and osmotic adjustment in *Dactylis glomerata* L., *Lolium perenne* L. and *L. multiflorum* Lam. *Ann. Bot.*, 57:225, 1986.
108. Toft, N.L., McNaughton, S.J., and Georgiadis, N.J., Effects of water stress and simulated grazing on leaf elongation and water relation of an East African grass, *Eustachys paspaloides*. *Aust. J. Plant Physiol.*, 14:211, 1987.
109. Maggio, A. et al., Moderately increased constituitive proline does not alter osmotic stress tolerance. *Physiol. Plant.*, 101:240, 1997.
110. Flower, D.J., Rani, U.R., and Peacock, J.M., Influence of osmotic adjustment on the growth, stomatal conductance and light interception of contrasting sorghum lines in a harsh environment. *Aust. J. Plant Physiol.*, 17:91, 1990.
111. Babu, R.C. et al., *HVA1*, a LEA gene from barley confers dehydration tolerance in transgenic rice (*Oryza sativa* L.) via cell membrane protection. *Plant Sci.*, 166:855, 2004.
112. Stuart, N.W., Comparative cold hardiness of scion roots from fifty apple varieties. *Proc. Soc. Hort. Sci.*, 37:330, 1939.
113. Flint, H.I., Boyce, B.R., and Beattie, D.J., Index of injury — A useful expression of freezing injnury to plant tissues as determined by the electrolytic method. *Can. J. Plant Sci.*, 47:229, 1966.
114. King, M.M. and Ludford, P.M., Chilling injury and electrolyte leakage in fruit of different tomato cultivars. *J. Amer. Soc. Hort. Sci.*, 108:74, 1983.
115. Yadava, U.L. and Doud, S.L., Evaluation of different methods to assess cold hardiness of peach trees. *J. Amer. Soc. Hort. Sci.*, 103:318, 1978.
116. Yelenosky, G., Survival of young cold-hardened "Hamlin" orange trees at $-6.7°C$. *HortScience*, 25:98 1990.
117. Peck, K.M. and Wallner, S.J., Ecotypic differences in heat resistance of aspen leaves. *Hort. Science*, 1:52, 1982.
118. Shanahan, J.F., Edwards, I.B., and Quick, J.S., Membrane thermostability and heat tolerance of spring wheat. *Crop Sci.*, 30:247, 1990.
119. Premachandra, G.S., Saneoka, H., and Ogata, S., Nutrio-physiological evaluation of the polyethylene glycol test of cell membrane stability in maize. *Crop Sci.*, 29:1287, 1989.
120. Sairam, R.K. et al., Differences in antioxidant activity in response to salinity stress in tolerant and susceptible wheat genotypes. *Biol. Plant.*, 49:85, 2005.

121. Xiong, Z.T. and Wang, H., Copper toxicity and bioaccumulation in Chinese cabbage. *Environ. Toxic.*, 20:188, 2005.
122. Whitlow, T.H. et al., An improved method for using electrolyte leakage to assess membrane competence in plant tissues. *Plant Physiol.*, 98:198, 1992.
123. Heckman, N.L. et al., Trinexapac-ethyl influence on cell membrane thermostability of Kentucky bluegrass leaf tissue. *Scientia Hort.*, 92:183, 2002.
124. Marcum, K.B., Cell membrane thermostability and whole-plant heat tolerance of Kentucky bluegrass. *Crop Sci.*, 38:1214, 1998.
125. Russell, N.J. and Fukunga, N., A comparison of thermal adaptation of membrane lipids in psychrophilic and thermophilic bacteria. *FEMS Microbiol. Rev.*, 75:171, 1990.
126. Oliver, A.E., Crowe, L.M., and Crowe, J.H., Methods for dehydration-tolerance: Depression of the phase transition temperature in dry membranes and carbohydrate vitrification. *Seed Sci. Res.*, 8:211, 1998.
127. Behzadipour, M. et al., Phenotypic adaptation of tonoplast fluidity to growth temperature in the CAM plant *Kalanchoe daigremontiana* Ham. et Per. is accompanied by changes in the membrane phospholipid and protein composition. *J. Membr. Biol.*, 166:61, 1998.
128. Jiang, Y. and Huang, B., Drought and heat stress injury to two cool-season turfgrasses in relation to antioxidant metabolism and lipid peroxidation. *Crop Sci.*, 41:436, 2001.
129. Wallner, S.J., Becwar, M.R., and Butler, J.D., Measurement of turfgrass heat tolerance *in vitro*. *J. Amer. Soc. Hort. Sci.*, 107:608, 1982.
130. Blum, A. and Ebercon, A., Cell membrane stability as a measure of drought and heat tolerance in wheat. *Crop Sci.*, 21:43, 1981.
131. Tripathy, J.N. et al., QTLs for cell-membrane stability mapped in rice (*Oryza sativa* L.) under drought stress. *Theor. Appl. Genet.*, 100:1197, 2000.
132. Saneoka, H. et al., Nitrogen nutrition and water stress effects on cell membrane stability and leaf water relations in *Agrostis palustris* Huds. *Environ. Exp. Bot.*, 52:131, 2004.
133. Gebre, G.M., Kuhns, M.R., and Brandle, J.R., Organic solute accumulation and dehydration tolerance in 3 water-stressed populus-deltoides clones. *Tree Physiol.*, 14:575, 1994.
134. Taiz, L. and Zeiger, E., *Plant Physiology*. Sinauer Associates, Sunderland, 2002.
135. Lyons, J.M. and Raison, J.K., Oxidative activity of mitochondria isolated from plant tissues sensitive and resistant to chilling injury. *Plant Physiol.*, 45:386, 1970.
136. Morris, G.J. and Clarke, A., *Effects of Low Temperature on Biological Membranes*. Academic, New York, 1981.
137. Wang, C.Y., Physiological and biochemical responses of plants to chilling stress. *Hort. Science*, 17, 1982.
138. Wolfe, J., Chilling injury in plants: The role of membrane. *Plant Cell Environ.*, 1:241, 1978.
139. Wilkinson, R.E., *Plant–Environment Interactions*. Marcel Dekker, New York, 1994.
140. Lieberman, M. et al., Biochemical studies of chilling injury in sweet potatoes. *Plant Physiol.*, 33:307, 1958.
141. Creencia, R.P. and Bramlage, W.J., Reversibility of chilling injury to corn seedlings. *Plant Physiol.*, 47:389, 1971.
142. Guin, G., Chilling injury in cotton seedlings: changes in permeability of cotyledons. *Crop Sci.*, 11:101, 1971.

143. Wright, M. and Simon, E.W., Chilling injury in cucumber leaves. *J. Exp. Bot.*, 24:400, 1973.
144. Murata, T. and Tatsumi, Y., Ion leakage in chilled plant tissues. In *Low Temperature Stress in Crop Plants*, Lyons, J. M., Graham, D., and Raison, J. K., Eds., Academic, New York, 1979.
145. Furmanski, R.J. and Buescher, R.W., Influence of chilling on electrolyte leakage and internal conductivity of peach fruits. *HortScience*, 14:167, 1979.
146. Apostol, I., Heinstein, P.F., and Low, P.S., Rapid stimulation of an oxidative burst during elicitation of cultured plant cells: role in defense and signal transduction. *Plant Physiol.*, 90:109, 1989.
147. Zhang, J. and Kirkham, M.B., Enzymatic responses of the ascorbate-glutathione cycle to drought in sorghum and sunflower plants. *Plant Sci.*, 113:139, 1996.
148. Zhang, J. and Kirkham, M.B., Drought-stress-induced changes in activites of superoxide dismutase, catalase, and peroxidase in wheat species. *Plant Cell Physiol.*, 35:785, 1994.
149. Smirnoff, N., The role of active oxygen in the response of plants to water deficit and desiccation. *New Phytol.*, 125:27, 1993.
150. Sgherri, C.L.M. and Navari-Izzo, F., Sunflower seedlings subjected to increasing water deficit stress: Oxidative stress and defense mechanisms. *Physiol. Plant*, 93:25, 1995.
151. Gossett, D.R., Millhollon, E.P., and Lucas, M.C., Antioxidant response to NaCl stress in salt-tolerant and salt-sensitive cultivars of cotton. *Crop Sci.*, 34:706, 1994.
152. Hernandez, A. et al., Salt-induced oxidative stresses mediated by activated oxygen species in pea leaf mitochondria. *Physiol. Plant*, 89:103, 1993.
153. Mishra, R.K. and Singhal, G.S., Function of photosynthesis apparatus of intact wheat leaves under high light and heat stress and its relationship with peroxidation of thylakoid lipids. *Plant Physiol.*, 98:1, 1992.
154. Prasad, T.K. et al., Evidence for chilling-induced oxidative stress in maize seedlings and a regulatory role for hydrogen peroxide. *Plant Cell*, 6:65, 1994.
155. Schoner, S. and Krause, G.H., Protective systems against active oxygen species in spinach: Response to cold acclimation in excess light. *Planta*, 180:383, 1990.
156. Zhou, R. and Zhao, H., Seasonal pattern of antioxidant enzyme system in the roots of perennial forage grasses grown in alpine habitat, related to freezing tolerance. *Physiol. Plant*, 121:399, 2004.
157. Zhang, J. et al., Protoplasmic factors, antioxidant responses, and chilling resistance in maize. *Plant Physiol. Biochem.*, 33:567, 1995.
158. Cakmak, I. and Marschner, H., Enhanced superoxide radical production in roots of zinc-deficient plants. *J. Exp. Biol.*, 39:1449, 1988.
159. Halliwell, B. and Gutteridge, J.M.C., Protection against oxidants in biological systems: the super oxide theory of oxygen toxicity. In *Free Radicals in Biology and Medicine*, Halliwell, B. and Gutteridge, J.M.C., Eds., Clarendon, Oxford, 1989.
160. Knox, J.P. and Dodge, A.D., Singlet oxygen and plants. *Phytochemistry*, 24:889, 1985.
161. Babbs, C.F., Pham, J.M., and Coolbaugh, R.C., Lethal hydroxyl radical production in paraquat-treated plants. *Plant Physiol.*, 90:1267, 1989.
162. Asada, K., The water-water cycle in chloroplasts: Scavenging of active oxygens and dissipation of excess photons. *Annu. Rev. Plant Physiol. Plant Mol. Biol.*, 50:601, 1999.

163. Halliwell, B., Superoxide dismutase, catalase and glutathione peroxidase: Solutions to the problems of living with oxygen. *New Phytol.*, 73:1075, 1974.
164. Asada, K. and Takahashi, M., Production and scavenging of active oxygen in photosynthesis. In *Photoinhibition*, Kyle, D. J., Osmond, C. B., and Arntzen, C. J., Eds., Elsevier Science, Amsterdam, 1987.
165. Cadenas, S.E., Biochemistry of oxygen toxicity. *Annu. Rev. Biochem.*, 58:79, 1989.
166. Foyer, C.H., Lelandais, M., and Kunert, K.J., Photooxidative stress in plants. *Physiol. Plant.*, 92:698, 1994.
167. Winston, G.W., Physiochemical basis for free radical production in cells: Production and defenses. In *Stress Responses in Plants: Adaptation and Acclimation Mechanisms*, Alscher, R.G. and Cumming, J.R., Eds., Wiley-Liss, New York, 1990.
168. Bowler, C., Montagu, M.V., and Inzé, D., Superoxide dismutase and stress tolerance. *Annu. Rev. Plant Physiol. Plant Mol. Biol.*, 43:83, 1992.
169. Gupta, A.S. et al., Overexpression of superoxide dismutase protects plants from oxidative stress. *Plant Physiol.*, 103, 1993.
170. Jagtap, V. and Bhargava, S., Variation in the antioxidant metabolism of drought tolerant and drought susceptible varieties of *Sorghum bicolor* (L.) Moench. exposed to high light, low water and high temperature stress. *J. Plant Physiol.*, 145:195, 1995.
171. Lascano, H.R. et al., Antioxidant system response of different wheat cultivars under drought: Field and *in vitro* studies. *Aust. J. Agric. Res.*, 28:1095, 2001.
172. Price, A.H. and Hendry, G.A.F., Stress and the role of activated oxygen scavengers and protective enzymes in plants subjected to drought. *Biochem. Soc. Trans.*, 17:493, 1989.
173. Sairam, R.K., Srivastava, G.C., and Saxena, D.C., Increased antioxidant activity under elevated temperatures: A mechanism of heat stress tolerance in wheat genotypes. *Biol. Plant.*, 43:245, 2000.
174. Anderson, M.D., Prasad, T.K., and Stewart, C.R., Changes in isozyme profiles of catalase, peroxidase, and glutathione reductase during acclimation to chilling in mesocotyls of maize seedlings. *Plant Physiol.*, 109:1247, 1995.
175. Saruyama, H. and Tanida, M., Effect of chilling on activated oxygen-scavenging enzymes in low temperature-sensitive and tolerant cultivars of rice (*Oryza sativa* L.). *Plant Sci.*, 109:105, 1995.
176. Queiroz, C.G.S. et al., Chilling-induced changes in membrane fluidity and antioxidant enzyme activities in *Coffea arabica* L. roots. *Biol. Plant*, 41:403, 1998.
177. Cavalcanti, F.R. et al., Superoxide dismutase, catalase and peroxidase activities do not confer protection against oxidative damage in salt-stressed cowpea leaves. *New Phytol.*, 163:563, 2004.
178. Hodges, D.M. et al., Improving the thiobarbituric acid-reactive-substances assay for estimating lipid peroxidation in plant tissues containing anthocyanin and other interfering compounds. *Planta*, 207:604, 1999.
179. Heath, R.L. and Packer, L., Photoperoxidation in isolated chloroplasts. *Arch. Biochem. Biophys.*, 125:189, 1968.
180. Ali, M.B., Hahn, E., and Paek, K., Effects of temperature on oxidative stress defense systems, lipid peroxidation and lipoxygenase activity in *Phalaenopsis*. *Plant Physiol. Biochem.*, 43:213, 2005.
181. Wang, Z. and Huang, B., Physiological recovery of Kentucky bluegrass from simultaneous drought and heat stress. *Crop Sci.*, 44:1729, 2004.

182. Janero, D.R., Malondialdehyde and thiobarbituric acid-reactivity as diagnostic indices of lipid peroxidation and peroxidative tissue injury. *Free Rad. Biol. Med.*, 9:515, 1990.
183. Taulavuori, E. et al., Comparison of two methods used to analyse lipid peroxidation from *Vaccinium myrtillus* (L.) during snow removal, reacclimation, and cold acclimation. *J. Exp. Biol.*, 52:2375, 2001.
184. Du, Z. and Bramlage, W.J., Modified thiobarbituric acid assay for measuring lipid oxidation in sugar-rich plant tisuse extracts. *J. Agric. Food Chem.*, 40:1566, 1992.
185. Stafford, H., Anthocyanins and betalains: evolution of mutually exclusive pathways. *Plant Sci.*, 101:91, 1994.
186. Lee, H.J. et al., Transgenic rice plants expressing a *Bacillus subtilis* protoporphyrinogen oxidase gene are resistant to diphenyl ether herbicide oxyfluorfen. *Plant Cell Physiol.*, 41:743, 2000.
187. Blokhina, O.B., Fagerstedt, K.V., and Chirkova, T.V., Relationships between lipid peroxidation and anoxia tolerance in a range of species, during post-anoxic reaeration. *Physiol. Plant.*, 105:625, 1999.
188. Sairam, R.K., Rao, V.K., and Srivastava, G.C., Differential response of wheat genotypes to long term salinity stress in relation to oxidative stress, antioxidant activity and osmolyte concentration. *Plant Sci.*, 163:1037, 2002.
189. Luna, C. et al., Oxidative stress indicators as selection tools for salt tolerance in *Chloris gayana*. *Plant Breed.*, 119:341, 2000.
190. Shalata, A. and Tal, M., The effect of salt stress on lipid peroxidation and antioxidants in the leaf of the cultivated tomato and its wild salt-tolerant relative *Lycopersicon pennellii*. *Physiol. Plant*, 104:169, 1998.
191. Sairam, R.K. and Srivastava, G.C., Changes in antioxidant activity in sub-cellular fractions of tolerant and susceptible wheat genotypes in response to long term salt stress. *Plant Sci.*, 162:897, 2002.
192. Bor, M., Ozdemir, F., and Turkan, I., The effect of salt stress on lipid peroxidation and antioxidants in leaves of sugar beet *Beta vulgaris* L. and wild beet *Beta maritima* L. *Plant Sci.*, 164:77, 2003.
193. Lima, A.L.S. et al., Photochemical responses and oxidative stress in two clones of *Coffea canephora* under water deficit conditions. *Environ. Exp. Bot.*, 47:239, 2002.
194. Sairam, R.K., Deshmukh, P.S., and Saxena, D.C., Role of antioxidant systems in wheat gentoypes tolerance to water stress. *Biol. Plant.*, 41:387, 1998.
195. Moran, J.F. et al., Drought induces oxidative stress in pea plants. *Planta*, 194:346, 1994.
196. Sachs, M.M., Alteration of gene expression during environmental stress in plants. *Annu. Rev. Plant Physiol.*, 37:363, 1986.
197. Wang, W.X., Vinocur, B., and Altman, A., Plant responses to drought, salinity and extreme temperatures: Towards genetic engineering for stress tolerance. *Planta*, 218:1, 2003.
198. Key, J.L., Lin, C.Y., and Chen, Y.M., Heat shock proteins of higher plants. *Proc. Natl. Acad. Sci. USA*, 78:3526, 1981.
199. Nover, L., Neumann, D., and Scharf, K.D., *Heat Shock and Other Stress Response Systems of Plants*. Springer-Verlag, Berlin, 1989.
200. Vierling, E., The roles of heat shock proteins in plants. *Annu. Rev. Plant Physiol.*, 42:579, 1991.
201. Sabehat, A., Weiss, D., and Lurie, S., Heat-shock proteins and cross-tolerance in plants. *Physiol. Plant*, 103:437, 1998.

202. Ellis, J., Proteins as molecular chaperones. *Nature*, 328:378, 1987.
203. Sun, W., Montagu, M.V., and Verbruggen, N., Small heat shock proteins and stress tolerance in plants. *Biochim. Biophys. Acta*, 1577:1, 2002.
204. Howarth, C., Molecular responses of plants to an increased incidence of heat shock. *Plant Cell Environ.*, 14:831, 1991.
205. Alamillo, J. et al., Constitutive expression of small heat shock proteins in vegetative tissues of the resurrection plant *Craterostigma plantagineum*. *Plant Mol. Biol.*, 29:1093, 1995.
206. Lee, J.H., Hubel, A., and Schoffl, F., Derepression of the activity of genetically engineered heat shock factor causes constitutive synthesis of heat shock proteins and increased thermotolerance in transgenic *Arabidopsis*. *Plant J.*, 8:603, 1995.
207. Hamilton E.W. III, and Heckathorn, S.A., Mitochondrial adaptations to NaCl. Complex I is protected by anti-oxidants and small heat shock proteins, whereas complex II is protected by proline and betaine. *Plant Physiol.*, 126:1266, 2001.
208. Sangster, T.A. and Queitsch, C., The HSP90 chaperone complex, an emerging force in plant development and phenotypic plasticity. *Curr. Opin. Plant Biol.*, 8:86, 2005.
209. Senthil-Kumar, M. et al., Screening of inbred lines to develop a thermotolerant sunflower hybrid using the temperature induction response (TIR) technique: A novel approach by exploiting residual variability. *J. Exp. Bot.*, 54:2569, 2003.
210. Srikanthbabu, V. et al., Identification of pea genotypes with enhanced thermotolerance using temperature induction response technique (TIR). *J. Plant Physiol.*, 159:535, 2002.
211. Bartels, D. and Sunkar, R., Drought and salt tolerance in plants. *Crit. Rev. Plant Sci.*, 24:23, 2005.
212. Dure, I.L., Greenway, S., and Galau, G.A., Developmental biochemistry of cotton seed embryogenesis and germination XIV. Changing mRNA populations as shown *in vitro* and *in vivo* protein synthesis. *Biochemistry*, 20:4162, 1981.
213. Galau, G.A., Hughes, D.W., and Dure, I.L., Abscisic acid induction of cloned cotton late embryogenesis-abundant (Lea) mRNAs. *Plant Mol. Biol.*, 7:155, 1986.
214. Goyal, K., Walton, L.J., and Tunnacliffe, A., LEA proteins prevent protein aggregation due to water stress. *Biochem. J.*, 388:151, 2005.
215. Figueras, M. et al., Maize Rab17 overexpression in *Arabidopsis* plants promotes osmotic stress tolerance. *Ann. Appl. Biol.*, 144:251, 2004.
216. Bernacchia, G. and Furini, A., Biochemical and molecular responses to water stress in resurrection plants. *Physiol. Plant*, 121:175, 2004.
217. Jayaprakash, T.L. et al., Genotypic variability in differential expression of lea2 and lea3 genes and proteins in response to salinity stress in fingermillet (Eleusine coracana Gaertn) and rice (Oryza sativa L.) seedlings. *Ann. Bot.*, 82:513, 1998.

12 Breeding and Genomic Approaches to Improving Abiotic Stress Tolerance in Plants

Stacy A. Bonos and Bingru Huang

CONTENTS

12.1 Introduction 357
12.2 Variability and Complexity in Genetics and Physiology
 of Abiotic Stress Tolerance 358
12.3 Classical Breeding Methods for Improving Abiotic
 Stress Tolerance 359
12.4 Application of QTLs for Abiotic Stress Tolerance 360
  12.4.1 QTLs across Environments and Populations 361
  12.4.2 QTL Analysis of Physiological Traits 362
  12.4.3 Limitations of QTL Analysis 363
12.5 Genomic Approaches to Understanding and Breeding
 for Abiotic Stress Tolerance 364
  12.5.1 Synteny and Comparative Genomics 364
  12.5.2 Gene Cloning from QTLs 365
  12.5.3 Microarrays 365
  12.5.4 Forward and Reverse Genetics 367
   12.5.4.1 Mutants 367
   12.5.4.2 Transgenic Approach 368
  12.5.5 Integrated Genomic Approaches 369
12.6 Summary and Prospects 371
References 372

12.1 INTRODUCTION

Abiotic stresses such as extremes of temperature, drought, flooding, and salinity are major factors limiting plant growth and productivity for most of the world's agricultural crops [1]. As climate changes including global warming continue, and water becomes scarcer, yield reductions due to these stresses will continue to be a major concern in agricultural production. Therefore, improving plant tolerance to abiotic stresses will become increasingly important, and the development of cultivars with improved abiotic stress tolerance will be an integral part of any strategic plans for high yield production [1–4].

Several approaches may be taken to improve plant stress tolerance, including traditional breeding and molecular and genomic techniques. Much of the past breeding efforts have aimed primarily to generate cultivars with high productivity under favorable environmental conditions. In recent years, improving abiotic stress tolerance through genetic modification using biotechnology or traditional breeding approaches has made significant progress, but has also been limited by many factors such as lack of genetic variability or germplasm and genes for stress tolerance, lack of understanding mechanisms of stress tolerance, and the complexity of abiotic stresses.

With recent advances in genomic tools, plant physiologists, biochemists, molecular biologists, and breeders are beginning to understand the complex array of genes involved in abiotic stress tolerance. The genomic advancements include the development of molecular markers for gene mapping, complete sequencing of plant genomes including *Arabidopsis*, rice (*Oryza sativa*), and maize (*Zea mays*); synteny and comparative mapping; development of EST libraries; genetic engineering technologies; the production of T-DNA or transposon-tagged mutagenic populations; and the development of forward and reverse genetics tools that can be used in gene function analysis [5]. Most important will be the application of this information toward breeding plants with better abiotic stress tolerance.

This chapter reviews some genetic, molecular, and genomic approaches that are being utilized or have the potential to be utilized in the improvement of plant tolerance to abiotic stresses. Additionally, the application of those different approaches in understanding plant tolerance to water and temperature stresses is discussed.

12.2 VARIABILITY AND COMPLEXITY IN GENETICS AND PHYSIOLOGY OF ABIOTIC STRESS TOLERANCE

Abiotic stresses are extremely unpredictable and can vary dramatically in duration and intensity in diverse environments [6]. Taking soil moisture as an example, it is highly heterogeneous in time (over hours, days, seasons, and years) and space (across soil profile horizontally or vertically or between and within sites). Soil water status ranges from flooding to drought stress due to excessive and deficit rainfall, respectively, and such stresses may last from days to months, depending on soil and climatic conditions. Additionally, it is often difficult to separate individual abiotic stresses as some of the stresses may occur at the same time [3]. For example, it is common for plants to suffer from the combined stress of drought, high temperature, and salinity in many areas. Heat stress is almost always associated with water deficit stress due to increases in soil evaporation and plant transpiration. Salinity stress can occur either as a direct result of dissolved minerals in irrigation water, added fertilizers and soil amendments, salts carried from the root zone by a high water table, flooding, and salt water intrusion [7] or indirectly through changes in soil water status and other physical properties, especially in hot environments. The extreme environmental variation that exists during abiotic stress makes identification and selection of resistant genotypes difficult.

Abiotic stress tolerance of plants is not only influenced by the variation in environmental conditions, it also depends on plant responses to the stress factors. Plant tolerance to abiotic stresses involves numerous mechanisms at the molecular, cellular, and whole-plant levels, as discussed in previous chapters in this book. Certain mechanisms such as whole-plant photosynthesis, osmotic adjustment (OA), cell membrane stability (CMS), antioxidant metabolism, and the production of stress-protection proteins such as heat shock proteins (HSP) and late embryogenesis abundant proteins (LEA) have been shown to play important roles in abiotic stress tolerance of various plant species. Those stress-tolerance mechanisms have been utilized to evaluate the variability in stress tolerance in plants (see Chapter 11). The change in a single physiological process in plant response to stress is rare, if it occurs at all. More likely it is a combination of multiple mechanisms that occur simultaneously in response to a particular type or level of stress.

Most abiotic stresses are considered to be highly complex traits with low habitability, due to the interaction between various environmental factors, the variability in plant responses to stresses, and the influence of many different genes [8–15]. Understanding the genetic control of multigenic (quantitative) traits is often difficult because the plant phenotype is the product of the genotype and the environment. The assessment of a desired genotype is highly dependent on the proper environmental conditions. The unpredictability and variable forms of abiotic stress in different environments make identifying promising individual plants and breeding for tolerance difficult [6,15]. The interdependence of genetics and physiology is important in assessing the phenotype in stressful conditions [16]. It is used to determine how much of the phenotypic variation is controlled by genotype (G), how much is due to environmental effects (E), and how much is a result of an interaction between the two (G × E) [16].

12.3 CLASSICAL BREEDING METHODS FOR IMPROVING ABIOTIC STRESS TOLERANCE

Classical breeding methods have been used for years in the selection of germplasm with improved stress tolerance. CIMMYT has been breeding tropical maize for drought tolerance since 1975 [2]. Typically, progenies are evaluated in replicated trials under well-watered conditions and at one or two drought stress levels during a rain-free period using irrigation. Breeders select for an index that will maintain time from sowing to anthesis; maintain or increase grain yield under well-watered conditions; increase grain yield under drought; and decrease anthesis-silking interval (ASI), the rate of leaf senescence, and leaf rolling under drought [2]. Depending on the selection scheme and selection intensity, yield increased from 59 to 233 kg/ha^{-1} cycle^{-1}, however, little change was observed in any trait indicative of plant water status (i.e., predawn or noon water potential, osmotic adjustment, canopy temperatures, or water extraction profiles) [2]. The cause of this may be a reflection of the restrictions imposed by their selection approach. For a trait to be considered it has to be fast and easy to measure under field conditions [2]. Physiological measurements such as osmotic adjustment are time consuming and not easily conducted in the field.

Selection for improved yield under well-watered conditions unfortunately does not correlate to yield under drought stress conditions [17,18] indicating that careful site selection for screening germplasm needs to be considered [4]. These germplasm evaluations require multiple seasons and a large number of locations to develop confidence in the performance of a hybrid/cultivar in terms of stress response, yield potential, and yield stability [4]. In fact, yield stability (or the ability of a genotype to show minimal interaction with the environment) seems to be the key to past success in breeding for drought tolerance) [4,17,19].

Although classical breeding methods utilized to investigate the genetic control of quantitative traits, such as drought stress, are valuable, they do not provide information on (1) the chromosome regions regulating the variation of each trait; (2) the simultaneous effects of each chromosome region on other traits and the genetic control (pleiotropy or linkage) of the associated effects; and (3) the interpretation of possible cause–effect (upregulated/downregulated) relationships among genes [15,16]. Some of these constraints can now be partially overcome by utilizing molecular markers, which can allow for the identification of quantitative trait loci (QTL) controlling a chosen phenotype and also enable scientists to assess the effects of the same QTL region on other traits [15]. Breeders can also utilize QTLs for marker-assisted selection (MAS) to improve selection efficiency. For example, if gene effects explain a large portion of the variation, then genetic markers can be used to select for desired phenotypes [16]. Recent advances in genetics including the identification of molecular markers have improved the ability of physiologists and breeders to study the genetic control of abiotic stress tolerance in plants.

12.4 APPLICATION OF QTLS FOR ABIOTIC STRESS TOLERANCE

The ultimate goals of QTL analysis are to dissect the complex inheritance of quantitative traits into "Mendelian-like" factors amenable to selection through the analysis of flanking molecular markers and to clone the genes underlying the QTLs. In order to identify QTLs, an experimental population segregating for the trait of interest is created [15] and a linkage map based on molecular markers is developed. In maize, the first linkage map using restriction fragment length polymorphism (RFLP) markers was created in 1986 [20]. The continued advancement in marker technology including the introduction of PCR-based markers such

Breeding and Genomic Approaches

as microsatellites (SSRs, simple sequence repeats) and AFLPs (amplified fragment length polymorphisms) [21] greatly reduced the time-consuming process of map construction and subsequent QTL analysis [15]. The analysis of QTLs is based on the association of phenotypic differences for the trait of interest with genetic markers located at specific positions on the chromosomes [22]. The ultimate goal is to identify QTLs that increase yield under abiotic stress or at least increase yield stability under abiotic stress [19].

12.4.1 QTLS ACROSS ENVIRONMENTS AND POPULATIONS

The identification of QTLs associated with drought tolerance has been the topic of many research projects in many agronomic crops [18,19,23–25]. The study of QTLs for drought tolerance has revealed some expected and promising results that will affect plant breeding programs for the future. Ribaut et al. [18] identified QTLs associated with yield components of tropical maize under drought conditions. They identified QTLs for grain yield, ear number, and kernel number using composite interval mapping. No major QTLs expressing more than 13% of the phenotypic variance were detected for any of these traits and there were inconsistencies in their genomic position across watering regimes. They found no positive correlation between drought tolerance index and yield under well-watered conditions but by using composite interval mapping they were able to identify "stable" QTLs across drought environments. They proposed to use a combination of the stable QTLs for marker-assisted selection that represents the largest phenotypic variance involved in the expression of traits correlated with yield such as ASI.

Frova et al. [23] evaluated QTL for yield components in a recombinant inbred population of maize between inbred lines B73 and H99 under both well-watered and water-stressed conditions [23]. In contrast to previous research, they found a high positive correlation between well-watered and water-stressed environments for all traits measured and more than 50% of the QTL detected were the same in both. Further research with this population by Sari-Gorla et al. [26] identified QTL associated with plant height and flowering in response to drought. They found that for male flowering time and plant height, most of the QTLs detected were the same under control and stress conditions, but QTLs for female flowering time and ASI appeared to be expressed either under control conditions or under stress. All of the QTLs conferring tolerance to drought were located in a different chromosome position compared to the map position of factors controlling the traits. This research suggested that plant tolerance to water stress may not be attributable to the presence of favorable allele combinations controlling the trait, but is based on physiological characteristics not directly associated with the control of the character [26].

Results of these studies indicate that QTLs for drought tolerance varied with the environment and populations evaluated. The importance of identifying stable QTLs across several growth conditions should be valuable for the discovery of chromosomal regions and ultimately genes that control these quantitative traits [27]. They also provide usefulness and insight into their potential use as markers for selection.

12.4.2 QTL ANALYSIS OF PHYSIOLOGICAL TRAITS

As discussed earlier, OA, relative water content (RWC), and CMS have been widely used as stress-tolerance indicators for various environmental stresses, especially drought, heat, and salinity. QTL analysis for root growth, osmotic adjustment, relative water content, and cell-membrane stability have been the topic of many research projects [8,25,27,28–34]. In rice, rooting traits and OA were found to be associated with improved drought tolerance [8], however, these traits are rarely selected for because phenotypic selection for these traits is difficult and labor intensive. Considering these limitations to efficient selection, molecular marker technology is a powerful tool for selecting such traits [8]. Price and Tomos [35] evaluated QTLs for root growth in rice. One QTL for root length on chromosome 11 explained 30% of the variation and appeared to be additive, indicating the phenotypic selection could be utilized in breeding programs. One QTL for root volume and two QTLs for adventitious roots were also detected. Further study in this area with different crosses under different environments revealed that within each experiment a large number of relatively small QTLs were detected, but the pattern of QTLs varied between type of rooting, environments, and between different treatments [19]. They suggest that the lack of co-location of drought avoidance and root morphological QTLs casts doubt on the notion that improving roots of rice will improve drought tolerance. In contrast, two other studies [8,36] found a positive association between root QTL and drought tolerance across different genetic backgrounds. Both studies indicate the benefits of utilizing these QTLs for marker-assisted selection breeding programs.

QTLs have also been identified for OA. Lilley et al. [29] found one major QTL associated with OA. This QTL and two others associated with dehydration tolerance were also associated with root morphology although the correlation was negative. They suggest that to develop cultivars with extensive root systems and high OA, linkage between these traits will have to be broken. Zhang et al. [36], Babu et al. [8], and Robin et al. [30] also identified QTLs for OA in rice. The QTL on chromosome 8 identified by Babu et al. [8] and the QTLs located on chromosomes 1 and 8 identified by Robin et al. [30] were located in the same genomic region as a QTL for OA identified by Zhang et al. [36]. Additionally, in the same region a QTL for stomatal behavior was detected in a rice F_2 mapping population [24]. These combined results suggest that genes in this genomic region might have been conserved in plant response to drought [8].

QTLs for RWC and OA have also been studied extensively in barley (*Hordeum vulgare* L.). Teulat et al. [31] evaluated 187 barley recombinant inbred lines for RWC, number of leaves on the main tiller (NL), number of tillers, and total shoot fresh and dry biomass under both well-watered and drought-stressed conditions. Under water stress, one QTL on chromosome 1 was involved in RWC and NL. They also found epistatic interactions between several QTLs detected only during the water-stress treatment, suggesting that some chromosomal regions might be involved in the regulation of the expressed traits under water stress. Another study conducted on the same population detected three QTLs for RWC, four QTLs for osmotic potential, and one QTL for OA under water-stress conditions [32]. Two chromosomal regions considered as regions controlling OA were found on chromosome 1 (7H) (similar to the region found by Teulat et al. [31]) and chromosome 6 (6H) [32].

QTLs for RWC and OA were also determined in another study conducted on the same population evaluated across several Mediterranean sites [27]. They found a total of nine chromosomal regions, four of which were considered stable across several growth conditions because they presented main effects across several field environments and previously identified under controlled conditions. This provides insight into potential markers to utilize for marker-assisted selection breeding programs as well as chromosomal regions to target for positional cloning of the QTL. The main region identified in this study, located on the long arm of chromosome 6H, was previously identified as controlling RWC, leaf osmotic potential under water stress and OA [27]. The latter region of chromosome 6H was also previously proven to contain a cluster of *dhn* genes including the barley *dhn4* and *dhn5* [37]. *Dhn* genes or dehydrin genes coding for LEA (late embryogenesis abundant) proteins are water-soluble lipid-associated proteins that accumulate in response to dehydration, low temperature, osmotic stress, or ABA, or during seed maturation [38]. Several QTL controlling tolerance traits, and particularly freezing tolerance, have been identified close to dehydrin genes (reviewed by [37]).

Cell membrane stability is a good indicator of stress tolerance, along with osmotic adjustment, however, CMS is difficult to use as a selection technique for breeding due to the time and labor required for the measurement. The development of QTLs for CMS would provide a more efficient selection technique compared to conventional phenotype selection, but it has not received as much attention as QTL studies on osmotic adjustment. QTLs for CMS have been identified in both maize [39] and rice [35]. In rice, nine putative QTLs for CMS were identified. In fact, one of the nine QTLs was mapped to the same locus as the QTL for OA on chromosome 8 identified by Zhang et al. [36], which was later confirmed by Babu et al. [8] and Robin et al. [30].

12.4.3 LIMITATIONS OF QTL ANALYSIS

It is generally accepted that QTL analysis is more useful in studying quantitative traits than classical genetics and provides a powerful magnifying lens for deciphering the chromosome regions regulating complex traits [15]. However, QTL analysis also has some major shortcomings. One of the most obvious limitations of QTL analysis is the inconsistency of QTL detection across environments and genetic backgrounds. The number, location, and effects of identified QTL vary according to the genetic background of the population and environmental conditions in which the experiment is being conducted [15]. QTLs also have large confidence intervals [40] and complete molecular dissection of the genetic basis underlying a QTL is not possible [15]. Confidence intervals can amount to between 5 and 30% of the chromosome [40]. As an example, a confidence interval of approximately 15 cM in a maize map of approximately 1500 cM long is expected to contain on average 400 genes [15]. Another problem with QTL analysis is the low resolution power in detecting the real number of QTLs regulating the expression of the investigated traits [15,40]. With such small sample sizes (100–200), it is believed that only a modest fraction of QTLs is discovered.

Perhaps the greatest limitation of QTL analysis is that it is time consuming and expensive [15]. The time from the initial cross to actual identification of QTLs is seldom less than three years and can be as long as six to eight years. The costs to do a traditional QTL study remain extremely expensive [15], especially if no linkage map is readily available for the species of study.

12.5 GENOMIC APPROACHES TO UNDERSTANDING AND BREEDING FOR ABIOTIC STRESS TOLERANCE

New developments over the past 15 years in the area of genomics have resulted in new tools available to scientists to uncover the complex mechanisms involved in abiotic stress tolerance. Whole genome sequences for *Arabidopsis* and rice combined with comparative mapping allow for the comparison and discovery of mechanisms across species. New techniques such as high-throughput analysis of expressed sequence tags (ESTs), targeted or random mutagenesis, large-scale analysis of gene expression, and improvements in genetic engineering provide the tools to improve our understanding of stress adaptation and will provide the basis for breeding cultivars and inbred lines with improved abiotic stress tolerance.

12.5.1 SYNTENY AND COMPARATIVE GENOMICS

The discovery of synteny or conservation of genetic content and order across the grasses [48] provides a powerful tool to study genes of numerous grasses as they relate to the rice sequence. All of the rice resources can be directly applied to the genetic analysis of wheat (*Triticum aestivum*), maize, barley, sorghum (*Sorghum bicolar*), pearl millet (*Pennisetum americanum*) [1], and even turfgrasses [42]. This information can also help facilitate fine mapping in the unsequenced genomes of many other grasses [43]. Unfortunately, syntenic relationships among *Arabidopsis* and dicot crops are not as well established as the grass model [1], but the genomic work on alfalfa (*Medicago*) should provide some useful information into the genome structure of broad-leafed plants.

Comparative mapping across related taxa can provide verification of QTL for traits tested across species. One example of this is the discovery of syntenic regions of rice, wheat, and barley associated with QTL for OA. The one QTL on rice chromosome 8 associated with OA identified by Lilley et al. [29] was found to be homologous with a segment of wheat chromosome 7 where a locus for OA was identified. Van Deynze et al. [44] confirmed this through the evaluation of co-segregation of molecular markers in both species. Teulat et al. [32] subsequently found that the QTL they had identified for OA on chromosome 1 (7H in rice) in barley shares homology with the rice chromosome 8 identified by Lilley et al. [29] as being associated with RWC. As stated previously, Zhang et al. [36] and Robin et al. [30] both identified the same chromosomal region associated with OA in different rice populations and both commented on the impact of this conserved region for map-based cloning of genes controlling the resistance to drought in plants [36].

12.5.2 GENE CLONING FROM QTLS

After identifying the QTL or chromosomal region associated with the trait of interest, the next step is to clone the gene and prove its function. When alignment maps of the species of interest and model plants (i.e., rice, *Arabidopsis*) are available candidate genes in the area of the QTL can be identified directly from the model DNA sequence [1]. As stated previously, QTLs have large confidence intervals and with approximately 30 to 50 genes per map unit, it is usually necessary to refine the map location of the QTL [1]. BAC (Bacterial Artificial Chromosomes) or YAC (Yeast Artificial Chromosomes) contigs, which are overlapping linear series of large insert clones, can be made of the species of interest and linked together using DNA markers from both the species molecular map and the model sequence that cover the QTL region [1,15]. Once these YAC or BAC libraries are in place, DNA fragments of known genes in model plants can be used as a probe to identify genes in the species of interest. Choi and Close [45] utilized this technique to identify a new barley dehydrin gene, *Dhn12*, located on chromosome 6H.

Positional cloning or map-based cloning requires increased efforts in phenotyping progenies and mass screening for additional markers close to the QTL [15]. In addition, the development of BAC or YAC genomic libraries is extremely expensive. Despite the large amount of research effort on the discovery of QTLs for abiotic stress tolerance, very few studies have focused on the actual isolation of a gene identified by a QTL. Several techniques including the analysis of gene expression of ESTs using microrrays and the development of gain-of-function or loss-of-function mutants (knock-out) are useful in identifying candidate genes for cloning [22]. The combination of these methods is expected to reduce the tens of thousands of candidate genes to a few target genes [4].

12.5.3 MICROARRAYS

High-throughput EST sequencing coupled with comparative microarray analysis can facilitate the identification of possible candidate genes and improve QTL cloning [1,15]. With the advent of automated high-throughput DNA sequencing, the number of ESTs available is growing rapidly. Most of these are publicly available in the dbEST section of GenBank (http:// www. ncbi. nlm. nih. gov/db EST/dbEST_summary.html) [46]. Differential display was one of the earliest methods of parallel screening for differences in levels of cDNA (or EST) fragments generated from mRNA isolated from samples between different experimental treatments [4,47]. This technology depends on semi-random PCR using degenerate primers and gel-based separation of the amplified cDNA fragments and no sequencing is necessary [4]. This technique has been useful for studying gene expression in abiotic stress tolerance in species for which limited genomic work has been conducted. Jamaux et al. [28] identified three cDNA fragments differentially expressed in water-stressed versus nonstressed sunflower (*Helianthus annus* L.). Jain et al. [48] identified a drought-responsive transcript in peanut (*Arachis hypogaea* L.) using differential display.

Other methods subsequently developed using polynucleotides bound to miniature solid supports (i.e., GeneChips, cDNA microarrays) hybridized with complex probes either as one or both treatments simultaneously [49]. Because this method relies on available cDNA clones and sequence information [4], it is currently only available in model crops (i.e., *Arabidopsis*, rice) and some important agricultural crops. Large-scale cDNA microarray analysis of expression profiles of genes involved in abiotic stress tolerance have been reported in important model crops [46] such as *Arabidopsis* [50–55], the ice plant (*Mesembryanthemum crystallinum* L.) [46], and other important crops such as tobacco (*Nicotiana tabacum*) [57], maize [58], and *Brassica oleracea* [59]. In addition, the *Arabidopsis* microarray has been used comparatively in species for which no microarray chips are available. The *Arabidopsis* microarray was used to evaluate the gene expression involved in salt tolerance in salt cress (*Thellungiella halophila*), an *Arabidopsis*-related halophyte [60], and used to study water-stress tolerance in sugar beet (*Beta vulagaris*) [61].

Microarray analysis is beneficial because it can be used to identify genes that are upregulated and those which are downregulated or turned off in stressed tissues [1]. Recent studies comparing the expression profiles of genes induced by different types of abiotic stresses are beginning to shed light on the level of crosstalk between different signaling pathways and the specific genes induced in response to specific stresses [51,54,57]. Takahashi et al. [54] evaluated the expression profiles of genes induced by hyperosmotic, high salinity, and oxidative stress and abscisic acid treatment using an *Arabidopsis* cDNA microarray. Their research revealed 11 genes induced significantly by mannitol, NaCl, and ABA, indicating crosstalk among these signaling pathways. Kreps et al. [51] evaluated changes in expression profiles in *Arabidopsis* in response to salt, osmotic, and cold stress. They found that 30% of the transcriptome was sensitive to regulation by common stress conditions, but the majority of changes were stimulus-specific.

Smith-Espinoza et al. [62] evaluated expression profiles of LEA genes in response to dehydration and salt in the resurrection plant (*Craterostigma plantagineum*). Expression analysis in their experiment revealed dehydration-specific profiles of LEA genes, which were different than the patterns of the same genes under salt stress. Rizhsky et al. [57] evaluated gene expression during drought stress and heat shock in tobacco. They found that genes typically upregulated during either drought or heat were downregulated when both stresses occurred at the same time. Furthermore, they determined that other transcripts not associated with either stress were upregulated when a combination of both stresses was applied. These results demonstrate that the response of plants to simultaneous drought and heat shock, similar to commonly occurring conditions in many field situations, is different from the response of plants to each of the stresses applied individually [57].

This emphasizes the importance of selecting tolerant germplasm in natural environmental conditions [63] in order to breed for improved abiotic stress tolerance.

Microarray analysis alone could reveal interesting variation between lines of organisms and reveal regulatory variation in genes for complex traits, but it would not link the variation to a particular phenotype [64]. By combining QTL mapping and fine mapping with arraying, scientists should be able to identify positional candidate genes for a phenotype of interest. This integrated approach could successfully identify candidate loci or TQL (transcript quantity loci) associated with the phenotype of interest and verify their coincidence with QTLs [15,64]. Jamaux et al. [28] isolated specific cDNAs differentially expressed during drought stress. They developed a molecular marker for that sequence using RFLPs and STS (sequence tagged site) markers, which then enabled the researchers to identify a putative QTL for osmotic adjustment and relative water content. The application of this technique in breeding for abiotic stress tolerance is important because it could potentially dramatically improve selection efforts for phenotypic variation in the field and make selection quicker, easier, and more reliable than current selection techniques.

Although microarray analysis provides invaluable information about gene regulation and expression under stress, it provides little information about the function of those transcripts. Furthermore, the correlation between the level of mRNA and the products of their translation or their biological importance can be low [15]. Therefore, microarray analyses are usually conducted in conjunction with other genomic approaches to validate the function of differentially expressed transcripts. A combination of both reverse and forward genetics methods including knock-out mutant analysis and transformation are being conducted to elucidate gene function and validate results of microarray analysis.

12.5.4 FORWARD AND REVERSE GENETICS

12.5.4.1 Mutants

The increasing information in DNA sequences has provided a unique opportunity to investigate the function of plant genes. The approach that determines the function of genes first defined by DNA sequence analysis is called reverse genetics. These reverse genetics approaches allow for the identification of lines in which any gene of interest is disabled [1]. Various T-DNA or transposon tagged populations are available in rice and *Arabidopsis*. These lines can be investigated to identify phenotypes that may give clues to the function of the gene. A recently developed genomic tool, TILLING (targeting induced local lesions in genomes), allows the production of targeted knock-outs and also the creation of allelic series in any gene [1]. TILLING is an integration of chemical mutagenesis and mutation screens of pooled PCR products, which results in the isolation of missense and nonsense mutant alleles of the targeted genes.

Mutants have been utilized to determine gene function of important genes involved in abiotic stress tolerance. Verslues and Bray [65] developed a mutant screening technique in *Arabidopsis* to create mutants impaired in low-water-potential responses. Two mutants designated low-water-potential response1, *lwr1*, and *lwr2* were characterized in detail. *Lwr1* had higher proline accumulation, higher total solute content, greater osmotic adjustment at low water potential, altered ABA content, and altered growth and morphology. *Lwr2* had low proline content and less osmotic adjustment leading to greater water loss at low water potential. They concluded the *lwr1* and *lwr2* affected multiple aspects of cellular osmoregulation [65]. The authors are using mapping approaches to identify the gene products involved in the *lwr1* and *lwr2* mutants [65]. Kim et al. [66] utilized a targeted gene approach to isolate *Arabidopsis* mutants with altered cold-responsive gene expression. They identified a gene, ACG1, as a negative regulator of the C-repeat/dehydration-responsive element (C/DRE) pathway. This gene also affected flowering time and the authors suggest the regulation of both pathways may have an evolutionary advantage for plants by increasing their survival rates.

Mutants have been proven important either for establishing biochemical pathways or to identify the physiological role of the mutated genes and processes [22]. Cloning of the corresponding genes has often provided important, and sometimes unique, information on the mode of action of these genes and offers the possibility of reintroducing these cloned genes into plants thereby overexpressing or underexpressing the genes [22].

12.5.4.2 Transgenic Approaches

In fact, many recent studies include a combination of mutant evaluation followed by transformation to prove the function of the gene of interest and how it is regulated. Hong and Vierling [67] developed a screening technique based on elongation of hypocotyls for mutants of *Arabidopsis* that are unable to acquire thermotolerance to high-temperature stress. They characterized one mutant in particular, *hot1*, which had a point mutation in the heat shock protein 101 (HSP101) gene. *Hot1* seedlings were unable to acquire thermotolerance and seeds had greatly reduced basal thermotolerance. Complementation of *hot1* plants by transformation with the wild-type HSP101 gene restored the *hot1* plants to the wild-type phenotype.

The development of transgenic plants containing genes to potentially improve abiotic stress tolerance is one of the most popular genomic techniques utilized to improve abiotic stress tolerance in plants. Genes known to be involved in stress tolerance, including super oxide dismutase (SOD) [68,69], and osmoprotectants such as betain aldehyde dehydrogenase (BADH) [25,70] have been incorporated into plants to try to improve stress tolerance.

McKersie et al. [68] evaluated water-deficit tolerance and field performance of transgenic alfalfa overexpressing SOD. They reported that yield and survival of transgenic plants were significantly improved over the wild-type [68]. However, when these plants were tested for winter survival, transgenic plants only had 1°C more freezing tolerance than the control and had no apparent improvement in freezing tolerance [69]. Both McKersie et al. [68] and Rathinasabapathi [70] acknowledge the importance of improving our understanding of the mechanisms involved in oxidative stress and the pathways involved in osmoprotectant synthesis.

Transgenic approaches can also be used to identify the biological importance of certain compounds to abiotic stress tolerance. Pilon-Smits et al. [71] investigated the possible functional significance of fructans in drought tolerance by developing transgenic tobacco (*Nicotiana tabacum*) plants that accumulated fructans. They found that fructan-producing tobacco plants had more extensive roots and performed significantly better under polyethylene-glycol mediated drought stress than wild-type tobacco. They concluded that introducing fructans into a nonfructan-producing species can enhance resistance to drought stress [71]. Nanjo et al. [72] developed antisense transgenic *Arabidopsis* plants to determine the role of proline in osmotic stress tolerance. They identified several transgenic plants that accumulated significantly less proline than the wild-type, exhibited morphological alterations in the leaves, and had decreased osmotolerance. Their results indicated that proline is not just an osmoregulator in stressed plants but also affects morphogenesis as a major constituent of cell wall structural proteins [72].

Most of the transgenic approaches often use single genes and constitutively active promoters, which lead only to marginal stress improvement in most cases [73] especially under field conditions [43,74]. The limited success with single gene transfers is not surprising considering that multiple genes are involved in abiotic stress tolerance. The application of combined genomic approaches (microarrays, mutagenesis, molecular markers, etc.) will allow for the identification of pathways and sets of genes involved in abiotic stress tolerance [72]. This information will help in the identification of the right genes to choose for biotechnological approaches to manipulate stress tolerance [73].

12.5.5 INTEGRATED GENOMIC APPROACHES

Integrated genomic approaches may be more powerful than using a single individual approach (discussed above) in improving abiotic stress tolerance in plants as abiotic stress tolerance is considered to be controlled by multiple genes. Studies with DREB transcription factors provide a good example for the application of combining genomic approaches to improve abiotic stress tolerance. The transcription factor DREB/CBF (dehydration-responsive element/C-repeat binding) specifically interacts with dehydration-responsive elements (DRE/C-repeat (CRT) *cis*-acting element (A/GCCGAC)) and controls the expression of many stress-inducible genes in *Arabidopsis* [75,76].

The transformation of *Arabidopsis* with the DREB1A gene under the control of the constitutive 35S cauliflower mosaic promoter gave rise to strong constitutive expression of stress-inducible genes and a marked increase in tolerance to freezing, water, and salinity stress [77], but under unstressed conditions "dwarfed" phenotypes were produced presumably due to the overexpression of DREB1A [78].

When the expression of DREB1A was under the control of rd29A promoter in transgenic *Arabidopsis* the plants exhibited improved stress tolerance and improved growth under unstressed conditions [77]. Six DRE target genes induced by DREB1A were identified, but it was not well understood how overexpression of DREB1A cDNA in transgenic plants increased stress tolerance [79]. Therefore, a cDNA microarray was utilized to identify novel DREB1A target genes. They found that the target stress-inducible genes encoded enzymes for the biosynthesis of osmoprotectants such as proline and sugar, membrane proteins, LEA proteins, detoxification enzymes, chaperones, and enzymes involved in phospholipid metabolism, protein kinases, and transcription factors [79].

Maruyama et al. [76] identified cold-inducible downstream genes of *Arabidopsis* DREB1A/CBF3 transcription factor using two microarray systems (7000 full-length cDNA microarray and 8000 Affymetric GeneChip array). They identified approximately 26 genes downstream from DREB1A. They searched for conserved regions in the promoter regions of the 26 genes. The functions of the identified genes were classified into two groups: (1) proteins that probably function in stress tolerance (LEA proteins, antifreeze proteins, hydrophilic proteins, RNA-binding proteins, etc.); and (2) protein factors involved in further regulation of signal transduction and gene expression. The wide range of function of the downstream genes indicates the DREB1A regulates a complex gene expression network in response to cold stress [76].

Similar work with DREB transcription factors is also being conducted in rice. Oh et al. [80] developed transgenic rice that constitutively expressed CBF3/D REB1A and ABF3 (ABA-dependent transcription factor). The transgenic plants exhibited improved tolerance to drought and high salinity, but not cold stress, and did not suffer growth inhibition [80]. Dubouzet et al. [81] identified DREB homologues in rice (OsDREB1A, OsDREB1B, OsDREB1C, OsDREB1D, and OsDREB2A). The OsDREB1A and OsDREB1B specifically bound to DRE and activated the transcription of the GUS reporter gene in rice protoplasts. Transgenic *Arabidopsis* overexpressing the rice OsDREB1A resulted in plants with higher tolerance to drought, and high salt and freezing stresses, indicating that OsDREB1A is functionally similar to the *Arabidopsis* DREB1A. Microarray and RNA gel-blot analysis indicated that some stress-inducible target genes of the DREB1A proteins were not overexpressed in the OsDREB1A transgenic *Arabidopsis* [81].

Additional research in this area using a combination of genomic approaches has identified genes that are involved in the upregulation of DREB [50,55]. Spermidine synthase, an enzyme involved in polyamine biosynthesis, was identified by Kasukabe et al. [50] to enhance tolerance to multiple environmental stresses and upregulate DREB transcription factors. Transgenic plants overexpressing spermidine synthase exhibited increased spermidine content and enhanced tolerance to various stresses including chilling, freezing, salinity, hyperosmosis, and drought. A cDNA microarray analysis of the transgenic plants revealed that several genes were more abundantly transcribed under chilling stress, including the transcription factor DREB1 [50].

Another gene SRK2C, a SNF1-related protein kinase 2, involved in protein phosphorylation, was found to improve drought tolerance by controlling stress-responsive gene expression, including DREB transcription factors [55]. Knockout mutants of SRK2C exhibited drought hypersensitivity in their roots. Additionally, transgenic plants with CaMV35S promoter::SRK2C-GFP displayed higher overall drought tolerance than control plants. Microarray analysis revealed that the drought tolerance observed in the transgenic plants coincided with upregulation of many stress-responsive genes including DREB1A/CBF3 [55]. From these results, they concluded that the SRK2C is capable of mediating signals initiated during drought stress, resulting in appropriate gene expression. The osmotic stress-activated SnRK2s were also identified in tobacco and soybean (*Glycine max*), indicating this system may be highly conserved among higher plants [55].

This combined genomic approach is a powerful tool for improving our knowledge of the highly complex mechanisms and pathways involved in abiotic stress tolerance. This research should provide applications for biotechnology and molecular breeding efforts to improve abiotic stress tolerance in plants [55].

12.6 SUMMARY AND PROSPECTS

Abiotic stress is a major constraint to food production and one that will continue to grow as water becomes increasingly limited [1]. Plant breeding for abiotic stress tolerance through conventional approaches has resulted in limited success to date. Due to the complicated and integrated response of plants to abiotic stress, a multidisciplinary approach involving genetics, biochemistry, physiology, plant breeding, and crop science will be necessary to ultimately develop plants with proven improved stress tolerance in the field [82]. Attention is being directed toward a better understanding of the genetic basis of abiotic stress tolerance by combining genomic approaches and evaluating the expression levels of different enzymes and proteins in different biochemical pathways involved in abiotic stress tolerance [82]. Genomic approaches including QTL analysis of important traits, genome scale EST sequencing, cDNA microarray analysis, mutagenesis studies, and transgenic applications have been summarized in this chapter. It is evident that progress is being made toward our understanding of the mechanisms involved in abiotic stress tolerance but more research is needed. Approaches with proteomics will be necessary to clarify the structural predictions of genome sequence information and to assess the protein modifications and protein–ligand interactions that are relevant to stress-tolerant phenotypes [46]. With this knowledge it will be possible to identify candidate genes to rationally manipulate and optimize tolerance traits for improved crop productivity [46,73].

A comprehensive breeding strategy for abiotic stress tolerance should include: (1) conventional breeding and germplasm selection; (2) elucidation of the specific molecular control mechanisms in tolerant and sensitive phenotypes through genomic approaches; (3) biotechnology-oriented improvement including improvement of selection and breeding procedures through functional genomics analysis, use of molecular probes and markers for selection among natural and breeding populations, and transformation with specific genes; and (4) improvement and adaptation of current agricultural practices [74]. By using an integrated approach it may be possible to develop plants that exhibit dramatic improvements in abiotic stress tolerance. This will be a necessity for the world, especially developing countries, as the population continues rising and available resources continue to decline.

REFERENCES

1. Gale, M., Applications of molecular biology and genomics to genetic enhancement of crop tolerance to abiotic stress — A discussion document, CGIAR interim Science Council, Food and Agriculture Organization of the United Nations, 2003, http://www.fao.org/WAIRDOCS/TAC/Y5198E/y5198e02.htm retrieved June 21, 2005.
2. Banziger, M., Mugo, S., and Edmeades, G.O., Breeding for drought tolerance in tropical maize-conventional approaches and challenges to molecular approaches. In *Molecular Approaches for the Genetic Improvement of Cereals for Stable Production in Water-Limited Environments*, Ribaut, J.M. and Poland, D., Eds., A Strategic Planning Workshop held at CIMMYT, El Batán, Mexico, June 21–25, 1999, 2000, pp. 154–155.
3. Bennet, J., Annex 1 Status of breeding for tolerance of abiotic stresses and prospects for use of molecular techniques. Consultative Group on International Agricultural Research Technical Advisory Committee. Food and Agriculture Organization of the United Nations. Retrieved Dec. 23, 2004. http://www.fao.org/WAIRDOCS / TAC / Y 5198E/y5198e03.htm. 2001.
4. Bruce, W.B., Edmeades, G.O., and Barker, T.C., Molecular and physiological approaches to maize improvement for drought tolerance. *J. Exp. Bot.*, 53:13, 2002.
5. Blumwald, E., Grover, A., and Good, A.G., Breeding for abiotic stress resistance: Challenges and opportunities. In *New Directions for a Diverse Planet: Proceedings for the 4th International Crop Science Congress*, Fischer, T. et al., Eds., Brisbane, Australia, September 26–October , 2004. Available online at: www.crops ci en ce.org.au.
6. Visser, B., Technical aspects of drought tolerance. *Biotech. Develop. Monitor,* 18:5, 1994.
7. Carrow, R.N. and Duncan, R.R., *Salt Affected Turfgrass Sites: Assessment and Management.* Ann Arbor Press, Chelsea, MI, 1998, p. 185.
8. Babu, R.C. et al., Genetic analysis of drought resistance in rice by molecular markers: Association between secondary traits and field performance. *Crop Sci.,* 43:1457, 2003.
9. Bohnert, H.J., et al., A genomics approach towards salt stress tolerance. *Plant Physiol. Biochem.* 39:295, 2001.
10. Flowers, T.J., Improving crop salt tolerance. *J. Exp. Bot.,* 55:307, 2004.
11. Lanceras, J.C. et al., Quantitative trait loci associated with drought tolerance at reproductive stage in rice. *Plant Physiol.,* 135:384, 2004.
12. Maestri, E.N. et al., Molecular genetics of heat tolerance and heat shock proteins in cereals. *Plant Mol. Biol.,* 48:667, 2002.
13. Neumann, P., Salinity resistance and plant growth revisited. *Plant Cell Environ.,* 20:1193, 1997.

14. Shannon, M.C., Breeding, selection, and the genetics of salt tolerance. In *Salinity Tolerance in Plants: Strategies for Crop Improvement*, Staples, R.C. and Toenniessen, G.H., Eds., John Wiley and Sons, New York, 1984, Chap. 13, p. 23.
15. Tuberosa, R.S., Mapping QTLs regulating morpho-physiological traits and yield: Case studies, short comings, perspectives in drought stress maize. *Ann. Bot.*, 89:941, 2002.
16. Forster, B.P. et al., The development and application of molecular markers for abiotic stress tolerance in barley. *J. Exp. Bot.*, 51:19, 2000.
17. Ivandic, V. et al., Phenotypic responses of wild barley to experimentally imposed water stress. *J. Exp. Bot.*, 51:2021, 2000.
18. Ribaut, J.M. et al., Identification of quantitative trait loci under drought conditions in tropical maize. 2. Yield components and marker assisted selection strategies. *Theor. Appl. Gene.*, 94:887, 1997.
19. Price, A.H. et al., Linking drought-resistance mechanisms to drought avoidance in upland rice using a QTL approach: Progress and new opportunities to integrate stomatal and mesophyll responses. *J. Exp. Bot.* 53:989, 2002.
20. Helentjaris, T., Wright, S., and Weber, D., Construction of a genetic linkage map in maize using restriction fragment length polymorphisms. *Maize Genetics Cooperation Newsletter*, 60:118, 1986.
21. Vos, P.R. et al., AFLP: A new technique for DNA fingerprinting. *Nucleic Acid Res.* 23:4407, 1995.
22. Korneef, M., Alonso-Blanco, C., and Peeters, A.J.M, Genetic approaches in plant physiology. *New Phytol.*, 137:1, 1997.
23. Frova, C. et al., Genetic analysis of drought tolerance in maize by molecular markers. I. Yield components. *Theor. Appl. Gene.*, 99:280, 1999.
24. Price, A.H., Young, E.M., and Tomos, A.D., Quantitative trait loci associated with stomatal conductance, leaf rolling and heading date mapped in upland rice (*Oryza sativa*). *New Phytol.*, 137:83, 1997.
25. Zhang, G. et al., Transformation of triploid bermudagrass (*Cynodon dactylon* x *C. transvaalensis* cv. TifEagle) with *BADH* gene for stress tolerance. *Abstract of ASA-CSSA-SSSA International Annual Meeting*, Oct. 21–25, Charlotte, NC, 2001.
26. Sari-Gorla, M. et al., Genetic analysis of drought tolerance in maize by molecular markers. II. Plant height and flowering. *Theor. Appl. Gene.*, 99:289, 1999.
27. Teulat, B. et al., QTL for relative water content in field grown barley and their stability across Mediterranean environments. *Theor. Appl. Gene.*, 108:181, 2003.
28. Jamaux, I., Steinmetz, A., and Belhassen, E., Looking for molecular and physiological markers of osmotic adjustment in sunflower. *New Phytol.*, 137:117, 1997.
29. Lilley, J.M. et al., Locating QTL for osmotic adjustment and dehydration tolerance in rice. *J. Exp. Bot.*, 47:1427, 1996.
30. Robin, S. et al., Mapping osmotic adjustment in an advanced back-cross inbred population of rice. *Theor. Appl. Gene.*, 107:1288, 2003.
31. Teulat, B. et al., Relationships between relative water content and growth parameters under water stress in barley: a QTL study. *New Phytol.*, 137:99, 1997.
32. Teulat, B. et al., Several QTL involved in osmotic-adjustment trait variation in barley (*Hordeum vulgare* L.). *Theor. Appl. Gene.*, 96:688, 1998.
33. Teulat, B., Borries, C., and This, D., New QTL identified for plant water status, water soluble carbohydrate and osmotic adjustment in a barley population grown in a growth-chamber under two water regimes. *Theor. Appl. Gene.*, 103:161, 2001.

34. Tripathy, J.N. et al., QTLs for cell-membrane stability mapped in rice (*Oryzae sativa* L.) under drought stress. *Theor. Appl. Gene.*, 100:1197, 2000.
35. Price, A.H. and Tomos, A.D., Genetic dissection of root growth in rice (*Oryza sativa* L.) II: Mapping quantitative trait loci using molecular markers. *Theor. Appl. Gene.*, 95:143, 1997.
36. Zhang, J., et al., 2001. Locating genomic regions associated with components of drought resistance in rice: Comparative mapping within and across species. *Theor Appl Gene.*, 103:19, 2001.
37. Campbell, S.A. and Close, T.J., Dehydrins: genes, proteins, and associations with phenotypic traits. *New Phytol.*, 137:61, 1997.
38. Close, T.J., Kortt, A., and Chandler, P.M., A cDNA-based comparison of dehydration-induced proteins (dehydrins) in barley and corn. *Plant Mol. Biol.*, 13:95, 1989.
39. Ottaviano, E. et al., Molecular markers (RFLPs and HSPs) for the genetic dissection of thermotolerance in maize. *Theor. Appl. Gene.*, 81;713, 1991.
40. Kearsey, M.J., QTL analysis: Problems and (possible solutions). In *Quantitative Genetics, Genomics and Plant Breeding*, Kang, M.S. Ed., CABI, Baton Rouge, LA, 2002, p. 45.
41. Gale, M.D. and Devos, K.M., Comparative genetics in the grasses. *Proc. Nat. Acad. Sci.*, 95:1971, 1998.
42. Jung, G. et al., Comparative genome analysis of recently domesticated turfgrasses (*Agrostis* and *Lolium*) with rice, wheat, and oat. W191. *Plant and Animal Genomes XIII Conference*, January 15–19, San Diego, 2005.
43. Tester, M. and Bacic, A., Abiotic stress tolerance in grasses. From model plants to crop plants. *Plant Physiol.*, 137:791, 2005.
44. Van Deynze, A.E., Comparative mapping in grasses. Wheat relationships. *Mol. General Gene.*, 248:744, 1995.
45. Choi, D.W. and Close, T.J., A newly identified barley gene, Dhn12, encoding a YSK2 DNH, is located on chromosome 6H and has embryo-specific expression. *Theor. Appl. Gene.*, 100:1274, 2000.
46. Cushman, J.C. and Bohnert, H.J., Genomic approaches to plant stress tolerance. *Curr. Opin. Plant Biol.*, 3:117, 2000.
47. Liang, P. and Pardee, A.B., Differential display of eukaryotic messenger RNA by means of the polymerase chain reaction. *Science*, 257:967, 1992.
48. Jain, A.K., Basha, S.M., and Holbrook, C.C., Identification of drought-responsive transcripts in peanut (*Arachis hypogaea* L.), *Electronic J. Biotech.*, 4:59, 2001.
49. Chee, M. et al., Accessing genetic information with high-density DNA arrays, *Science*, 274:610, 1996.
50. Kasukabe, Y. et al., Overexpression of spermiding synthase enhances tolerance to multiple environmental stresses and up-regulates the expression of various stress-regulated genes in transgenic *Arabidopsis thaliana*. *Plant Cell Physiol.*, 45:712, 2004.
51. Kreps, J.A. et al., Transcriptome changes for *Arabidopsis* in response to salt, osmotic, and cold stress. *Plant Physiol.*, 130:2129, 2002.
52. Liu, F. et al., Global transcription profiling reveals comprehensive insights into hypoxic response in *Arabidopsis*. *Plant Physiol.*, 137:1115, 2005.
53. Sze, H. et al., Expression patterns of a novel AtCHX gene family highlight potential roles of osmotic adjustment and K^+ homoeostasis in pollen development. *Plant Physiol.*, 136:2532, 2004.

54. Takahashi, S. et al., Monitoring the expression profiles of genes induced by hyperosmotic, high salinity, and oxidative stress and abscisic acid treatment in *Arabidopsis* cell culture using a full-length cDNA microarray. *Plant Mol. Biol.,* 56:29, 2004.
55. Umezawa, T. et al., SRK2, a snf1-related protein kinase 2, improves drought tolerance by controlling stress-responsive gene expression in *Arabidopsis thaliana. Proc. Natl. Acad. Sci.,* 101:1306, 2004.
56. Bohnert, H.J., and Cushman, J.C., The ice plant cometh: lessons in abiotic stress tolerance. *J. Plant Growth Reg.,* 19:334, 2000.
57. Rizhsky, L., Liang, H., and Mittler, R., The combined effect of drought stress and heat shock on gene expression in tobacco. *Plant Physiol.,* 130:1143, 2002.
58. Yu, L. and Setter, T.L., Comparative transcriptional profiling of placenta and endosperm in developing maize kernels in response to water deficit. *Plant Physiol.,* 131:568, 2003.
59. Soeda, Y., Gene expression programs during *Brassica oleracea* seed maturation, osmopriming, and germination are indicators of progression of the germination process and the stress tolerance level. *Plant Physiol.,* 137:354, 2005.
60. Taji, T. et al., Comparative genomics in salt tolerance between *Arabidopsis* and *Arabidopsis*-related halophyte salt cress using *Arabidopsis* microarray. *Plant Physiol.,* 135:1697, 2004.
61. Bagatta, M. et al., A physiological and molecular description of the water stress in sugar beet. Poster Abstract. D.19. *Proceedings of the XLVIII Italian Society of Agricultural Genetics — SIFV SIGA Joint Meeting,* Lecce, Italy 15–18 Sept., 2004.
62. Smith-Espinoza, C. et al., Dissecting the response of dehydration and salt (NaCl) in the resurrection plant *Craterostigma plantagineum. Plant Cell Environ.,* 26:1307, 2003.
63. Hazen, S.P., Wu, Y., and Kreps, J.A., Gene expression profiling of plant responses to abiotic stress. *Funct. Integr. Genomics,* 3:105, 2003.
64. Wayne, M.L. and McIntyre, L.M., Combining mapping and arraying: An approach to candidate gene identification. *Proc. Natl. Acad. Sci.,* 99:14903, 2002.
65. Verslues, P.E. and Bray, E.A., LWR1 and LWR2 are required for osmoregulation and osmotic adjustment in *Arabidopsis. Plant Physiol.,* 136:2831, 2004.
66. Kim, H. et al., A genetic link between cold responses and flowering time through FVE in *Arabidopsis thaliana. Nature Genetics,* 36:167, 2004.
67. Hong, S. and Vierling, E., Mutants of *Arabidopsis thaliana* defective in the acquisition of tolerance to high temperature stress. *Proc. Natl. Acad. Sci.,* 97:4392, 2000.
68. McKersie, B.D. et al., Water-deficit tolerance and field performance of transgenic alfalfa overexpressing superoxide dismutase. *Plant Physiol.,* 111;1177, 1996.
69. McKersie, B.D., Bowley, S.R., and Jones, K.S., Winter survival of transgenic alfalfa overexpressing superoxide dismutase. *Plant Physiol.,* 119:839, 1999.
70. Rathinasabapathi, B., Metabolic engineering for stress tolerance: Installing osmoprotectant synthesis pathways. *Ann. Bot.,* 86:709, 2000.
71. Pilon-Smits, E.A.H. et al., Improved performance of transgenic fructan-accumulating tobacco under drought stress. *Plant Physiol.,* 107:125, 1995.
72. Nanjo, T. et al., Biological functions of proline in morphogenesis and osmotolerance revealed in antisense transgenic *Arabidopsis thaliana. Plant J.,* 18:185, 1999.
73. Ramanjulu, S. and Bartels, D., Drought and dessication-induced modulation of gene expression in plants. *Plant Cell Environ.,* 25:141, 2002.

74. Wang, W., Vinocur, B., and Altman, A., Plant response to drought salinity and extreme temperatures: Towards genetic engineering for stress tolerance. *Planta*, 218, 2001.
75. Ingram, J. and Bartels, D., The molecular basis of dehydration tolerance in plants. *Ann. Rev. Plant Physiol. Plant Mol. Biol.*, 47:377, 1996.
76. Maruyama, K. et al., Identification of cold-inducible downstream genes of the *Arabidopsis* DREB1A/CBF3 transcriptional factor using two microarray systems. *Plant J.*, 38:982, 2004.
77. Kasuga, M. et al., Improving plant drought, salt, and freezing tolerance by gene transfer of a single stress inducible transcription factor. *Nature Biotech.*, 17:287, 1999.
78. Mundree, S.G. et al., Physiological and molecular insights into drought tolerance. *Afr. J. Biotech.*, 1:28, 2002.
79. Yamaguchi-Shinozaki, K., Identification of target genes of the DREB1A transcription factor controlling abiotic-stress tolerance-responsive gene expression using a full-length cDNA microarray. *JIRCAS Annual Report*, 2001, p. 46-47.
80. Oh, S. et al., *Arabidopsis* CBF3/DREB1A and ABF3 in transgenic rice increased tolerance to abiotic stress without stunting growth. *Plant Physiol.*, 138:341, 2005.
81. Dubouzet, J. et al., OsDREB genes in rice, *Oryza sativa* L. encode transcription activators that function in drought, high-salt and cold-responsive gene expression. *Plant J.*, 33:751, 2003.
82. Mitra, J., Genetics and genetic improvement of drought resistance in crop plants. *Curr. Sci.*, 80:758, 2001.

Index

Page numbers followed by f denote figures; t, tables.

A

A/Ci curve (carbon dioxide assimilation to partial pressure curve), 327–329
AAA protease subfamily, in acquired thermotolerance, 34–35
ABA (abscisic acid), 102–103
  in dehydration stress, 103–110, 104f, 282–284, 294
  gene expression induced by, 110–111
  in guard cell phosphatidic acid production, 108–109
  in guard cell signaling, 106–109
    compared to mesophyll cell signaling, 109–110
  in osmoprotectant metabolism, 146
  in plant growth, 111–113
  and reactive oxygen species interaction, 293–294
  in root growth and physiology, 275, 278f, 282–285, 286, 288
  in salinity stress, 130, 136, 137
  in stomatal closure, 106–109, 107f, 286–287
ABA-activated serine-threonine protein kinase (AAPK), 108
ABA inducible genes, 110–111
ABA responsive elements and binding proteins/factors, 110–111, 131, 131f, 136, 153
ABA specific glucosyltransferase gene *(AOG)*, 106
ABH1, in negative abscisic acid regulation, 107–108
ABI2 protein phosphatase 2C, in SOS pathway regulation, 137
Abscisic acid, *see* ABA (abscisic acid)
Abscisic acid aldehyde oxidase (AAO3), 104f, 105
ACC (1-aminocyclopropane-1-carboxylic acid) production
  in dehydration, 283
  in soil compaction, 284–285
  in waterlogging, 285–286
Acquired acclimation, systematic, 79
Adaptive/protective mechanisms (*see also* Stress tolerance)
  in abiotic stress, 149, 151–153
    calcium sensor proteins in, 151–153 (*see also* Calcium ion flux, cytosolic)
    genes and transcriptional regulation of, 151–155
    LEA proteins in, 149–151 (*see also* Late embryogenesis abundant (LEA) proteins)
    membrane lipids in, 1–25 (*see also* Membrane lipids)
    signaling pathways in, 155–157 (*see also* Signal transduction)
  in cold stress and freezes, 47–67 (*see also* Cold acclimation; Cold stress; Freeze tolerance)
  in drought, 101–119 (*see also* Drought/dehydration stress; Drought tolerance)
  in heat stress, 27–45 (*see also* Heat stress; Heat tolerance)
  in light intensity change, 69–99 (*see also* Photoacclimation)
  in low phosphorus availabililty, 209–242 (*see also* Low phosphorus availability)
  in salinity changes, 121–149 (*see also* Salinity stress; Salinity tolerance)
  in waterlogging, 177–207 (*see also* Waterlogging tolerance)
ADC expression, 139
Adventitious root growth
  and oxygen limitations, 184
  in phosphorus deficiency, 213
Aerenchyma
  cortical, 277–278
  and metabolic efficiency in root exploration, 218–221

377

in oxygen limitations, 183, 276–280, 277f
   secondary, 278–279
Alanine synthesis, ATP production in, 191
Anaerobic metabolism, 186–192
   ATP production and enzymes involved in,
      190–191
   gene expression and enzyme synthesis in,
      186–193
   glycolytic and fermentative enzymes in, 190
   injury and recovery in, 185–186, 193–194,
      196, 275–278, 292–293
Anaerobic proteins (ANPs), 187
Anthoycyanins, 72
Antioxidant enzymes and metabolism, 147,
      290–291, 338, 359
   in abscisic acid induced stomatal closure, 108
   in chloroplasts, 87
   excess light energy and production of, 71
   freeze tolerance and production of, 57–58
   hypoxic/anoxic damage caused by, 193–194,
      196, 275–278, 292–294
   in salinity stress, 146–149, 288, 291–292
   in stress signaling pathways, 87, 155–156,
      290–294
   transgenic overexpression of, 148–149
AP2 domain, 53, 187
AP2 domain proteins, 54, 60
APX, see Ascorbate peroxidase
Aquaporins, 38
Arabidopsis, stress adaptations in
   desiccation, 11–12
   heat and cold, 4–5, 8–10
   osmotic, 15–16
Arginine decarboxylase activity, and putrescine
   accumulation, 139
Ascorbate peroxidase (APX), 338
   in salinity and oxidative stress tolerance, 149,
      291–292
Assimilation quotient (AQ), 327
Assimilation (CO_2) to partial pressure (CO_2)
   curve (A/Ci curve), 327–329
ATHK4, 102–103
AtHKT1 expression, 137, 298
   in osmoprotectant metabolism, 146
AtMYB2, 194
AtNHX1 expression, 130, 135–136, 298
ATP generation
   diminished, as hypoxia sensor, 180–181,
      184–186
   in hypoxic cellular conditions, 190–191, 195,
      276
   in soil dehydration, 283
ATP supply, and demand, 185
Auxins, 102–103, 286

B

β carotene quenching, 74
BAC (Bacterial Artificial Chromosomes)
   genomic libraries, 365
Breeding strategies, for stress tolerance,
      157–158, 196–197, 321–322, 357–372
   (see also Genetics; Genomics;
      Quantitative trait loci)
   classical, and use of molecular markers,
      359–364
   genomic, 364–371
   references on, 372–376
bZIP (basic lucine-zipper) transcription factors,
   153, 187

C

C-repeat binding proteins, 154–155
Ca^{++}, see calcium ion entries
Cadaverine, as heat shock signal, 138
Cadmium toxicity, 257–259
Calcineurin, 133
Calcium dependent protein kinases, 131
Calcium ion dependent protein kinases
   (CDPKs), 151–153
Calcium ion flux, cytosolic
   abscisic acid in, 107–109
   and cold gene activation, 49, 50
   in root response to abiotic stress, 287, 294
   in salt stress and sodium uptake, 130–135
Calcium sensor proteins, 151–153
Calmodulins, 131, 151–153, 287
CAM (crassulacean acid metabolism), 129
Carbon dioxide, diffusion and fixation of, 327
Carbon dioxide assimilation to partial pressure
   curve (A/Ci curve), 327–329
Carboxylate excretion, root, 225–226
Carotenoids, salinity stress and effect on, 125
Casparian strip, 132, 282
Catalase (CAT), 338
CBF pathway, in plant metabolism, 51–54,
      154–155, 369–370
CBF regulated genes
   activation of, in cellular collapse, 58
   functions of, 53–54
   overexpression of, and freeze tolerance,
      51–52, 59–61
CBF regulon, 53, 54
cDNA microarray analysis, 366
CDPKs (calcium ion dependent protein
   kinases), 151–153
Cell death, in oxygen limitation, 184–186
Cell membrane, see membrane entries

Index

Chlamydomonas npq5, 75
Chloride uptake, root, 298
Chlorophyll
  per leaf area, 79
  salinity stress and effect on, 125, 127
Chlorophyll accumulation bioassay, and gene identification for heat tolerance, 35
Chlorophyll fluorescence, 323–326
Chlorophyll level, and light level, 81
Chloroplast(s)
  alterations in a/b ratio, 79
  in *Arabidopsis*, movement of, 71
  lipids in, 6
  redox status signaling in, 84–87
  responses of, in photoacclimation, 72–76
  and signal transduction to nucleus, 83–88
Chromium, physiologic effects of, 259–260
Chromium toxicity, 260
Cluster roots, 226
Cobalt, physiologic effects of, 253–255
Cold acclimation, 4, 47–48
  genes activated in, 52–54
  membrane lipids in, 56–59
  metabolic shift in, 54
  references on, 62–67
  signal transduction in, 4, 48–50
  water stress and cold stress in, 50–52
Cold stress
  chlorophyll fluorescence as indicator of, 325–326
  and cold acclimation, 50–52
  membrane adaptation to, 7–8, 49–50, 336–337
  plant adaptation to, 47–62 (*see also* Cold acclimation)
  transcription activators in, 51–52, 53–54
Comparative genomics, 364–365
Copper, physiologic effects of, 247–249
Copper toxicity, 248–249
COR (cold regulated/responsive) genes, 49, 50, 51–56
  in freeze tolerant plants, 57–58 (*see also* Freeze tolerance)
  overlapping signal pathways for, 51–52
  proteins produced by, 55–56
  transcriptional regulation of, 151, 154–155
COR (cold regulated/responsive) proteins, 55–56
Cotton lines, genotypic diversity in heat tolerance, 37–38
Crassulacean acid metabolism (CAM), 129
Cre1, as sensor in dehydration stress, 102–103
Cryoprotectants, in freeze tolerance, 59–62, 337
Cyclic nucleotides, and sodium ion influx, 133

Cyclic sugar alcohols, as osmoprotectants, 140
CYP707A1-A4, upregulation in dehydration stress, 106
Cytochrome C, in membrane damage, 16
Cytokinin(s), in root response in waterlogging, 286–287
Cytokinin biosynthesis pathway, 195–196
Cytokinin receptor, as sensor in dehydration stress, 102–103, 283
Cytoplasmic acidosis, 191
Cytosolic ion flux, *see* Calcium ion flux; Ion homeostasis
Cytosolic pH, in response to hypoxia, 181, 191

D

Dehydration, *see* Drought/dehydration stress
Dehydrin LEA proteins, 150
DGDG (digalactosyldiacylglycerol), 6, 11–12
DGDG synthase, mutation of, and acquired thermotolerance, 33–34
DNA landmarks, 157–158
DRE (dehydration responsive elements), 51–52
DREB transcription factors, 51–52, 154–155, 369–370 (*see also* CBF pathway)
Drought/dehydration stress, 113f (*see also* Drought tolerance)
  abscisic acid in response to, 103–111, 104f (*see also* ABA (abscisic acid))
  chlorophyll fluorescence as indicator of, 325
  in cold acclimation, 50–52
  and cold stress response, 49–50
  and freeze tolerance, 51–52, 54
  heat shock proteins in, 29–30
  hormones in, 102–103, 282–284
  membrane adaptation in, 11–13, 336
  photosynthetic gas exchange in, 328
  plant growth in, 111–113
  references on, 114–119
  relative water content as indicator of, 330
  root growth and physiology in, 272–274
    nutrient ion uptake in, 294–295
    osmotic adjustment in, 288–290
  signal transduction in, 282–284
  signaling mechanisms in, 12–13, 103, 106–110, 282–284
  stomatal closure in, regulation of, 106–109
Drought responsive genes, 53
Drought tolerance, 11–13, 101–102, 113f (*see also* Drought/dehydration stress)
  abscisic acid in regulation of, 103–111
  hormones in regulation of, 102–103
  membrane lipids in, 11–13

plant growth in, 111–113
quantitative trait loci in, 361
references on, 114–119
root growth and physiology in, 272–274
  osmotic adjustment in, 288–290
  signal transduction in, 282–284
signal transduction in, 109–110, 282–284
stomatal closure and signal transduction in, 106–109

E

EG (endo-1,4-β-D-glucanase), 38
Electrolyte leakage, in membrane damage, 17, 335–337
Electrolyte leakage analysis, 335
Elements
  essential, for plant growth, 243–244
  trace, 247
Environmental stress, 2–3
Epinastic leaf curvature, 286–287
ERA1 mutation, in hypersensitivity to abscisic acid, 107
ERF-like transcription factors, 187
ESTs, *see* Expressed sequence tags
Ethanol fermentation, 195
Ethanol production, in anoxia, 190
Ethylene
  in aerenchyma formation, 278, 280
  in root morphology, 284–285
  in water stress signaling, 283
  in waterlogging signaling, 286–287
Ethylene evolution among cultivars, differences in, 112–113
Ethylene-forming-enzyme complex, 284–285
Expansins, and cell enlargement, 38
Expressed sequence tags (ESTs), 364, 365
  high throughput analysis of, 365–367

F

fad8 ω-3 desaturase gene, 9
Fermentative enzymes, in hypoxic ATP production, 190–191, 195
Fluidity, of membrane lipids, 4–5
Freeze tolerance, 48, 337
  bioengineering for, 59–62
  CBF regulated genes in, 51–52, 59–61
  membrane lipids in, 56–59
  phospholipase Dα1 deficiency and, 57–59
  and production of oxidative species, 57–58
  references on, 62–67
  water stress and, 51–52, 54

FRY1 mutation, and inositol polyphosphate accumulation, 109–110, 130
FtsH protease, mutation of, and acquired thermotolerance, 34–35
Full turgor adjustment method, 333
Fv/Fm ratio, 324

G

Galactinol synthase gene, 52, 56
Gamma amino butyric acid (GABA) synthesis, and cytosolic pH, 191, 287
Gas exchange, photosynthetic, as indicator of stress tolerance, 326–329
GenBank, 365
Gene cloning, 158, 368
  from quantitative trait loci, 365
GeneChips, 366
Genetic maps, molecular, 157–158
Genetics, in breeding for stress tolerance, 149–150, 194–197, 358–360
  forward and reverse, 367–368
  genomic approach in, 364–371 (*see also* Genomics)
  quantitative trait loci in, 360–364
  references on, 372–376
  transcriptional regulation in, 151–155, 152f
  transgenic, 368–369
Genomics, in breeding for stress tolerance, 364–371
  gene cloning in, 365
  integrated, 369–371
  microarray analysis in, 365–367
  references on, 372–376
  reverse genetic analysis in, 367–369
  synteny and comparative, 364–365
Germination, and stand establishment, heat tolerance development in, 29–32
Gibberellins, in root response in waterlogging, 286–287
Glutathione peroxidase (GPX), and glutathione-S-transferase (GST)
  overexpression of, 148
  regulation of, 155
Glycine-betaine
  as osmoprotectant, 141–142
  transgenic production of, for salinity tolerance, 143t
Glycolytic enzymes, in hypoxic ATP production, 190–191, 194–195
Glycolytic flux, as hypoxia sensor, 180–181, 190–191
Glyoxylase system overexpression, and salinity tolerance, 149

Index

Growth inhibition, in oxygen limitation, 184–186
Growth regulation, in phosphorus deficiency, 212–213
GTP binding proteins, 180
Guard cell regulation (*see also* Stomatal closure)
   signal transduction in, 108–109, 130–131
GUN1-GUN5 (genomes uncoupled) mutants, 84

H

HAL1, *HAL3* expression, in yeast, 133
HAP1 (heme-activating protein), 84
Heat shock proteins (HSPs), 28–29, 340–342, 359
   developmentally induced, 29–30
   in drought tolerance, 29–30
   genetic diversity and induction of, 32–38
   in heat tolerance, 30–41
Heat stress
   chlorophyll fluorescence as indicator of, 324–325
   membrane adaptation to, 4–7, 336
   plant adaptation to, 27–29 (*see also* Heat tolerance)
   relative water content as indicator of, 331
Heat tolerance, 27–29
   development of, in germination and stand establishment, 29–32
   and expansive cell growth, 38–40
   heat shock proteins in, 30–38
   inducible processes associated with, 32–35
   references on, 41–46
   in reproductive phase, 35–38
   seed reserve mobilization in, 31–32
   in vegetative phase, 38–40
Heavy metals
   availability of, factors affecting, 245–247
   physiologic effects of, 243–245
      cadmium, 257–259
      chromium, 259–260
      cobalt, 253–255
      copper, 247–249
      future research in, 264–265
      iron, 249
      lead, 260–262
      manganese, 249–251
      mercury, 262–263
      molybdenum, 251–252
      nickel, 255–256
      references on, 265–269
      thallium, 263

      vanadium, 256–257
      zinc, 252–253
Hexokinase, in signaling pathways, 86–87
High affinity potassium ion uptake, 132–133
High temperature, *see* Heat stress
Histidine kinase (HK), 102–103, 131f, 194–195
   in hypoxic ATP production, 190–191
   in sodium transport from shoot to root, 136–137
HKT1, 131f, 136–137
HKT1 expression
   and *AtHKT1* expression, comparisons of, 137
   and sodium ion transport, 132–133
HSP70, induction of, in pollen, 36
HSP90, and genetic variability for thermotolerance, 342
HSP101 gene, 368
HSPs, *see* Heat shock proteins
Hydrogen peroxide
   in cold response signaling, 50
   in stress response signaling, 155
Hydrotropism, 102–103
Hypoxia, response in
   anaerobic metabolism in, 186–192
   ATP and metabolic sensors in, 180–181
   ATP production and fermentative enzymes in, 190–191, 195
   cytosolic pH in, 181, 191
   downregulated biosynthetic pathways in, 188
   gene expression in, 186–188
   glycolytic and fermentative enzymes in, 190
   growth inhibition and cell death in, 184–186
   in mitochondria, 179–180
   morphologic and anatomic changes in, 183–184, 275–280
   osmotic adjustments in, 288–290
   phytohemoglobin in, 180, 191–192
   reactive oxygens species and antioxidant metabolism in, 292–293
   recovery after, 193–194
   root to shoot signaling in, 182–183, 285–287
   second messenger molecules in, 182
   sugar transport and degradation in, 188–190

I

Inorganic ions, and membrane stability, 15
Inositol polyphosphate accumulation, *FRY1* mutation and, 109–110, 130
Intercellular air space, carbon dioxide partial pressure in, 327
Ion homeostasis

regulation, in SOS (salt overly sensitive) pathway, 131, 131f
transgenic manipulation of, 137
Ion leakage, in membrane damage, 17, 335–337
Iron, physiologic effects of, 249

K

Knockout mutants, 365, 367–369

L

Lactate fermentation, 195
Late embryogenesis abundant (LEA) genes
expression profiles of, 366–367
and transcriptional regulation, 151–153
Late embryogenesis abundant (LEA) proteins, 29, 149–151, 340–341, 342–343, 359
cellular protection associated with, 149–151
group classification of, 150
Lead, physiologic effects of, 260–262
Light absorbing compounds, 71–72
Light activated stomatal opening, abscisic acid inhibition of, 108
Light energy, excess, adaptive mechanisms for, 70–72
Light energy dissipation, and quenching, 72–74
Light gradient, 79
Light harvesting complexes (LHCs), 70–71, 75
antennae of, 81–82
antennae size in, molecular regulation of, 82–83
and state transitions, 75–76
Lipid metabolism
in abiotic stress, 2–3, 337–340 (*see also* Membrane lipids)
damage mechanisms in, 16–17 (*see also* Lipid peroxidation; Lipid stability)
references on, 18–25
in desiccation, 11–13 (*see also* Drought/dehydration stress; Drought tolerance)
in salinity stress, 13–16 (*see also* Salinity stress; Salinity tolerance)
in temperature stress, 3–11 (*see also* Cold acclimation; Cold stress; Freeze tolerance; Heat stress; Heat tolerance)
Lipid peroxidation, 16, 337–338 (*see also* Reactive oxygen species)
analysis of, 338–340
as stress indicator, 340
Lipid peroxidation marker, 193 (*see also* Malondialdehyde)
Lipid stability, membrane, 2–18, 335–337

Low oxygen concentration, *see* Anaerobic metabolism
Low phosphorus availability, 209–212
root growth in, 212–214
aerenchyma in, 218–221
alternative respiration in, 222–223
carboxylate excretion in, 225–226
clustered, 226
as ecological issue, 232–233
etiolation in, 221–222
genetic interaction in, 228–230
genotype and competition for, 231–232
metabolic costs of, 215–218
mycorrhizal symbiosis in, 226–227
nurtrient distribution and, 230–231
phenologic response to, 227–228
references on, 234–242
root hairs in, 223–225
senescence in, 223
topsoil exploration and, 214–215
trait interactions in, 228–229
Low temperatures, *see* Cold stress
Lucine-zipper family transcription factors, 153

M

Maize pollen development, and heat shock protein induction, 36
Malondialdehyde (MDA), 193
Malondialdehyde (MDA) content analysis, 338–340
Manganese, physiologic effects of, 249–251
Manganese toxicity, 250–251
Mannitol
as osmoprotectant, 140
transgenic production of, for salinity tolerance, 144t
MAPKs, *see* Mitogen activated protein kinases
McHAK transporter genes, 132
MDA, *see* Malondialdehyde
Membrane lipids, 3
in cold acclimation, and freeze stress, 56–59
in cold stress, 7–8, 49–50
damage of, mechanisms of, 16–17
in drought tolerance, 11–13
fluidity of, 4–5
in heat stress, 4–7
and membrane permeability, 17, 335–337 (*see also* Ion leakage)
osmolytes and, 32, 56–59, 332–333
peroxidation of, 16, 337–340 (*see also* Reactive oxygen species)
quantitative trait loci associated with, 362–363

references on, 18–25
in salinity tolerance, 13–16
saturated and unsaturated, 5–10
stress damage mechanisms in, 16–17
structure and stability of, 2–18, 335–337, 359
temperature sensors and signaling in, 10–11, 49–50
in temperature stress tolerance, 8–11, 336–337
Mercury, physiologic effects of, 262–263
Mesembrayanthemum cyrstallinum, as model for molecular response to salinity stress, 129
Mg-protoporphyrin IX concentration, in chloroplast signaling, 84
MGDG (monogalactosyldiacylglycerol), 6, 11–12
Microarray analysis, high throughput, 365–367
Micronutrient(s), 247
Mitochondria, in hypoxic response, 179–180
Mitogen activated protein kinases (MAPKs)
and reactive oxygen species in stress signaling, 293–294
in signaling temperature stress, 10–11
in stress signaling, 155–156, 157f
Mitotic cyclin expression, in salinity stress, 124
Modulated light, 324
Molecular chaperones, 28–29
Molecular markers, 157–158
in classical breeding strategies, 359–363
lipid peroxidation, 193 (*see also* Malondialdehyde)
of quantitative trait loci, 360–364
of waterlogging tolerance, 191
Molybdenum, physiologic effects of, 251–252
Monogalactosyldiacylglycerol (MGDG), 6, 11–12
Mutants, gain of function or loss of function (knockout), 365, 367–369
MYB/MYC type transcription factors, 153–154, 187
Mycorrhizal symbiosis, 226–227
Myristolation mutations, in salinity tolerance, 134–135
N-Myristoylation motif, 133

N

Na$^+$, *see* sodium ion entries
Na$^+$/H$^+$ antiporter expression, 135–136, 297–298
NCEDs (9-*cis*-epoxycarotenoid dioxygenases), 104f, 105
NHX expression, 135–136

NHX1, 131f
Nickel, physiologic effects of, 255–256
Nickel toxicity, 255–256
Nitrogen starvation, in root, 279
Nodal roots, and oxygen limitations, 184
Nonphotochemical quenching (NPQ), 71, 74–75, 323
in drought/dehydration stress, 325
and trimerization of light harvesting complexes, 81

O

D-Ononitol, transgenic production of, for salinity tolerance, 144t
Optimal temperature, of plant species, 28
OsHKT1 expression, in salt tolerant rice, 132–133, 298
Osmolyte metabolism, 145f
Osmolytes, 289, 331
in cold stress and water stress, 52, 58
functions of, in cold acclimation, 55–56
genetic control of, 140
and membrane lipid stabilization, 32, 56–59, 331–333
and sodium compartmentation, 135–136
Osmoprotectant metabolism, regulation of, 145–146
Osmoprotectants (*see also* Cryoprotectants)
and compatible solutes, 139–142, 331–332
metabolic engineering of, 142–145, 143t–144t
Osmotic adjustment, 135–136, 331–332, 359
determination of, 332–333
genetic capacity for, 334
as indicator of stress tolerance, 333–334
quantitative trait loci associated with, 362–363, 364–365
in roots, 288–290, 332
Osmotic stress, *see* Osmolytes; Osmotic adjustment; Salinity tolerance
Oxidative species, *see* Reactive oxygen species
Oxygen limitation, *see* Hypoxia, response in
Oxygen loss, root barriers to, 184

P

Partial pressure, carbon dioxide, in intercellular air space, 327
Permeability, membrane, 17, 335–337
Peroxidase, 338
Peroxidation, lipid, 16, 337–340 (*see also* Reactive oxygen species)
analysis of, 338–340

Phaseic acid (PA), in abscisic acid synthesis, 105–106
Phenologic response, to low phosphorus availability, 227–228
Phenotype of interest, identifying candidate genes for, 365–367
Phosphatidic acid (PA, PtdOH), 56, 57
  abscisic acid stimulation of, in guard cells, 108–109
  accumulation of, and transcription of phospholipase D, 110
Phosphatidylcholine (PC), 56, 57
Phosphatidylethanolamine (PE), 56, 57
Phosphatidylglycerol (PG), 6, 7
Phosphoinositide specific phospholipase C (PI-PLC) isoforms, 12–13
Phospholipase(s)
  in abiotic stress response, 9–16, 56–57 (see also specific lipase)
  activation of, 10–11
  in salt stress response, 130–131
Phospholipase D (PLD), 9–10, 13, 15–16, 57
  and abscisic acid in stomatal closure in water stress, 108–110
  in cold acclimation, and freeze stress, 56–61
  in drought tolerance, 57–59
  genes encoding classes of, 57
Phospholipid hydrolysis, and environmental stress, 9–10, 13, 15–16
Phospholipids, 3 (see also Membrane lipids)
Phosphorus bioavailability (see also Low phosphorus availability)
  factors in, 209–211
  global, 210f
Phosphorus efficient genotypes, 231–232
  ecologic benefits of, 232–233
  genetic interaction in, 228–230
  and metabolic cost of root exploration, 216–218
Phosphorus starvation, in root, 279
Photoacclimation, 69–72
  cellular and chloroplast responses in, 72–76
  developmental attributes in, 78–79
  energy dissipation and quenching in, 72–74
  light harvesting antennae in, and molecular regulation, 81–83
  long term responses in, 77–83
  metabolic responses in, 77–78
  photoinhibition in, 76
  photoreceptors in, 87–88
  photosystem stoichiometry in, 80
  references on, 88–99

short and long term response integration in, 77f
  signal transduction in, 83–88
  species specific, 79–80
  state transitions in, 75–76
  thermal dissipation in, 74–75
Photochemical quenching, 72–74, 323
Photoinhibition, 76
Photoreceptors, 87–88
Photosynthesis
  salinity stress and effect on, 126–128
  as stress indicator, 70–71, 322–329 (see also Chlorophyll fluorescence; Gas exchange)
Photosystem(s) (PSI, PSII), 70–71
  fluorescence in, 72–74, 325–326
  light buffering in, 72
  salt stress and effects on, 127–128
  state transitions in, 75–76
  stoichiometry of, in photoacclimation, 80
  variable fluorescence, temperature dependence of, 28
Phytohemoglobin expression, in hypoxic or anoxic plant tissue, 180, 191–192, 196
Phytohormones, in plant growth and development, 112–113
PI-PLC (phosphoinositide specific phospholipase C) isoforms, 12–13
Plant breeding, molecular genetics in, 157–158 (see also Genetics; Genomics; Transgenic plants)
Plant transformation technology, 158
Plasma membrane(s), 3 (see also Membrane lipids)
Plastoquinone (PQ) pool, 85–86
Pollen dehiscence, in elevated temperatures, 37–38
Pollen sensitivity, to heat, genetic regulation of, 36–38
Polyamines
  DNA stabilization by, 139
  in repair of salinity damage, 138–139
Potassium loss, in membrane damage, 17
PQ (plastoquinone) pool, 85–86
Proline
  as osmoprotectant, 141
  transgenic production of, for salinity tolerance, 143t
Protective mechanisms, see Adaptive/protective mechanisms
Protein(s), cold regulated, 53, 55–56 (see also COR (cold regulated/responsive) proteins)
Protein stability

Index

in heat tolerance development, 29–31
hypoxic degradation of, 192
osmolytes promoting, 32 (*see also* Osmoprotectants)
in salinity stress, 125–126
in stress conditions, 16, 340–343 (*see also* Heat shock proteins; Late embryogenesis abundant proteins)
PSI, PSII, *see* Photosystem(s)
PtdIns(3,5) P_2 synthesis, 15–16
PtdOH, *see* Phosphatidic acid
Pulse modulated fluorometers, 324
Putrescine
  accumulation of, in abiotic stress, 138–139
  DNA stabilization by, 139

Q

Quantitative trait loci (QTL), 360
  analysis of in breeding for stress tolerance, 360–364
    limitations of, 363–364
  comparative mapping of, 364–365
  gene cloning using, 365
  physiological traits and, 362–363
  populations and environment and, 361

R

Radial oxygen loss, in root, 184, 278, 279
  barriers to, 279–280
RAV1, 54
RD29B, AREB mediated activation of, 111
Reactive oxygen species (ROS), 290–291
  in abscisic acid induced stomatal closure, 108
  excess light energy and production of, 71
  freeze tolerance and production of, 57–58
  hypoxic/anoxic damage caused by, 193–194, 196, 275–278, 292–294
  in salinity stress, 146–149, 288, 291–292
  and signaling in chloroplasts, 87
  in stress signaling pathways, 155–156, 290–294
Redox status signaling, in chloroplasts, 84–87
Reduced specific root length, 221
Rehydration method, 333
Relative water content (RWC)
  analysis of, 329–330
  as indicator of stress tolerance, 330–331
  plant maturity and use of, 331
  quantitative trait loci associated with, 362–363

regression analysis of, and osmotic adjustment determination, 332–333
Reproductive phase, heat tolerance in, 35–38
Resistance mechanisms, 2–3
Respiratory gas exchange
  analysis of, 326–327
  as indicator of stress tolerance, 326–329
  in root, 218
  in low phosphorus soils, 222–223
Rhizosphere alteration, 225–226
Root etiolation, 221–222
Root foraging duration, 227–228
Root growth and physiology
  in abiotic stress, 271–272
    morphologic adaptations, 272–282
    nutrient ion uptake, 294–298
    osmotic adjustments, 288–290
    physiologic adaptations, 282–298
    reactive oxygen species metabolism, 290–294
    references on, 299–319
    signal transduction, 278f, 282–288
  in drought, 272–274, 282–284, 294–295 (*see also* Drought/dehydration stress; Drought tolerance)
  in low phosphorus availability, 212–234 (*see also* Low phosphorus availability)
  in salinity stress, 280–282, 287–288, 297–298 (*see also* Salinity stress; Salinity tolerance)
  in soil compaction, 274–275, 284–285, 295–296
  in waterlogging, 275–280, 285–287, 296–297 (*see also* Hypoxia; Waterlogging)
Root hairs
  in dehydration stress, 273–274
  in low phosphorus availability, 223–225
Root-length density, 273
Root porosity, and deep rooting, correlation between, 280
Root respiration, 218
Root senescence, in low phosphorus soils, 223
ROS, *see* Reactive oxygen species
Rubisco, in photoacclimation, 77–78, 80
RWC, *see* Relative water content

S

Salinity stress, 123–128
  and cell cycle, 124
  damage and repair mechanisms in, 137–146
  and effect on photosynthesis, 126–128
  ion homeostasis in, 129–131

photosynthetic gas exchange in, 328–329
physiology of, 123–125
and reactive oxygen species production, 146–149, 290–292
references on, 159–175
root growth and physiology in, 280–282
  nutrient ion uptake, 297–298
  osmotic adjustments, 288–290
  reactive oxygens species, 290–292
  signal transduction, 287–288
signaling mechanisms in, 129–131, 287–288
sodium uptake in, 132–133
Salinity tolerance, 122, 123
and antioxidant enzyme overexpression, 148–149, 291–292
calcium ion binding mutations in, 134–135
genetic encoding of, 149–155
genetic mutations conferring, 133–135
genetic regulation of, 128–131
membrane lipids in, 13–16
metabolic engineering of osmoprotectants in, 142–145
references on, 159–175
regulation of osmoprotectant metabolism in, 145–146
sodium compartmentation in, 135–136
sodium influx and efflux in, 132–135
sodium transport from shoot to root in, 136–137, 280–281
Salt overly sensitive, see SOS (salt overly sensitive) gene expression; SOS (salt overly sensitive) pathway
Saturated and unsaturated membrane lipids, 5–10
regulation of, 8–10, 12
SDR1 (short-chain dehydrogenase reductase), in *Arabidopsis,* 105
Serine-threonine protein kinase
ABA activated, 108
SOS3 activated, 133, 139, 298
Signal transduction, 155–157
in cold acclimation, 48–50
in dehydration stress, 12–13, 103, 106–110, 282–284
in hypoxia in waterlogging, 178–182, 285–287
in membrane lipid maintenance, 10–11, 49–50
in photoacclimation, 83–88
reactive oxygen species in, 155–156, 290–294
in root response to abiotic stress, 278f, 282–288
in salinity stress and tolerance, 15–16, 129–131, 287–288

in soil compaction stress, 284–285
in stomatal closure in water stress, 106–109
in temperature stress, 10–11
SOD, see Superoxide dismutase
Sodium ion compartmentation, in salinity tolerance, 135–136
Sodium ion transport
salinity stress and effect on, 132–135
from shoot to root, 136–137, 287–288
Soil compaction stress
nutrient ion uptake in, 295–296
osmotic adjustments in, 288–290
root growth and physiology in, 274–275
signal transduction in, 284–285
Soil water saturation, 178–179 (*see also* Waterlogging)
Sorbitol
as osmoprotectant, 141
transgenic production of, for salinity tolerance, 144t
SOS (salt overly sensitive) gene expression, 129–131
SOS (salt overly sensitive) pathway, 129–131
and *AtHK1* expression, 137
and *AtNHX1* expression, 135–136
ion homeostasis regulation in, 131, 131f
in sodium ion efflux, 133–135
targets of, 129–131
SOS1
in sodium transport from shoot to root, 135–136, 297–298
as target in SOS (salt overly sensitive) pathway, 134–135
SOS3-like calcium binding proteins, 131, 133–134, 136, 151
Specific root length, 273
reduced, 221
SRK2C gene, 371
State transitions, 75–76
Stomatal closure, abscisic acid in, 106–109, 107f, 286–287
Stress genes, transcriptional regulation of, 149–155 (*see also* Late embryogenesis abundant (LEA) proteins)
Stress signaling pathways, 155–157 (*see also* Signal transduction)
Stress tolerance, (*see also* Adaptive/protective mechanisms)
breeding strategies for, 157–158, 196–197, 321–322, 357–372
(*see also* Breeding strategies; Genetics; Genomics; Quantitative trait loci)

indicators of, 321–355
  chlorophyll fluorescence, 323–326
  gas exchange, 326–329
  heat shock proteins, 341–342
  late embryogeneis abundant proteins, 342–343
  lipid peroxidation, 337–340
  membrane stability, 335–337
  osmotic adjustment, 331–335
  references on, 345–355
  water content, 329–331
Sugar concentration, and signaling in chloroplasts, 86–87
Sugar transport, and degradation, in hypoxia, 188–190
Superoxide dismutase (SOD), 291, 338
  overexpression of, 368–369
Synteny, 364–365

T

T-DNA insertional mutants, in metabolic genes, 35
Temperature stress
  membrane lipids in, 3–11
  and photosynthetic gas exchange, 327–328
  signaling mechanisms of, 10–11
Tetrapyrrole biosynthetic pathway, 83–84
Thallium toxicity, 263
Thermal dissipation, in photoacclimation, 74–75
Thermal kinetic window (TKW), 27–29
Thermotolerance, acquired, 28–29
  mutational analysis of, 32–35
Thiobarbituric acid reactive substances (TBARS), 338
Thylakoid membrane
  photoacclimation in, short term, 73–74
  sensitivity to cold stress, 7–8
  sensitivity to desiccation, 12
  sensitivity to heat stress, 5
TILLING (targeting induced local lesions in genomes), 367–368
Topsoil foraging, in low phosphorus availabililty, 214–215
Trace elements, 247
Transgenic plants, 368–369
  antioxidant enzyme overexpression in, 148–149
  heat tolerance in, 33–35
  ion homeostasis in, 137
  lipid unsaturation in, and cold acclimation, 8
  osmoprotectants in, 142–145
  trienoic acid levels in, 9

Trehalose, transgenic production of, for salinity tolerance, 144t
Turgor loss (*see also* Osmotic adjustment; Salinity stress)
  and exposure to low temperature, 49–50
  root, 288–290

V

Vacuolar sodium ion sequestration, 131f, 135–136
Vanadium, physiologic effects of, 256–257
VCX1, 131f
Vegetative growth, heat tolerance in, 38–40
Violaxanthin, in abscisic acid synthesis pathway, 103–105, 104f
Violaxanthin de-epoxidase (VDE), 74, 75

W

Water binding LEA proteins, 150
Water content, relative
  analysis of, 329–330
  as indicator of stress tolerance, 330–331
  plant maturity and use of, 331
  quantitative trait loci associated with, 362–363
  regression analysis of, and osmotic adjustment determination, 332–333
Water homeostasis, sensors of change in, 178–179
Water stress, *see* Drought/dehydration stress; Drought tolerance
Waterlogging, 178–179
  anoxia and injury in, 193–194, 196, 275–278, 292–294
  and manganese toxicity, 250
  root growth and physiology in, 275–280
    nutrient ion uptake, 296–297
    osmotic adjustments, 288–290
    signal transduction, 285–287
Waterlogging tolerance, 177–178
  breeding plants for, 196–197
  evaluation of, 197–199
  genetic control of, 194–197
  genetic manipulation for, 194–196
  injury and recovery in, 193–194, 292–294
  metabolic adaptations in, 186–192 (*see also* Anaerobic metabolism; Hypoxia, response in)
  molecular markers of, 197
  morphologic and anatomic mechanisms of, 183–184, 275–280
  osmotic adjustment in, 288–290

references on, 200–208
sensors and signal transduction in, 178–184, 285–287
Whole leaf alteration, 79
WRKY related genes, 187

X

Xanthophyll cycle, functional site of, 75
Xanthophyll cycle quenching, 74
Xanthoxin, in abscisic acid synthesis pathway, 104f
XET (xyloglucan endotransglycosylase), 38

Y

YAC (Yeast Artificial Chromosomes) genomic libraries, 365

Z

ZAT12, 54, 187
Zeaxanthin epoxidase (ZEP), in abscisic acid synthesis pathway, 103–105, 104f
Zinc, physiologic effects of, 252–253